T0314273

LINEAR CIRCUIT TRANSFER FUNCTIONS

LINEAR CIRCUIT TRANSFER FUNCTIONS

AN INTRODUCTION TO FAST ANALYTICAL TECHNIQUES

Christophe P. Basso

ON Semiconductor, Toulouse, France

WILEY

This edition first published 2016

Registered office
John Wiley & Sons Ltd, The Atrium, Southern Gate, Chichester, West Sussex, PO19 8SQ, United Kingdom

For details of our global editorial offices, for customer services and for information about how to apply for permission to reuse the copyright material in this book please see our website at www.wiley.com.

Library of Congress Cataloging-in-Publication Data

Names: Basso, Christophe P., author.
Title: Linear circuit transfer functions : an introduction to fast analytical techniques / Christophe Basso.
Description: Chichester, West Sussex ; Hoboken, NJ : Wiley, 2016. | Includes index.
Identifiers: LCCN 2015047967 | ISBN 9781119236375 (cloth) | ISBN 9781119236351 (epub)
Subjects: LCSH: Transfer functions. | Electric circuits, Linear.
Classification: LCC TA347.T7 B37 2016 | DDC 621.3815–dc23 LC record available at http://lccn.loc.gov/2015047967

A catalogue record for this book is available from the British Library.

ISBN: 9781119236375

Set in 9.5/11.5 pt TimesLTStd-Roman by Thomson Digital, Noida, India

1 2016

Contents

About the Author

Christophe Basso is a Technical Fellow at ON Semiconductor in Toulouse, France, where he leads an application team dedicated to developing new offline PWM controller's specifications. He has originated numerous integrated circuits among which the NCP120X series has set new standards for low standby power converters.

Further to his 2008 book *Switch-Mode Power Supplies: SPICE Simulations and Practical Designs*, published by McGraw-Hill, he released a new title in 2012 with Artech House, *Designing Control Loops for Linear and Switching Power Supplies: a Tutorial Guide*. He holds 17 patents on power conversion and often publishes papers in conferences and trade magazines including How2Power and PET.

Christophe has over 20 years of power supply industry experience. Prior to joining ON Semiconductor in 1999, Christophe was an application engineer at Motorola Semiconductor in Toulouse. Before 1997, he worked at the European Synchrotron Radiation Facility in Grenoble, France, for 10 years. He holds a BSEE equivalent from the Montpellier University (France) and a MSEE from the Institut National Polytechnique of Toulouse (France). He is an IEEE Senior member.

When he is not writing, Christophe enjoys snowshoeing in the Pyrenees.

Preface

First as a student and later as an engineer, I have always been involved in the calculation of transfer functions. When designing power electronics circuits and switch mode power supplies, I had to apply my analytical skills on passive filters. I also had to linearize active networks when I needed the control-to-output dynamic response of my converter. Methods to determine transfer functions abounded and there are numerous textbooks on the subject. I started in college with mesh-node analysis, and at some point ended up using state variables. If all paths led to the correct result, I often struggled rearranging equations to make them fit a friendly format. Matrices were useful for immediate numerical results but, when trying to extract a meaningful symbolic transfer function, I was often stuck with an intractable result. What matters with a transfer function formula is that you can immediately distinguish poles, zeros and gains without having to rework the expression. This is the idea behind the term *low-entropy*, a concept forged by Dr. Middlebrook.

Simulation gives you an idea where poles and zeros hide by interpreting the phase and magnitude plots with minimum-phase functions. However, inferring which terms really affect a pole or a zero position from a Bode plot is a different story. Fortunately, if the transfer function is written the right way, then you can immediately identify which elements contribute to the roots and assess how they impact the dynamic response. As some of these parasitics vary in production or drift with temperature, you have to counteract their effects so that reliability is preserved during the circuit's life. The typical example is when you are asked to assess the impact of a parasitic term variation on a product you have designed: if a new capacitor or a less expensive inductor is selected by the buyers, will production be affected? Is there a chance that stability will be jeopardized in some operating conditions? Implementing the classical analysis method will surely deliver a result describing the considered circuit, but extracting the information you need from the final expression is unlikely to happen if the equations you have are disorganized or in a *high-entropy* form.

This is where Fast Analytical Circuit Techniques (FACTs) come into play. The acronym was formed by Dr. Vatché Vorpérian, who formalized the technique you are about to discover here. Before him, Dr. Middlebrook published numerous papers and lectured on his Extra-Element Theorem (EET), later generalized to the N extra-element theorem by one of his alumni. Since Hendrik Bode in the 40's, authors have come up with techniques aiming to simplify linear circuit analysis through various approaches. All of them were geared towards determining the transfer function at a pace quicker than what traditional methods could provide. Unfortunately, while traveling and visiting customers world-wide, I have found that, despite all the available documentation, FACTs were rarely adopted by engineers or students. When describing examples in my seminars and showing the method at work in small-signal analysis, I could sense interest from the audience through questions and comments. However, during the discussions I had later on with some of the engineers or students, they confessed that they tried to acquire the skill but gave up because of the intimidating mathematical formalism and the complexity of the examples. If one needs to be rigorous when tackling electrical analysis, perhaps a different approach and pace could make people feel at ease when learning the method. This is what I strived to do with this new book, modestly shedding a different light on the subject by progressing with simple-to-understand examples and clear explanations. As a student, I too struggled to apply these fast analytical circuits techniques to real-world problems; as such, I identified the obstacles and worked around them with success. Thus, the seeds for this book were sown.

This book consists of five chapters. The first chapter is a general introduction to the technique, explaining what transfer functions are and how time constants characterize a circuit. The second chapter digs into transfer function definitions and polynomial forms, introducing the low-Q approximation, and how to organize 2nd and 3rd-order denominators or numerators. The third chapter uses the superposition theorem to gently introduce the extra-element theorem. Numerous examples are given to illustrate its usage in different 1st-order configurations. The fourth chapter deals with the 2-extra element theorem, generalized and applied to 2nd-order networks. Numerous examples illustrated with Mathcad® and SPICE punctuate the explanations. Finally, the fifth chapter tackles 3rd- and 4th-order circuits, all illustrated with examples. Each chapter ends with 10 fully documented problems. There is no secret; mastering a technique requires patience and practice, and I encourage you to test what you have learned after each chapter through these problems.

I have adopted the same casual writing style already used in my previous books, as readers' comments show that the way I present things better explains complex matters. Please let me know if my approach still applies here and if you enjoy reading this new book. As usual, feel free to send me your comments or any typos you may find at cbasso@wanadoo.fr. I will maintain an errata list in my personal webpage as I did for the previous books (http://cbasso.pagesperso-orange.fr/Spice.htm). Thank you, and have fun determining transfer functions!

Christophe Basso
May 2015

Acknowledgement

A book like this one could not have been written and published without the help of many contributing friends. My warmest thanks and love first go to my sweet wife Anne who endured my ups and downs when determining some of the book transfer functions: equations time is over and we can now enjoy the long and warm evenings of summer to come!

I was fortunate to share my work with my ON Semiconductor colleagues and friends who played a crucial role in reviewing my pages and challenging the method. Stéphanie Cannenterre reviewed and practiced numerous book exercises. She now masters the method: well done! Dr. José Capilla raced with me several times to determine a transfer function with his Driving Point Impedance method and I recognize his skills in doing so. Special thanks go to my friend Joël Turchi with whom I spent endless hours debating the method or discussing the validity of an equation. Merci Joël for your kindness and invaluable support for this book!

Two people did also accompany me from the beginning of the writing process. Mon ami Canadien Alain Laprade from ON Semiconductor in East Greenwich who developed an addicted relationship to the FACTs and kindly reviewed all my work. Monsieur Feucht from Innovatia did also a tremendous work in correcting my pages but also kindly polished my English. I am not exactly a novelist and cannot hide my French origins Dennis!

I want to warmly thank the following reviewers for their kind help in reading my pages during the 2015 summer: Frank Wiedmann (Rhode & Schwarz), Thierry Bordignon, Doug Osterhout (both are with ON Semiconductor), Tomas Gubek – děkuji! (FEI), Didier Balocco (Fairchild), Jochen Verbrugghe, Bart Moeneclaey (both are with Ghent University), Bruno Allard (INSA Lyon), Vatché Vorpérian (JPL), Luc Lasne (Bordeaux University) and Garrett Neaves (Freescale Semiconductor).

Last but not least, I would like to thank Peter Mitchell at Wiley & Sons UK for giving me the opportunity to publish my work.

1

Electrical Analysis – Terminology and Theorems

This first chapter is an introduction to some of the basic definitions and terms you must understand in order to perform electrical analysis with efficiency and speed. By electrical analysis, I imply finding the various relationships that characterize a particular electrical network. To excel in this field, as in any job, you need to master a few tools. Obviously, they are innumerable and I am sure you have learned a plethora of theorems during your student life. Some names now seem distant simply because you never had a chance to exercise them. Or you actually did but implementation was so obscure and complex that you left quite a few of them aside. This situation often happens in an engineer's life where real-case experience helps clean up what you have learned at school to only retain techniques that worked well for you. Sometimes, when what you know fails to deliver the result, it is a good opportunity to learn a new procedure, better suited to solve your current case. In this chapter, I will review some of the founding theorems that I extensively use in the examples throughout this book. However, before tackling definitions and examples, let us first understand what the term *transfer function* designates.

1.1 Transfer Functions, an Informal Approach

Assume you are in the laboratory testing a circuit encapsulated in a box featuring two connectors: one for the input, the second for the output. You do not know what is inside the box, despite the transparent case in the picture! You now inject a signal with a function generator to the input connector and observe the output waveform with an oscilloscope. Using the right terminology, you *drive* the circuit input and observe its *response* to the stimulus. The input waveform represents the *excitation* denoted u and it generates a *response* denoted y. In other words, the excitation variable propagates through the box, undergoes changes in phase, amplitude, perhaps induces distortion etc. and the oscilloscope reproduces the response on its screen.

The waveform displayed by the oscilloscope is a *time-domain* graph in which the horizontal axis x is graduated in seconds while the vertical axis y indicates the signal *amplitude* (positive or negative). Its dimension depends on the observed variable (volts, amperes and so on). The input waveform is denoted in lower case as it is an *instantaneous* signal, observed at a time – the *instant* t – $u(t)$. A similar notation applies to the output signal, $y(t)$. In Figure 1.1, you see a low duty ratio square-wave injected in the box engendering a rather distorted waveform on the output.

Linear Circuit Transfer Functions: An Introduction to Fast Analytical Techniques, First Edition. Christophe P. Basso.

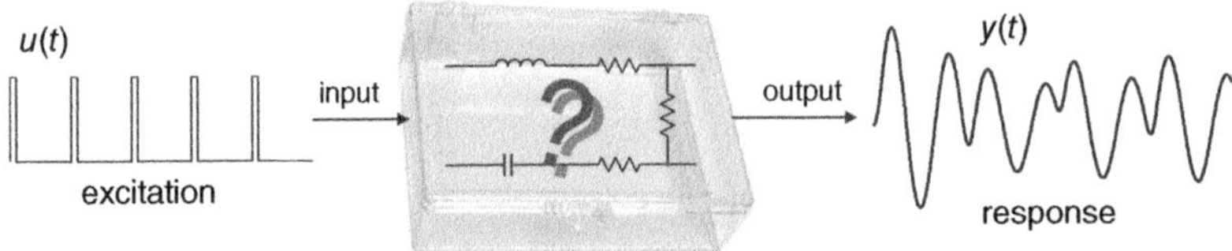

Figure 1.1 A black box featuring an input and an output signal. What is the relationship linking output and input waveforms?

This ringing signal tells us that the box could associate resonant elements, probably capacitors and inductors but not much more than that. If we change the excitation, what type of shape will we obtain? Knowing what is inside the box will let us predict its response to various types of excitation signals.

There are several available ways to characterize an electrical linear circuit. One of them is called *harmonic analysis*. The input signal is replaced by a sinusoidal waveform and you observe how the stimulus propagates through the box to form the response. This is shown in Figure 1.2:

The excitation level must be of reasonable amplitude – understand *small* – so that the response signal is not distorted. The input signal dc bias must also be set accounting for the physical constraints imposed by the active circuit so that upper- or lower-rail saturation is avoided. In other words, the box internal circuitry is not *overdriven* and remains *linear* during the analysis. Linearity is confirmed if the output signal is sinusoidal with the same frequency as the input sine and only varies in amplitude and phase while you ac-sweep the network. This is a so-called *small-signal* analysis. In the Laplace domain, you perform such harmonic analysis when you set $s = j\omega$ in which $\omega = 2\pi f$ represents the angular frequency expressed in radians per seconds (rads/s). Laplace analysis with $s = j\omega$ applies to linear circuits only.

Should you increase the input signal amplitude or change the operating bias point, slewing or clipping may happen. In this case, you explore the box *large-signal* or *nonlinear* response. This is a characterization different than the small-signal approach and it offers another insight into the circuit operation. Let us keep linear and once the right input amplitude is found, i.e. a signal of comfortable amplitude is observed on the oscilloscope screen, the frequency is varied step by step while output amplitude/phase couples are recorded in an array. At each frequency point f, we store the ratio of the response amplitude $Y(f)$ in volts to the excitation amplitude $U(f)$ in volts also. At each frequency point f, we save the phase information linking both input and output waveforms. As U and Y are complex variables affected by a magnitude and a phase, we can write:

$$A_v(s) = \frac{Y(s)}{U(s)} \tag{1.1}$$

A_v represents a *transfer function*, a mathematical relationship linking a response signal Y to an excitation signal U. Please note that the excitation signal U resides in the transfer function

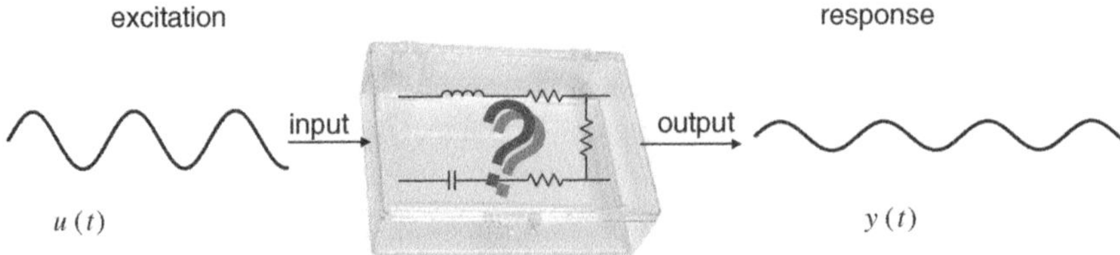

Figure 1.2 The black box is now driven by a sinusoidal stimulus for a small-signal analysis.

denominator while the response Y sits in the numerator. It will always be this way throughout the book.

The transfer function is a complex variable characterized by a magnitude noted $|A_v(f)|$ and an argument, $\angle A_v(f)$ also noted $\arg A_v(f)$. The ratios $Y(f)/U(f)$ we have stored correspond to the transfer function magnitude (also called *modulus*) observed at a frequency f while the phase difference between Y and U represents the transfer function argument or phase at the considered frequency. The transfer function magnitude dimension depends on the observed variables as we will later see. Here, because volts are involved for both variables, the transfer function magnitude is *dimensionless* or *unitless*. Furthermore, $|A_v|$ can only be greater than or equal to zero. It is what makes the difference between an amplitude which can take on any value, positive, null or negative and a magnitude which can only be zero or positive. If it is 0, there is no output signal. If $|A_v|$ is less than 1, we talk about *attenuation*. Now, if $|A_v|$ is greater than 1, it is designated as a *gain*. If the magnitude can only be a null or positive number, what about a gain of −2 then? It simply characterizes a stage offering a gain of 2, lagging or leading the excitation signal phase by 180°.

1.1.1 Input and Output Ports

It is convenient to represent our box as a two-port circuit. A *port* is a pair of connections that can input or output signals such as voltage and current. Figure 1.3 shows an illustration of this principle where you see two connecting ports, one input and one output.

Under some conditions, a port can take on the input and output roles at the same time. Imagine you want to measure the output impedance of the box. To realize this measurement, you classically implement Figure 1.4 where a current across the output terminals is injected while the voltage across the same terminals is observed This is what is called a *single injection*, i.e. one stimulus and one response. In this experiment, the box input port is shorted (see Appendix 1A). The excitation variable is the current $I_{out}(s)$ injected into the port while the response is the voltage $V_{out}(s)$ collected across the port's terminals. The output impedance Z obtained from the ratio of the port voltage to the injected

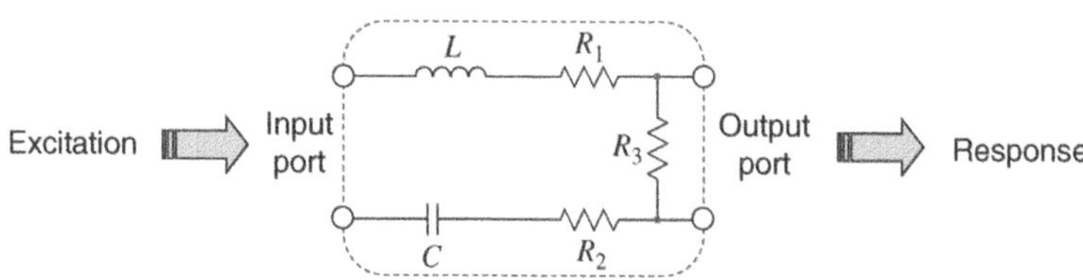

Figure 1.3 The input port receives the excitation signal while the output port delivers the response.

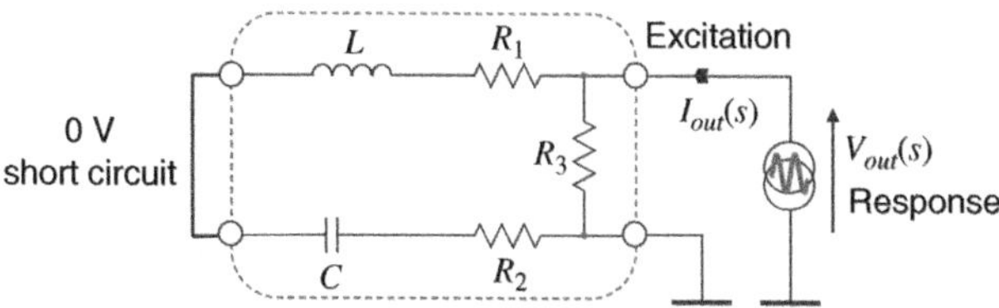

Figure 1.4 A port can be both an input and an output at the same time. Here, an output impedance measurement.

current is a transfer function. It has the dimension of an impedance expressed in ohms:

$$Z_{out}(s) = \frac{V_{out}(s)}{I_{out}(s)} \tag{1.2}$$

I_{out}, the excitation signal lies in the denominator while the response, V_{out}, stands in the numerator. We will come back on this important peculiarity.

If input and output connectors are fixed, physical ports, which let you respectively inject and observe signals, nothing prevents one from creating other observation ports as needed. Simply remove a resistor, a capacitor or an inductor and its connecting points become a new port. This port can now be used as a new input stimulus or as an output variable you want to observe. As already mentioned, this newly created port can also play the role of an input and output port at the same time. In that case, the box originally featuring one input and one output, becomes a two-input/two-output system as illustrated in Figure 1.5 in which the inductor has been removed. Using adequate terminology, we analyze the system by performing a *double-injection*: two stimuli – inputs 1 and 2 – giving two responses, outputs 1 and 2.

In this example, the voltage across the removed inductor terminals is the response while the injected current is the excitation signal. By dividing the port voltage by the injected current, we have the resistance offered by the port terminals when the element initially connected has been removed. In other words, we 'look' at the resistance offered by the inductor port as shown in Figure 1.6 where the symbol R? and the arrow imply this exercise. Expressed in a different manner, we find the equivalent *output resistance* exhibited by the port when 'driving' the inductor, hence the name *driving point* resistance or *driving point impedance* abbreviated as DPI. Combining resistance and inductance gives us a time constant τ ('tau') associated with this inductive element:

$$\tau = \frac{L}{R} \tag{1.3}$$

To conduct this exercise and find the resistance R, you can directly look at the sketch and infer the resistive series-parallel arrangement without solving a single equation. This exercise is called *network inspection*: you simply observe the network in certain conditions (for instance in dc, or when V_{in} is set to 0) and find resistance values by observing how components are connected together. For example, in Figure 1.6, what resistance do you 'see' looking into the inductor port while capacitor C is

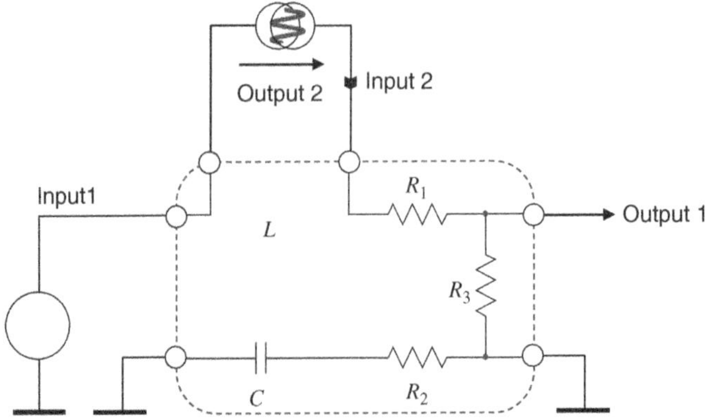

Figure 1.5 If you remove a component from this circuit, its connections become a connecting port. You can bias this port and consider it as a new input, or as a new output, or both of them at the same time.

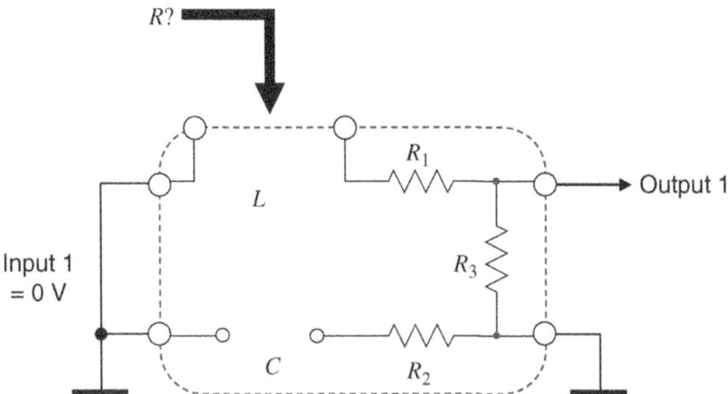

Figure 1.6 Removing the inductor lets you look at the port output resistance that drives the inductor. Associating the port resistance and the inductance leads to a time constant. Here, the resistance seen at the inductor port is $R_1 + R_3$.

disconnected for the exercise? R_1 appears first and then R_3 in series goes to ground and returns to the inductor left terminal via the shorted input source. R_2 is open and plays no role:

$$R = R_1 + R_3 \tag{1.4}$$

Applying (1.3) with (1.4) gives the definition for the time constant involving L:

$$\tau_1 = \frac{L}{R_1 + R_3} \tag{1.5}$$

A similar exercise can be conducted with the capacitor to also unveil the resistance R that drives this element. In this case, the time constant associated with the capacitance is simply:

$$\tau = RC \tag{1.6}$$

Assuming a shorted inductance in this particular illustration, what resistance value do you see in Figure 1.7 when looking into the capacitor port? The left terminal is grounded while the second

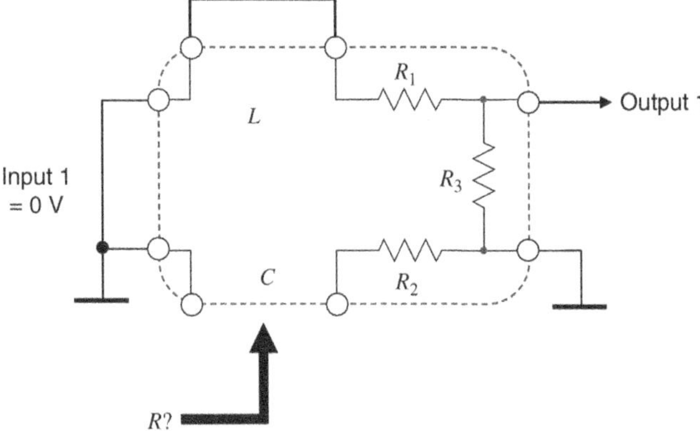

Figure 1.7 Removing the capacitor lets you conduct a similar exercise to unveil the time constant associated with this component. In this case, the resistance seen at the capacitor port is R_2.

terminal also goes to ground via R_2. R_1 and R_3 play no role since their series combination goes from one ground to the other one. Therefore:

$$R = R_2 \tag{1.7}$$

The time constant involving the capacitor is simply:

$$\tau_2 = R_2 C \tag{1.8}$$

We have two storage elements, C and L, and there are two time constants. For each storage element, there is an associated time constant.

Rather than looking into a capacitive or an inductive port, we could also remove a resistor and define what resistance drives it, the exercise remains the same. Sometimes, looking into the port to 'see' the resistance is not as straightforward, especially when controlled sources are involved. In this case, you need to add a test current generator as in Figure 1.5 and define the voltage generated across the considered terminals. The resistance offered by the port being the port voltage divided by the test current generator. This test generator will later be labeled I_T and the voltage across its terminals V_T.

What we just described is part of the technique foundations we will later describe: find resistances offered across the connecting terminals of resistive, capacitive or inductive elements once they have been temporarily removed from the circuit under certain conditions. Breaking a complex passive or active circuit into a succession of *simple* configurations where time constants are unveiled will help us characterize a network featuring poles and zeros. The Extra Element Theorem (EET) and later, the n Extra Element Theorem (nEET), make an extensive usage of these methods and it is important to understand this prerequisite. Appendix 1A will refresh our memory regarding available methods to derive output impedances while Appendix 1B collects several examples to let you exercise your skills at finding these resistances.

1.1.2 Different Types of Transfer Function

Depending where you inject the excitation and where you observe the response, you can define six types of transfer functions as detailed in [1]. For the sake of simplicity, input and output ports are ground-referenced but could also be differential types. The first one, is the voltage gain A_v already encountered in the above lines and it appears in Figure 1.8 together with an operational amplifier (op amp) in an inverting configuration. In all the following illustrations, the op amp is considered a perfect element (infinite open-loop gain, infinite bandwidth, zero output and infinite input impedances). You sweep the input voltage with a sinusoid, the stimulus, and observe the voltage at the op amp output, the response. In Laplace notation, you compute A_v as:

$$A_v(s) = \frac{V_{out}(s)}{V_{in}(s)} \tag{1.9}$$

A_v is dimensionless, sometimes expressed in [V]/[V].

The second one is the current gain, A_i, this time involving input and output currents as shown in Figure 1.9. The excitation signal is now the input current I_{in} while the observed variable is the output current I_{out}:

$$A_i(s) = \frac{I_{out}(s)}{I_{in}(s)} \tag{1.10}$$

A_i is dimensionless, sometimes expressed in [A]/[A].

The third transfer function is called a *transadmittance* – short name for transfer admittance – and is denoted Y_t. You observe the output current while the input is excited by a voltage source.

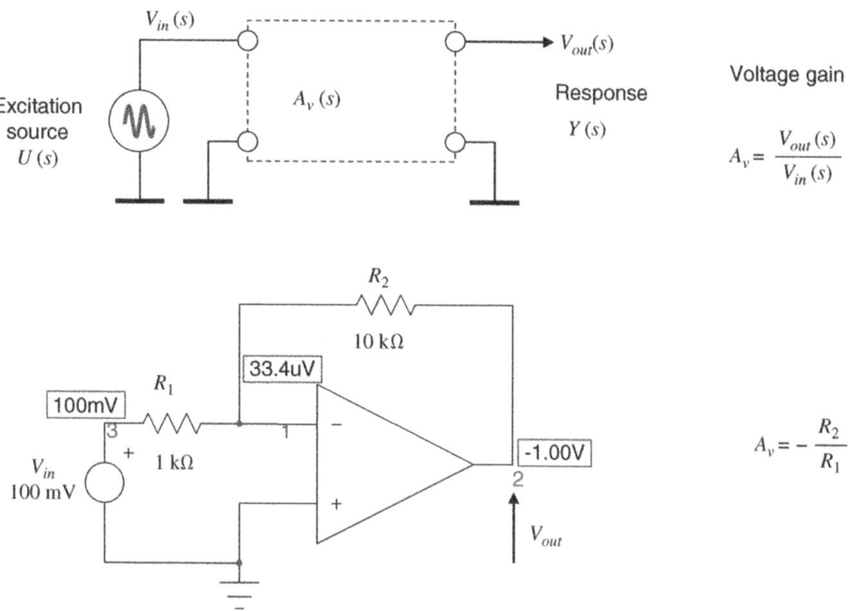

Figure 1.8 The voltage gain A_v is the first transfer function and links the output voltage to the input voltage.

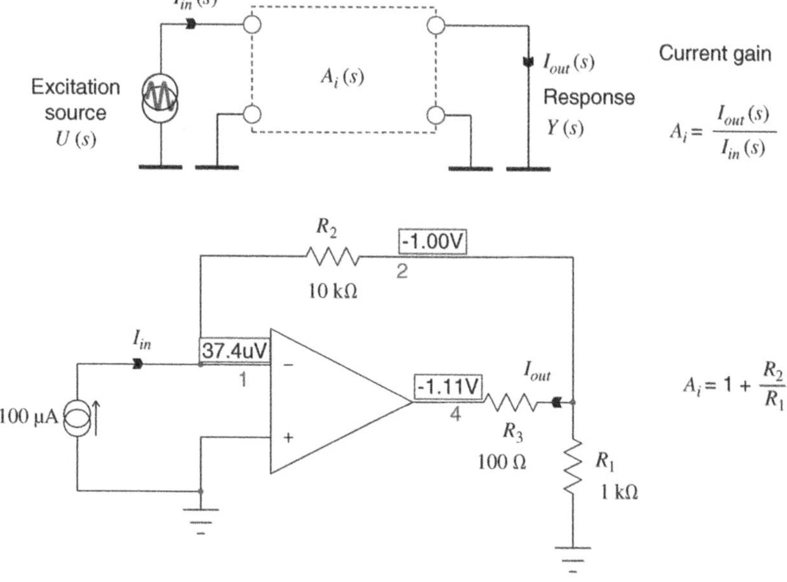

Figure 1.9 The current gain A_i is the second transfer function and links the output current to the input current.

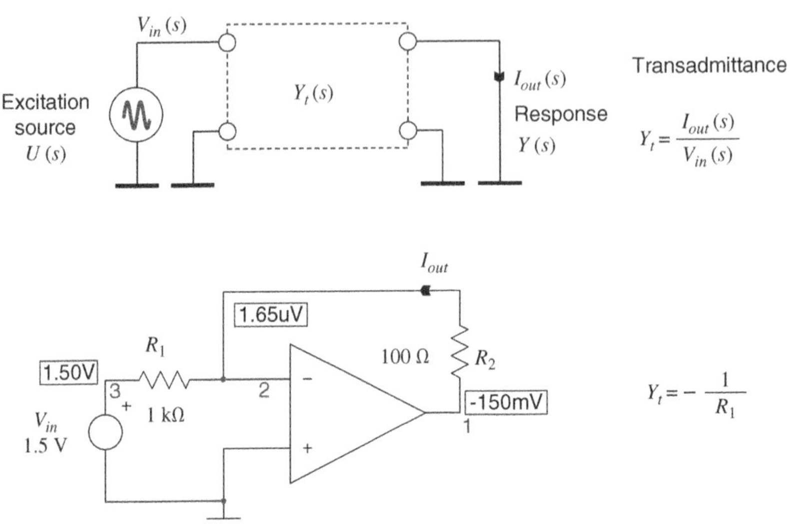

Figure 1.10 The transadmittance Y_t links the output current to the input voltage. Here the current in R_2 is imposed by V_{in} and reaches 1.5 mA. The transadmittance gain is −0.001 A/V or −1 mS.

The measurement configuration is shown in Figure 1.10. The definition is as follows:

$$Y_t(s) = \frac{I_{out}(s)}{V_{in}(s)} \tag{1.11}$$

If the two preceding gains were dimensionless, the transadmittance is expressed in ampere per volt, [A]/[V] or siemens [S]. Similarly, we can define the fourth transfer function in which, this time, the input is excited by a current source while the output voltage is the response (Figure 1.11). The ratio of these two variables is designated as a *transimpedance* – short name for transfer impedance – denoted Z_t and expressed in volt per ampere, [V]/[A] or ohm [Ω]:

$$Z_t(s) = \frac{V_{out}(s)}{I_{in}(s)} \tag{1.12}$$

Transimpedance amplifiers are often used in case you want to amplify a photodiode current for instance. You will find in [2] a design example of such a circuit.

In the four previous transfer functions, the involved quantities – excitation and response signals – appear at two different places in the network. We conveniently considered the box input and output terminals for the examples, but definitions apply equally for relationships between any ports in the network. For the two remaining transfer functions, impedance Z and admittance Y, excitation and response signals are observed at the same port terminals. It is therefore important to distinguish how we create the excitation signal and what is considered the response signal. You can argue that it is not a problem to reverse excitation and response because impedance and admittances are reciprocal to each other. However, if we want to stick to our transfer function definition in which the excitation waveform lies in the denominator while the response appears in the numerator, then, for a *driving point impedance* (DPI) function $Z_{dp}(s)$, the excitation signal is a current source and for a driving point admittance function $Y_{dp}(s)$, the stimulus is a voltage source.

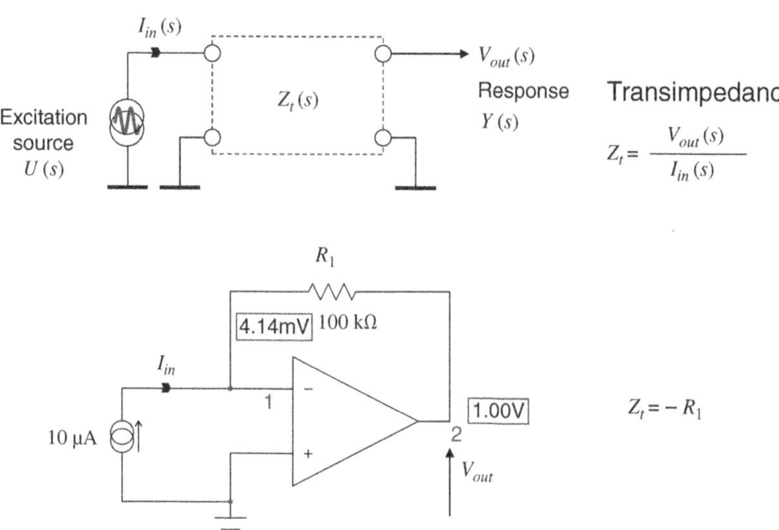

Figure 1.11 The transimpedance Z_t links the output voltage to the input current. In the op amp example, resistor R_1 brings a transimpedance gain of −100 kV/A.

The 5th transfer function is thus the port input impedance $Z(s)$ whose generalized transfer function is given below:

$$Z_{dp}(s) = \frac{V_1(s)}{I_1(s)} \tag{1.13}$$

If you consider V_{in} and I_{in} or V_{out} and I_{out}, you respectively measure the network input and output impedances by injecting a test current in the port and measuring the voltage across the port terminals. Figure 1.12 shows sources arrangement for this specific measurement. The dimension of an impedance is ohm, [Ω].

Finally, the 6th transfer function is the *admittance*, the inverse of an impedance. You measure an admittance by exciting the concerned port with a voltage source which produces a current, the response (Figure 1.13). The generalized transfer function of an admittance is:

$$Y_{dp}(s) = \frac{I_1(s)}{V_1(s)} \tag{1.14}$$

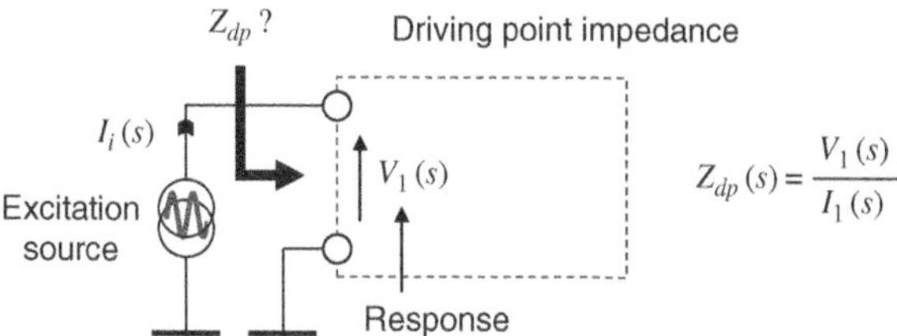

Figure 1.12 Impedances have the dimension of ohms. The excitation signal is a current.

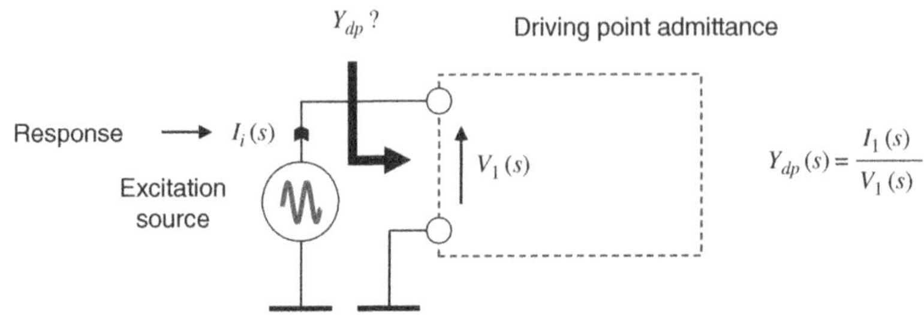

Figure 1.13 Admittances have the dimension of Siemens. The excitation signal is a voltage.

If you consider I_{in} and V_{in} or I_{out} and V_{out}, you respectively measure the network input and output admittances.

Admittances are expressed in siemens, abbreviated [S]. Old notations such *mhos*, ℧ or Ω^{-1} are no longer in use in the International System of units (SI, after the French *Système International d'unités*).

As explained, when determining a port impedance, the excitation signal is a current source. In certain configurations, it is sometimes more convenient to actually calculate the admittance instead by exciting the circuit with a voltage source. The final result is simply reversed to obtain the impedance we are looking for. We will see an application of this principle in an example later on. Figure 1.14 below summarizes the 6 transfer functions we just described.

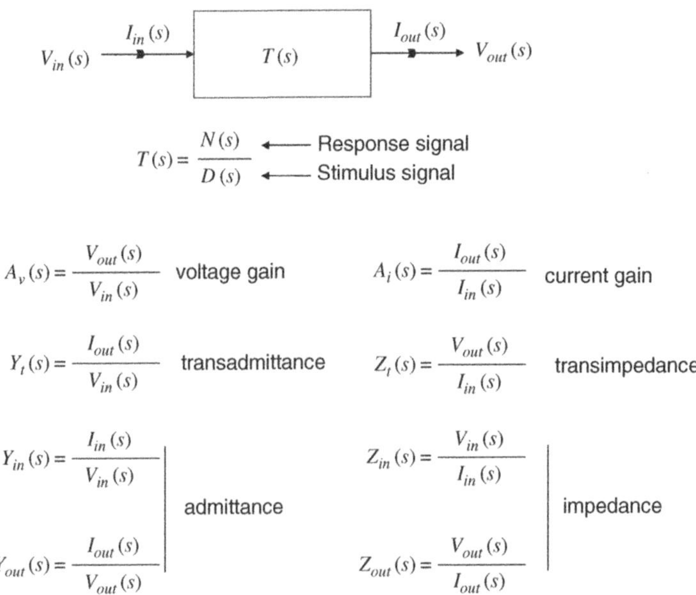

Figure 1.14 There are six different transfer functions, 4 of them have a stimulus and a response at different locations – different ports – while two of them, Z_{dp} and Y_{dp}, have stimulus and response at the same port.

1.2 The Few Tools and Theorems You Did Not Forget . . .

In the litany of theorems and analysis tools I had been taught during my university years, there are a few I did not forget because I exercise them almost every day in my engineer's job. Voltage and current dividers are the first in the tools list. They are of tremendous help when it comes to simplifying circuits and a quick refresh is given below. Among theorems, the first one is Thévenin's theorem, after the French electrical engineer, Charles Léon Thévenin, in 1883. The second is the dual of Thévenin's theorem, Norton's theorem, after the American electrical engineer, Edward Lawry Norton who described the theorem in his 1926 technical memorandum. The third one is obviously the superposition theorem whose extension will lay the foundations for the EET and, later, the *n*EET. Superposition and the EET are thoroughly detailed in Chapter 3.

Let's have a look at a few examples applying these tools, showing how Thévenin and Norton can help us simplify circuits in a quick and efficient way.

1.2.1 The Voltage Divider

This is one of the most useful tools I employ when analyzing electrical circuits. It works with all passive elements in dc or ac (direct or alternating voltages/currents) and the Thévenin theorem makes an extensive use of it. Figure 1.15 shows its simple representation.

The circulating current I_1 is the input voltage V_{in} divided by the total resistive path, $R_1 + R_2$:

$$I_1 = \frac{V_{in}}{R_1 + R_2} \tag{1.15}$$

The voltage across R_2 is the resistance value multiplied by current I_1:

$$V_{out} = I_1 R_2 \tag{1.16}$$

Substituting (1.15) in (1.16), we have:

$$V_{out} = V_{in} \frac{R_2}{R_1 + R_2} \tag{1.17}$$

If we divide both sides of the equation by V_{in}, we have the transfer function linking V_{out} to V_{in}:

$$\frac{V_{out}}{V_{in}} = \frac{R_2}{R_1 + R_2} \tag{1.18}$$

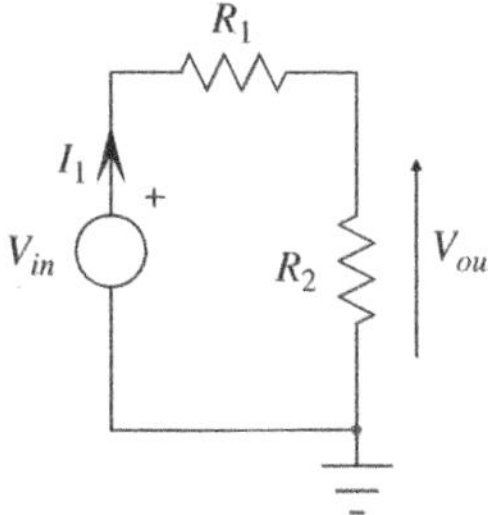

Figure 1.15 A resistive divider is a great tool to simplify circuits.

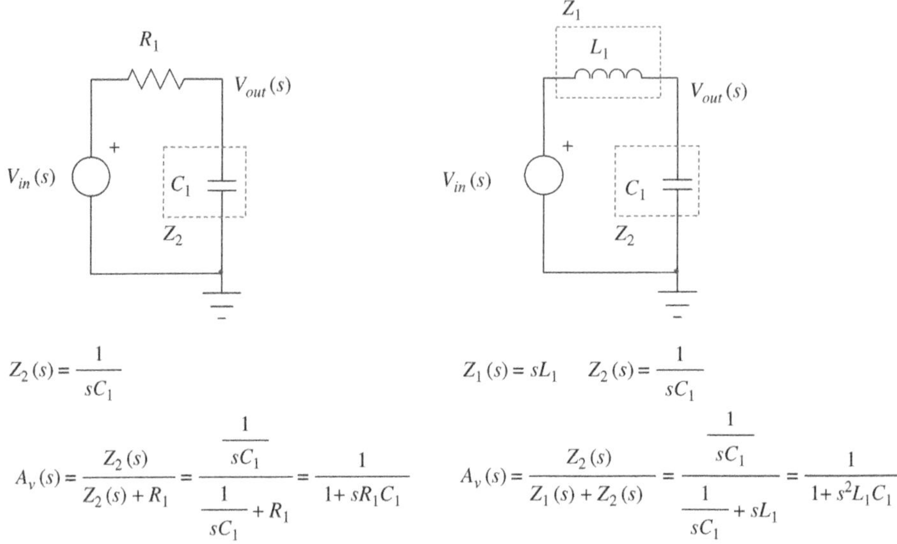

R_1
$V_{out}(s)$
$V_{in}(s)$
C_1
Z_2
Z_1
L_1
$V_{out}(s)$
$V_{in}(s)$
C_1
Z_2
$Z_2(s) = \frac{1}{sC_1}$
$Z_1(s) = sL_1 \quad Z_2(s) = \frac{1}{sC_1}$
$A_v(s) = \frac{Z_2(s)}{Z_2(s) + R_1} = \frac{\frac{1}{sC_1}}{\frac{1}{sC_1} + R_1} = \frac{1}{1 + sR_1C_1}$
$A_v(s) = \frac{Z_2(s)}{Z_1(s) + Z_2(s)} = \frac{\frac{1}{sC_1}}{\frac{1}{sC_1} + sL_1} = \frac{1}{1 + s^2L_1C_1}$

Figure 1.16 The divider equation works with passive elements such as capacitors and inductors.

When you see networks such as those of Figure 1.16, you can immediately apply (1.18) without writing a single line of algebra. In this example, (1.18) is updated with impedances rather than resistances:

$$A_v(s) = \frac{V_{out}(s)}{V_{in}(s)} = \frac{Z_2(s)}{Z_1(s) + Z_2(s)} \tag{1.19}$$

Please note that (1.18) and (1.19) only work if R_2 or Z_2 are unloaded. Should you have another circuit connected across R_2 or Z_2 respectively in Figure 1.15 and Figure 1.16, (1.18) and (1.19) no longer work.

1.2.2 The Current Divider

This is another example of a very useful tool often involved in electrical analysis. Consider Figure 1.17a circuit in which you need to find the current flowing in R_3.

The total current I_1 is V_{in} divided by the resistive path connected to the source:

$$I_1 = \frac{V_{in}}{R_1 + R_2||R_3} \tag{1.20}$$

In this expression, the '||' operator refers to the paralleling of R_2 and R_3:

$$R_2||R_3 = \frac{R_2R_3}{R_2 + R_3} \tag{1.21}$$

Mathematically, the parallel operator has precedence over the addition: $R_2||R_3$ is first computed and then added to R_1.

The original sketch can then be updated to a simpler one as shown in Figure 1.17b. Kirchhoff's current law (KCL) tells us that the sum of the currents entering a junction equals the sum of currents

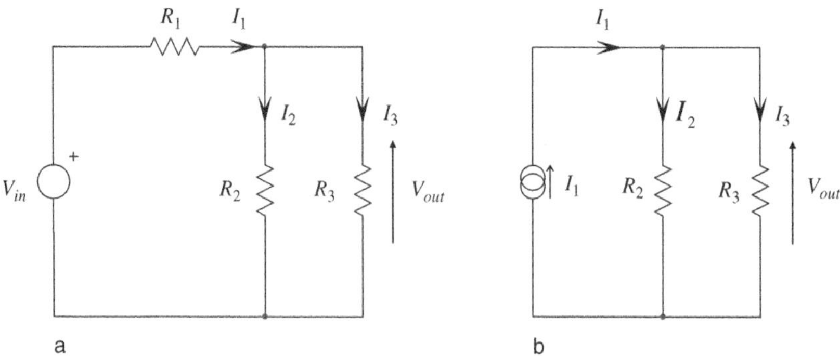

Figure 1.17 The current divider is another great simple tool.

leaving it. Thus:

$$I_1 = I_2 + I_3 \tag{1.22}$$

Currents I_2 and I_3 are defined by the voltage across their terminals, V_{out}:

$$I_3 = \frac{V_{out}}{R_3} \tag{1.23}$$

$$I_2 = \frac{V_{out}}{R_2} \tag{1.24}$$

Extracting V_{out} from (1.23) and (1.24) then equating results gives another relationship linking I_3 and I_2:

$$R_3 I_3 = R_2 I_2 \tag{1.25}$$

Extracting I_2 from (1.22) and substituting it in (1.25) leads to:

$$R_3 I_3 = R_2(I_1 - I_3) \tag{1.26}$$

Rearranging and factoring leads to the relationship linking I_3 and I_1:

$$I_3 = I_1 \frac{R_2}{R_2 + R_3} \tag{1.27}$$

This is the current divider expression which helps us get the current into R_2 or R_3 when I_1 splits between these elements. Figure 1.18 gives another representation. The current flowing in R_2 equals

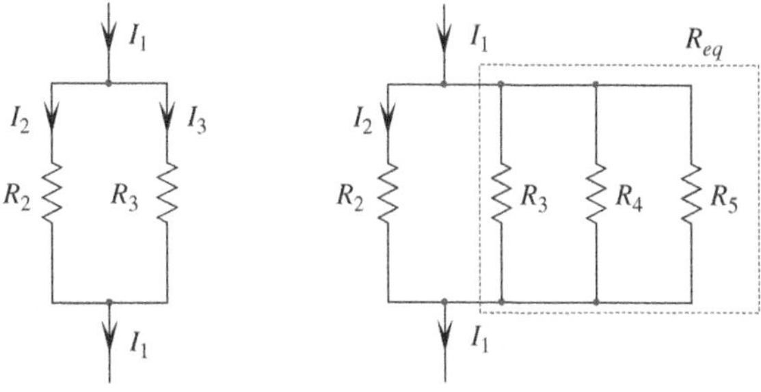

Figure 1.18 The current divider is easily generalized to paralleled resistors.

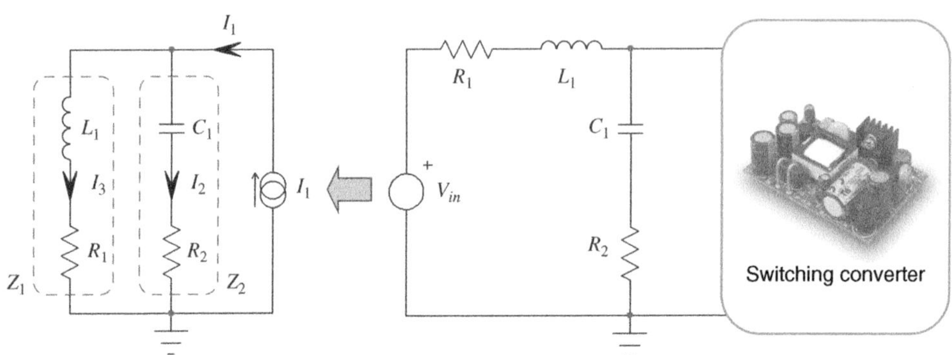

Figure 1.19 Passive elements arranged to form a filter: how much current flows in L_1?

the main current I_1 multiplied by the resistance 'facing' R_2 (thus R_3) and divided by the sum of resistances, R_2+R_3. The right side of Figure 1.18 generalizes the concept where more resistors are connected in parallel with R_3. If $R_{eq} = R_3||R_4||R_5$ then the current in R_2 is simply:

$$I_2 = I_1 \frac{R_{eq}}{R_{eq} + R_2} \tag{1.28}$$

This technique works equally well with energy-storing components as represented in Figure 1.19. This is a typical Electromagnetic Interference (EMI) filter found in switching converters. I_1 illustrates the converter current signature – its high-frequency input current – C_1 is the front-end capacitor while L_1 is the filtering inductor. With a perfect filter, all the alternating current would flow in C_1 while only direct current flows in L_1, providing the dc source with the right isolation to the switching current. Reality differs and what you need is the current really flowing in L_1 and check what attenuation this configuration brings. Apply the current divider expression to Figure 1.19 circuit and you have

$$\frac{I_3(s)}{I_1(s)} = \frac{Z_2(s)}{Z_1(s)+Z_2(s)} = \frac{R_2 + \frac{1}{sC_1}}{R_2 + \frac{1}{sC_1} + R_1 + sL_1} = \frac{1 + sR_2C_1}{1 + sC_1(R_1+R_2) + s^2L_1C_1} \tag{1.29}$$

We did not write a single equation to derive this transfer function, we just *inspected* the figure and applied the current division law. This technique is called solving for a transfer function by *inspection*.

1.2.3 Thévenin's Theorem at Work

Any 2-port *linear* system made of resistors, capacitors, inductors, dependent/independent current/voltage sources can be represented by an equivalent Thévenin model. This equivalent circuit is made of a complex generator V_{th} associated with a complex output impedance Z_{th}. When solving complex networks transfer functions, or if the current or voltage at a given point is needed, the idea is to apply Thévenin's theorem and break the complex circuit into a simpler representation with a Thévenin equivalent circuit in place. This idea behind Thévenin's approach is to model the I-V characteristics 'seen' by the load. You remove the load and model the equivalent source that drives it, affected by a certain output impedance/resistance. As such, Thévenin's and Norton's equivalent circuits do not

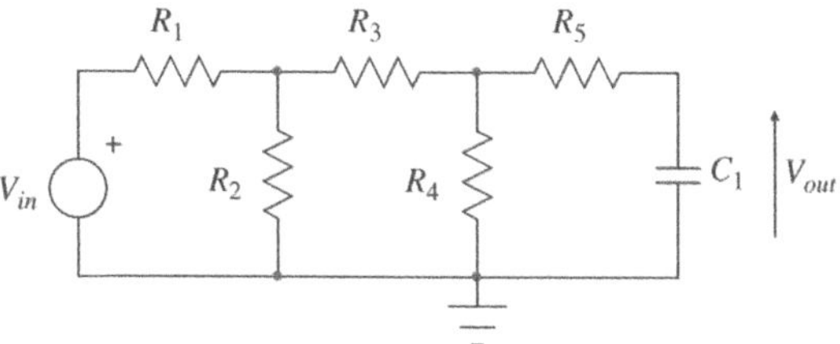

Figure 1.20 In this circuit, five resistors drive capacitor C_1. Rather than going through KCL and KVL, use the Thévenin's generator approach.

reflect the power dissipated by the network they replace. Use them carefully when evaluating powers or currents at certain points in the circuit.

Assume you need to calculate the transfer function $V_{out}(s)/V_{in}(s)$ of the circuit in Figure 1.20. This is a classical case to which we purposely added more resistors than in examples you can find in the web. The goal of using Thévenin is to reduce this complex circuit into a simple structure from which you can immediately deduce the transfer function by inspection. The first option is to use Kirchhoff's voltage and current laws (KCL and KVL) and write mesh and nodes equations. It is very likely that you will obtain the result but the chance also exists that you will make mistakes while writing these expressions. This is the so-called brute-force analysis. The second option uses Thévenin and represents a step towards *fast analytical circuit techniques* whose acronym is FACTs. We must find a place in the network where the insertion of our equivalent generator will simplify the analysis. Let's proceed step by step. We first cut the circuit after R_2 as shown in Figure 1.21 to isolate a first equivalent generator.

The Thévenin voltage is the voltage appearing across R_2 while separated from the rest of the circuit. We can apply the voltage divider law as R_2 is unloaded after the separation. Looking at the upper side of Figure 1.21, this voltage is simply:

$$V_{th1} = V_{in}(s)\frac{R_2}{R_1 + R_2} \tag{1.30}$$

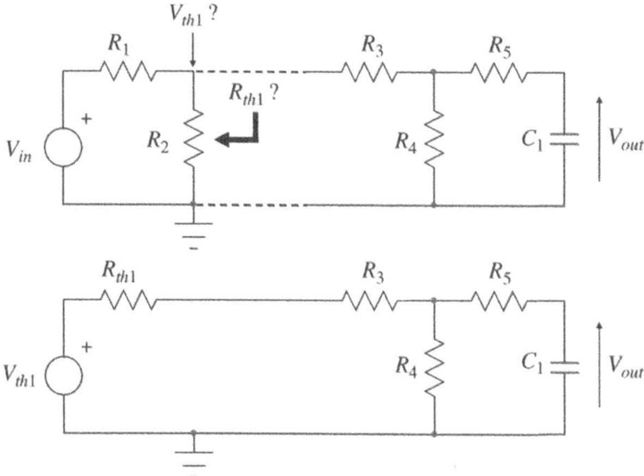

Figure 1.21 You must find a place in the circuit to identify a Thévenin equivalent generator.

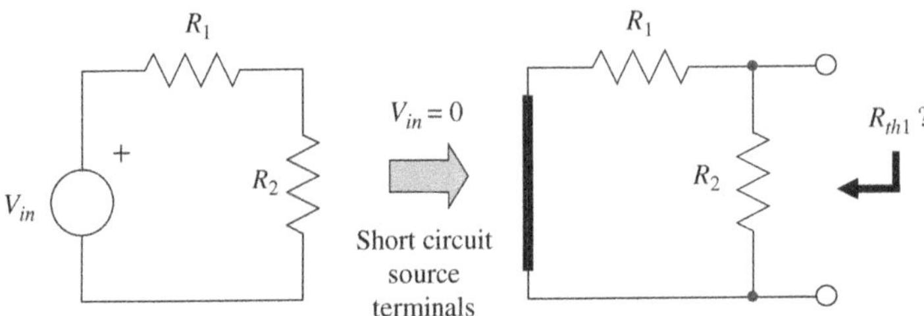

Figure 1.22 The output resistance is found by looking into the output port, across R_2 while the voltage generator is set to 0 V.

We have the Thévenin generator expression so what is its output resistance in this case? The output resistance, as explained in Appendix 1A, is found by setting the input voltage to 0 V and find the resistance seen across R_2's terminals (Figure 1.22).

The resistance is immediate:

$$R_{th1} = R_1 \| R_2 \tag{1.31}$$

We can now replace the input source associated with R_1 and R_2 by its equivalent Thévenin's generator. It appears in the upper side of Figure 1.23. We have an equivalent circuit that mimics the I-V characteristic driving the circuit made of R_3 and the rest of the elements. Coming back on our note regarding caution in using Thévenin (or Norton), you can see that the generator in Figure 1.22 dissipates power when unloaded $-V_{in}^2/(R_1+R_2)-$ while the equivalent model involving V_{th} and R_{th} does not. Using Thévenin to calculate power levels or efficiency figures would lead to a wrong result.

Simplifying further on, before reaching the capacitor, another resistive divider is present. We can update the previous Thévenin generator by accounting for the presence of these elements. The voltage divider approach is still useful:

$$V_{th2} = V_{th1}\frac{R_4}{R_4+R_3+R_{th1}} = V_{in}\frac{R_2}{R_1+R_2}\frac{R_4}{R_4+R_3+R_{th1}} \tag{1.32}$$

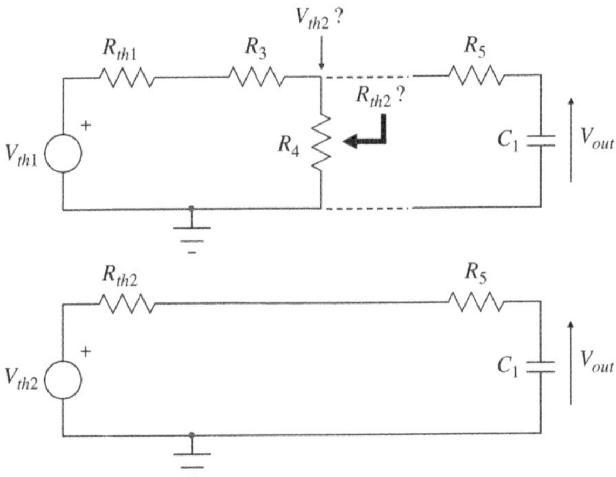

Figure 1.23 The Thévenin output resistance is found by looking into the output port, across R_4:

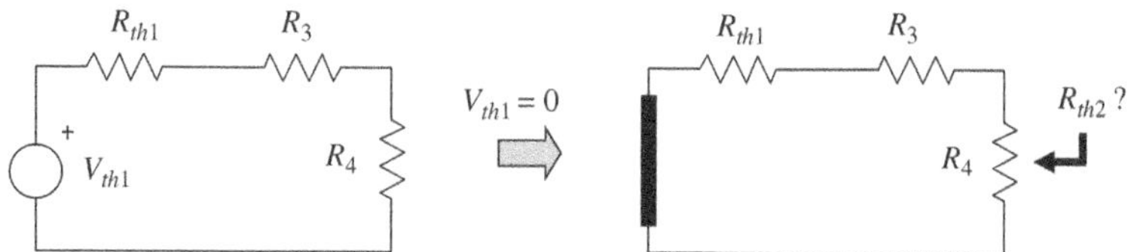

Figure 1.24 The first Thévenin generator is updated with the presence of R_3 and R_4:

The output resistance is found by setting V_{th1} to 0 and looking into R_4 terminals to obtain the resistance while R_4 remains in place (Figure 1.24).

$$R_{th2} = (R_{th1} + R_3) \| R_4 \tag{1.33}$$

The final circuit appears in the low side of Figure 1.23 where, again, a simple voltage divider appears. Its transfer function is that of the low-pass filter in the lower left corner of Figure 1.16 where the resistance is the sum of R_{th2} and R_5:

$$\frac{V_{out}(s)}{V_{th2}(s)} = \frac{1}{1 + s(R_{th2} + R_5)C_1} \tag{1.34}$$

V_{th2} must be replaced by its definition in (1.32) with V_{in} and $R_{th1/2}$ expressions brought back in (1.34). After the update, the final expression is:

$$\frac{V_{out}(s)}{V_{in}(s)} = \frac{R_2}{R_1 + R_2} \cdot \frac{R_4}{R_4 + R_3 + R_1 \| R_2} \cdot \frac{1}{1 + s\left[\left(R_1 \| R_2 + R_3\right) \| R_4 + R_5\right] C_1} \tag{1.35}$$

It is a rather large equation but we don't know if it is correct yet. Let's try a different approach. If we look at (1.34), the denominator expression includes a term in which C_1 is multiplied by a resistance, $R_{th2} + R_5$. The resulting RC term is a time constant. If we refer to our first steps when looking into ports (Figure 1.7), we said that the resistance 'seen' by the capacitor when looking into its port is further associated with the capacitor to form a time constant denoted τ. Well, let's try to do the same with our complex circuit from Figure 1.20. When calculating output impedance/resistance, the excitation source V_{in} plays no role and can be turned off. Turning off a voltage source is equivalent to replacing it with a short circuit, a 0-V source. We will see later a more rigorous explanation for this fact, let's accept it now. Once V_{in} is replaced by a strap and capacitor C_1 removed from the circuit, Figure 1.25 appears.

With a simple drawing like that, inspection is child's play. If we start from the left, we see R_1 paralleled with R_2, then in series with R_3 and the whole is paralleled with R_4. This total resistance is in series with R_5. Finally:

$$R = R_5 + \left(R_1 \| R_2 + R_3\right) \| R_4 \tag{1.36}$$

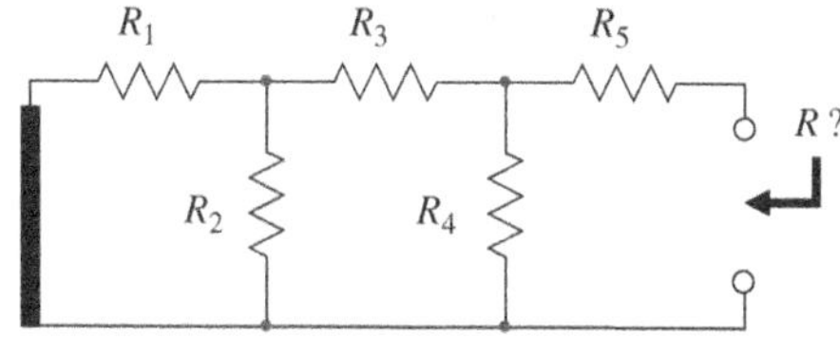

Figure 1.25 What resistance is 'driving' capacitor C_1?

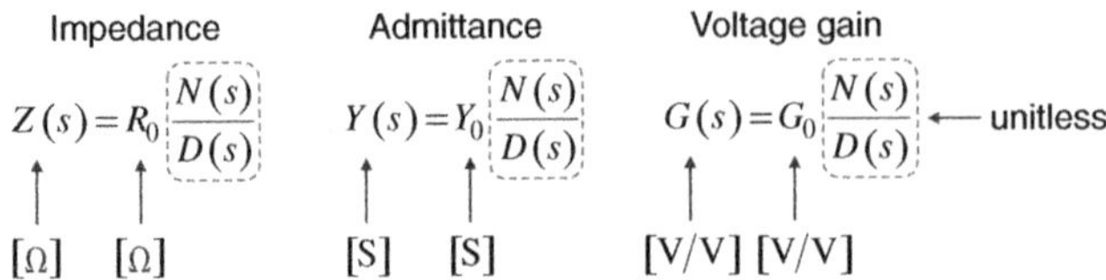

Figure 1.26 In a properly-written transfer function, the leading term carries the unit (if any) while the numerator and the denominator are unitless.

This is the exact same definition we have for the resistive portion of (1.35) time constant. No Thévenin, no complex manipulations were involved to build this expression.

In (1.35), when s equals 0, we talk about dc condition, the 0-Hz response. In some transfer function equations we will derive in this book, the 0-Hz response can be null, infinite or finite as in our case. In (1.35), if you replace s by 0, the right-side denominator becomes 1 and the equation's left term remains alone: this is a dc term whose unit must be that of the transfer function under study. Here, we calculate a gain in [V]/[V] and it has no unit. Should we calculate an impedance, its dimension would be ohms. You mark this dc term by a subscripted 0 when you write it. It is usually accepted that its letter is the same as that of the transfer function you study: A_0 for $A(s)$, H_0 for $H(s)$, G_0 for $G(s)$ and so on. Exception is for R_0 when calculating an impedance $Z(s)$. Figure 1.26 illustrates this fact, valid regardless of the transfer function order.

Below appears the first generalized transfer function of a circuit described by (1.35) that we call H:

$$H(s) = H_0 \frac{1}{1 + s\tau} \tag{1.37}$$

in which

$$H_0 = \frac{R_2}{R_1 + R_2} \frac{R_4}{R_4 + R_3 + R_1 \| R_2} \tag{1.38}$$

and

$$\tau = \left(R_5 + \left(R_1 \| R_2 + R_3\right) \| R_4\right) C_1 \tag{1.39}$$

What physically happens in a circuit under dc condition or at a 0-Hz excitation? A capacitor becomes an infinite resistance (no current flows in it) and an inductor is replaced by a short circuit. When you analyze a circuit under dc conditions, you thus open all capacitors and short all inductors. This is, by the way, what SPICE does when it calculates a bias point prior to starting a simulation whether it is a .TRAN or .AC analysis. You have the corollary that at very high frequencies or infinite frequency, capacitors becomes short circuits and inductors become open circuits. When you analyze a circuit for s approaching infinity, then you short all capacitors and open all inductors from the network. We will come back to these important points, but let's focus on Figure 1.20 where C_1 has been removed. The new diagram appears in Figure 1.27.

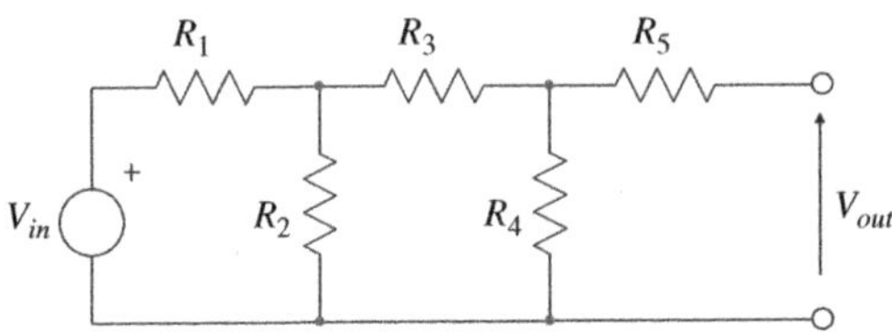

Figure 1.27 In dc conditions, capacitor C_1 is removed as no current flows in it.

In this drawing, R_5 does not play a role as no current flows through it. The dc transfer function is what we already derived in (1.32) when applying Thévenin two times. It is H_0 of (1.38). As a preliminary conclusion, we could have derived (1.35) in two steps, first by considering $s=0$ and obtaining H_0, then by setting the excitation source to 0 V and looking into the capacitor ports to get the associated time constant.

In this approach, what we calculate is a simple gain involving resistors only (where capacitors or inductors are respectively open or shorted) further followed by an output resistance calculation, the one seen from the capacitor terminals. Assembling these elements according to (1.37) gave the transfer function. These are the first steps towards fast analytical circuit techniques also known as FACTs.

1.2.4 Norton's Theorem at Work

Any 2-port *linear* system made of resistors, capacitors, inductors, dependent or independent current or voltage sources can be represented by an equivalent Norton model. This equivalent circuit is made of a complex current generator I_{th} associated with a complex output impedance Z_{th}. Thévenin and Norton can be used interchangeably depending on the circuit you need to analyze. With similar output impedance Z_{th} in both approaches, I_{th} and V_{th} are linked by the simple formula $I_{th} = V_{th}/Z_{th}$.

Assume the filter shown in Figure 1.28 in which you see an inductor associated with three resistors. r_L symbolizes the inductor equivalent series resistance (ESR), its ohmic losses.

To obtain the transfer function, let's cut the circuit after R_2 and transform the input source involving R_1 and R_2 into a Norton generator. The result appears in Figure 1.29. First, the Norton current is found. This current is either equal to V_{th}/R_{th} or to the short circuit current when a strap is applied across R_2. In this case, the current I_{th} is simply V_{in}/R_1. The output resistance has already been evaluated in previous examples and is equal to the parallel arrangement of R_1 and R_2. Once the Norton transformation is done, you can place the equivalent generator into the circuit as proposed in Figure 1.30. In this case, the output voltage is simply:

$$V_{out}(s) = I_{th}(s)Z_1(s) = \frac{V_{in}(s)}{R_1}\left[R_{th}\|(r_L + sL_1)\right] \tag{1.40}$$

If you now develop this expression and rearrange the terms, you should obtain an expression similar to (1.41):

$$\frac{V_{out}(s)}{V_{in}(s)} = \frac{r_L R_2}{R_1 R_2 + r_L(R_1 + R_2)} \frac{1 + s\dfrac{L_1}{r_L}}{1 + s\dfrac{L_1(R_1 + R_2)}{R_1 R_2 + r_L(R_1 + R_2)}} \tag{1.41}$$

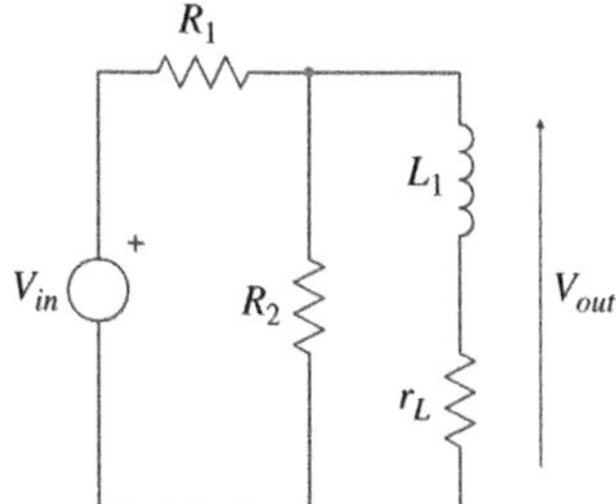

Figure 1.28 Norton's theorem can be applied to obtain the transfer function quickly.

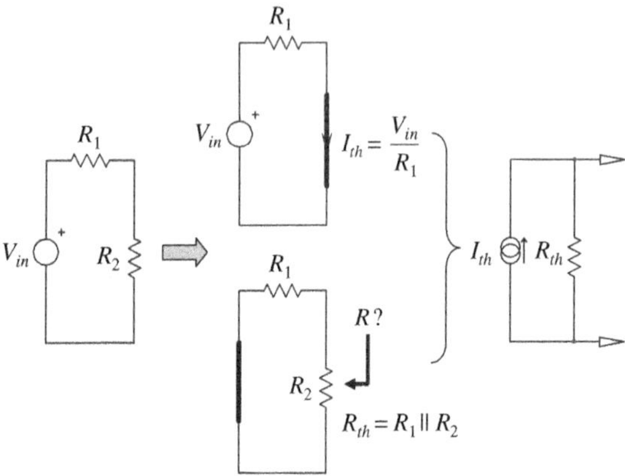

Figure 1.29 The Norton current generator sources current to a series-parallel arrangement involving inductor L_1.

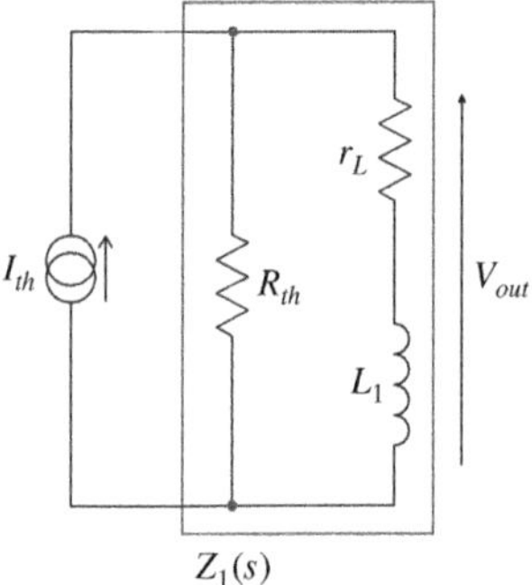

Figure 1.30 The output voltage is simply the current source times impedance Z_1.

We now see two time constants, one is in the numerator while the second lies in the denominator. We can rewrite this transfer function capitalizing on the notation introduced with (1.37):

$$H(s) = H_0 \frac{1 + s\tau_1}{1 + s\tau_2} \tag{1.42}$$

in which

$$H_0 = \frac{r_L R_2}{R_1 R_2 + r_L(R_1 + R_2)} \tag{1.43}$$

$$\tau_1 = \frac{L_1}{r_L} \tag{1.44}$$

and

$$\tau_2 = \frac{L_1}{R_{eq}} \tag{1.45}$$

where $R_{eq} = \dfrac{R_1 R_2 + r_L(R_1 + R_2)}{R_1 + R_2}$.

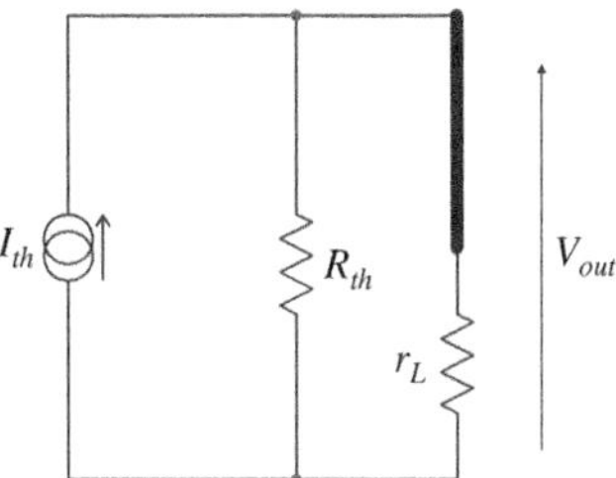

Figure 1.31 In dc conditions, the inductor is a short circuit: you replace it by a strap.

Now, rather applying ohm's law as in (1.40), let's see if we can already apply what we learned in the Thévenin example. First, the easiest thing, set s to 0 and solve the dc transfer function H_0. If a capacitor is an open circuit at dc, an inductor becomes a short circuit. In Figure 1.30, short the inductor and you have Figure 1.31. From this figure, the dc gain is immediate:

$$\frac{V_{out}(0)}{I_{th}(0)} = R_{th} \| r_L \tag{1.46}$$

Now substitute R_{th} and I_{th} definitions in (1.46) to obtain:

$$H_0 = \frac{(R_1 \| R_2) \| r_L}{R_1} \tag{1.47}$$

No special development or rearrangement was necessary here. If you check, (1.47) is the same as (1.43). Should you start from Figure 1.28 instead and short the inductor in dc, you would find another definition for H_0, equal to (1.47) but expressed differently:

$$H_0 = \frac{r_L \| R_2}{(r_L \| R_2) + R_1} \tag{1.48}$$

In these above expressions, (1.47) and (1.48), resistors appear in an ordered series-parallel arrangement. This is not the case for (1.43) in which resistors are combined with each other without a noticeable relationship between them. An ordered arrangement helps gain insight into the formula without rearranging elements. For instance, in (1.47), you see that if r_L goes to infinity, the dc gain reduces to that of a simple voltage divider involving R_1 and R_2. Even simpler in (1.48) where H_0 simplifies immediately to $R_2/(R_1 + R_2)$. Equation (1.43) does not offer the same immediate insight; you would need to factor r_L and make it infinite to simplify the formula. In other words, you would need more effort to rearrange the complex formula and get the response you want. In that respect, (1.47) and (1.48) are designated as *low-entropy* expressions by analogy to thermodynamic laws.

The entropy of a system qualifies its degree of internal disorder: to produce the work the system has been designed for, you need to bring less external energy when its entropy is low. In our equations, with well-organized, well-ordered constitutive elements, insight is immediate and no further work is required to unveil gains, poles or zeros positions. On the other hand, in a *high-entropy* equation, where elements are in disordered form, you need to spend more energy to rearrange terms and reveal key relationships. We will see that FACTs naturally deliver low-entropy expressions whereas brute-force analysis often produces a correct but abstruse result.

Now that we have the dc gain H_0, let's go back to (1.42). Compared to (1.37), this time, we have two time constants. One is in the numerator while the other lies in the denominator. In Chapter 2, we will learn that transfer functions are combining gains, poles and zeros. Without disclosing too many details now, zeros appear in the transfer function numerator N while poles are in the denominator D. In other words, τ_1 in (1.42) corresponds to the zero time constant while τ_2 characterizes the pole time constant. As already highlighted, both time constants involve a resistive term driving the considered element (C or L) that we sometimes can find by inspection.

The mathematical definition of a zero in a function $f(x)$ is the value of x for which f returns 0. In a transfer function, a zero noted s_z represents the root of the numerator N. When a network featuring zeros is evaluated at $s = s_z$, the numerator N of the corresponding transfer function cancels:

$$N(s_z) = 0 \tag{1.49}$$

For instance, in (1.41), when $s = -\frac{r_L}{L}$, a real value in the complex plane, the numerator equals 0. In this condition, the transfer function linking the response signal to the driving signal also returns 0:

$$\frac{V_{out}(s_z)}{V_{in}(s_z)} = \frac{N(s_z)}{D(s_z)} = \frac{0}{D(s_z)} = 0 \tag{1.50}$$

If the transfer function is 0 at $s = s_z$, then despite the presence of a driving signal V_{in}, the response V_{out} is also 0. From this simple observation, we can infer that the presence of a zero in a transfer function implies that the response is *nulled* when the *transformed* network is examined at $s = s_z$. Figure 1.32 illustrates this fact through a simple drawing.

The word *transformed* means that all energy-storing elements are replaced by their impedance expressed in the Laplace domain as shown in Figure 1.33. If the response is a null while a driving signal exists, it means that the excitation does not reach the output and is lost somewhere in the *transformed* network examined at $s = s_z$. Figure 1.34a and b illustrates two cases leading to $V_{out} = 0$ V in this particular condition.

A null in the response implies that no current circulates in resistor R_1 hence the label $I_{out}(s_z) = 0$. If you observe Figure 1.34a, the absence of current in R_1, despite the presence of a driving signal, is due to the series network Z_1 becoming a *transformed open* for $s = s_z$. The presence of this series infinite impedance blocks all current circulation and induces an output null at $s = s_z$. In Figure 1.34b, a current circulates in resistor R_2 but a *transformed short circuit* diverts all of it from resistor R_1 to ground, nulling the output. By observing the conditions for which an output null is created in the transformed circuit, we have the possibility to obtain the transfer function zeros just by inspecting the network.

Back to Figure 1.30 or Figure 1.28, what circuit association could bring a null to the output when the network is evaluated at $s = s_z$? R_{th} is fixed and frequency-invariant. However, the series association of L_1 and r_L could perhaps be a transformed short circuit at a certain s? The impedance of this network is:

$$Z(s) = r_L + sL_1 \tag{1.51}$$

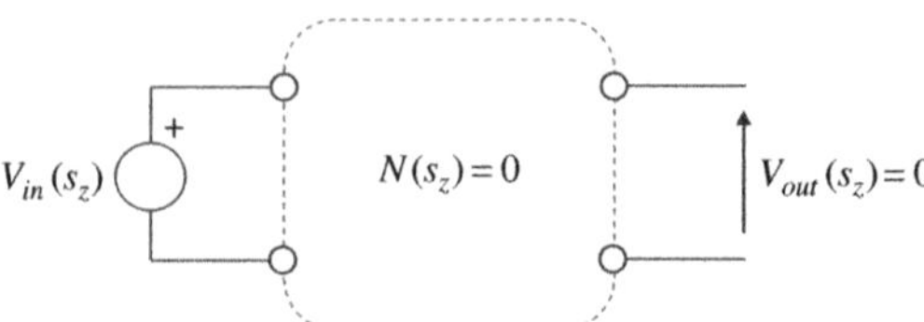

Figure 1.32 For $s = s_z$ the numerator of a transfer function featuring a zero cancels and the response is nulled.

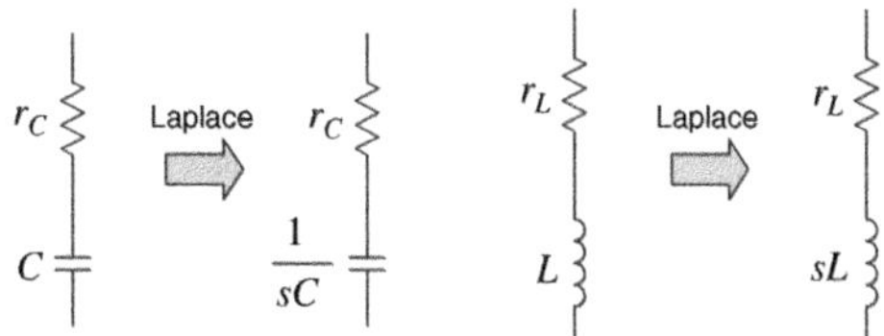

Figure 1.33 In the transformed world, capacitive and inductive elements must be replaced by their Laplace impedance expressions.

For what value of s will this expression be 0? In other words, what is the root of this equation? You have to solve:

$$0 = r_L + sL_1 \tag{1.52}$$

which leads to

$$s_z = -\frac{r_L}{L_1} \tag{1.53}$$

This is a complex root whose magnitude is:

$$\omega_z = |s_z| = \frac{r_L}{L_1} \tag{1.54}$$

In the next chapters, we will learn that:

$$\tau_1 = \frac{1}{\omega_z} = \frac{L_1}{r_L} \tag{1.55}$$

Now, if you go to the laboratory and solder a resistor in series with an inductor then drive the obtained element with an ac current source, there is no way you will cancel the response (the voltage across the network). Actually, as (1.53) suggests with this real zero, the only way to make the transformed impedance equal a short circuit would be to consider s in the entire complex plane and not only along the imaginary axis as we do when we set $s = j\omega$. The method offered above is thus an abstraction which translates the mathematical definition of a zero into the Laplace-transformed world. Despite its lack of physical significance, it is an extremely useful way to identify zeros and will be heavily used in what we will later call the *Null Double Injection* (NDI). By the way, an output null can be physically produced at a 0-Hz frequency when you have a zero at the origin as the origin is common to both real

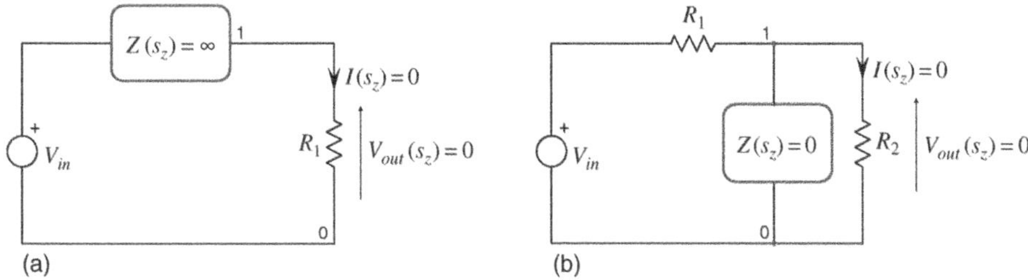

Figure 1.34 In the transformed circuit, when s equals s_z, a series infinite impedance or a transformed short circuit to ground prevents the driving signal from reaching the output and creates a null in the response: $V_{out}(s_z) = 0$

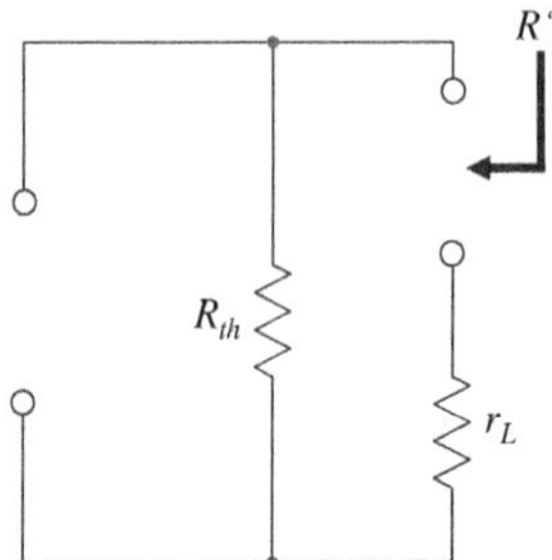

Figure 1.35 When the excitation signal is a current source, turning it off is similar to removing it from the circuit.

and imaginary axes. This is the case in a circuit where a capacitor lies in series with the signal path, for instance, or an inductor is paralleled with the response signal. The other case in which a null can be physically obtained is with a highly underdamped notch filter. As the transfer function numerator quality factor increases, the zero-pair approaches the imaginary axis and zeros become pure imaginary conjugates. When you excite this filter at a frequency f_z where both zeros are located, you truly observe a null in the output.

To find the second time constant τ_2, we can apply what we already learned: we suppress the excitation signal (no role in output resistances definitions) and look at the resistance that drives L_1. If we apply this concept to Figure 1.30, the output resistance seen at the inductor terminals does not depend on the current source I_{th}. We can then turn it to 0 A. Turning an independent current source off is similar to removing it from the circuit thus leaving an open circuit in place. This is what Figure 1.35 suggests.

What resistance do you see from the inductor terminals? r_L in series with R_{th}. If you replace R_{th} by its definition from Figure 1.29, you have:

$$R = r_L + R_{th} = r_L + \left(R_1 \| R_2\right) \tag{1.56}$$

The second time constant is thus:

$$\tau_2 = \frac{L_1}{r_L + \left(R_1 \| R_2\right)} \tag{1.57}$$

Should you do the same in Figure 1.28 but shorting V_{in} instead (excitation is 0 V) and look at the inductor port resistance you see, you will find (1.56).

Now associating (1.48), (1.55) and (1.57), we can write the transfer function describing Figure 1.28 in a normalized form. This is truly a low-entropy expression:

$$\frac{V_{out}(s)}{V_{in}(s)} = \frac{r_L \| R_2}{\left(r_L \| R_2\right) + R_1} \frac{1 + s\frac{L_1}{r_L}}{1 + s\frac{L_1}{\left(R_1 \| R_2\right) + r_L}} \tag{1.58}$$

We can even rearrange it in a more readable format, where a zero and a pole now appear:

$$\frac{V_{out}(s)}{V_{in}(s)} = H_0 \frac{1 + \frac{s}{\omega_{z_1}}}{1 + \frac{s}{\omega_{p_1}}} \tag{1.59}$$

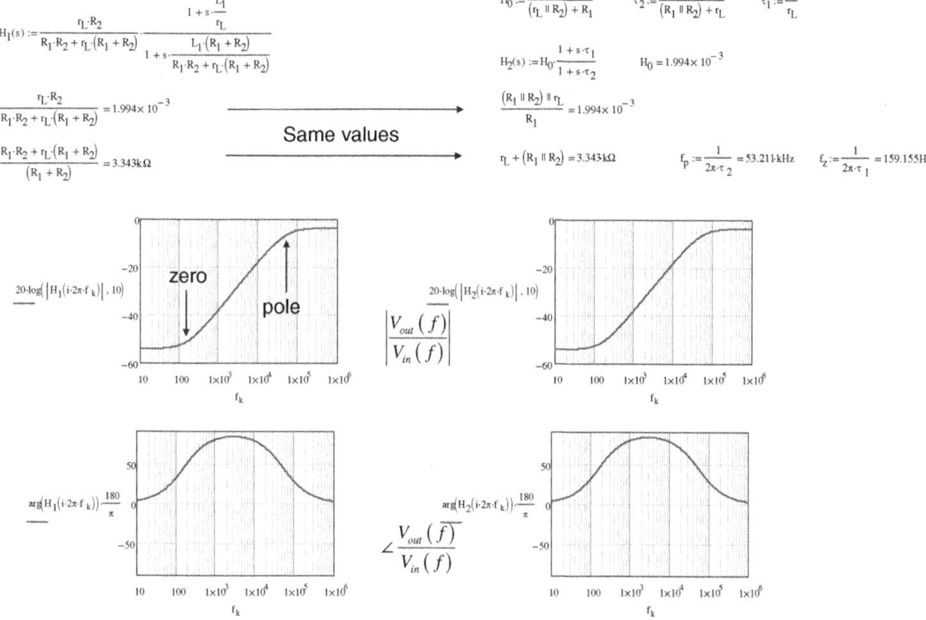

Figure 1.36 Ac responses from the low and high entropy equations are identical.

where

$$H_0 = \frac{r_L \| R_2}{(r_L \| R_2) + R_1} \tag{1.60}$$

$$\omega_{z_1} = \frac{1}{\tau_1} = \frac{r_L}{L_1} \tag{1.61}$$

$$\omega_{p_1} = \frac{1}{\tau_2} = \frac{(R_1 \| R_2) + r_L}{L_1} \tag{1.62}$$

To verify our calculations, we have captured these equations into a Mathcad® sheet and plotted the ac response. Results appear in Figure 1.36.

In the upper left corner is the high-entropy expression whose ac response is plotted below. The right side shows the low-entropy version involving the time constants we quickly obtained by inspection. Values returned either way are strictly identical.

1.3 What Should I Retain from this Chapter?

In this first chapter, we have learned key information that are summarized below:

1. A transfer function is a mathematical relationship linking an excitation signal (the input) to a response signal (the output). Excitation and response can appear at different terminals or ports but can also be observed across a common port. This is the case for impedance and admittance transfer functions.

2. A transfer function is usually made of a numerator N and a denominator D but not always. When written in the form of a fraction, the zeros of the transfer function are the numerator roots while poles are the denominator roots.
3. A network featuring storage elements such as capacitors and inductors involve time constants. These time constants imply a resistive term R that 'drives' the concerned capacitor or inductor. This resistance can be observed, in certain conditions, by 'looking' into the considered element terminals while the said element is removed from the circuit. A time constant involving a capacitive term is $\tau = RC$ while a time constant characterizing an inductive term is $\tau = L/R$.
4. When the port output resistance is evaluated, we have seen that the input source does not play a role in the resistance expression. When evaluating a port output resistance, the excitation voltage source is turned off (set to 0 V) and is replaced by a short circuit (a strap). For the dual case, if the excitation source is a current generator, it must be set to 0 A or become an open circuit.
5. Fast Analytical Circuits Techniques (FACTs) consist of expressing a transfer function with the above time constants and gains in a clear and ordered form. This form is said to be of *low entropy* if you can tell where poles, zeros, and gains are located without having to rework the equation.
6. There are several important analysis techniques that you must know and be at ease with to start manipulating complex networks: the voltage divider, the current divider and Thévenin's/Norton's theorems. Superposition sets the foundations for the Extra Element Theorem we will discover in the next chapter.
7. By applying some of the simple techniques explored in this chapter, we were able to derive a transfer function without writing a single equation. In other words, we derived the transfer function by inspection. When circuits are not too complex, writing the transfer function by inspection is a real pleasure!

References

1. V. Vorpérian. *Fast Analytical Techniques for Electrical and Electronic Circuits*. Cambridge University Press, 2002, pp. 15–17.
2. Understand and apply the transimpedance amplifier. Planet Analog blog, http://www.planetanalog.com/document.asp?doc_id=527534&site=planetanalog (last accessed, 12/12/2015).

1.4 Appendix 1A – Finding Output Impedance/Resistance

As exemplified in the introductory figures, finding time constants associated with capacitors or inductors will often involve the derivation of the resistive term that 'drives' the considered capacitor or inductor. Besides capacitors and inductors, the exercise can also involve a simple resistor for which finding the resistance or impedance seen from its terminals is important. In other terms, what resistance is offered from the terminals the considered element is connected to?

There are several known methods to find the output impedance or resistance of a given network. They are reviewed in the appendix below. In the examples, we will use SPICE notations for the sake of simplicity: $1k = 10^3$, $1Meg = 10^6$, $1p = 10^{-12}$, $1n = 10^{-9}$, $1m = 10^{-3}$ and $1u = 10^{-6}$.

The Voltage Output is Divided by Two

For the first option, assume you have a resistive divider made of two resistors driving a capacitor. You have removed the capacitor as in Figure 1.7 and the circuit involving R_1 and R_2 appears in the left side of Figure 1.37. If you load that circuit with a resistance R, the output voltage takes a certain value V_{out}, lower than V_{in}. The drop is incurred to the output resistance we want and the current delivered to the load. If you now load the same circuit with a resistance equal to the circuit output resistance R_{th}, as in the right side of Figure 1.37, you obtain an output value exactly half of the voltage obtained with no load ($I_{out} = 0$):

$$V_{in}\frac{R_2\|R_{th}}{R_1 + R_2\|R_{th}} = \frac{V_{th}}{2} \tag{1.63}$$

This voltage, V_{th}, is the Thévenin voltage and R_{th} the Thévenin output resistance we want.

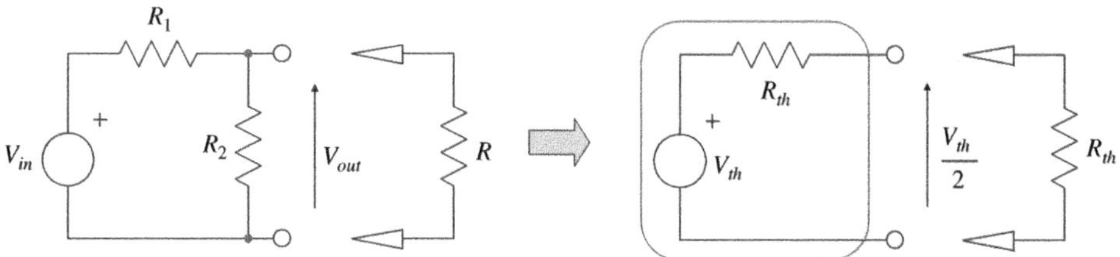

Figure 1.37 If you load a circuit with a resistance R equal to the output resistance R_{th} of that circuit, the output voltage is divided by 2.

Capitalizing on (1.63), you can write:

$$V_{in}\frac{R_2\|R_{th}}{R_1 + R_2\|R_{th}} = \frac{V_{in}}{2}\frac{R_2}{R_1 + R_2} \tag{1.64}$$

In this expression, the input voltage V_{in} does not play a role and disappears from both sides. If you solve (1.64) for R_{th}, you obtain the output resistance we want:

$$R_{th} = \frac{R_1 R_2}{R_1 + R_2} = R_1\|R_2 \tag{1.65}$$

A Dynamic Output Resistance

A second method consists of calculating the output voltage in relationship to a current injected by the generator I_{out}, $V_{out} = f(I_{out})$. This is the method already introduced in Figure 1.4. Assuming the same two-resistor circuit in Figure 1.38, the output voltage across R_2 can be defined as follows:

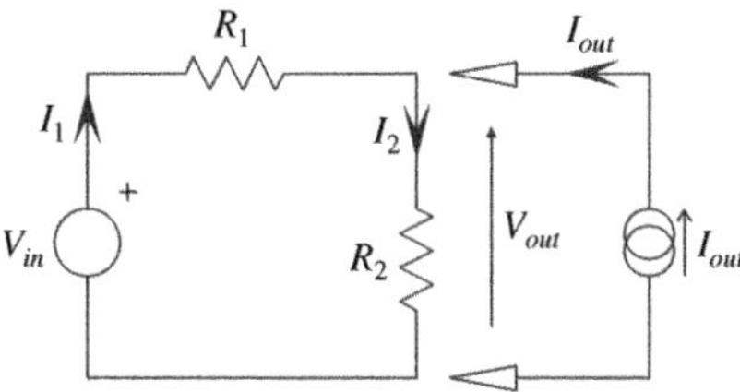

Figure 1.38 If you load the resistive network by a current source, the output voltage drops by a certain quantity, proportional to I_{out} and the network output resistance:

$$V_{out} = V_{in} - R_1 I_1 \tag{1.66}$$

I_1 is made of the output current I_{out} and I_2:

$$I_1 = \frac{V_{out}}{R_2} - I_{out} \tag{1.67}$$

Substituting (1.67) into (1.66):

$$V_{out} = V_{in} - R_1\left(\frac{V_{out}}{R_2} - I_{out}\right) = V_{in} - \frac{R_1}{R_2}V_{out} + R_1 I_{out} \tag{1.68}$$

Rearranging and factoring V_{out} in the left side leads to:

$$V_{out}(I_{out}) = \frac{V_{in} + R_1 I_{out}}{1 + \frac{R_1}{R_2}} \tag{1.69}$$

The incremental or small-signal output resistance is found by differentiating (1.69) with respect to I_{out}:

$$\frac{dV_{out}(I_{out})}{dI_{out}} = \frac{R_1 R_2}{R_1 + R_2} = R_1 \| R_2 \tag{1.70}$$

The word *incremental* refers to measurements involving small voltage (dV) and current (dI) variations around a defined operating point. We talk about small variations so that the system remains *linear* when measurements are performed. To that respect, (1.70) is also referred to as the *small-signal* output resistance: the stimulus signal I_{out} is purposely kept of small amplitude so that V_{out}, the response, remains undistorted.

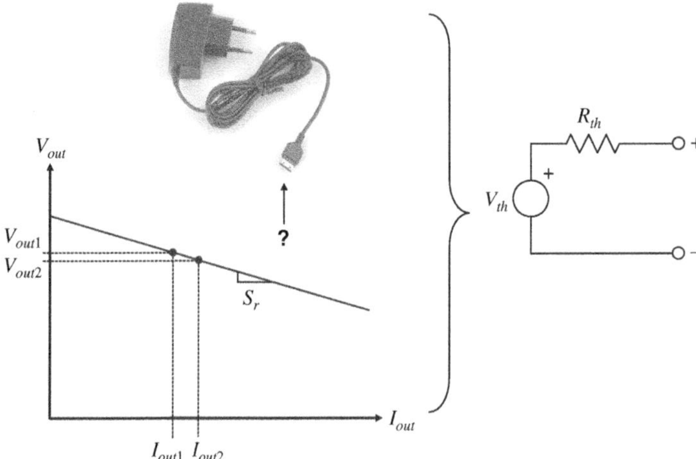

Figure 1.39 The slope of this I-V characteristic is negative. The unit is ohm and is obtained at a given operating point, I_{out1}.

You can find a practical application of (1.70). Assume you want to characterize the dynamic output resistance of a given power supply, such as the simple ac-dc charger in Figure 1.39. This measurement must be performed at the selected operating point of interest. For instance, what is the output resistance at a 12-V output and a 1-A current? First, measure the power supply voltage at current I_{out1} and record the output value, V_{out1}. Then slightly increase the current to I_{out2}, and record the new output voltage V_{out2}. The output resistance at the given operating point is simply:

$$R_{out} = \frac{V_{out1}(I_{out1}) - V_{out2}(I_{out2})}{I_{out1} - I_{out2}} \tag{1.71}$$

To make sure you keep the power supply in a linear zone, the output current variation at which the two output voltages are recorded must remain small. In other words, I_{out2} must be close to I_{out1}, perhaps 1 A and 1.1 A. Mathematically speaking, you performed a differentiation of the output voltage with respect to the output current at $I_{out} = I_{out1}$:

$$\lim_{I_{out2} \to I_{out1}} \frac{V_{out}(I_{out2}) - V_{out}(I_{out1})}{I_{out2} - I_{out1}} = \frac{dV_{out}(I_{out})}{dI_{out}} \tag{1.72}$$

What you obtain with (1.72) is the linear or the small-signal dc or static output resistance of your converter measured at a given output current, 1 A in this example. Please note that (1.71) returns a negative value simply because I_{out} leaves the output port rather than entering it as in Figure 1.38. However, having a voltage drop across R_{th} that is proportional to the output current is the result of a positive resistance.

Setting the Source Contribution to Zero Volts

All elements involved in Figure 1.37 are linear. The source itself, V_{in}, is considered a perfect voltage generator with a zero output resistance. When we load this circuit, V_{in} remains constant. Again, considering the differentiation, we can write $\frac{dV_{in}(I_{out})}{dI_{out}} = 0$: the source contribution to calculating the small-signal output resistance or impedance is 0. Therefore, when we calculate the output resistance (or impedance) of a linear circuit in which a generator is involved, we can set the generator to 0 V or replace it by a strap. Should we need to repeat the exercise with a current generator in the circuit, setting the current to 0 A would be the same as disconnecting the current source from the circuit. Please note that controlled sources must not be put to 0 for this type of analysis.

This third method is exemplified in Figure 1.40 where V_{in} is set to 0 V and replaced by a short. The output resistance is found by looking at the resistance across R_2 terminals, while R_2 remains in place:

$$R_{th} = R_1 \| R_2 \tag{1.73}$$

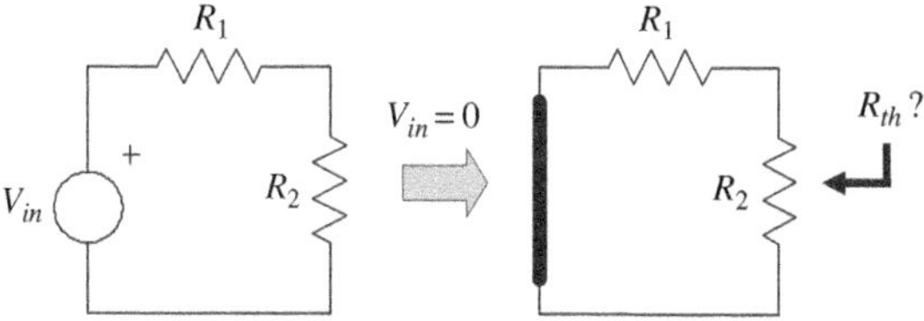

Figure 1.40 The Thévenin output resistance is found by looking into the output port, across R_2.

A similar result to that of (1.65) and (1.70).

It works nicely also when storage elements are added to the circuit as in Figure 1.41a. The output impedance is found by setting the source to 0 V as shown in Figure 1.41b. The impedance is made of

R_1 paralleled with C_1:

$$Z_{th}(s) = R_1 \| \frac{1}{sC_1} \tag{1.74}$$

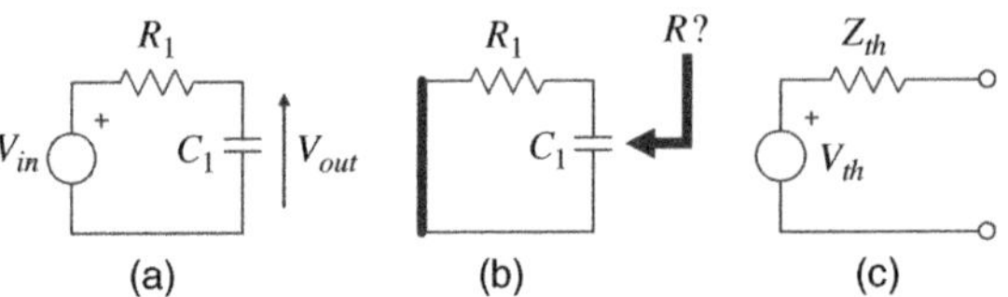

Figure 1.41 A simple 1st-order network featuring a capacitor offers a frequency-dependent output impedance.

The equivalent Thévenin circuit appears in the right side of the figure, its generator value is:

$$V_{th}(s) = V_{in}(s)\frac{\frac{1}{sC_1}}{\frac{1}{sC_1}+R_1} = \frac{1}{1+sR_1C_1} \tag{1.75}$$

The Short Circuit Current

The fourth and last method involves the short circuit current. In Figure 1.42, a Thévenin equivalent circuit is drawn. If you short circuit its output terminals, you have a current equal to:

$$I_{sc} = \frac{V_{th}}{R_{th}} \tag{1.76}$$

From an unknown circuit, if you have V_{th} and calculate or measure the short circuit current I_{sc}, you can find the Thévenin resistance. In Figure 1.37, the output voltage across R_2 is the circuit Thévenin voltage equal to:

$$V_{th} = V_{in}\frac{R_2}{R_1+R_2} \tag{1.77}$$

If you now short circuit the output as shown in Figure 1.43, the current is simply:

$$I_{sc} = \frac{V_{in}}{R_1} \tag{1.78}$$

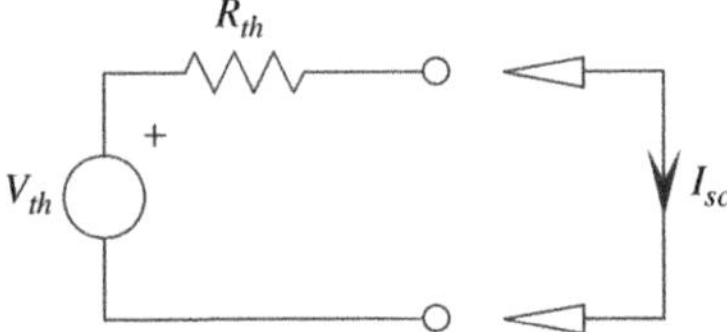

Figure 1.42 The short circuit current of a Thévenin equivalent circuit is the Thévenin voltage V_{th} divided by resistance R_{th}.

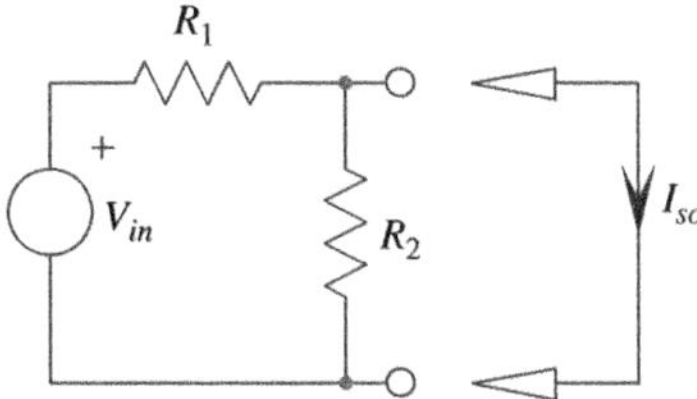

Figure 1.43 The Thévenin output resistance can also be found by calculating the short circuit current.

By applying (1.76), the Thévenin resistance is derived as:

$$R_{th} = \frac{V_{th}}{I_{sc}} = \frac{V_{in}\dfrac{R_2}{R_1 + R_2}}{\dfrac{V_{in}}{R_1}} = \frac{R_1 R_2}{R_1 + R_2} = R_1 \| R_2 \tag{1.79}$$

Dependent Sources

So far we have drawn circuits in which one single independent source was used, the V_{in} generator. Assume Figure 1.44 example in which a dependent source now appears in parallel with R_3. The term 0.19 has the dimension of siemens. What resistance drives capacitor C_1?

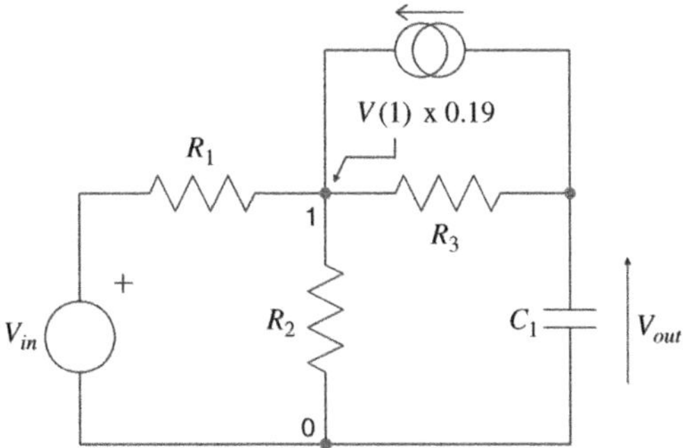

Figure 1.44 The presence of the dependent current source does not modify the circuit analysis: only the independent source is set to 0.

To find the answer, we can still set V_{in} to 0 but the controlled current source remains untouched as we previously said: it depends on $V(1)$, not V_{in}. Should it depend on V_{in} instead then it would be another *dependent* source and putting V_{in} to 0 would naturally remove that current source from the circuit.

The updated circuit appears in Figure 1.45. To find the resistance offered by the capacitor terminals, we connect a current test generator as suggested in Figure 1.46 and Figure 1.47. When doing so, we realize that the excitation signal is our current source while the response is the voltage developed across the considered terminals. It complies with our impedance transfer function definition introduced in the beginning of this chapter – see (1.13). The resistance offered by the port in

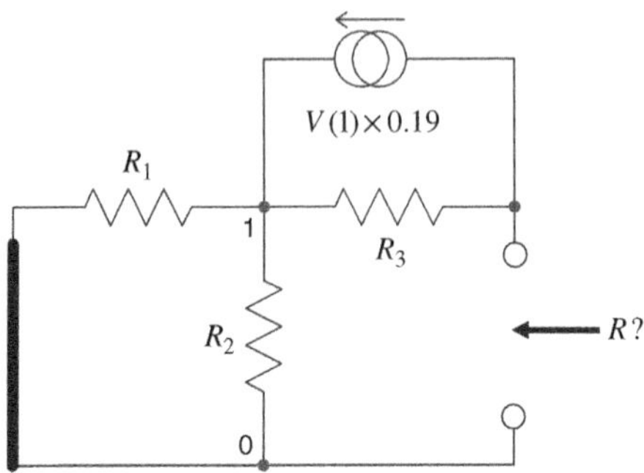

Figure 1.45 The independent source V_{in} is classically set to 0 V but the dependent source remains in the sketch:

Figure 1.47 is thus

$$R = \frac{V_T}{I_T} \tag{1.80}$$

From Figure 1.46 right side, we can write that the current flowing in the paralleled arrangement of R_1 and R_2 (R_{eq}) is:

$$I_{Req} = I_3 + 0.19 \cdot V(1) \tag{1.81}$$

with

$$R_{eq} = R_1 \| R_2 \tag{1.82}$$

The current flowing in R_{eq} is node (1) voltage divided by R_{eq}. Updating (1.81), we have:

$$\frac{V(1)}{R_{eq}} - 0.19 \cdot V(1) = I_3 \tag{1.83}$$

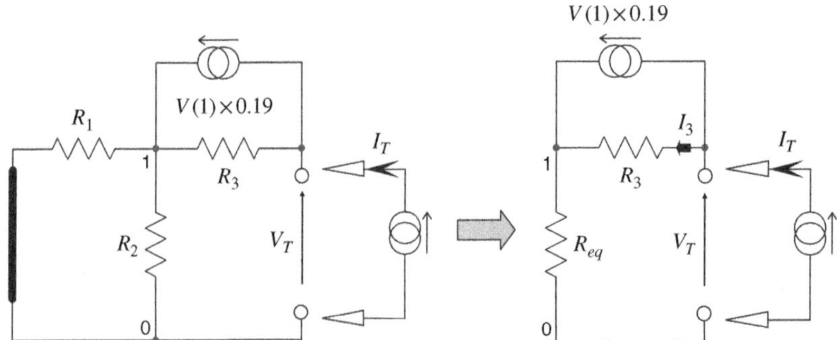

Figure 1.46 A current test generator is now biasing the capacitor terminals to determine its driving resistance.

The test voltage V_T is equal to the voltage across R_{eq}, V(1), plus the drop across R_3:

$$V_T = V(1) + I_3 R_3 = V(1) + V(1) R_3 \left(\frac{1}{R_{eq}} - 0.19 \right) \tag{1.84}$$

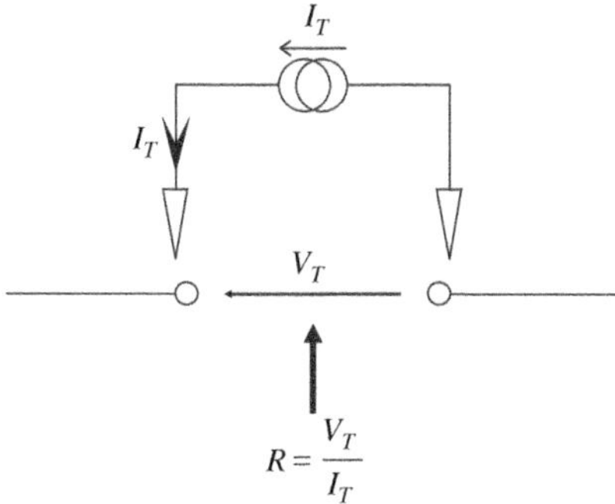

Figure 1.47 The test generator helps find the resistance offered by the terminals at which current (excitation) and voltage (response) are simultaneously observed.

$V(1)$ is the voltage across R_{eq} whose current is the test generator I_T. Therefore, $V(1)$ is equal to:

$$V(1) = I_T R_{eq} \tag{1.85}$$

Updating (1.84) with (1.85), leads to the definition of V_T:

$$V_T = I_T R_{eq}\left[1 + R_3\left(\frac{1}{R_{eq}} - 0.19\right)\right] \tag{1.86}$$

The resistance seen by the capacitor terminals is obtained by rearranging (1.86) in a transfer function form:

$$R = \frac{V_T}{I_T} = R_3 + (R_1 \| R_2)(1 - 0.19R_3) \tag{1.87}$$

The time constant associated with capacitor C_1 is thus:

$$\tau = \left[R_3 + (R_1 \| R_2)(1 - 0.19R_3)\right] C_1 \tag{1.88}$$

This example shows that a network involving controlled sources requires the use of KVL and KCL compared to other networks whose characteristics were derived by inspection only.

A Transistor-Based Circuit

A transistorized circuit is another typical example involving controlled sources. A simple amplifying circuit appears in Figure 1.48. Here, the exercise will consist of finding the resistance driving

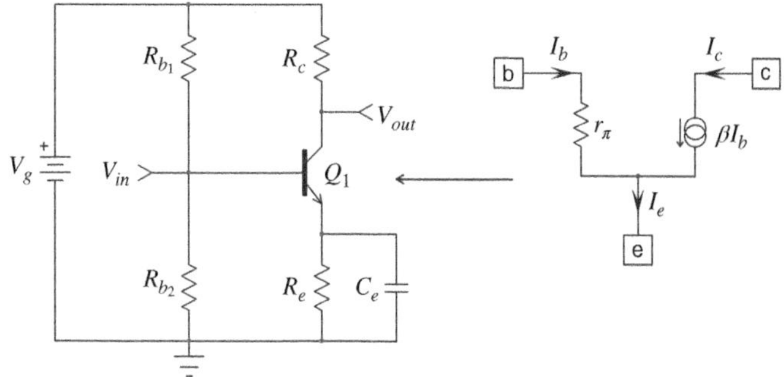

Figure 1.48 A simple amplifier built around a bipolar circuit.

capacitor C_e when that capacitor is removed from the circuit. Before proceeding, Q_1 is replaced by its (simplified) small-signal equivalent circuit, the hybrid-π model, that appears in the right side of the figure.

The excitation signal is the V_{in} source applied at the bias bridge made of R_{b1} and R_{b2}. To find the resistance seen at capacitor C_e terminals, V_{in} is turned off and its terminals are strapped. V_g is a dc supply and in this example, its ac contribution to the circuit is 0. You can imagine that V_g is decoupled by an infinite value capacitor so that R_{b1} and R_c upper terminals are at the ground level in ac: V_g is also replaced by a strap for the ac analysis. The updated circuit is shown in Figure 1.49.

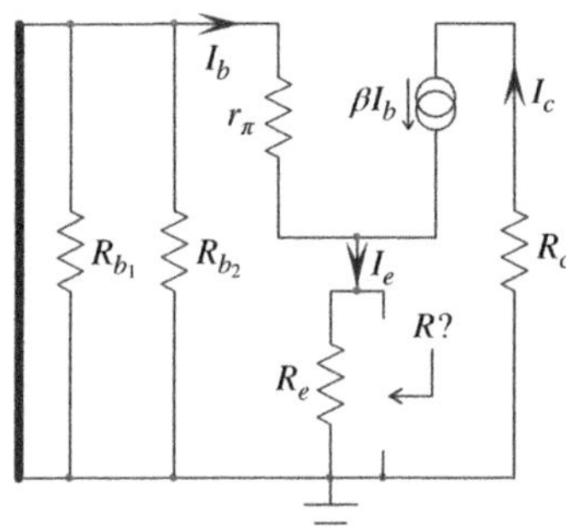

Figure 1.49 With a 0-V excitation voltage, the equivalent circuit nicely simplifies.

Further rearranging the network leads to a rather simple circuit as drawn in Figure 1.50. Here, you can observe that resistance R_e is connected in parallel with the capacitor terminals. It can therefore be temporarily removed from the analysis and, once the resistance seen from the terminals is identified, the final expression will bring R_e back in parallel with the intermediate result.

First, let's express the base current I_b equal to

$$I_b = -\frac{V_R}{r_\pi} \tag{1.89}$$

I_1 depends on the collector current as:

$$I_1 = (\beta + 1)I_b = -I_T \tag{1.90}$$

The intermediate resistance seen at the capacitor terminals is simply the ratio of V_R by the test current I_T:

$$R_{\text{int}} = \frac{V_R}{I_T} = \frac{r_\pi}{\beta + 1} \tag{1.91}$$

Bringing R_e back in the circuit gives the resistance seen at the capacitor terminals:

$$R = R_e \| \left(\frac{r_\pi}{\beta + 1} \right) \tag{1.92}$$

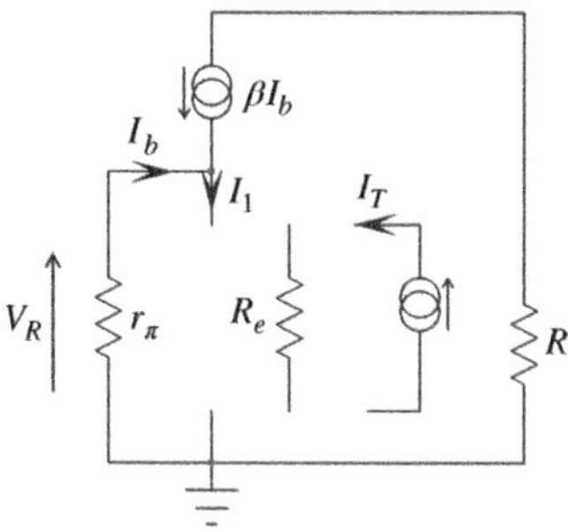

Figure 1.50 The rearranged circuit reduces to a simple network.

The time constant of this circuit is now immediately available:

$$\tau = C_e\left[R_e \| \left(\frac{r_\pi}{\beta+1}\right)\right] \tag{1.93}$$

SPICE Can Help Verify Results

In some complex circuits, the output resistance you have derived can be made of several series-paralleled combinations and a mathematical solver such as Mathcad® can be of great help to obtain the result quickly. However, how do you know if your derivation is correct? A SPICE simulator can help verify results quickly.

Assume the circuit in Figure 1.51 in which an inductor is inserted into a resistive circuit. What we want is the resistance 'seen' from the inductor terminals. Otherwise stated, what is the output resistance driving the inductor when the source is set to 0 V? We can see that resistance r_L is in series with L, this is the first term. Then, the left inductor terminal goes to ground via the paralleled combination of R_1 and R_2 and returns to the right terminal through R_3:

$$R = r_L + R_2 \| R_1 + R_3 \tag{1.94}$$

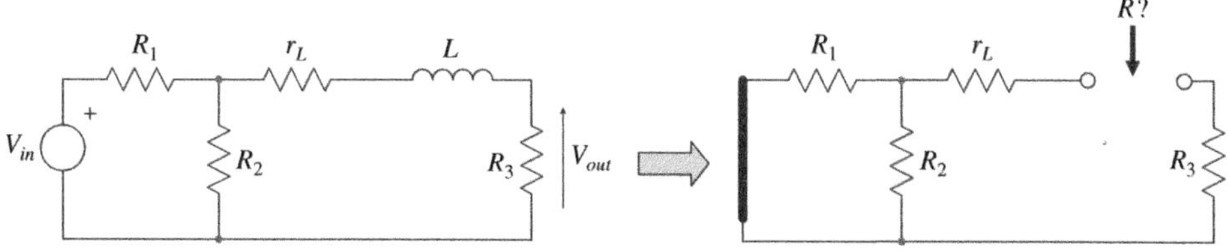

Figure 1.51 Find the resistance offered by the inductor terminals when the component is removed.

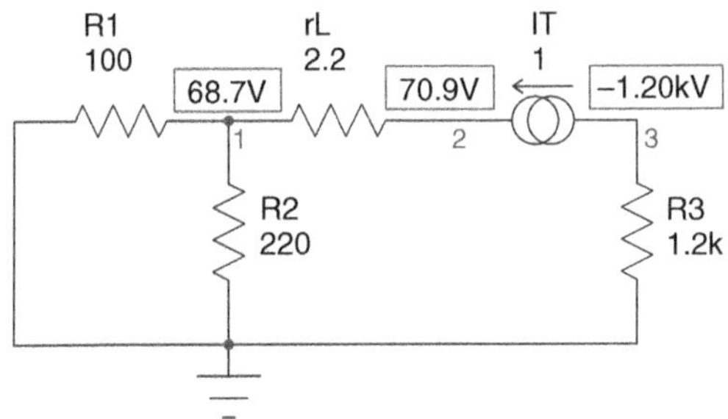

Figure 1.52 A SPICE dc simulation will give you the resistance seen from the inductor terminals.

If we apply Figure 1.52 values to (1.94) we find:

$$R = 2.2 + \frac{220 \times 100}{220 + 100} + 1.2k = 1270.95\,\Omega \tag{1.95}$$

The simulation circuit uses a 1-A test source I_T injecting current into the inductor terminals when the source is set to 0 V. Involving a dc operating point simulation, the figure gives a voltage difference equal to $70.95 - (-1.2k) = 1270.95\ V$. Considering the 1-A generator, the resistance seen from the terminals is $1270.95\,\Omega$ as calculated by SPICE.

A similar exercise can be run on Figure 1.45 circuit where I_2, the test generator, injects 1 A into the capacitor terminals. With component values such as those labeled in Figure 1.53, the resistance seen from the capacitor port is equal to:

$$R = 2.2 + (22\|60)(1 - 0.19 \times 2.2) = 2.2 + 16.09756 \times 0.582 = 11.56878\,\Omega \tag{1.96}$$

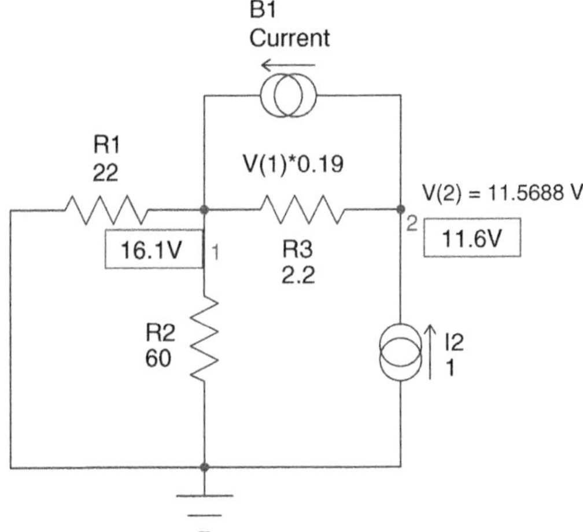

Figure 1.53 The SPICE simulation of the circuit involving the controlled current source confirms hand-calculated results.

The value rounded by the simulator in Figure 1.53 is 11.5688 V.

The technique involving SPICE is extremely useful when you deal with large and complex networks, especially those with a lot of dependent sources. Externally biasing the storage element terminals with the 1-A current generator and running a quick .OP bias point analysis has proven to be extremely efficient to trap errors and flaws in the analysis. Highly recommended!

1.5 Appendix 1B – Problems

We have gathered simple to more complex sketches in which you are asked to determine the dc transfer function and the resistance driving the storage element. Answers are at the end.

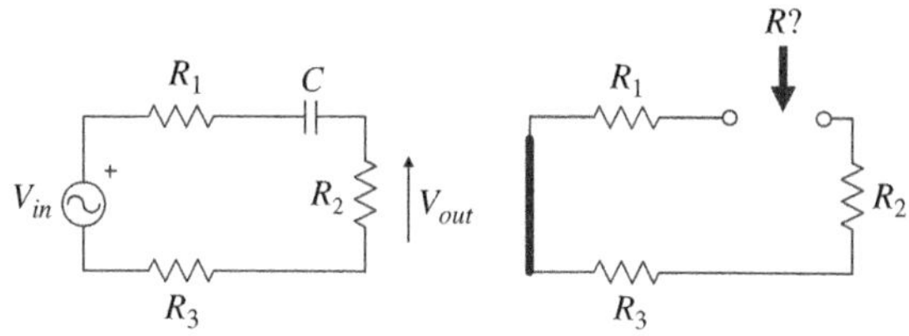

Figure 1.54 – Problem 1

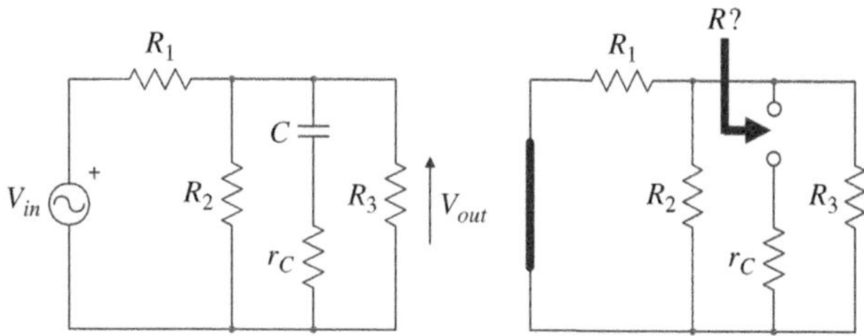

Figure 1.55 – Problem 2

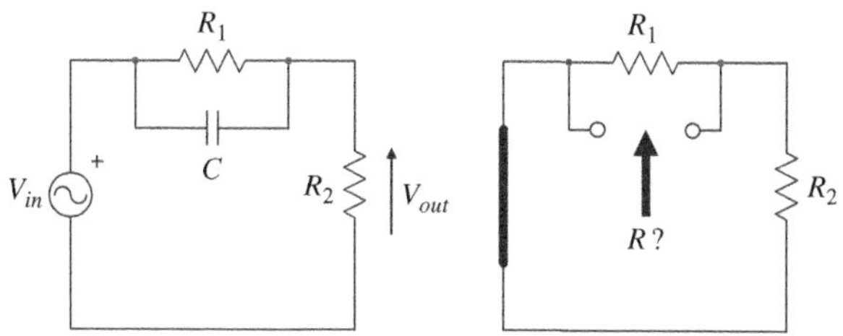

Figure 1.56 – Problem 3

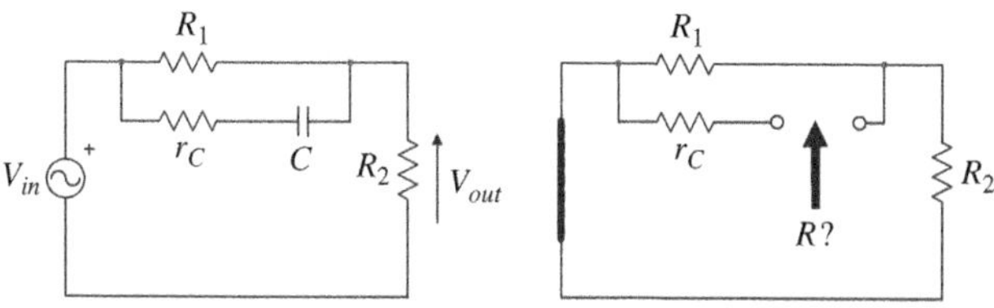

Figure 1.57 – Problem 4

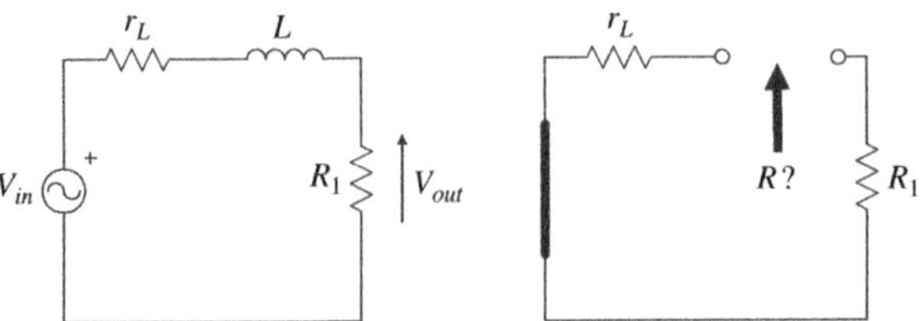

Figure 1.58 – Problem 5

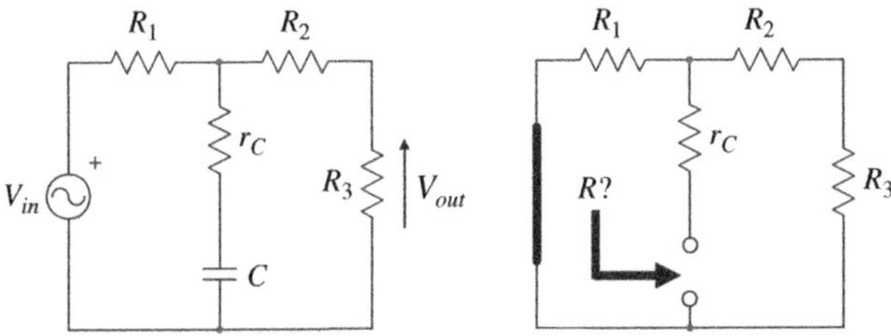

Figure 1.59 – Problem 6

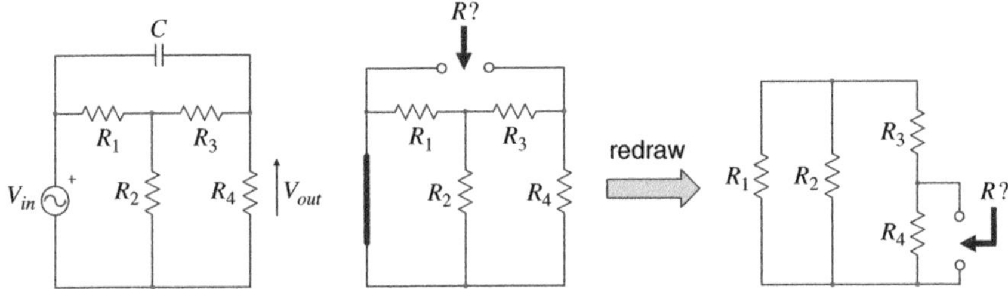

Figure 1.60 – Problem 7

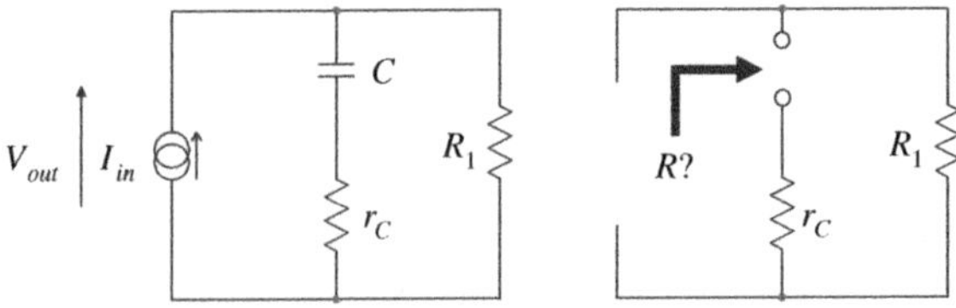

Figure 1.61 – Problem 8

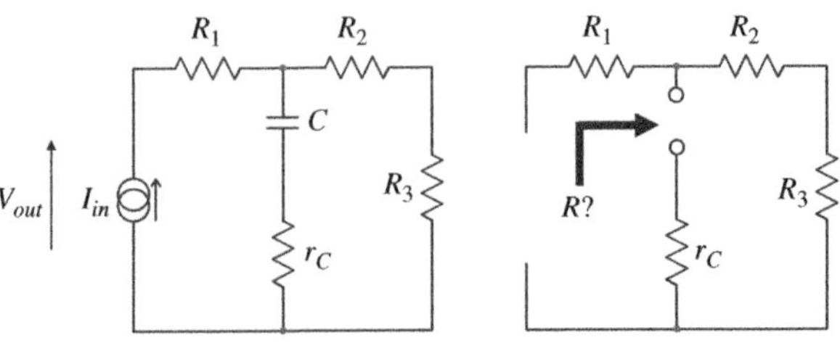

Figure 1.62 – Problem 9

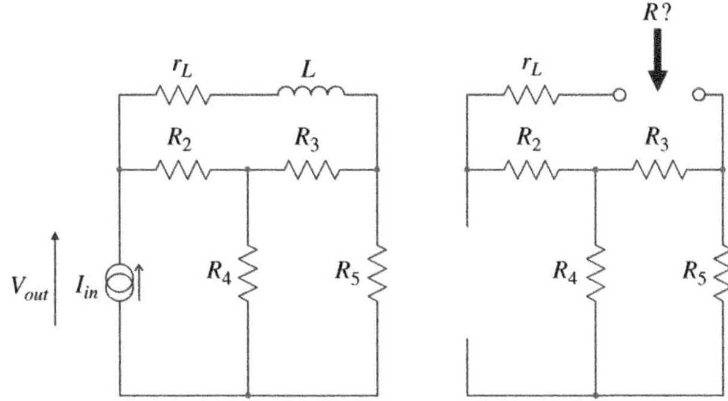

Figure 1.63 – Problem 10

Answers

Problem 1:

$$R = R_1 + R_2 + R_3$$

$$\tau = (R_1 + R_2 + R_3)C$$

$$H_0 = 0$$

Problem 2:

$$R = \left(R_1 \| R_2 \| R_3\right) + r_C$$

$$\tau = \left[\left(R_1 \| R_2 \| R_3\right) + r_C\right]C$$

$$H_0 = \frac{R_2 \| R_3}{R_2 \| R_3 + R_1}$$

Problem 3:

$$R = R_1 \| R_2$$

$$\tau = \left(R_1 \| R_2\right)C$$

$$H_0 = \frac{R_2}{R_1 + R_2}$$

Problem 4:

$$R = R_1 \| R_2 + r_C$$

$$\tau = \left(R_1 \| R_2 + r_C\right)C$$

$$H_0 = \frac{R_2}{R_1 + R_2}$$

Problem 5:

$$R = r_L + R_1$$

$$\tau = \frac{L}{r_L + R_1}$$

$$H_0 = \frac{R_1}{R_1 + r_L}$$

Problem 6:

$$R = r_C + R_1 \| (R_2 + R_3)$$

$$\tau = \left[r_C + R_1 \| (R_2 + R_3)\right] C$$

$$H_0 = \frac{R_3}{R_1 + R_2 + R_3}$$

Problem 7:

$$R = R_4 \| \left(R_3 + R_1 \| R_2\right)$$

$$\tau = \left[R_4 \| \left(R_3 + R_1 \| R_2\right)\right] C$$

$$H_0 = \frac{R_4}{R_4 + R_3 + R_1 \| R_2} \frac{R_2}{R_1 + R_2}$$

Solving by inspection is not possible, use Thévenin's theorem for R_1 and R_2 driving R_3.

Problem 8:

$$R = r_C + R_1$$

$$\tau = (r_C + R_1) C$$

$$R_0 = R_1$$

Problem 9:

$$R = r_C + R_2 + R_3$$

$$\tau = (r_C + R_2 + R_3) C$$

$$R_0 = R_1 + R_2 + R_3$$

Problem 10:

$$R = r_L + R_2 + (R_5 + R_4) \| R_3$$

$$\tau = \frac{L}{r_L + R_2 + (R_5 + R_4) \| R_3}$$

$$R_0 = ?$$

It cannot be found by inspection. Solution in Chapter 3!

2

Transfer Functions

In the first chapter, we learned how a transfer function defines a relationship linking a response to a stimulus. While propagating through the considered electrical network, the excitation waveform undergoes amplifications, attenuations and phase distortions. These characteristics can be mathematically described by organizing gains, poles and zeros in a frequency-dependent function. If tools exist to write such a function in a rather quick way, we must learn how to organize the result in an intelligible form so that the response in the frequency domain can be inferred by reading the equation. This is the principle of *low-entropy* expressions in which the proper factorization of poles and zeros gives immediate insight into the frequency response. This chapter explores transfer functions by first defining what a linear system is and how time constants shape the response of the analyzed circuit.

2.1 Linear Systems

A system is said to be *linear* if it satisfies the properties of *additivity* and *proportionality*. Assume a system encapsulated in a box as in Figure 2.1. A stimulus u_1 is applied and delivers a response y_1. Then a stimulus u_2 with a different amplitude is applied and delivers y_2. If we now bias the input with the sum of the two stimuli, $u_1 + u_2$ and observe the corresponding signal output y, the *additivity* property is respected if that output signal equals $y_1 + y_2$.

Mathematically, you would write:

$$y(u_1 + u_2) = y(u_1) + y(u_2) \tag{2.1}$$

The second property is *proportionality* or *homogeneity* as shown in Figure 2.2. If a stimulus u_1 is applied at the system input and delivers an amplitude y_1, then growing the stimulus by a factor k will deliver an output equal to $k \cdot y_1$. Mathematically, you write this principle as:

$$y(k \cdot u) = k \cdot y(u) \tag{2.2}$$

These two properties form the so-called *superposition* principle and must be satisfied by a system before you can apply the Laplace transform or any of the tools that we have studied (Norton's or Thévenin's theorems for instance). What type of function satisfies this principle? Assume a system which divides the input voltage u by 2 then adds 2 V as a fixed offset. The function y characterizing this system would be defined as:

$$y(u) = 0.5u + 2 \tag{2.3}$$

Linear Circuit Transfer Functions: An Introduction to Fast Analytical Techniques, First Edition. Christophe P. Basso.

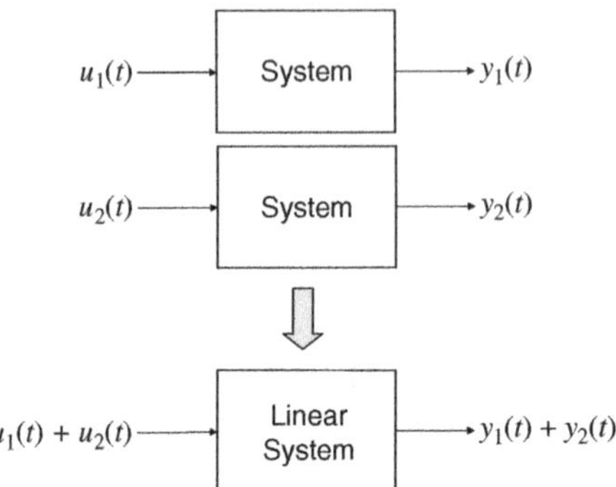

Figure 2.1 The property of additivity assumes that the response of two added stimuli is equal to the added response of the two individual stimuli responses.

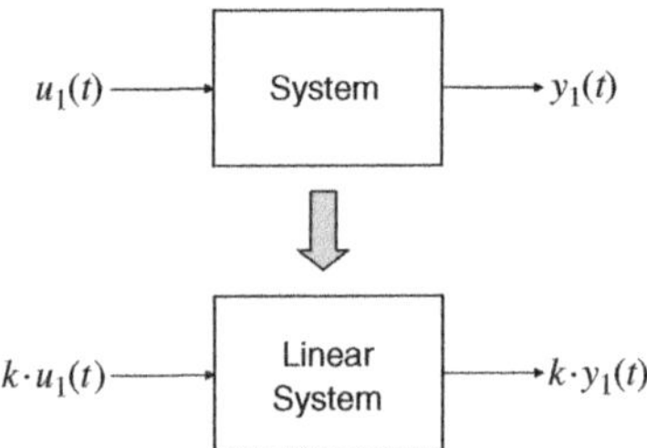

Figure 2.2 The property of *homogeneity* or *proportionality* states that if we measure the response to a stimulus then grow the stimulus by a factor k, the response is k times the original response.

This is the shifted straight line shown in Figure 2.3. Does this function characterize a linear system per previous properties?

Let's see the proportionality principle first. What value does (2.3) deliver for two u?

$$y(2u) = u + 2 \tag{2.4}$$

Then, what does $2 \cdot y(u)$ give?

$$2 \cdot y(u) = u + 4 \tag{2.5}$$

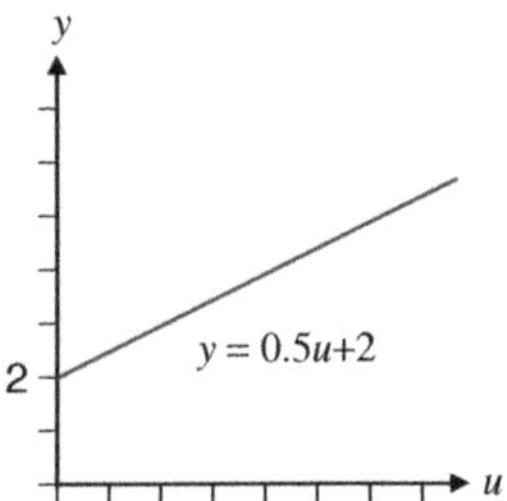

Figure 2.3 This system divides the input voltage u by 2 and adds a 2-V offset.

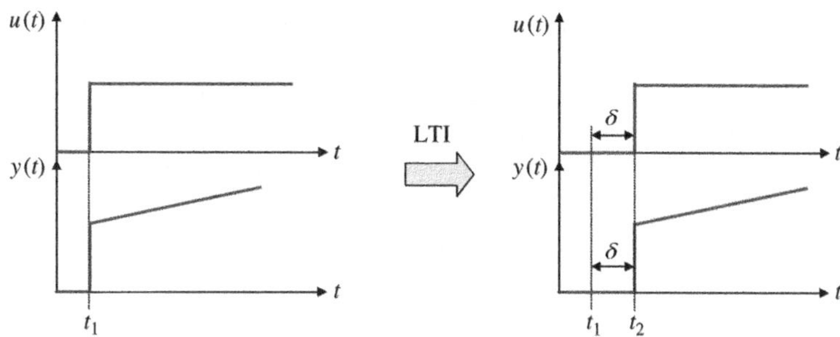

Figure 2.4 If an input signal u is applied to a system that gives an output y, the same stimulus applied with a delay δ, gives the same output y but shifted in time by δ.

(2.5) is different than what (2.4) gave. At this point, we can already state that it is a non-linear system as per our definition. Let's check additivity now with two different input levels:

$$u = u_1 \text{ then } y_1 = 0.5u_1 + 2$$
$$u = u_2 \text{ then } y_2 = 0.5u_2 + 2$$

If we sum up y_1 and y_2 to form y, we find $y = 0.5(u_1 + u_2) + 4$. Now if the stimulus is made of the two independent stimuli summed up $u = u_1 + u_2$, then $y = 0.5(u_1 + u_2) + 2$ which, again, does not satisfy the principle of additivity. As a conclusion, the graph displayed in Figure 2.3 does not represent a linear system.

2.1.1 A Linear Time-invariant System

Besides the superposition principle, a second characteristic is required to define what is a called a *Linear Time-Invariant* (LTI) system. If we apply a signal u to our system and obtain a signal y, then applying u with a delay of δ seconds will produce the same response y but delayed (time-shifted) by δ.

This principle is illustrated in Figure 2.4 in which you see that the stimulus u applied at two different instants t_1 or t_2 delivers the same output y but shifted by the time delay δ.

Does (2.3) characterize a time-invariant system? We first start by injecting a delayed stimulus $u(t-\delta)$ and we calculate the corresponding output signal y:

$$y_1(t) = 0.5u(t-\delta) + 2 \tag{2.6}$$

Now we calculate the response $y_2(t)$ to $u(t)$ but shifted by a delay δ:

$$y_2(t-\delta) = 0.5u(t-\delta) + 2 \tag{2.7}$$

Both equations return the same value, therefore, the system characterized by equation (2.3) is time-invariant. However, as it does not satisfy the superposition principle, it is not linear. More details on LTI systems will be found in [1].

2.1.2 The Need for Linearization

If we now think of Figure 2.3 graph as a transfer function characterizing an electrical system, we could say that output y is linked to the input u by a ratio of 0.5 to which a fixed 2-V dc voltage is added. To linearize this system (so that Thévenin's, Norton's, the superposition theorem or the Laplace transform could be applied), one widely adopted principle is called *perturbation*. Each variable (u and

y) is perturbed by adding to it a small ac modulation. The modulating waveform is assumed of sufficiently-low amplitude so that nonlinearities (saturation for instance) are negligible in the network under analysis. The alternating component is designated by a small hat or caret (^) placed on top of the considered lower-cased variable ($\hat{v}_{out}$ for instance). The direct component (if any) is usually upper-cased but it can sometimes be identified by the presence of an added subscripted 0 as below. In (2.3) we have two variables y and u, so we perturb both:

$$y = \hat{y} + y_0 \tag{2.8}$$

$$u = \hat{u} + u_0 \tag{2.9}$$

Now substituting (2.8) and (2.9) into (2.3), we have:

$$\hat{y} + y_0 = 0.5(\hat{u} + u_0) + 2 \tag{2.10}$$

We develop and collect ac and dc terms to form a set of two equations. All dc terms multiplied by an ac (^) term become an ac component while ac cross-products (e.g. $\hat{y}\hat{u}$) are eliminated because they form nonlinear terms of negligible contribution. Here, the expression is quite simple and there is no problem to sort out ac and dc components:

Ac: $\hat{y} = 0.5\hat{u}$ this is the small-signal or *incremental* transfer function

Dc: $y_0 = 0.5u_0 + 2$ this is the *static operating point* value, also called *bias point*.

The sum of the above definitions forms the *large-signal* response also called the *total* variable. To study the frequency response of that system, we are only interested by the small-signal or incremental output $\hat{y}$ not by the static offset y_0 which is a fixed value, independent of frequency. This dc component would be used to calculate a static operating point for instance. We can now redraw Figure 2.3 with separated dc and ac components as shown in Figure 2.5 for $u_0 = 0$.

The function describing this curve is written $\hat{y} = 0.5\hat{u}$ and if you check, it now satisfies the superposition principle and the time-invariant definition: this is a linear function describing a linear time-invariant system.

2.2 Time Constants

Figure 2.6 depicts a simple network associating two linear elements, a resistor R and a capacitor C. Let's calculate the relationship linking y and u. Applying KCL and KVL, we can immediately write that u equals y plus the voltage drop across R:

$$u(t) = Ri(t) + y(t) \tag{2.11}$$

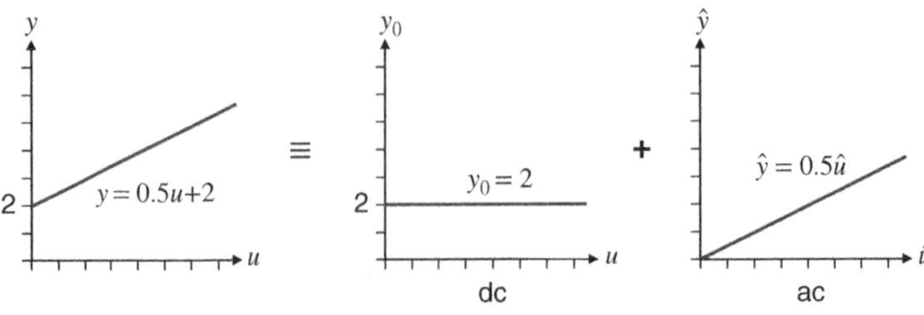

Figure 2.5 The idea is to perturb the non-linear equation and have it split into dc and an ac expressions. The ac expression is linear.

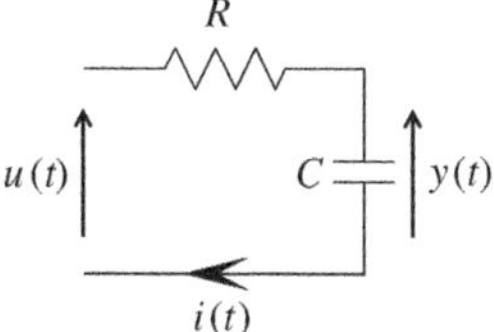

Figure 2.6 A simple *RC* network.

The current *i* circulating in the capacitor depends on the instantaneous voltage across its terminals, v_C:

$$i(t) = C\frac{dv_C(t)}{dt} \tag{2.12}$$

Substituting (2.12) into (2.11) we have:

$$y(t) = u(t) - RC\frac{dv_C(t)}{dt} = u(t) - RC\frac{dy(t)}{dt} \tag{2.13}$$

To solve this 1st-order differential equation, we can take the Laplace transform of (2.13) remembering that a differentiation is a multiplication by s while an integration is a division by s.

Applying this principle, we obtain the following Laplace expression, considering an initial voltage in the capacitor labeled V_0:

$$Y(s) = L\{y(t)\} = U(s) - RC(sY(s) - V_0) \tag{2.14}$$

If we rearrange this expression by factoring some of the terms, we obtain:

$$Y(s) = \frac{U(s)}{1 + sRC} + \frac{RCV_0}{1 + sRC} \tag{2.15}$$

Now assume the input is a step function, a pulse starting at 0 V and stepping up immediately to V_1. To extract the time-domain response of (2.15), we simply replace $U(s)$ by V_1/s (the Laplace transform of a step featuring a peak value V_1) and we calculate the inverse Laplace transform of the result:

$$L_s^{-1}\{Y(s)\} = L_s^{-1}\left\{\frac{V_1}{s}\frac{1}{1 + sRC}\right\} + L_s^{-1}\left\{\frac{RCV_0}{1 + sRC}\right\} \tag{2.16}$$

Applying inverse Laplace transform tables, the result is made of two terms:

$$y(t) = V_1\left(1 - e^{-\frac{t}{\tau}}\right) + V_0 e^{-\frac{t}{\tau}} \tag{2.17}$$

in which τ is the time constant equal to *RC*. These two terms form the *overall* response of the system under study, our simple *RC* network. Mathematically, (2.17) can be rewritten as:

$$y(t) = r_f(t) + r_n(t) \tag{2.18}$$

in which r_f is the *forced* or *steady-state* response and r_n the *natural* or *transient* response. In the literature, these responses are also respectfully denoted as *zero-state* and *zero-input* responses.

$$r_f(t) = V_1\left(1 - e^{-\frac{t}{\tau}}\right) \tag{2.19}$$

$$r_n(t) = V_0 e^{-\frac{t}{\tau}} \tag{2.20}$$

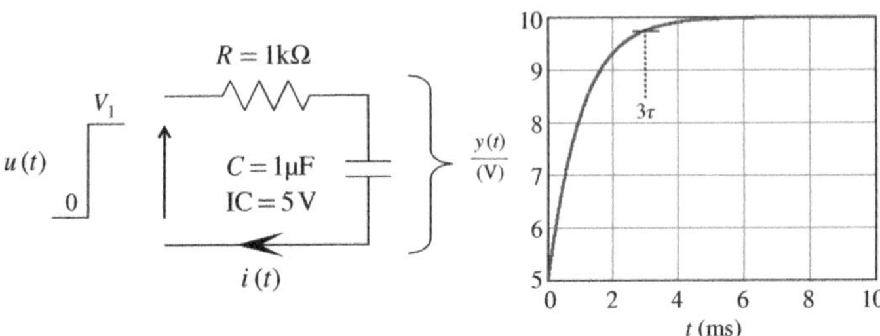

Figure 2.7 The time-domain response reveals an exponential shape reaching a *forced* value as t approaches infinity. The term IC = 5 V instructs the simulator to use a voltage Initial Condition at $t=0$:

The forced response, $r_f(t)$, is the result of an externally applied stimulus, the *force*, V_1 in our case which is of 10-V amplitude. When t approaches infinity, the output reaches the final value of 10 V. However, one usually considers the output to be at the forced value - or the capacitor to be fully charged – after a delay equal to three times the time constant or 3τ. In this example, the time constant value is 1 ms so the capacitor is considered charged after 3 ms as shown in Figure 2.7. The forced response depends on the input stimulus.

On the other hand, the natural or zero-input response $r_n(t)$ is only due to the initial conditions, V_0 in our example. There is no contribution from the source or excitation. To obtain the natural response, we turn the excitation to 0 (strap the input voltage source to set it to 0 V, or remove the current generator and open circuit its terminals to make it a 0-A source). The new circuit appears in Figure 2.8 with its individual response plotted on the right side.

In Figure 2.8, if you remove the capacitor and look at the resistance seen from its terminals, you 'see' the 1-kΩ resistance. The time constant is $RC = 1\ \mu\text{F} \times 1\ \text{k}\Omega = 1$ ms.

In (2.15), if we consider the capacitor to be fully discharged at $t=0$, $v_C(0) = 0$, then the equation simplifies to:

$$\frac{Y(s)}{U(s)} = \frac{1}{1+sRC} = \frac{1}{1+s\tau} \tag{2.21}$$

For $s=0$, the transfer function returns 1. As s increases, the transfer function brings attenuation to finally reach 0 as s approaches infinity: this is a low-pass filter whose pole ω_p is the inverse of

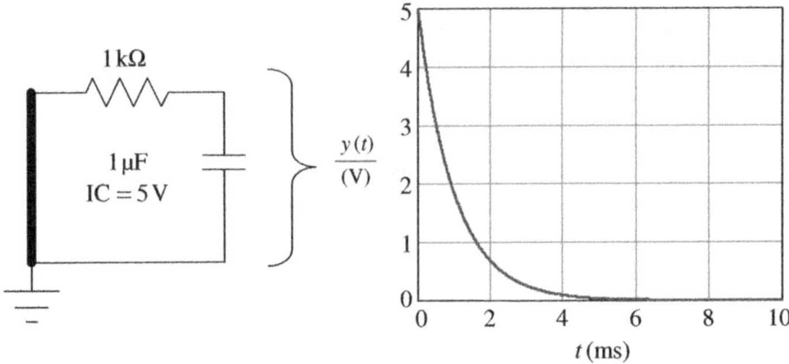

Figure 2.8 The natural response of the network is studied with the excitation source turned off. Here we have a voltage generator so turning it off means replacing it by a strap.

the time constant:

$$\omega_p = \frac{1}{\tau} \tag{2.22}$$

Using this notation, we can rewrite (2.21) in a way where the above pole appears:

$$\frac{Y(s)}{U(s)} = \frac{1}{1 + \frac{s}{\omega_p}} \tag{2.23}$$

This is the proper way of writing the transfer function characterizing a 1st-order low-pass filter.

2.2.1 Time Constant Involving an Inductor

Similarly to what we did with the *RC* network, we can write nodal and mesh equations for the *RL* network shown in Figure 2.9. Applying KCL and KVL, we can immediately write that *u* equals *y* plus the voltage drop across the inductor *L*:

$$u(t) = L\frac{\mathrm{d}i(t)}{\mathrm{d}t} + y(t) \tag{2.24}$$

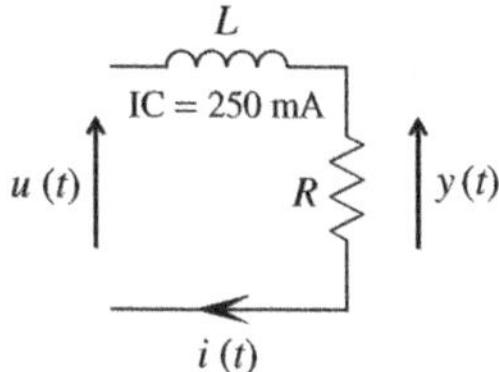

Figure 2.9 A simple *RL* network.

Rearranging (2.24), we have:

$$y(t) = u(t) - L\frac{\mathrm{d}i(t)}{\mathrm{d}t} \tag{2.25}$$

Considering an initial inductor current I_0, we can solve this 1st-order differential equation by calculating the Laplace transform of (2.25):

$$Y(s) = L\{y(t)\} = U(s) - L(sI(s) - I_0) \tag{2.26}$$

The current $I(s)$ is the output voltage $Y(s)$ divided by R thus:

$$Y(s) = U(s) - L\left(s\frac{Y(s)}{R} - I_0\right) \tag{2.27}$$

Factoring and rearranging, we obtain the final Laplace equation characterizing our *RL* network:

$$Y(s) = \frac{U(s)}{1 + s\frac{L}{R}} + \frac{LI_0}{1 + s\frac{L}{R}} \tag{2.28}$$

Now assuming a step function similar as the one we used, 0 to 10 V, then the response of this *RL* network to an input step is calculated as follows:

$$L_s^{-1}\{Y(s)\} = L_s^{-1}\left\{\frac{V_1}{s}\frac{1}{1 + s\frac{L}{R}}\right\} + L_s^{-1}\left\{\frac{LI_o}{1 + s\frac{L}{R}}\right\} \tag{2.29}$$

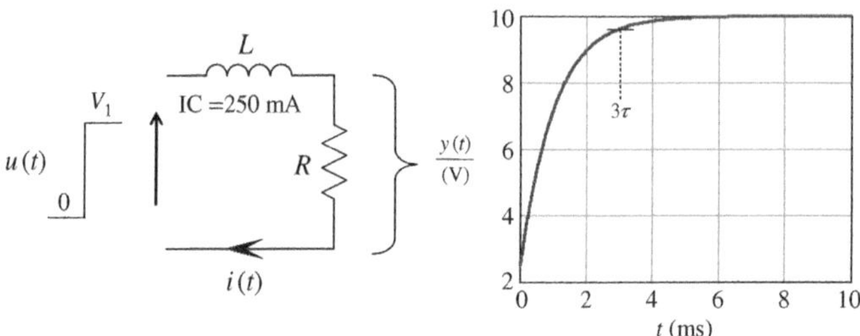

Figure 2.10 The time-domain response reveals an exponential shape reaching a *forced* value as t approaches infinity. The term IC = 250 mA instructs the simulator to use a current initial condition at $t = 0$.

It gives the following time-domain response:

$$y(t) = V_1\left(1 - e^{-\frac{t}{\tau}}\right) + RI_0e^{-\frac{t}{\tau}} \tag{2.30}$$

again made of two terms, a forced response imposing V_1 as t approaches infinity and a natural response only considering the initial condition:

$$r_f(t) = V_1\left(1 - e^{-\frac{t}{\tau}}\right) \tag{2.31}$$

$$r_n(t) = RI_0e^{-\frac{t}{\tau}} \tag{2.32}$$

In these expression, the time constant is now defined as $\tau = \frac{L}{R}$.

The forced response depends on the excitation (the 10-V source) as described by (2.31). The natural response depends on the initial condition only, the 250-mA current in the energized inductor. This current creates a 2.5-V drop across the 10-Ω resistance which dies out to zero volt as time elapses. To plot this response, simply turn the excitation to 0 V and compute the voltage according to (2.32). The result appears in Figure 2.11.

In the situation where the stimulus is turned off, if you remove the inductor and look into its terminals, you see the 10-Ω resistor: the time constant is $\tau = \frac{10\ \text{mH}}{10\ \Omega} = 1$ ms.

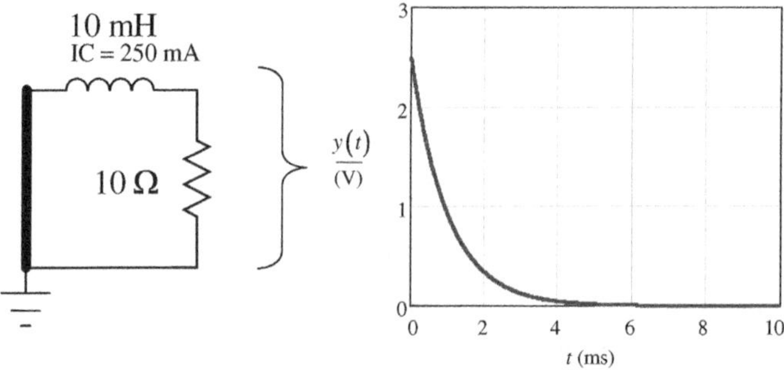

Figure 2.11 The initial response depends solely on initial conditions because the excitation is turned off. Here, the voltage source is set to 0 V and replaced by a short circuit.

In (2.28), if we consider the inductor to be fully de-energized at $t = 0$, $i_L(0) = 0$, then the equation simplifies to:

$$\frac{Y(s)}{U(s)} = \frac{1}{1 + s\frac{L}{R}} = \frac{1}{1 + s\tau} \tag{2.33}$$

It is exactly the same format as equation (2.21) derived for the *RC* filter in which the time constant appears at the denominator. This is another form of a 1st-order low-pass filter.

From the above examples, we have seen that the natural response of a network does not depend on its excitation signal. Besides initial conditions, the natural response solely depends on the network structure, how the storage elements, *C* and *L*, are arranged together with resistive terms to produce time constants. Therefore, we confirm the point already tackled in Chapter 1 and now set as a definition. To find the time constants of the network under study, turn the excitation waveform off. If the stimulus is a voltage source, set it to 0 V or strap its terminals in the electrical diagram. If the stimulus is a current source, turn it to 0 A or open circuit it in the schematic. Figure 2.12 illustrates this principle.

Voltage and current sources cannot be applied inconsistently to the circuit under study: a voltage source must be applied in series with an existing branch while a current source can only be applied in parallel to a component or branch. The reason is to revert the circuit into its original structure when you turn the excitation off as shown in Figure 2.12. We will come back to this important notion after the introduction of transfer functions and poles/zeros.

2.3 Transfer Functions

Transfer functions have been the object of many text books and are covered in details in [2] and [3] for example. In this paragraph, we will briefly review some of the important aspects of a transfer function. However, the important point is learning how to write and organize Laplace expressions in a swift and efficient way. By efficient I refer to the *low-entropy* formalism described in Chapter 1: immediately identify the presence of poles/zeros and gains by looking at the expression without reworking it.

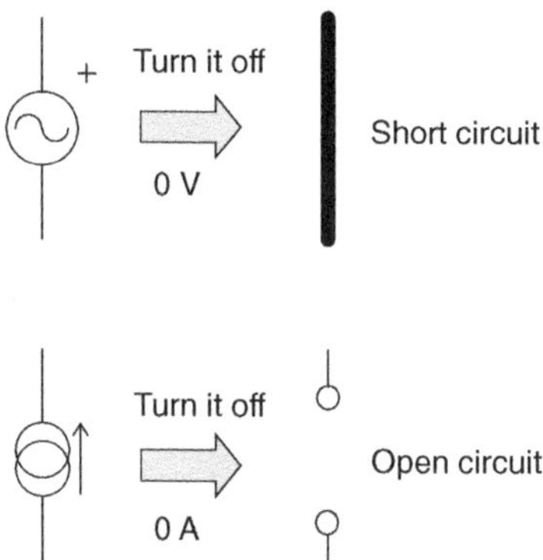

Figure 2.12 You turn off a voltage generator by setting it to 0 V and you replace the source symbol in the circuit by a short circuit. A current generator is turned off by setting it to 0 A and you replace the source symbol in the circuit by an open circuit.

The transfer function of a LTI system without delays can be defined by the ratio of two polynomials:

$$H(s) = \frac{N(s)}{D(s)} \tag{2.34}$$

The numerator $N(s)$ hosts the *zeros* of the transfer function while the denominator $D(s)$ holds the *poles*. The zeros are the roots of the equation $N(s) = 0$ while the poles are the roots of the *characteristic equation*, $D(s) = 0$. Mathematically, a zero in a function $f(x)$ is a value of x for which the function has the value of 0. Alternatively, a pole in a function is a point at which the function magnitude becomes infinite. It is a singularity. As already discussed in Chapter 1, a Laplace-transformed circuit featuring a zero and examined at $s = s_z$ brings an output null, implying that the input waveform – the stimulus – does not generate an output waveform, the response. Somewhere in the circuit, a transformed short circuit or a transformed open in the signal path block the stimulus and the ac output response is a *null*: $\hat{v}_{out} = 0$ V or $V_{out}(s_z) = 0$. In the transformed circuit observed at $s = s_p$, the pole value, the denominator of the transfer function cancels to 0 and the output magnitude becomes infinite.

A 1st-order zero in the numerator of a transfer function involving a capacitor and a resistor would follow the form:

$$N(s) = 1 + sR_1C_1 = 1 + s\tau_1 \tag{2.35}$$

To obtain the zero frequency, we solve $N(s_z) = 0$ leading to the following value for the root:

$$s_z = -\frac{1}{R_1C_1} = -\frac{1}{\tau_1} \tag{2.36}$$

Sometimes, the zero can be at the origin, meaning that in dc, $s = 0$, the output voltage is a null:

$$N(s) = s\tau_1(1 + \ldots) \tag{2.37}$$

The root here is $s_z = 0$, hence the term *at the origin*. With a zero at the origin, the 0-Hz gain of the considered network is 0. Sometimes, there can be more than a zero at the origin. A s^2 in (2.37) would indicate a double zero, s^3 a triple zero and so on. When you see a capacitor in series with the excitation signal or in series somewhere in the path to the output, then the dc component is blocked and it is likely that you find one or several zeros at the origin, depending how many dc blocks you can spot. The same applies for an inductor when in dc it shorts the signal path or the load to ground.

In our harmonic analysis, s is equal to $j\omega$, a complex imaginary frequency. s_z is also a complex number but affected by real and imaginary parts. In (2.36), there is no imaginary part and we say the root is *real* and negative. This is a left half-plane zero (LHPZ), a zero placed in the left half of the s-plane. If the root is positive, the zero lies in the right half-plane (it becomes a RHPZ) and exhibits different properties than a LHPZ. The angular frequency ω_z at which the zero occurs is obtained by calculating the magnitude of (2.36):

$$\omega_z = \left| -\frac{1}{R_1C_1} + 0j \right| = \frac{1}{R_1C_1} \tag{2.38}$$

With this definition and (2.36) in mind, we can re-write (2.35) in a slightly different way:

$$N(s) = 1 + \frac{s}{\omega_z} \tag{2.39}$$

which is the normalized form of the 1st-order numerator. Figure 2.13 plots the frequency response of a typical zero located at 1.6 kHz and a unity denominator (no pole):

$$H(s) = \frac{N(s)}{D(s)} = 1 + \frac{s}{\omega_z} \tag{2.40}$$

This is the typical response of a LHPZ where the phase lead increases with frequency to become asymptotic with 90°. Should we plot a RHPZ instead, $N(s) = 1 - \frac{s}{\omega_z}$, the magnitude graph would be similar but the phase would lag (rather than lead) down to 90°. A RHPZ, in the literature, is sometimes referred to as an *unstable* zero.

Looking at the graph, we do not see a response becoming null as the frequency reaches the zero position, 1.6 kHz. Does it conflict with our mathematical definition of a zero? The root of $N(s) = 0$ is a real negative value defined by (2.36). When we set $s = j\omega$ we only consider values along the imaginary axis as σ is purposely set to 0 in harmonic analysis. In this mode, there is no way we could cancel the numerator as s never matches (2.36). There are two cases, however, where a transfer function numerator is truly canceled and produces a nulled response. The first case happens with a simple *RC* differentiator where a series capacitor blocks the dc component. The zero is located at the origin, a point common to both imaginary and real axes. When s equals 0 then the response is really a null. The second case is with a double underdamped zero-pair. When the quality factor approaches infinity (no damping), then both zeros become imaginary conjugates and for $s = s_{z_1} = s_{z_2}$, the numerator cancels and the output is a null. The physical implementation of this particular circuit is a notch filter.

For a 1st-order pole, the writing is similar. Back to our expressions in (2.21) and (2.33), the denominator $D(s)$ for a 1st-order system can be written the same way, here involving a simple *RC* network:

$$D(s) = 1 + sR_2C_2 = 1 + s\tau_2 \tag{2.41}$$

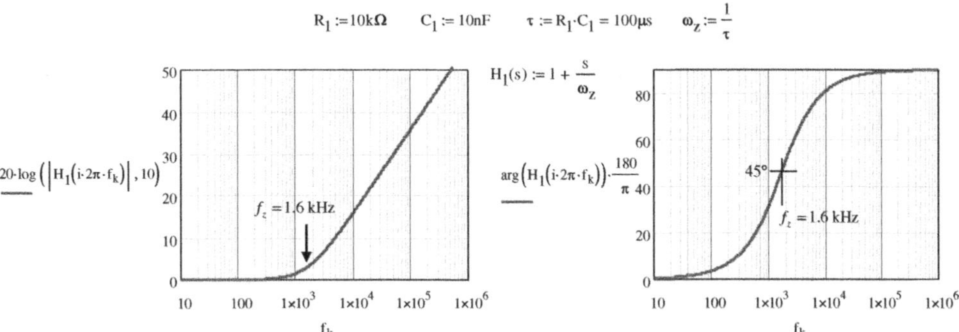

Figure 2.13 The dynamic response of a left half-plane zero shows a gain increase and a phase hitting 90° beyond the zero position.

To obtain the pole frequency, we solve $D(s_p) = 0$ leading to the following value for the root expression:

$$s_p = -\frac{1}{R_2C_2} = -\frac{1}{\tau_2} \tag{2.42}$$

Sometimes, the pole can be at the origin, meaning that in dc, $s = 0$, the transfer function is infinite since the denominator goes to 0:

$$D(s) = s\tau_1(1 + \ldots) \tag{2.43}$$

The root here is $s_p = 0$, hence the term at the origin. A system featuring an origin pole offers an infinite quasi-static gain. In reality, this gain is bounded by the operational amplifier (op amp) open-loop gain or any other sort of amplifier. Sometimes, there can be more than one pole at the origin. s^2 in (2.43) would indicate a double pole, s^3 a triple pole and so on.

The negative sign in (2.42) indicates a root placed in the left half-plane (LHP pole). As we did for the zero angular frequency extraction, we calculate (2.42) magnitude and obtain:

$$\omega_p = \left| -\frac{1}{R_2C_2} + 0j \right| = \frac{1}{R_2C_2} \tag{2.44}$$

Rearranging (2.41) in a more familiar way, we have:

$$D(s) = 1 + \frac{s}{\omega_p} \tag{2.45}$$

Figure 2.14 plots the frequency response of a typical pole located at 1.6 kHz with a unity numerator (no zero):

$$H(s) = \frac{1}{D(s)} = \frac{1}{1 + \frac{s}{\omega_p}} \tag{2.46}$$

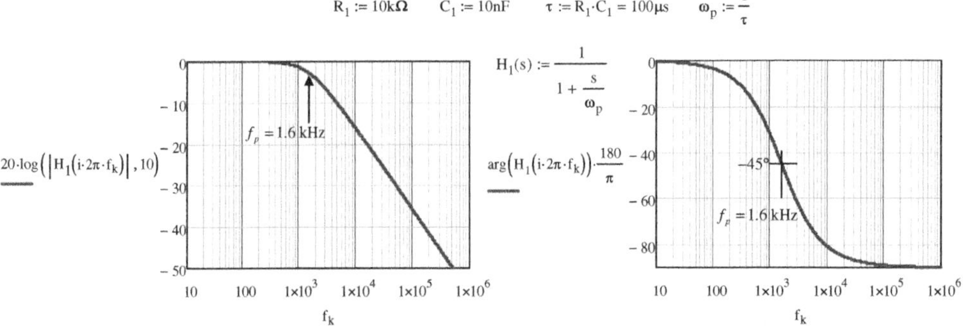

Figure 2.14 The dynamic response of a pole shows a gain decrease and a phase hitting −90° beyond the pole position.

This is the typical response of a left half-plane pole (LHPP) where the phase lag increases with frequency to asymptotically hit −90°. Should we plot a right half-plane pole (RHPP) instead $D(s) = 1 - \frac{s}{\omega_p}$, the magnitude graph would be similar but the phase would lead up to 90°. A RHPP, in the literature, is sometimes referred to as an *unstable* pole.

Again, we do not observe the transfer function magnitude becoming infinite as *f* approaches the pole location. The explication we gave for the zero is still valid for the pole: with $s = j\omega$ you cannot cancel the denominator at $s = s_p$ which is a real negative value defined by (2.42). The first case where the denominator could truly be set to zero with $s = j\omega$ is if the circuit features an origin pole. An integrator built with an op amp for instance exhibits a pole at the origin. For $s = 0$, the gain is not infinite but very high, physically bounded by the op amp open-loop gain for instance. The second case is with an under-damped second-order denominator. With an infinite quality factor, the poles are imaginary conjugates and $D(s)$ truly cancels as s approaches $s_{p_1} = s_{p_2}$. An underdamped 2nd-order low-pass filter can be physically realized to produce a really high output voltage at resonance.

If we now combine the frequency response of the zero and the pole, the transfer function becomes:

$$H(s) = \frac{1 + \frac{1}{\omega_z}}{1 + \frac{1}{\omega_p}} \tag{2.47}$$

Figure 2.15 graphs the resulting magnitude and phase responses for a zero and a pole distant by a ratio *k* of 20. As the zero kicks in, the phase increases towards 90°. The pole appears later and brings the phase down again.

As shown in [4], the phase peaks at a frequency equal to $\sqrt{f_z f_p}$. The phase bump, also called a *phase boost*, is typical of compensators used in control systems.

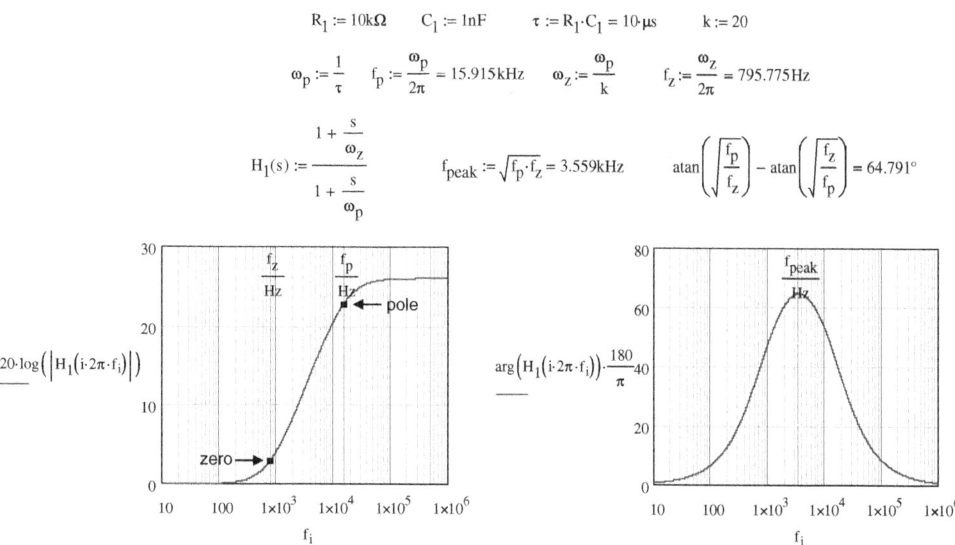

Figure 2.15 The combined response of a pole and a zero shows a phase peaking and returning to zero.

2.3.1 Low-entropy Expressions

Building on the notations we have seen, it is possible to extend the concept to higher-order transfer functions by factoring zeros and poles. Below is one possible factored pole-zero form widely used in technical papers, for instance written by Dr. Middlebrook [5]:

$$H(s) = H_0 \frac{\left(1+\frac{s}{\omega_{z_1}}\right)\left(1+\frac{s}{\omega_{z_2}}\right)\left(1+\frac{s}{\omega_{z_3}}\right)\cdots}{\left(1+\frac{s}{\omega_{p_1}}\right)\left(1+\frac{s}{\omega_{p_2}}\right)\left(1+\frac{s}{\omega_{p_3}}\right)\cdots} \tag{2.48}$$

In this expression, H_0 is the dc or static gain obtained for s equals 0. However, as we will later see in different expressions, the leading term can be a gain selected at s approaches infinity or even a gain at resonance. Why reorganize poles and zeros to follow these particular frames? For always the same motivation: we want a *low-entropy* expression from which gains, poles, zeros can be immediately identified and check how many of them are involved. Assume you have derived the following expression:

$$H(s) = -\frac{R_4 + sR_2R_4C_1}{R_3 + sR_6R_3(C_1 + C_3)} \tag{2.49}$$

Can you tell immediately when looking at this transfer function if there is a static gain, where the poles and zeros are, if any? I personally could not. Rework the numerator by factoring R_4 and rearrange the denominator by factoring R_3:

$$H(s) = -\frac{R_4(1 + sR_2C_1)}{R_3[1 + sR_6(C_1 + C_3)]} \tag{2.50}$$

Then separate R_4/R_3 from the main fraction to obtain:

$$H(s) = -\frac{R_4}{R_3}\frac{1 + sR_2C_1}{1 + sR_6(C_1 + C_3)} \tag{2.51}$$

You can now rewrite the expression following a familiar and friendly form:

$$H(s) = H_0 \frac{1+\frac{s}{\omega_z}}{1+\frac{s}{\omega_p}} \tag{2.52}$$

in which you have

$$H_0 = -\frac{R_4}{R_3} \tag{2.53}$$

a zero located at

$$\omega_z = \frac{1}{R_2C_1} \tag{2.54}$$

and a pole placed at

$$\omega_p = \frac{1}{R_6(C_1 + C_3)} \tag{2.55}$$

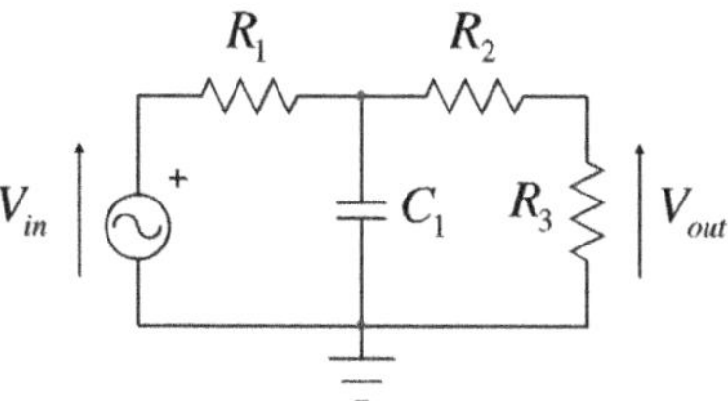

Figure 2.16 A simple 1st-order circuit involving three resistors and a capacitor.

Please note that when writing a transfer function, the leading term (identified by a subscripted 0 or ∞ respectively for a low- or high-frequency asymptote) and the transfer function must be of similar dimension (unit), meaning that the rest of the expression must be dimensionless. If we assume that $H(s)$ describes a voltage transfer function, then its dimension is [V]/[V] and H_0 (if different than 0) is unitless. Suppose you derived an impedance expression $Z(s)$, its dimension is [Ω]. Therefore, the right way of writing the final expression should be:

$$Z(s) = R_0 \frac{1 + \frac{s}{\omega_z}}{1 + \frac{s}{\omega_p}} \ldots \tag{2.56}$$

in which R_0 can be the resistance offered for $s = 0$ and has the dimension of ohms. Should you derive an admittance, $Y(s)$, then Y_0 dimension would be in siemens, [S], and so on.

Low-entropy expressions are naturally obtained by doing circuit inspection. Observe Figure 2.16 for example.

If we use brute-force analysis, we can calculate the transfer function V_{out}/V_{in} by cutting the circuit at R_2's left terminal and modeling an equivalent Thévenin generator as shown in Figure 2.17. The transfer function is thus defined as:

$$\frac{V_{out}(s)}{V_{in}(s)} = \frac{R_3}{R_2 + \left(\frac{\frac{1}{sC_1} R_1}{\frac{1}{sC_1} + R_1}\right) + R_3} \frac{\frac{1}{sC_1}}{R_1 + \frac{1}{sC_1}} \tag{2.57}$$

If you develop the fraction, you end up with a nice meaningless equation such as (2.58): you have squared s terms and various unfactored coefficients that are not organized at all. Good luck to see the

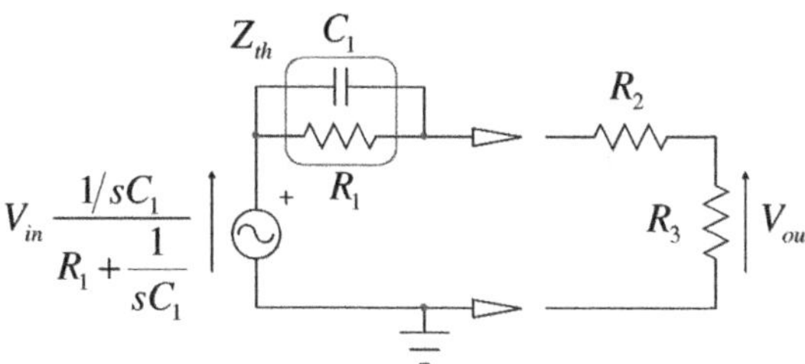

Figure 2.17 By revealing a Thévenin generator, we can apply the resistive divider theorem.

pole and a zero if any:

$$\frac{V_{out}(s)}{V_{in}(s)} = \frac{R_3}{R_2 + R_3 + sC_1R_1(R_2 + R_3) + \dfrac{sR_1C_1}{R_1{C_1}^2s^2 + sC_1} + \dfrac{s{R_1}^2C_1}{1 + sR_1C_1}} \tag{2.58}$$

Rather than applying the classical method, use what we learned in Chapter 1. What is the dc transfer function H_0 that you can infer by looking at Figure 2.16? C_1 disappears from the picture in dc (the capacitor offers an infinite impedance at 0 Hz, while an inductor would be replaced by a short circuit in similar conditions) and you are left with a simple resistive divider made of the series combination of R_1 and R_2 driving R_3:

$$H_0 = \frac{R_3}{R_1 + R_2 + R_3} \tag{2.59}$$

Then, in presence of a 1st-order network, calculate the time constant involving C_1 and the resistance it 'sees' from its terminals while the excitation source is set to 0. Here, this is a voltage source so we simply short its terminals as shown in Figure 2.18.

Looking into C_1 terminals, you see R_1 in parallel with the series arrangement of R_2 and R_3:

$$R = R_1 \| (R_2 + R_3) \tag{2.60}$$

The time constant of this system is simply $\tau = C_1\left[R_1 \| (R_2 + R_3)\right]$.

Is there a zero in this network? In other words, can we identify a condition for which the excitation would not bring a response? Besides shorting C_1 (which occurs for s approaches infinity, however, $s \to \infty$ does not count as a salient frequency to define a pole or zero by inspection), the rest of the elements are constant so there is no zero in this circuit. The transfer function obeys the form developed with (2.33), here with a static term, H_0:

$$H(s) = H_0\frac{1}{1 + s\tau} = \frac{R_3}{R_1 + R_2 + R_3}\frac{1}{1 + sC_1\left[R_1 \| (R_2 + R_3)\right]} = H_0\frac{1}{1 + \dfrac{s}{\omega_p}} \tag{2.61}$$

in which H_0 has been defined by (2.59). The pole ω_p is immediate, it is the inverse of the natural time constant:

$$\omega_p = \frac{1}{C_1\left[R_1 \| (R_2 + R_3)\right]} \tag{2.62}$$

Expressions (2.61) and (2.58) are similar however, (2.58) yields absolutely no insight into the transfer function response. Not only can going from (2.57) to (2.58) lead to errors but rearranging it into a

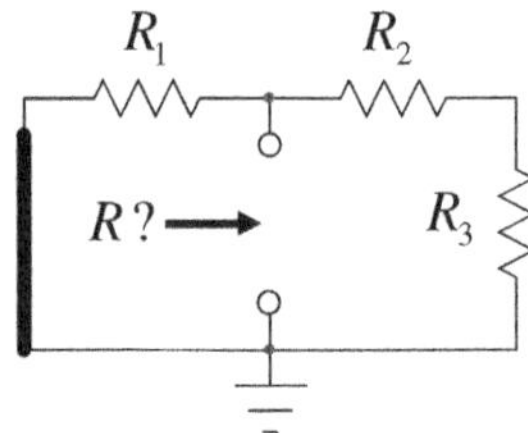

Figure 2.18 The time constant is obtained by shorting the excitation voltage and looking into the capacitor terminals to get the resistance $R?$ which drives it.

factored and ordered form requires further manipulations. On the other hand, deriving (2.61) took us less than one minute and the result is meaningful at once. Having series-parallel arrangements lets you immediately see what element dominates or lets you figure what happens to the expression if that element goes to zero or is becoming extremely large? In (2.62) if R_2 is much smaller than R_3, the pole will mainly depend on R_1 paralleled with R_3 and the dc gain H_0 will increase. I could not see that in (2.58) without spending more time rearranging the whole expression.

How do we know by the way if both equations are equal? We can capture these expressions in Mathcad® and plot their frequency response in the same graph. If equations are equal, curves must perfectly superimpose in magnitude and phase. You can also plot the difference of both equations (magnitude and phase) along the frequency axis to check an almost 0-response. The slightest deviation indicates the presence of an error.

Figure 2.19 shows how both curves nicely superimpose. If you discover that an error was made while deriving H_0 or the time constant, you can see how easy it would be to fix it. Should it happen after you wrote (2.58), you restart from scratch.

Regarding Mathcad®, I highly recommend that you apply units to all the elements you manipulate. The software checks the dimensions of the derived formulas and immediately points out units inconsistency. For defining the paralleling of components, I use the symbol || found in the 'Custom characters' toolbar. You display it in the View submenu and define what paralleling x and y means. The || operation uses the 'xfy' operator found in the Evaluation toolbar. It can be further used in expressing paralleled components in a quick and straightforward way. I also use subscripts in all my labels, making the final equation more readable: R_1 versus R1 for instance. You subscript a number by pressing the '.' key after you typed the label letter. Finally, in all the plots we have presented so far, the x-axis features a logarithmic scale. To mitigate the number of points per decade (as SPICE does in an . AC analysis), I use a simple routine developed by my colleague Dr. Capilla. It nicely distributes the points along the axis in relationship to the total amount you have selected per decade. For a standard

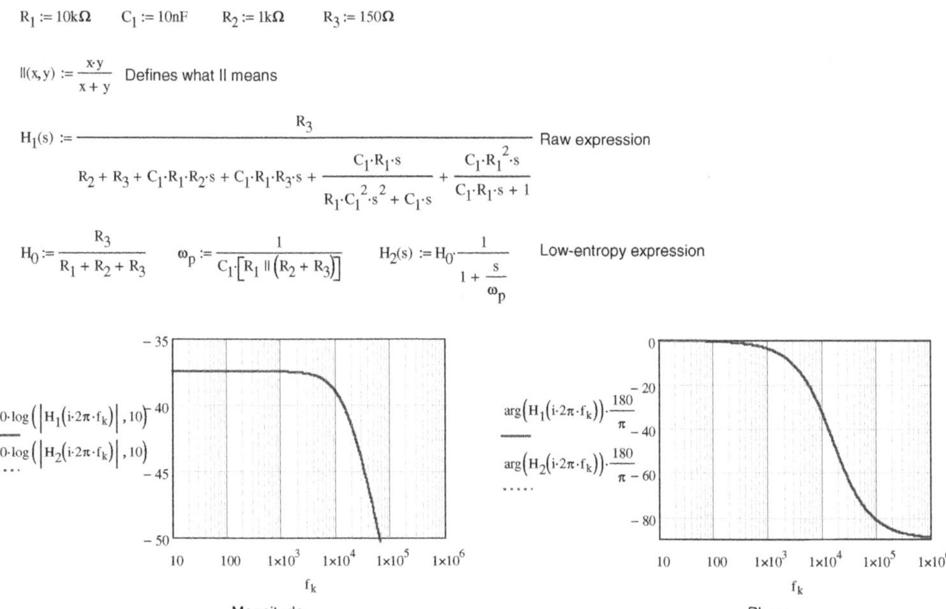

Figure 2.19 Mathcad® or SPICE help plot various types of expressions and check for expressions integrity.

Bode plot, 100 points per decade are enough but if you do not want to miss a sharp peaking, increasing the points per decade to 1000 or more is necessary. In the below code, the subscript in the counter variable k is not obtained by pressing '.' in this case. As it refers to a position in the storing matrix f, you have to press '[' after typing the letter.

$\text{Start\_Freq} := 10^1$

$\text{Stop\_Freq} := 10^6$

$\text{Points\_per\_decade} := 1000$

$\text{Number\_of\_decades} := (\log(\text{Stop\_Freq}) - \log(\text{Start\_Freq}))$

$\text{Number\_of\_points} := \text{Number\_of\_decades} \cdot \text{Points\_per\_decade} + 1$

$k := 0..\ \text{Number\_of\_points}$

$$f_k := 10^{\log(\text{Start\_Freq}) + k \cdot \frac{\text{Number\_of\_decades}}{\text{Number\_of\_points}}} \cdot \text{Hz}$$

Please note that a built-in function also exists and is designated as *logspace*. You pass the start and end frequencies plus the total number of points. A routine using *logspace* could look like this:

$\text{Number\_of\_points} := 1000$

$k := 0..\ \text{Number\_of\_points}$

$f := \text{logspace}\left(10, 10^6, \text{Number\_of\_points}\right) \cdot \text{Hz}$

And now, the icing on the cake: what if a small resistance r_C is added in series with capacitor C_1 as in Figure 2.20? Can we reuse what we have already done in (2.61) without restarting from scratch? Absolutely!

Is H_0 affected? No, as C_1 disappears at 0 Hz, adding a series term such as r_C does not modify H_0. Now consider the time constant τ. Without r_C, looking into C_1 terminals while the source was shorted, we saw a series/parallel resistive arrangement. Now that r_C is added, it simply comes in series with the previous term:

$$\tau = C_1 \left[R_1 \| (R_2 + R_3) + r_C\right] \tag{2.63}$$

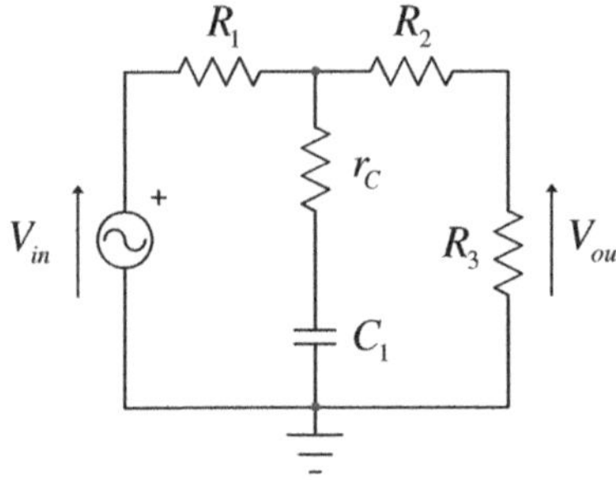

Figure 2.20 Adding a small resistance in series is not a problem with a low-entropy approach.

and the pole definition is updated to be:

$$\omega_p = \frac{1}{C_1\left[R_1 \| (R_2 + R_3) + r_C\right]} \tag{2.64}$$

With C_1 alone, we said that besides a short circuit at infinite frequency, the excitation always produces a response. Now that we have installed a series resistance, is it always true? In other words, can the series arrangement of r_C and C_1 (in a transformed circuit where C_1 is replaced by $1/sC_1$) become a short circuit at a certain angular frequency s_z, the zero?

$$r_C + \frac{1}{sC_1} = \frac{1 + sr_CC_1}{sC_1} = 0 \tag{2.65}$$

Solving for the numerator root, we find a zero for the angular frequency $\omega_z = \frac{1}{r_CC_1}$. We can then quickly update (2.61) by reusing on what we found by adding r_C:

$$H(s) = \frac{R_3}{R_1 + R_2 + R_3} \frac{1 + sr_CC_1}{1 + sC_1\left[R_1 \| (R_2 + R_3) + r_C\right]} = H_0 \frac{1 + \frac{s}{\omega_z}}{1 + \frac{s}{\omega_p}} \tag{2.66}$$

in which H_0 is still defined by (2.59) and the pole by (2.64). The zero is expressed as:

$$\omega_z = \frac{1}{r_CC_1} \tag{2.67}$$

There is no way we could have derived the first and second examples in such a small amount of time without using the fast analytical techniques.

2.3.2 Higher Order Expressions

The degree of the denominator defines the degree of the network under study. So far we have looked at first-order expressions in which the degree of the denominator was 1. As such, we can combine (2.35) or (2.45) under a polynomial form to express a transfer function:

$$H(s) = \frac{a_0 + a_1s}{b_0 + b_1s} \tag{2.68}$$

Factoring a_0 and b_0 we have:

$$H(s) = \frac{a_0}{b_0} \frac{1 + \frac{a_1}{a_0}s}{1 + \frac{b_1}{b_0}s} \tag{2.69}$$

In this linear expression, all coefficients are real and a_0/b_0 represents the low-frequency static gain ($s=0$) that we will note H_0. Rearranging (2.69) we have:

$$H(s) = H_0 \frac{1 + s\frac{a_1}{a_0}}{1 + s\frac{b_1}{b_0}} \tag{2.70}$$

We said earlier that writing the equation the proper way implies that H_0 and $H(s)$ share a similar dimension. If $H(s)$ is a voltage or current gain, then clearly H_0 is unitless. If $Z(s)$ instead is derived (an impedance), then the dc term must be a resistance noted R_0. If the leading term carries the unit, then the numerator and the denominator are dimensionless. In this case, then sa_1/a_0 and sb_1/b_0 are also dimensionless in (2.70). As s has the dimension of a frequency in Hz, a_1/a_0 or b_1/b_0 must have the dimension of Hz^{-1} or time: they are time constants, those appearing in (2.35) and in (2.41). For a 1st-order system, b_1/b_0 is the circuit time constant evaluated when the excitation is removed: RC or L/R respectively in Figure 2.6 and Figure 2.9. Similarly, a_1/a_0 is also a time constant but obtained in a different way as we shall see later.

We can describe higher order networks simply by increasing the order of the polynomials:

$$H(s) = \frac{a_0 + a_1 s + a_2 s^2 + a_3 s^3 + \ldots + a_m s^m}{b_0 + b_1 s + b_2 s^2 + b_3 s^3 + \ldots + b_n s^n} \tag{2.71}$$

(2.71) is another way of writing (2.48). Actually, high-order expressions derived using FACTs will first come out as in (2.71). This result can already be exploited to plot a frequency response but it still does not fit the format of (2.48) in which poles and zeros are collected in a clear and ordered form. Part of the skill you will acquire with this book is how to shape a ratio of polynomials such as in (2.71) and rewrite it in a low-entropy form.

The degree of the denominator n sets the order of the network. It also sets the number of poles: if you deal with a 3rd-order network, there are three poles (three roots for a third order polynomial). Same for the denominator, if you have a 2nd-order form in the numerator, you have two zeros. These poles and zeros can be real, imaginary or conjugates. Again, if a_0/b_0 is factored, we obtain a slightly different expression:

$$H(s) = H_0 \frac{1 + \frac{a_1}{a_0}s + \frac{a_2}{a_0}s^2 + \frac{a_3}{a_0}s^3 + \ldots + \frac{a_m}{a_0}s^m}{1 + \frac{b_1}{b_0}s + \frac{b_2}{b_0}s^2 + \frac{b_3}{b_0}s^3 + \ldots + \frac{b_n}{b_0}s^n} \tag{2.72}$$

To maintain a unitless ratio, a_1/a_0 and b_1/b_0 must still have the dimension of Hz^{-1} or time as in (2.70). In higher order networks, a_1/a_0 and b_1/b_0 are *the sums* of the network time constants obtained in certain conditions. If b_1/b_0 gathers the time constants found when the excitation is set to 0, a_1/a_0 will be unveiled in a different way, when the output is a null. a_2/a_0 and b_2/b_0 are multiplied by s^2 which has the dimension of Hz^2. Therefore, a_2/a_0 and b_2/b_0 add products of two time constants to form the dimension of time^2 or Hz^{-2}. If we carry on, a_3/a_0 and b_3/b_0 are multiplied by s^3 which has the dimension of Hz^3. Therefore, a_3/a_0 and b_3/b_0 add products of three time constants to form a dimension of time^3 or Hz^{-3}. You can continue up to the nth element. These explanations may sound obscure for now but step-by-step examples will help make things clearer in the coming chapters.

2.3.3 *Second-order Polynomial Forms*

It is not easy to identify poles and zeros in (2.71). Assume the denominator D of our 2nd-order transfer function has been derived in the following form in which b_0 equals 1:

$$D(s) = 1 + b_1 s + b_2 s^2 \tag{2.73}$$

This is the typical expression our fast analytical techniques will let you derive in an amazingly quick way. To unveil the poles positions, you must remember a few typical polynomial forms. One of the most popular is that of a 2nd-order low-pass filter. Its denominator is characterized by a quality factor Q (or a damping ratio ζ) and a resonant frequency ω_0:

$$H(s) = \frac{1}{1 + \frac{s}{\omega_0 Q} + \left(\frac{s}{\omega_0}\right)^2} = \frac{1}{1 + 2\zeta\frac{s}{\omega_0} + \left(\frac{s}{\omega_0}\right)^2} \tag{2.74}$$

When you come up with a formula such as in (2.73), whether it is in the denominator or in the numerator (you deal with double zeros then) you must rearrange it to fit (2.74) denominator format. To do so, identify the terms by equating (2.73) with (2.74) denominator:

$$1 + b_1 s + b_2 s^2 = 1 + \frac{s}{\omega_0 Q} + \left(\frac{s}{\omega_0}\right)^2 \tag{2.75}$$

and identify coefficients

$$b_1 = \frac{1}{\omega_0 Q} \tag{2.76}$$

$$b_2 = \frac{1}{\omega_0^2} \tag{2.77}$$

From (2.77) we have:

$$\omega_0 = \frac{1}{\sqrt{b_2}} \tag{2.78}$$

which substituted in (2.76) gives the quality factor definition:

$$Q = \frac{\sqrt{b_2}}{b_1} \tag{2.79}$$

This 2nd-order form can be rearranged in different ways as illustrated in Figure 2.21. Assume you derive a 2nd-order transfer function like in (2.71) but you don't know the type of filter it looks like. Just plot its frequency response using a software like Mathcad® for instance and check how it compares to Figure 2.21 and Figure 2.22 curves. Then identify Q and ω_0 exactly as we did with (2.76) and (2.77). Rewrite the expression fitting the identified format in Figure 2.21 or in Figure 2.22.

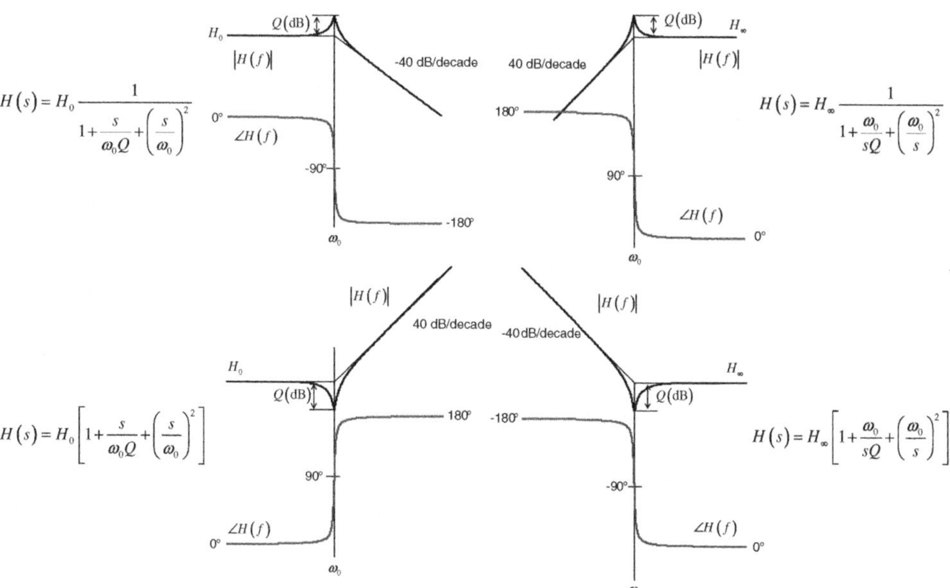

Figure 2.21 Various 2nd-order networks frequency responses. The x-axis is $Log(f)$.

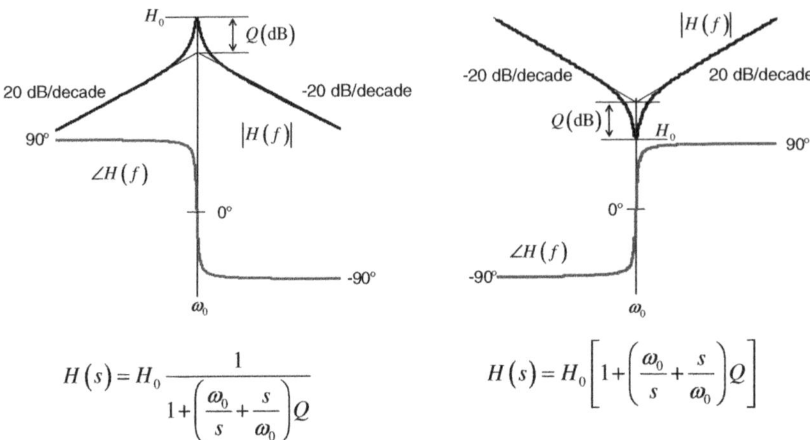

Figure 2.22 Frequency responses of 2nd-order networks, here a band-pass filter (left) and a notch (right).

2.3.4 *Low-Q Approximation for a 2nd-order Polynomial*

In (2.74), checking for the denominator roots – the poles positions – you would solve:

$$1+\frac{s}{\omega_0 Q}+\left(\frac{s}{\omega_0}\right)^2=0 \tag{2.80}$$

leading to a pair of roots defined as:

$$s_{p_1}, s_{p_2}=\frac{\omega_0}{2Q}\left(\pm\sqrt{1-4Q^2}-1\right) \tag{2.81}$$

Depending on the quality factor value, the roots can be real (no imaginary part), conjugate (real and imaginary parts) or conjugate imaginaries (no real part). For Q less than 0.5 the roots are real and the response to an input step is not oscillating at all. For Q equals 0.5, roots are real and coincident. The response is fast and there are no oscillations. For Q above 0.5, roots become conjugate and you start observing an oscillatory response but the real parts represent damping (energy dissipation) and oscillations eventually die out. As Q increases, the real parts shrink and the response becomes more and more oscillatory with a pronounced overshoot. When Q is infinite, the real parts are 0 and you have permanent oscillations. These transient responses appear in Figure 2.23 for different Q values. They are obtained from a simple *RLC* filter driven by a 1-V voltage step whose R is adjusted in relationship to the desired damping.

If you now place s_1 and s_2 obtained from (2.81) for the various Q values on a s-plane chart, you have a graph as shown in Figure 2.24. In sketch (a), Q is low, well below 0.5 and the poles are split. Root s_{p1} is close to the vertical axis and dominates the low-frequency ac response while s_{p2} concerns higher frequencies. In (b), Q equals 0.5 and the roots are *coincident*: poles occur at the same frequency but the transient response is ringing free because there is no imaginary part in s_{p1} and s_{p2}. As Q increases beyond 0.5 as in (c), roots become conjugate and the response is a damped oscillatory sinusoidal waveform. The real parts are all the losses that dissipate energy and damp the system. In (d), no energy is dissipated cycle by cycle, Q is infinite. The system is undamped and oscillates freely. Please note that all roots in (a), (b) and (c) sit in the left half-plane, they are designated as LHP poles: the response is exponentially decaying. In other terms, it is *bounded*, it does not diverge. Should one of these poles

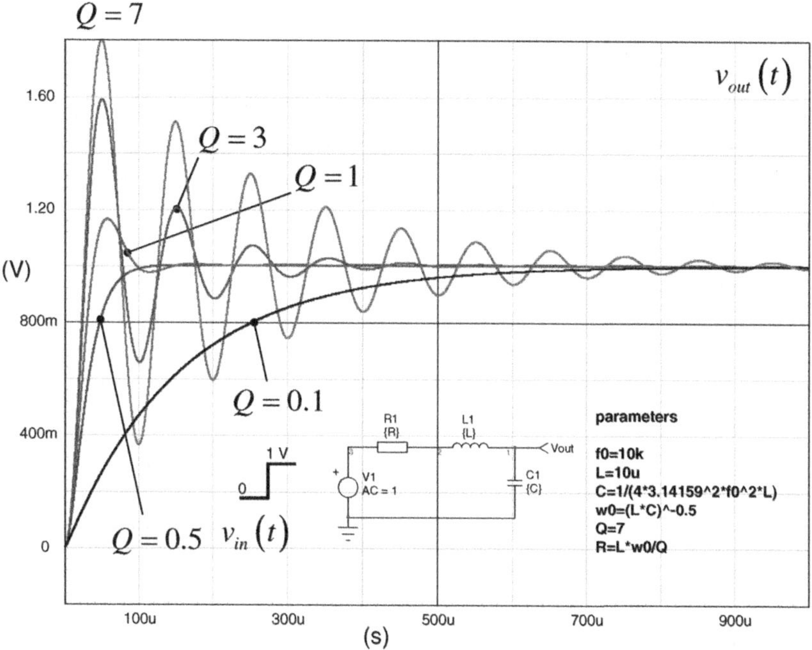

Figure 2.23 The transient response of a 2nd-order network varies depending on the quality factor Q.

jump to the right side of the map and become a positive root (RHPP), the response would diverge and become *unbounded*.

If we now plot the frequency responses of the *RLC* network for different Q values, they appear in Figure 2.25. For high Qs, the response peaks: a Q of 7 corresponds to a 17-dB excursion as shown in the graph. Please note that the 2nd-order network magnitude response produces a −2 slope (a 40-dB

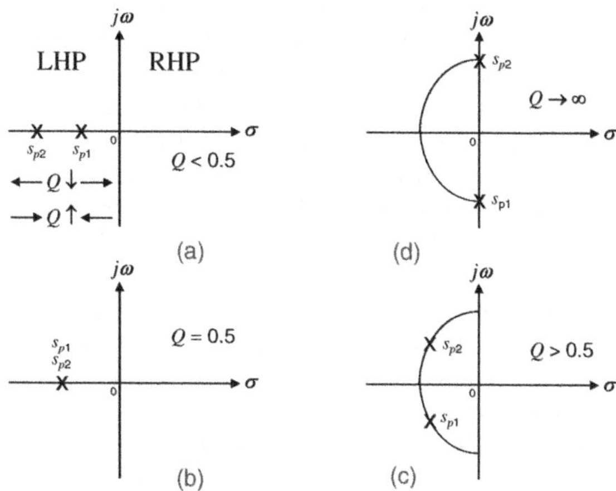

Figure 2.24 Roots can be placed in an Argand chart or s-plane. Here we have left half-plane poles (LHPP). The roots position changes as Q also moves. Poles are represented by crosses (×). Should you want to represent zeros instead, you would use circles (O).

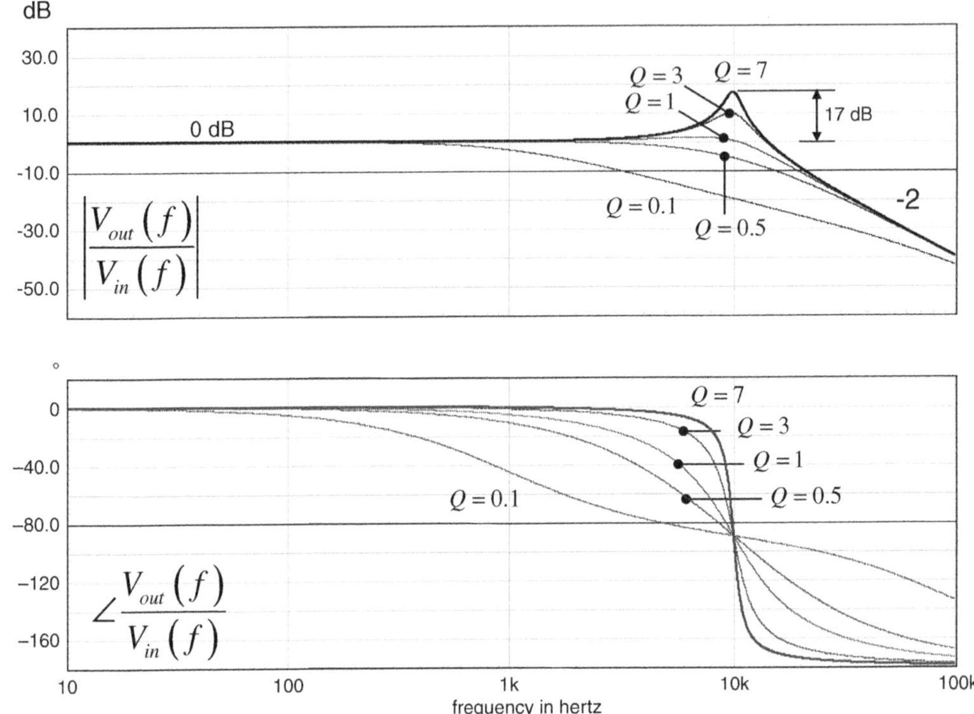

Figure 2.25 Frequency responses of the 2nd-order *RLC* network with different Q values.

attenuation per decade) beyond the peak. The phase sharply drops for high Q and asymptotically reaches $-180°$ at high frequencies. As Q lowers, the peak starts to weaken and disappears when Q passes below 0.5. The phase drops in a softer way. For the lowest Q, the amplitude really differs from the previous curves and the phase slowly falls as if it would like to pause around $-90°$ to fall again towards $-180°$ in higher frequencies. As we explained, for a low Q, the two poles are split and one of them dominates at low frequency. It is the one closer to the vertical axis in Figure 2.24a. The second acts later at higher frequencies. The first magnitude slope is a 1st-order slope (-1 or -20 dB per decade) and when the second pole kicks in, the slope turns into a -2 slope as with any 2nd-order network.

Figure 2.26 shows the ac response when the quality factor is way smaller than 1, 0.01 for instance. You can clearly see the poles at 100 Hz and 1 MHz identified at $-45°$ and $-135°$ phase points. As a conclusion, when the quality factor is low, well below 1, you can consider a 2nd-order denominator (or numerator, it also works for zeros) as two cascaded 1st-order filters.

To calculate the position of these roots and replace the complex 2nd-order form by two cascaded poles (or zeros) we can use the MacLaurin series to simplify the square root term in (2.81) assuming $Q \ll 1$:

$$(1+x)^n \approx 1 + nx \tag{2.82}$$

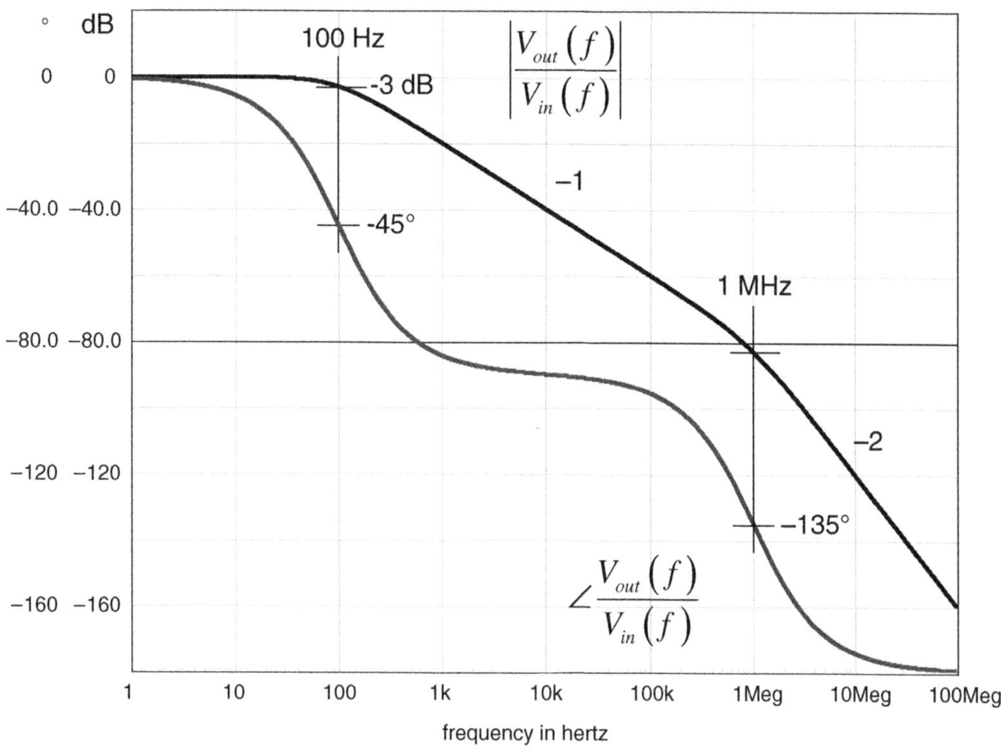

Figure 2.26 Frequency responses of the 2nd-order *RLC* network for a very low Q (0.01).

which implies that

$$\sqrt{1+x} \approx 1 + \frac{1}{2}x \tag{2.83}$$

If we introduce this term into (2.81) and simplify the final expressions, we have our definitions for the split poles:

$$\omega_{p_1} = \frac{\omega_0}{Q}\frac{1-\sqrt{1-4Q^2}}{2} \approx \frac{\omega_0}{Q}\frac{1-\left(1-2Q^2\right)}{2} = Q\omega_0 \tag{2.84}$$

$$\omega_{p_2} = \frac{\omega_0}{Q}\frac{1+\sqrt{1-4Q^2}}{2} \approx -\frac{\omega_0}{Q}\left(Q^2-1\right) \approx \frac{\omega_0}{Q} \tag{2.85}$$

In this case, for Q much lower than 1, the complex 2nd-order expression described in (2.75) can be replaced by a product of two poles whose position is defined by (2.84) and (2.85):

$$\text{For } Q_D \ll 1 \rightarrow D(s) = 1 + \frac{s}{\omega_0 Q_D} + \left(\frac{s}{\omega_0}\right)^2 \approx \left(1+\frac{s}{\omega_{p_1}}\right)\left(1+\frac{s}{\omega_{p_2}}\right) \tag{2.86}$$

Please note that 2 zeros can be rearranged in a similar way:

$$\text{For } Q_N \ll 1 \rightarrow N(s) = 1 + \frac{s}{\omega_{0N} Q_N} + \left(\frac{s}{\omega_{0N}}\right)^2 \approx \left(1+\frac{s}{\omega_{z_1}}\right)\left(1+\frac{s}{\omega_{z_2}}\right) \tag{2.87}$$

If the polynomial form is still in a raw format, then the simplification for a low-Q system uses (2.78)/(2.79):

$$\omega_{p_1} = \frac{1}{\sqrt{b_2}} \frac{\sqrt{b_2}}{b_1} = \frac{1}{b_1} \tag{2.88}$$

$$\omega_{p_2} = \frac{\frac{1}{\sqrt{b_2}}}{\frac{\sqrt{b_2}}{b_1}} = \frac{b_1}{b_2} \tag{2.89}$$

and leads to

$$D(s) = 1 + b_1 s + b_2 s^2 \approx (1 + b_1 s)\left(1 + \frac{b_2}{b_1} s\right) \tag{2.90}$$

For two zeros, we can proceed the same way:

$$\omega_{z_1} = \frac{1}{\sqrt{a_2}} \frac{\sqrt{a_2}}{a_1} = \frac{1}{a_1} \tag{2.91}$$

$$\omega_{z_2} = \frac{\frac{1}{\sqrt{a_2}}}{\frac{\sqrt{a_2}}{a_1}} = \frac{a_1}{a_2} \tag{2.92}$$

and leads to

$$N(s) = 1 + a_1 s + a_2 s^2 \approx (1 + a_1 s)\left(1 + \frac{a_2}{a_1} s\right) \tag{2.93}$$

For $Q_D = 0.5$, the poles are coincident and (2.86) reduces to:

$$\text{For } Q_D = 0.5 \to D(s) = 1 + \frac{s}{\omega_0 Q_D} + \left(\frac{s}{\omega_0}\right)^2 = \left(1 + \frac{s}{\omega_p}\right)^2 \tag{2.94}$$

in which

$$\omega_p = \frac{1}{\sqrt{b_2}} \tag{2.95}$$

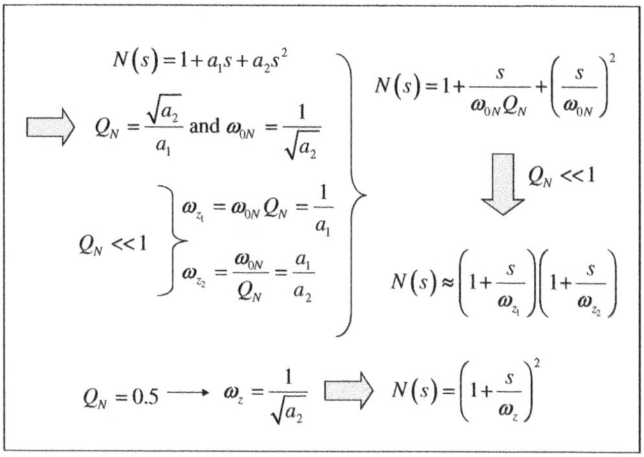

$$N(s)=1+a_1s+a_2s^2$$

$$\Rightarrow Q_N=\frac{\sqrt{a_2}}{a_1} \text{ and } \omega_{0N}=\frac{1}{\sqrt{a_2}}$$

$$Q_N \ll 1 \quad \begin{cases} \omega_{z_1}=\omega_{0N}Q_N=\dfrac{1}{a_1} \\ \omega_{z_2}=\dfrac{\omega_{0N}}{Q_N}=\dfrac{a_1}{a_2} \end{cases}$$

$$N(s)=1+\frac{s}{\omega_{0N}Q_N}+\left(\frac{s}{\omega_{0N}}\right)^2$$

$$\Downarrow \; Q_N \ll 1$$

$$N(s)\approx\left(1+\frac{s}{\omega_{z_1}}\right)\left(1+\frac{s}{\omega_{z_2}}\right)$$

$$Q_N=0.5 \longrightarrow \omega_z=\frac{1}{\sqrt{a_2}} \Rightarrow N(s)=\left(1+\frac{s}{\omega_z}\right)^2$$

Figure 2.27 Summary of operations around a 2nd-order transfer function numerator.

Considering the numerator, for $Q_N=0.5$, the zeros are coincident and (2.87) reduces to:

$$\text{For } Q_N=0.5 \rightarrow N(s)=1+\frac{s}{\omega_{0N}Q_N}+\left(\frac{s}{\omega_{0N}}\right)^2=\left(1+\frac{s}{\omega_z}\right)^2 \tag{2.96}$$

in which

$$\omega_z=\frac{1}{\sqrt{a_2}} \tag{2.97}$$

We will refer to these expressions quite often when rearranging expressions in the next part of the book. Figure 2.27 and Figure 2.28 summarize what we have discussed.

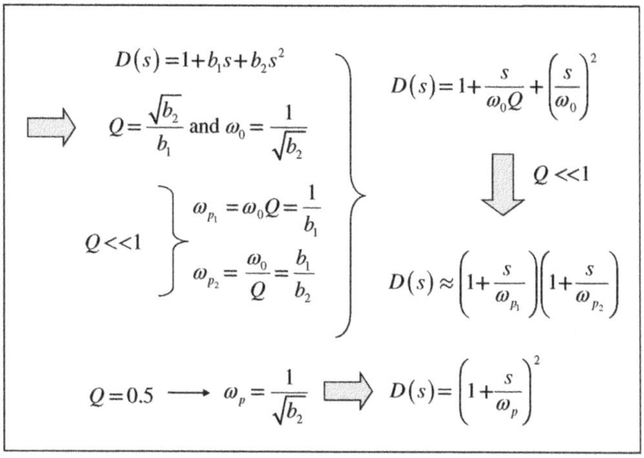

$$D(s)=1+b_1s+b_2s^2$$

$$\Rightarrow Q=\frac{\sqrt{b_2}}{b_1} \text{ and } \omega_0=\frac{1}{\sqrt{b_2}}$$

$$Q \ll 1 \quad \begin{cases} \omega_{p_1}=\omega_0 Q=\dfrac{1}{b_1} \\ \omega_{p_2}=\dfrac{\omega_0}{Q}=\dfrac{b_1}{b_2} \end{cases}$$

$$D(s)=1+\frac{s}{\omega_0 Q}+\left(\frac{s}{\omega_0}\right)^2$$

$$\Downarrow \; Q \ll 1$$

$$D(s)\approx\left(1+\frac{s}{\omega_{p_1}}\right)\left(1+\frac{s}{\omega_{p_2}}\right)$$

$$Q=0.5 \longrightarrow \omega_p=\frac{1}{\sqrt{b_2}} \Rightarrow D(s)=\left(1+\frac{s}{\omega_p}\right)^2$$

Figure 2.28 Summary of operations around a 2nd-order transfer function denominator.

2.3.5 Approximation for a 3rd-order Polynomial

A 3rd-order polynomial is described by the following form, here a denominator D in which b_0 is unity:

$$D(s) = 1 + b_1 s + b_2 s^2 + b_3 s^3 \tag{2.98}$$

If this expression is already organized, the best would be to put it under a polynomial form involving a single pole (or zero) followed by a 2nd-order polynomial form. That way, we can immediately see where poles (or zeros) occur:

$$D(s) = \left(1 + \frac{s}{\omega_p}\right)\left(1 + \frac{s}{\omega_0 Q} + \left(\frac{s}{\omega_0}\right)^2\right) \tag{2.99}$$

If we now develop (2.99) and identify (2.98) coefficients, it is possible to write:

$$\left(1 + \frac{s}{\omega_p}\right)\left(1 + \frac{s}{\omega_0 Q} + \left(\frac{s}{\omega_0}\right)^2\right) = 1 + s\left(\frac{1}{\omega_p} + \frac{1}{\omega_0 Q}\right) + s^2\left(\frac{1}{\omega_0{}^2} + \frac{1}{Q\omega_0\omega_p}\right) + \frac{s^3}{\omega_0{}^2\omega_p} \tag{2.100}$$

Then:

$$b_1 = \frac{1}{\omega_p} + \frac{1}{\omega_0 Q} \approx \frac{1}{\omega_p} \text{ if } \omega_0 Q \gg \omega_p \tag{2.101}$$

$$b_2 = \frac{1}{\omega_0^2} + \frac{1}{Q\omega_0\omega_p} \tag{2.102}$$

$$b_3 = \frac{1}{\omega_0^2\omega_p} \tag{2.103}$$

From the above equations, if we solve for ω_p, Q and ω_0, we should find:

$$\omega_p = \frac{1}{b_1} \tag{2.104}$$

$$Q = \frac{b_1 b_3\sqrt{\frac{b_1}{b_3}}}{b_1 b_2 - b_3} \tag{2.105}$$

$$\omega_0 = \sqrt{\frac{b_1}{b_3}} \tag{2.106}$$

In (2.99), we can see a factor $Q\omega_0$ which is the product of (2.105) by (2.106). If we expand this product, we have:

$$Q\omega_0 = \frac{b_1^2}{b_1 b_2 - b_3} \tag{2.107}$$

If we consider $b_3 \ll b_1 b_2$ then (2.107) can be rewritten as:

$$Q\omega_0 \approx \frac{b_1}{b_2} \tag{2.108}$$

Now using (2.106) and (2.108) definitions, we can rewrite the 2nd-order polynomial in (2.99):

$$1+\frac{s}{\omega_0 Q}+\left(\frac{s}{\omega_0}\right)^2 \approx 1+s\frac{b_2}{b_1}+s^2\frac{b_3}{b_1} \tag{2.109}$$

As a first result, in case the second and third poles are not separated (a resonating circuit for instance, such as subharmonic poles in a current mode switching converter), then (2.98) can be expressed as:

$$1+b_1 s+b_2 s^2+b_3 s^3 \approx (1+b_1 s)\left(1+s\frac{b_2}{b_1}+s^2\frac{b_3}{b_1}\right) \tag{2.110}$$

Please note that the response is dominated by $1/b_1$ at low frequency, before the double poles kick in. If you have a 3rd-order denominator in which one separate pole occurs at a higher position than the double poles, approximation from (2.110) no longer holds. Now, if the 2nd-order polynomial in (2.110) features a low Q (poles are well separated), then it can further be rearranged capitalizing on (2.90) recommendations:

$$\left(1+s\frac{b_2}{b_1}+s^2\frac{b_3}{b_1}\right) \approx \left(1+s\frac{b_2}{b_1}\right)\left(1+s\frac{b_3}{b_2}\right) \tag{2.111}$$

Now updating the complete expression in (2.98):

$$1+b_1 s+b_2 s^2+b_3 s^3 \approx (1+b_1 s)\left(1+\frac{b_2}{b_1}s\right)\left(1+\frac{b_3}{b_2}s\right) \tag{2.112}$$

This is the expression of a 3rd-order polynomial form describing a system having three well-separated poles or zeros with the same sign.

Figure 2.29 shows a 3rd-order transfer function assembled using all b coefficients under the label H_1. A first simplification using (2.110) is done under the H_2 label. Then, a third simplification based on (2.112) recommendation is performed under the H_3 label. When all three frequency responses are plotted in magnitude and phase, they look very similar. Please note the correct dimension for the various b coefficients, as discussed in several paragraphs above. Without the correct units in b, Mathcad® would point out a unit homogeneity problem.

Taking advantage of (2.112), the generalization for a nth-order polynomial with well-separated roots is now possible as shown below:

$$1+a_1 s+a_2 s^2+a_3 s^3+a_4 s^4+\ldots+a_n s^n \approx (1+a_1 s)\left(1+\frac{a_2}{a_1}s\right)\left(1+\frac{a_3}{a_2}s\right)\left(1+\frac{a_4}{a_3}s\right)\ldots\left(1+\frac{a_n}{a_{n-1}}s\right) \tag{2.113}$$

A summary of these calculations appears in Figure 2.30 and Figure 2.31.

2.3.6 How to Determine the Order of the System?

The order of a transfer function depends on the polynomial degree of its denominator. That degree itself relates to the number of independent *state variables* you can identify in the circuit you

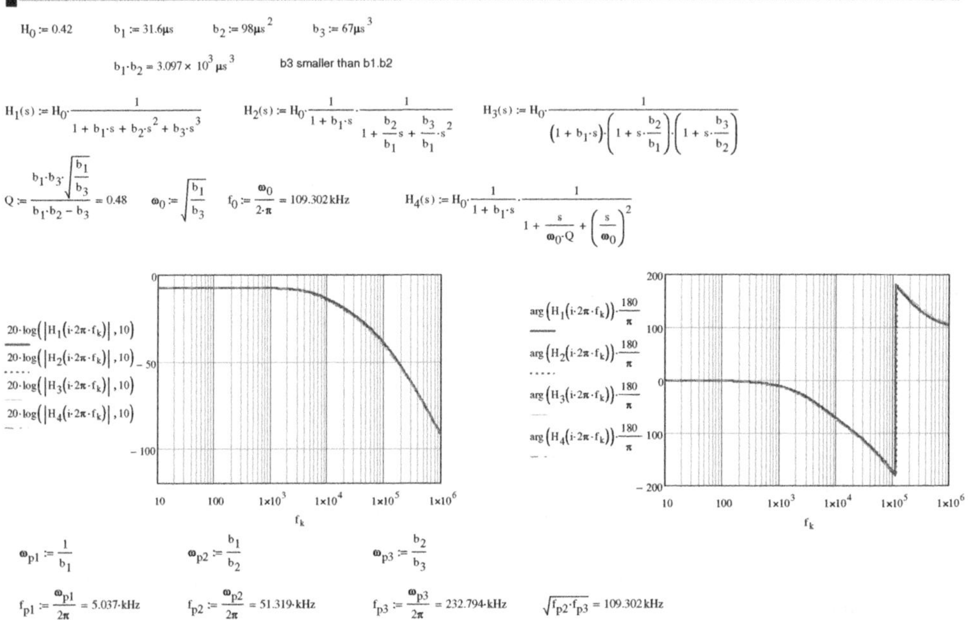

Figure 2.29 A 3rd-order polynomial – provided its roots are well separated and of similar signs – can be rewritten in the form of 3 cascaded poles.

$$N(s) = 1 + a_1 s + a_2 s^2 + a_3 s^3$$

$$\omega_z = \frac{1}{a_1} \qquad \omega_{0N} = \sqrt{\frac{a_1}{a_3}} \qquad Q_N = \frac{a_1 a_3 \sqrt{\frac{a_1}{a_3}}}{a_1 a_2 - a_3}$$

Dominant low-frequency zero
Double high-frequency zeros

$$N(s) \approx \left(1 + \frac{s}{\omega_z}\right)\left(1 + \frac{s}{\omega_{0N} Q_N} + \left(\frac{s}{\omega_{0N}}\right)^2\right)$$

$$a_3 << a_1 a_2$$

$$N(s) \approx (1 + s a_1)\left(1 + s\frac{a_2}{a_1} + s^2 \frac{a_3}{a_1}\right)$$

High-frequency 3rd zero

$$\omega_{z_1} = \frac{1}{a_1} \qquad \omega_{z_2} = \frac{a_1}{a_2} \qquad \omega_{z_3} = \frac{a_2}{a_3}$$

Well-spread zeros

$$N(s) \approx \left(1 + \frac{s}{\omega_{z_1}}\right)\left(1 + \frac{s}{\omega_{z_2}}\right)\left(1 + \frac{s}{\omega_{z_3}}\right)$$

$$1 + a_1 s + a_2 s^2 + a_3 s^3 + a_4 s^4 + \ldots + a_n s^n \approx (1 + a_1 s)\left(1 + \frac{a_2}{a_1}s\right)\left(1 + \frac{a_3}{a_2}s\right)\left(1 + \frac{a_4}{a_3}s\right)\ldots\left(1 + \frac{a_n}{a_{n-1}}s\right)$$

Figure 2.30 Summary of approximations for 3rd- and higher-order polynomial forms for the numerator.

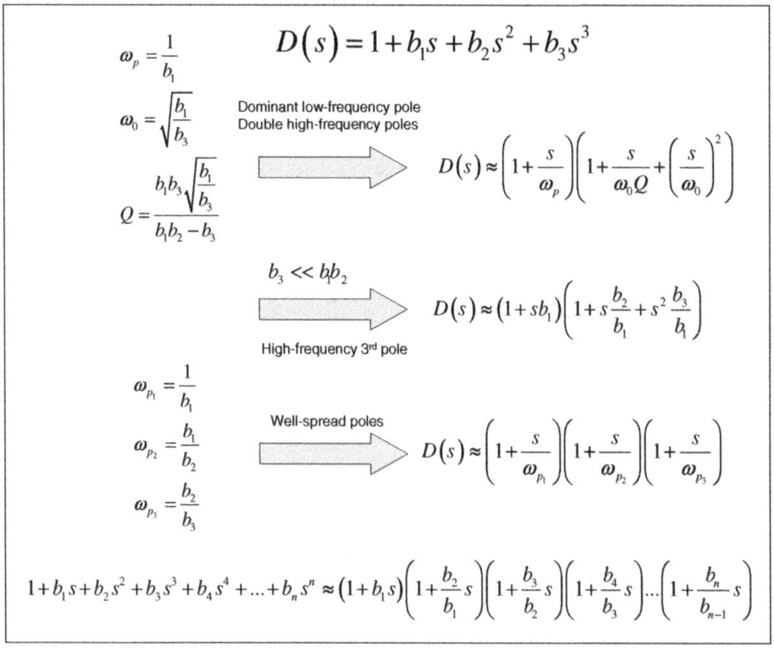

$$D(s) = 1 + b_1 s + b_2 s^2 + b_3 s^3$$

$$\omega_p = \frac{1}{b_1} \qquad \omega_0 = \sqrt{\frac{b_1}{b_3}} \qquad Q = \frac{b_1 b_3 \sqrt{\frac{b_1}{b_3}}}{b_1 b_2 - b_3}$$

Dominant low-frequency pole
Double high-frequency poles

$$D(s) \approx \left(1 + \frac{s}{\omega_p}\right)\left(1 + \frac{s}{\omega_0 Q} + \left(\frac{s}{\omega_0}\right)^2\right)$$

$b_3 \ll b_1 b_2$

$$D(s) \approx (1 + s b_1)\left(1 + s\frac{b_2}{b_1} + s^2\frac{b_3}{b_1}\right)$$

High-frequency 3rd pole

$$\omega_{p_1} = \frac{1}{b_1} \qquad \omega_{p_2} = \frac{b_1}{b_2} \qquad \omega_{p_3} = \frac{b_2}{b_3}$$

Well-spread poles

$$D(s) \approx \left(1 + \frac{s}{\omega_{p_1}}\right)\left(1 + \frac{s}{\omega_{p_2}}\right)\left(1 + \frac{s}{\omega_{p_3}}\right)$$

$$1 + b_1 s + b_2 s^2 + b_3 s^3 + b_4 s^4 + \ldots + b_n s^n \approx (1 + b_1 s)\left(1 + \frac{b_2}{b_1}s\right)\left(1 + \frac{b_3}{b_2}s\right)\left(1 + \frac{b_4}{b_3}s\right)\ldots\left(1 + \frac{b_n}{b_{n-1}}s\right)$$

Figure 2.31 Summary of approximations for 3rd- and higher-order polynomial forms for the denominator.

study. State variables are associated with energy-storage elements such as inductors and capacitors. These state variables are often designated with the lower case letter x like x_1 for the inductor current, x_2 for the capacitor voltage and so on. If you know the state of the circuit at $t = 0$ – actually what the initial conditions (IC in SPICE) are – then you can predict the circuit state for any time different than 0. This is what we showed in (2.17) and (2.30) where the voltage V_0 and the current I_0 are the respective state variables for the capacitor and the inductor at t equals 0. To determine the order of a given circuit, you count the number of capacitors and inductors to obtain the denominator degree.

Figure 2.32 shows a few network examples in which storage elements are present. Evaluate the number of storage elements and you have the network degree. In Figure 2.32a, b and d, there is just one storage element, a capacitor or an inductor: they are 1st-order networks. In Figure 2.32c, you have a capacitor and an inductor, this is a 2nd-order system. In Figure 2.32e, two capacitors with an inductor form a 3rd-order system.

In the above lines, I said that you must count the *independent* state variables. By that statement, I implied that a state variable was considered independent if its value did not uniquely depend on other state variables. This is typical of a loop involving capacitors only, or a mesh involving capacitors and voltage sources. The comment equally applies to a node having inductors only or inductors and current sources in series. In these configurations, one of the state variables is a linear combination of the others or is not independent (e.g. the inductor current depends on a current source or the capacitor voltage depends on a voltage source). Such arrangement is called a *degenerate case*. Obvious degenerate cases are when capacitors or inductors can be arranged in series or parallel to form a single storage element. When a degenerate case occurs in a circuit,

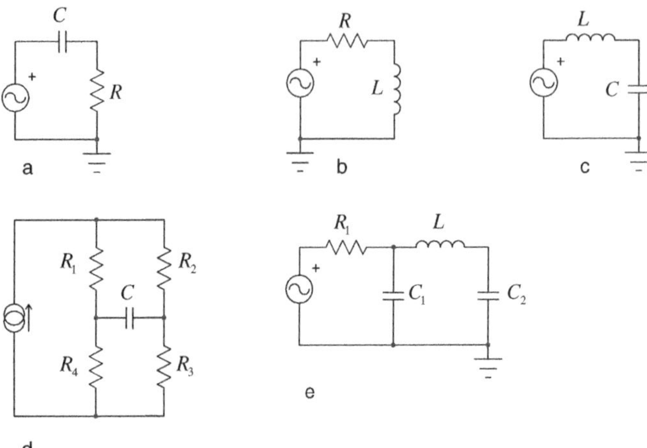

Figure 2.32 To obtain the order of a circuit, simply count the number of independent state variables.

for instance a mesh with three capacitors or a node with three inductors, the denominator order is reduced by 1 (a three-capacitor circuit would be a 3rd-order system but because of the degenerate case it is 2nd-order). If there is another degenerate case brought by a purely inductive node, then another order is lost. The generic formula defining the order n of a network having n_{LC} energy storage elements and in which you can identify n_C capacitive or n_L inductive degenerate cases is expressed as:

$$n = n_{LC} - n_C - n_L \tag{2.114}$$

Figure 2.33a shows a classical degenerate case for a 3-capacitor mesh. The voltage across C_2 terminals can be defined as:

$$x_2 = x_1 - x_3 \tag{2.115}$$

and illustrates a capacitive loop where state variable x_2 is uniquely defined by state variables x_3 and x_1. In that case, the network loses one order: there are four storage elements but the denominator will be of 3rd order. Should you add a resistor in series with C_2 as in Figure 2.33b, you break the loop and (2.115) becomes:

$$x_2 = x_1 - x_3 - I_1 R_2 \tag{2.116}$$

The state variable x_2 is no longer uniquely dependent of the other variables as R_2 introduces a voltage drop in the mesh. The network has now 4 independent state variables, this is a 4th-order system.

Figure 2.34 depicts a circuit in which a voltage source is now involved in the capacitive loop:

$$x_1 = V_{in} - x_2 \tag{2.117}$$

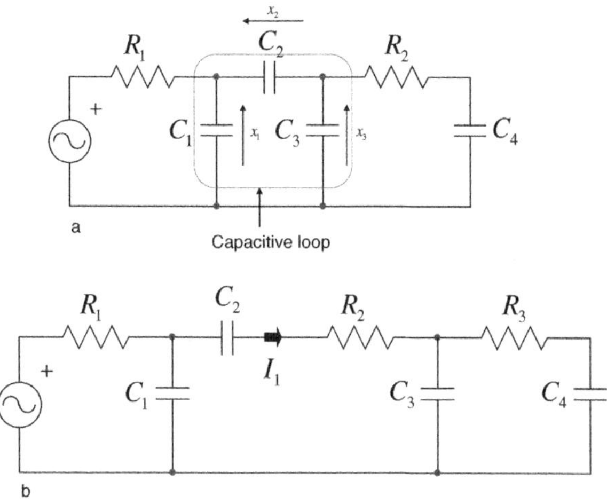

Figure 2.33 When state variables are not independent, the denominator polynomial expression loses a degree as in sketch (a). Should the loop be broken by a resistor such as R_2, the denominator degree increases to 4.

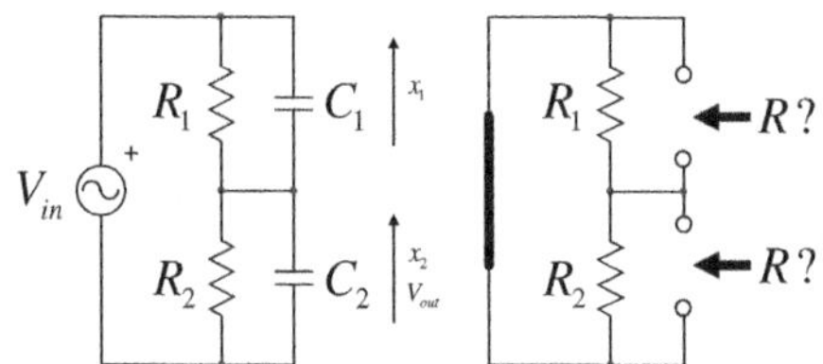

Figure 2.34 When capacitors are in parallel as the source is turned off, they are combined into one capacitor: it is a degenerate case.

This loop illustrates a degenerate case because the voltage source fixes the potential of the series arrangement of C_1 and C_2: their associated state variables are not independent and despite the presence of two storage elements, an order is lost. You could identify the loss in the polynomial order if you calculate the time constant of each capacitor: you strap V_{in} terminals while looking into C_1 and C_2 terminals. In this case, you see that R_1 comes in parallel with R_2 for both capacitors. The time constant of this circuit is thus:

$$\tau = \left(R_1 \| R_2\right)\left(C_1 + C_2\right) \tag{2.118}$$

and the denominator is $D(s) = 1 + s\tau_1$. This is a 1st-order system as shown in Problem 8.

In Figure 2.35a, we have three inductors connected to a node. This is a degenerate case as the current in inductor L_2 for instance, uniquely depends on the two other inductor currents:

$$x_2 = x_1 - x_3 \tag{2.119}$$

Despite five storage elements, this is a 4th-order network. Now, if you add a resistor in parallel with L_2, (2.119) needs an update:

$$x_2 = x_1 - x_3 - I_4 \tag{2.120}$$

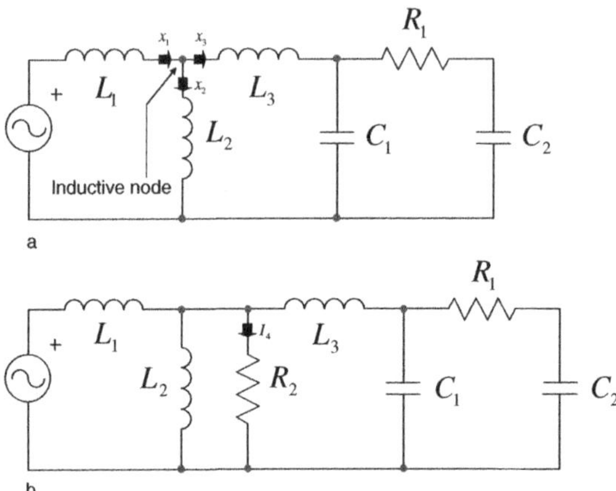

Figure 2.35 In this example, the three inductors connected at the same node form a degenerate case. This is a 4th-order network despite the presence of five storage elements. In sketch (b), resistor R_2 is added so the current in L_2 is no longer uniquely depending upon currents in L_1 or L_3. The circuit becomes 5th order.

and the current in inductor L_2 no longer uniquely depends on the other state variables x_1 and x_3. We now have a 5th-order system.

In Figure 2.36, you see a network made of two inductors and a resistor. You want to calculate the impedance offered by its connecting terminals. To derive this impedance (even if the result is obvious in this case), you connect a current source – the stimulus, I_T – and calculate what the response is. In this example, as the response is the voltage V_T across the network terminals, we want to determine a *driving point impedance* (DPI). You have two storage elements, you may think it is a 2nd-order circuit. However, the first inductor lies in series with the current source: its state variable x_1 is not independent. Only one independent state variable remains, the current in L_2, x_2. We have a degenerate case where an order is lost.

This is a 1st-order circuit whose impedance is quickly obtained by calculating the series impedance brought by L_1 and L_2 paralleled with R. The denominator is of 1st order as expected:

$$Z(s) = sL_1 + sL_2 \| R = \frac{sL_1R + sL_2R + s^2L_1L_2}{R + sL_2} \tag{2.121}$$

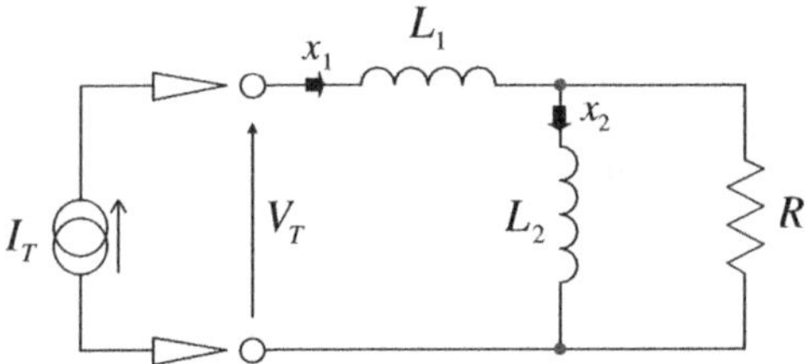

Figure 2.36 Here, we want to calculate the network impedance using a current source. We have a mesh involving the series connection of two inductors and a current source: it is a degenerate case as in Figure 2.35a.

Are we done here? Certainly not; it does not comply with our model defined in (2.56) for an impedance. First, we can factor R in the denominator and the numerator:

$$Z(s) = \frac{R}{R}\frac{s(L_1 + L_2) + s^2\frac{L_1L_2}{R}}{1 + s\frac{L_2}{R}} = \frac{s(L_1 + L_2) + s^2\frac{L_1L_2}{R}}{1 + s\frac{L_2}{R}} \tag{2.122}$$

Still, the numerator can be further rearranged if you purposely multiply and divide all terms by R:

$$Z(s) = R\frac{s\frac{L_1 + L_2}{R} + s^2\frac{L_1L_2}{R^2}}{1 + s\frac{L_2}{R}} = R_0\frac{\frac{s}{\omega_0 Q} + \left(\frac{s}{\omega_0}\right)^2}{1 + \frac{s}{\omega_p}} \tag{2.123}$$

In this expression, after identifying the various coefficients, we have:

$$R_0 = R \tag{2.124}$$

$$\omega_0 = \frac{R}{\sqrt{L_1L_2}} \tag{2.125}$$

$$Q = \frac{\sqrt{L_1L_2}}{L_1 + L_2} \tag{2.126}$$

$$\omega_p = \frac{R}{L_2} \tag{2.127}$$

(2.123) can still be rewritten by factoring $\frac{s}{\omega_0 Q}$:

$$Z(s) = R_0\frac{\frac{s}{\omega_0 Q}\left(1 + \frac{sQ}{\omega_0}\right)}{1 + \frac{s}{\omega_p}} = R_0\frac{\frac{s}{\omega_{z_1}}\left(1 + \frac{s}{\omega_{z_2}}\right)}{1 + \frac{s}{\omega_p}} \tag{2.128}$$

In which $\omega_{z_1} = \omega_0 Q = \frac{R}{L_1+L_2}$ and $\omega_{z_2} = \frac{\omega_0}{Q} = R\frac{L_1+L_2}{L_1L_2}$. In (2.128), we see a zero at the origin followed by a zero ω_{z2} and a pole ω_p. In case we have $L_2 \ll L_1$, then $\omega_{z_2} \approx \omega_p$ and (2.128) simplifies to:

$$Z(s) \approx R_0\frac{s}{\omega_{z_1}} \tag{2.129}$$

(2.128) tells us that when the pole and the zero are well spread, the impedance no longer increases with frequency: the pole breaks the first +1 slope – the zero at the origin – to a 0-slope. The magnitude pauses at R_0 until the second zero appears. Figure 2.37 confirms this behavior if R is 100 Ω or 40 dBΩ. It would be impossible to infer this resistive plateau from (2.121).

Sometimes, it is not obvious how to identify a degenerate case if the network is complex or arranged in a non-obvious way. In that case, you identify a time constant with each storage element (degenerate case or not by the way) but at the end of the exercise, you will simply find that the term b_n in (2.71) is zero. For instance, assume four identified time constants in a circuit hosting one degenerate case but you did not see it. You go through all calculation steps for $D(s)$ combining all time constants and at the end, the term b_4 will equal 0.

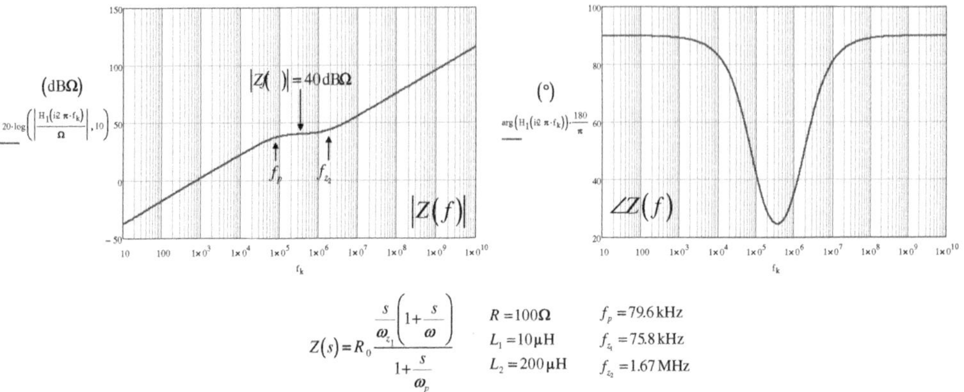

Figure 2.37 For well-spread pole and zero, the impedance pauses to R_0 until the second zero frequency occurs.

2.3.7 Zeros in the Network

As explained in Chapter 1, zeros of a transfer function correspond to conditions created in the transformed circuit for which a null is observed in the response signal. While the circuit is examined as $s = s_z$, a null in the response occurs because a transformed open (infinite series impedance) opposes the stimulus propagation or because a branch becomes a transformed short circuit to ground and shunts the stimulus.

There are several ways you can unveil zeros while analyzing a network. The easiest way is by inspection. To learn how to do it, look at Figure 2.38. What combination in the transformed circuit – meaning C is replaced by $1/sC$ and L by sL – examined at $s = s_z$ could prevent the stimulus from reaching the output? Let's start from the left where the stimulus is located. R_1 becomes infinite? Not

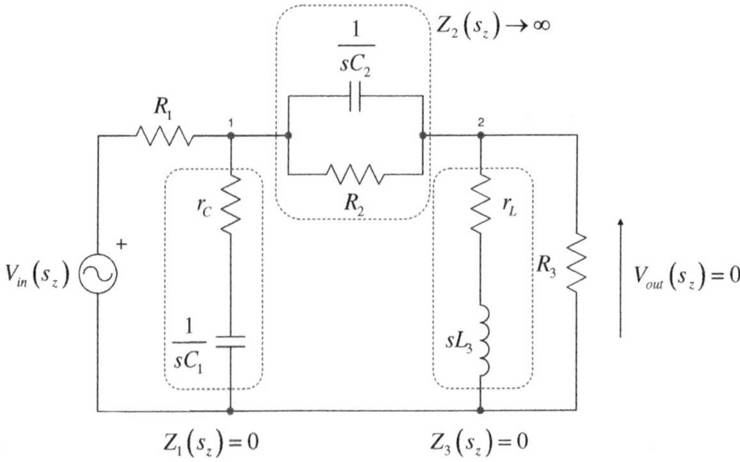

Figure 2.38 Conditions exist in the transformed network examined at $s = s_z$ which yield a null in the output response.

likely; it is a fixed element, independent of frequency. What if node 1 goes to ground via a transformed short circuit? Indeed, in this case, the input signal can no longer reach the output. The network that connects node 1 to ground is made of capacitor C_1 in series with its ESR r_C. The series impedance of C_1 and r_C is simply:

$$Z_1(s) = r_C + \frac{1}{sC_1} = \frac{1 + sr_CC_1}{sC_1} \tag{2.130}$$

What is the condition for which $Z_1(s_{z_1}) = 0$? Solve for the root of (2.130) when $1 + sr_CC_1 = 0$ and you find that the zero occurs for an angular frequency equal to:

$$s_{z_1} = -\frac{1}{r_CC_1} \text{ or } \omega_{z_1} = \frac{1}{r_CC_1} \tag{2.131}$$

No mathematics, no complex equations, just look at the figure.

A quick example will help you understand the concept. Look at Figure 2.39 in which a current source I_T – the excitation – ac-sweeps a simple network. For s different than s_z, current $\hat{i}_1$ exists and produces a response $\hat{v}_{out}$ across R_1. When $s = s_z$, the series connection of r_C and $1/sC$ forms a transformed short circuit across the current source. The response $\hat{v}_{out}$ disappears because all the excitation current I_T now flows in the shunt while $\hat{i}_1$ is a null.

Let's continue the exploration in Figure 2.38. Node 1 is connected to node 2 via the parallel arrangement of C_2 and R_2. To block the signal propagation, it would mean that Z_2 impedance evaluated at $s = s_{z2}$ becomes infinite. Is it possible? The impedance of the paralleled connection of C_2 and R_2 is defined as:

$$Z_2(s) = R_2 \| \frac{1}{sC_2} = \frac{R_2}{1 + sR_2C_2} \tag{2.132}$$

R_2 is fixed and cannot become infinite however, if we cancel the denominator, we have another zero for which Z_2 will become infinite:

$$1 + sR_2C_2 = 0 \tag{2.133}$$

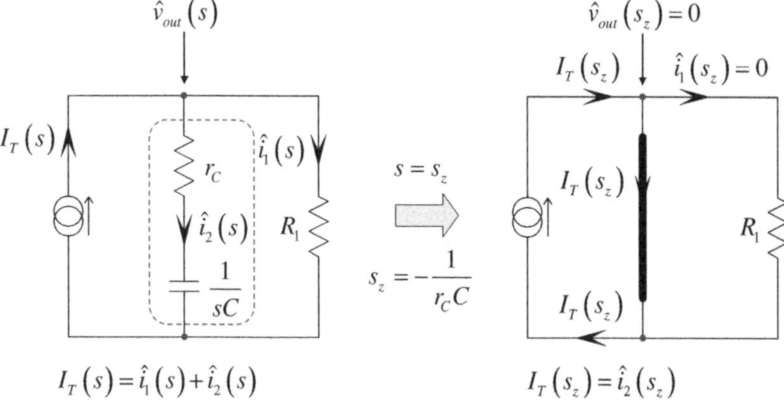

Figure 2.39 When the transformed circuit is examined at $s = s_z$, the current in R_1 is simply zero and the response is nulled.

implies that

$$s_{z_2} = -\frac{1}{R_2C_2} \text{ or } \omega_{z_2} = \frac{1}{R_2C_2} \tag{2.134}$$

In this intermediate step, the pole of (2.132) becomes the zero of the transfer function. Let's finish since we are now at node 2. In parallel with the load, we see the series connection of L_3 and its ESR labeled r_L. Can this series arrangement become a transformed short circuit for $s = s_{z3}$? Let's check:

$$Z_3(s) = sL_3 + r_L = 0 \tag{2.135}$$

the solution is simply:

$$s_{z_3} = -\frac{r_L}{L_3} \text{ or } \omega_{z_3} = \frac{r_L}{L_3} \tag{2.136}$$

This is it, we have found 3 zeros in the above circuit arrangement just by inspecting the drawing. Not bad for a 3rd-order network! With the help of (2.131), (2.134) and (2.136), we can immediately write the denominator polynomial:

$$N(s) = \left(1 + \frac{s}{\omega_{z_1}}\right)\left(1 + \frac{s}{\omega_{z_2}}\right)\left(1 + \frac{s}{\omega_{z_3}}\right) \tag{2.137}$$

This is a *low-entropy* form that we obtained without a complex analysis of the circuit. This method works fine when the network lends itself well to this kind of interpretation: looking at Figure 2.38, you can easily identify inductive and capacitive associations that could become either a short or an open circuit. It is usually easy with passive networks, without controlled sources. When inspection is not possible, you need another method.

2.4 First Step Towards a Generalized 1st-order Transfer Function

In more complex circuits, when controlled or dependent sources are involved, solving zeros by inspection becomes an almost impossible exercise. Besides inspection, how to easily see if there are zeros in the network under study and which energy-storing element contributes one or several zeros? In a 1st-order system featuring a capacitor or an inductor, the time constant τ associated with the storage element is either *RC* or *L/R*, with *R* the equivalent resistance driving *C* or *L* in certain conditions. In all the simple transfer functions we have derived so far, *L* or *C* always appear combined with *s* at the denominator and the numerator in presence of zeros [6]. Assuming a circuit featuring a capacitor and considering b_0 equal to 1, we can rewrite (2.68) as:

$$H(s) = \frac{H_0 + \alpha_1 C_1 s}{1 + \beta_1 C_1 s} \tag{2.138}$$

In this expression, α_1 and β_1 have the dimensions of ohms to form a time constant with C_1. They respectively represent the resistance offered by the capacitor terminals when the response is a null ($\hat{v}_{out} = 0$ which occurs at the zero frequency) and when the excitation is set to zero for the pole (see Figure 2.12).

When we calculate H_0, we consider the circuit in static conditions, at a 0-Hz frequency. In this mode, a capacitor offers a high impedance path and disappears from the analysis while an inductor is assimilated as a short circuit. Removing a capacitor from a circuit can be seen as bringing its capacitance to 0 F. Similarly, replacing an inductor by a strap means that the inductance has dropped to 0 H. Capacitors and inductors can thus be put in different states depending on what needs to be analyzed. The points below describe these different states and formalize the terminology:

- The dc state or the zero value state of a capacitor is an open circuit. If in the text you read that capacitor C_1 must be put in its dc state or is set to 0 farad, simply remove the capacitor from the circuit and redraw it without the capacitor.
- For an inductor, putting it in its dc state or having it zero-valued ($L=0$ henry) implies that the inductor symbol is replaced by a strap: it is a short circuit.
- For s approaching infinity, the capacitor in its high-frequency (HF) state can be replaced by a short circuit or a strap. It is similar to saying that the capacitor is infinite-valued.
- An inductor in its high-frequency state (or when infinite-valued) is equivalent to an open circuit: you simply remove the inductor from the circuit and redraw it.

Figure 2.40 illustrates this concept for the two cases, s is 0 or approaches infinity.

Now, how does (2.138) simplify if C_1's value approaches infinity rather than being put to 0 as we did to calculate H_0?

$$\lim_{C_1\to\infty}\frac{H_0+\alpha_1 C_1 s}{1+\beta_1 C_1 s}=\frac{\alpha_1}{\beta_1} \tag{2.139}$$

When capacitor C_1 becomes a short circuit – or is *infinite-valued* – it transforms the 1st order network into a different circuit involving α_1 and β_1. The gain of the new circuit is changed and is noted H^1 in which the '1' in the exponent refers to the element involved in τ_1 – C_1 in this case – set in its high-frequency state:

$$H^1 \equiv H|_{C_1\to\infty}=\frac{\alpha_1}{\beta_1} \tag{2.140}$$

If an inductor L_2 is involved instead of capacitor C_1, (2.138) will be expressed as:

$$H(s)=\frac{H_0+\frac{L_2}{\alpha_2}s}{1+\frac{L_2}{\beta_2}s} \tag{2.141}$$

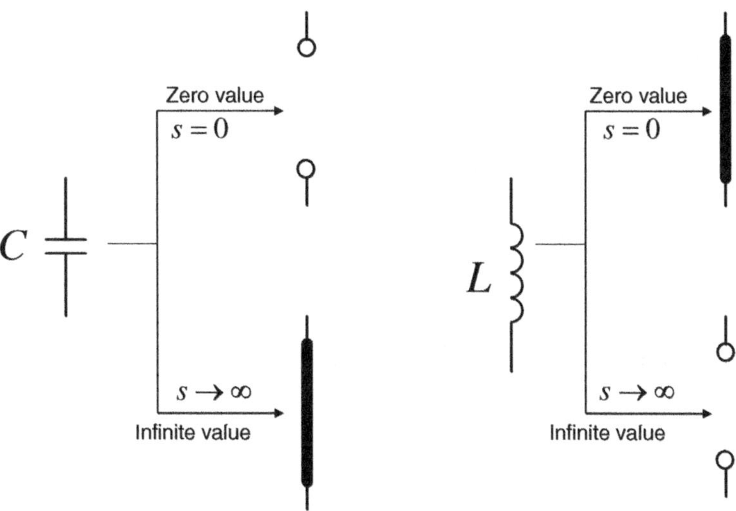

Figure 2.40 The different states of capacitors and inductors for $s=0$ (static conditions) or when s approaches infinity (high frequency).

Now, how does (2.141) simplify if L_2's value approaches infinity rather than being put to 0 as we did to assess H_0?

$$\lim_{L_2\to\infty}\frac{H_0+\dfrac{L_2}{\alpha_2}s}{1+\dfrac{L_2}{\beta_2}s}=\frac{\beta_2}{\alpha_2} \tag{2.142}$$

When L_2 is infinite-valued, it transforms the 1st order network into a different circuit also involving α and β but in a reversed expression compared to (2.139). The gain of the new circuit is changed and is noted H^2 in which the '2' in the exponent refers to the element involved in τ_2-L_2 in this case – set in its high-frequency state:

$$H^2\equiv H|_{L_2\to\infty}=\frac{\beta_2}{\alpha_2} \tag{2.143}$$

Quantities obtained in (2.140) and (2.143) are unitless. Both are a gain obtained when L or C (or both in higher order circuits) are infinite-valued: C is replaced by a strap while L is removed from the circuit. This new gain is noted H^i in which exponent i refers to the element involved in time constant i. When you start analyzing the circuit, it is a good habit to associate the time constant subscript with the component label. For instance, assume you deal with C_1, C_3 and L_6, then choose time constants as $\tau_1=f(C_1)$, $\tau_3=f(C_3)$ and $\tau_6=f(L_6)$. If τ_1 involves C_1, then H^1 means that C_1 is infinite-valued in this configuration (C_1 is replaced by a strap). Similarly, if τ_6 involves L_6, then H^6 means that L_6 is infinite-valued in this configuration (L_6 is replaced by an open circuit). In case of higher order circuits, as we will see later, the notation H^{235} simply means that storage elements associated with τ_2, τ_3 and τ_5 are infinite-valued for this gain calculation while the remaining energy-storing components in the circuit are zero-valued. This is an important concept and it must be fully acquired for the rest of the book. Figure 2.41 shows a graphic illustration of this approach in several examples and a few problems will exercise your skills at the end of this chapter.

What we calculate when setting C or L to infinity is a gain H^i which differs from H_0, obtained in dc conditions. Sometimes, this gain can be equal to 0 as shown in Figure 2.41b. From (2.140), we can extract the value of α_1 quite easily and obtain:

$$\alpha_1=H^1\beta_1 \tag{2.144}$$

If we now substitute (2.144) in (2.138), we have:

$$H(s)=\frac{H_0+H^1\beta_1C_1s}{1+\beta_1C_1s} \tag{2.145}$$

In this equation, β_1C_1 is the time constant τ_1 associated with C_1. (2.145) can thus be rewritten as:

$$H(s)=\frac{H_0+H^1\tau_1s}{1+\tau_1s} \tag{2.146}$$

If H_0 is different than 0, we obtain the generalized transfer function for a 1st order circuit built with a single capacitor:

$$H(s)=H_0\frac{1+\dfrac{H^1}{H_0}\tau_1s}{1+\tau_1s}=H_0\frac{1+\dfrac{s}{\omega_z}}{1+\dfrac{s}{\omega_p}} \tag{2.147}$$

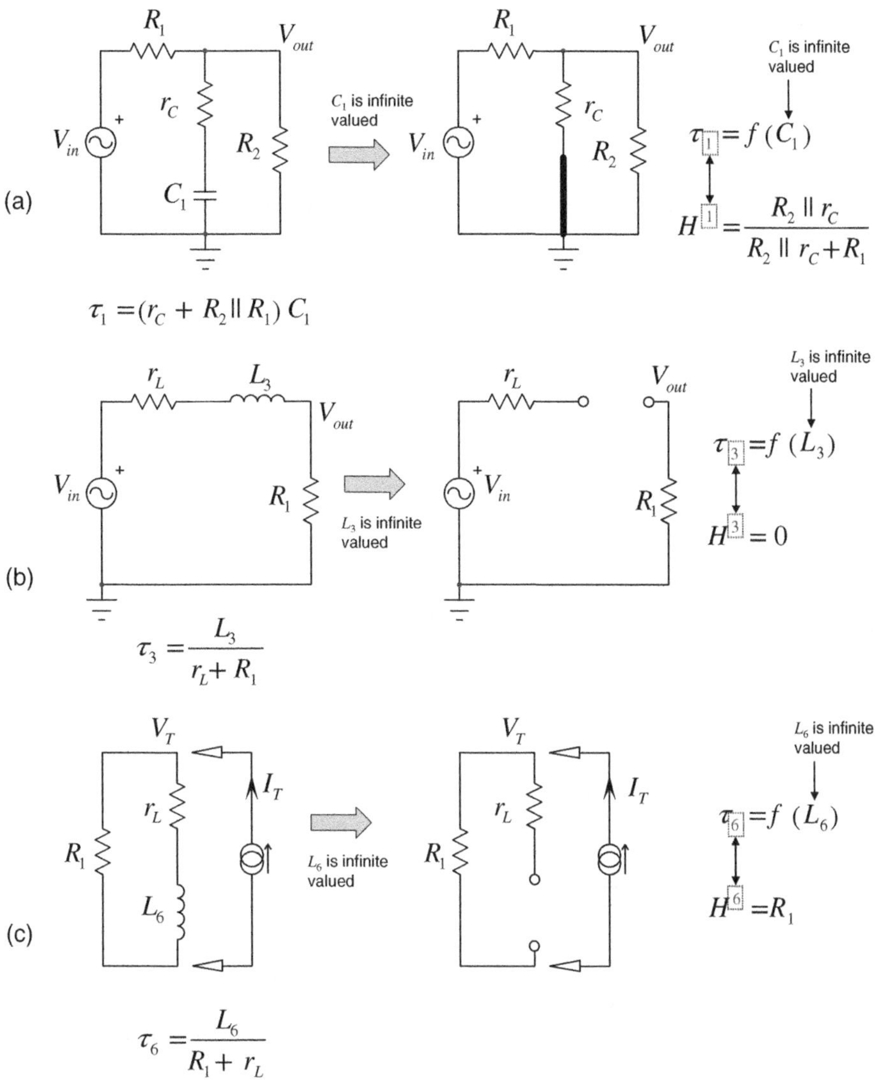

Figure 2.41 The high-frequency gain is obtained for infinite-valued C or L.

in which

$$\omega_z = \frac{H_0}{H^1 \tau_1} \tag{2.148}$$

and

$$\omega_p = \frac{1}{\tau_1} \tag{2.149}$$

We can run the same exercise when extracting α_2 from (2.143) and substituting it in (2.141):

$$H(s) = \frac{H_0 + L_2 \dfrac{H^2}{\beta_2} s}{1 + \dfrac{L_2}{\beta_2} s} \tag{2.150}$$

In this expression, L_2/β_2 is the time constant associated with inductor L_2. Substituting τ_2 in (2.150), gives a formula similar to that already obtained in (2.146):

$$H(s) = \frac{H_0 + H^2 \tau_2 s}{1 + \tau_2 s} \tag{2.151}$$

Considering H_0 different than 0 and rearranging, we have:

$$H(s) = H_0 \frac{1 + \dfrac{H^2}{H_0} \tau_2 s}{1 + s\tau_2} = H_0 \frac{1 + \dfrac{s}{\omega_z}}{1 + \dfrac{s}{\omega_p}} \tag{2.152}$$

in which

$$\omega_z = \frac{H_0}{H^2 \tau_2} \tag{2.153}$$

and

$$\omega_p = \frac{1}{\tau_2} \tag{2.154}$$

Expressions (2.146) and (2.151) are identical. For a single energy-storing element network, they deliver the exact transfer function of the system via three calculations:

1. The resistance seen from the capacitor or the inductor port is evaluated with the excitation source turned to 0 V or 0 A.
2. The first gain H_0 is computed with the single energy-storing element put in its dc state or zero valued: with a capacitor in the circuit, C is removed from the schematic diagram. If the circuit involves an inductor L, it is simply shorted.
3. The second gain H^i is obtained when the single energy-storing component is infinite-valued: C_i is replaced by a short circuit or L_i is removed from the circuit.

2.4.1 *Solving 1st-order Circuits with Ease, Three Examples*

Let's exercise this new formula through a few simple examples. The first one appears in Figure 2.42a. What is the transfer function linking V_{out} to V_{in}? Step 1 in the above list is described by sketch (b) where the excitation source is set to 0 V and we look into the capacitive port to obtain the resistance R:

$$R = R_1 + R_2 \tag{2.155}$$

Therefore, the time constant τ_1 associated with C_1 is:

$$\tau_1 = (R_1 + R_2)C_1 \tag{2.156}$$

The static gain is now obtained when C_1 is removed from the circuit as shown in Figure 2.42c. In this particular case, the signal path is interrupted, therefore $H_0 = 0$. The final step is when C_1 is infinite-

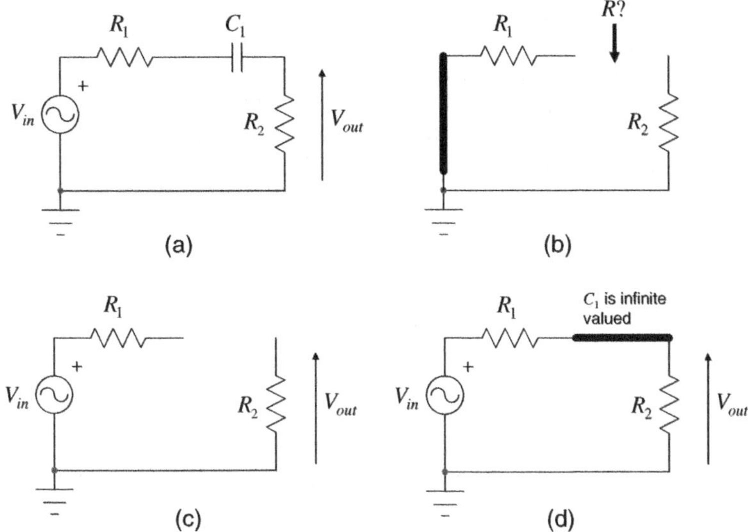

Figure 2.42 What is the transfer function of this simple 1st-order circuit?

valued. What is the circuit gain in Figure 2.42d?

$$H^1 = \frac{R_2}{R_1 + R_2} \tag{2.157}$$

This is it, we can now combine the pieces according to (2.146):

$$H(s) = \frac{H_0 + H^1 \tau_1 s}{1 + s\tau_1} = \frac{0 + s\frac{R_2}{R_1 + R_2}(R_1 + R_2)C_1}{1 + s(R_1 + R_2)C_1} \tag{2.158}$$

Rearranging and simplifying, we have:

$$H(s) = \frac{sR_2C_1}{1 + s(R_1 + R_2)C_1} = \frac{\frac{s}{\omega_{zo}}}{1 + \frac{s}{\omega_{p_1}}} \tag{2.159}$$

in which

$$\omega_{zo} = \frac{1}{R_2C_1} \tag{2.160}$$

and

$$\omega_{p_1} = \frac{1}{(R_1 + R_2)C_1} \tag{2.161}$$

In (2.159), there is a zero at the origin, meaning that at 0 Hz, the gain of H is 0. What (2.160) describes is what I call the 0-dB *crossover zero*. It is the cutoff frequency for which the magnitude of s/ω_{zo} is 1 or 0 dB, hence the name.

In this example, we have derived this transfer function just by drawing three extra simple sketches, without writing a single equation. When skilled in the art of fast analytical circuit techniques, you will be able to derive this transfer function in your head, without sketching anything!

Let's carry on with another example shown in Figure 2.43. This time, we have an inductor. Let's roll out the steps as we did before. The time constant configuration is described by sketch (b) in which the excitation is replaced by a short circuit. The resistance looking into the inductive port is:

$$R = R_1 \| (R_2 + R_3) \tag{2.162}$$

Therefore, the time constant τ_4 associated with L_4 is:

$$\tau_4 = \frac{L_4}{R_1 \| (R_2 + R_3)} \tag{2.163}$$

The static gain is now obtained when L_4 is replaced by a short as shown in Figure 2.43c. In this particular case, the signal path is shorted to ground at R_1's right terminal, therefore $H_0 = 0$. The final step is when L_4 is infinite-valued. What is the gain of the circuit in Figure 2.43d in this case?

$$H^4 = \frac{R_3}{R_1 + R_2 + R_3} \tag{2.164}$$

This is it, we can now combine the pieces according to (2.146) or (2.151):

$$H(s) = \frac{H_0 + H^4 \tau_4 s}{1 + s\tau_4} = \frac{0 + s\dfrac{L_4}{R_1 \| (R_2 + R_3)} \dfrac{R_3}{R_1 + R_2 + R_3}}{1 + s\dfrac{L_4}{R_1 \| (R_2 + R_3)}} \tag{2.165}$$

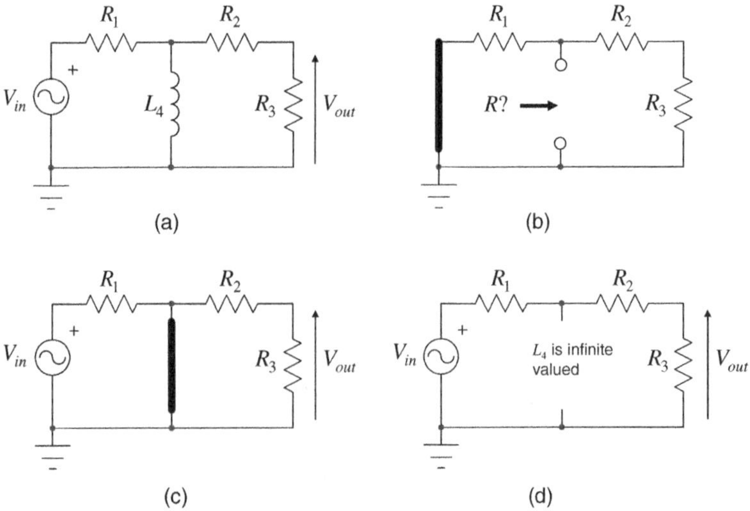

Figure 2.43 What is the transfer function of this simple 1st-order inductive circuit?

In the above expression numerator, if we develop $R_1\|(R_2+R_3)$, a simplification will occur with $R_1+R_2+R_3$. The final transfer function is then:

$$H(s)=\frac{s\dfrac{L_4}{\dfrac{R_1(R_2+R_3)}{R_3}}}{1+s\dfrac{L_4}{R_1\|(R_2+R_3)}}=\frac{\dfrac{s}{\omega_{zo}}}{1+\dfrac{s}{\omega_p}} \tag{2.166}$$

in which:

$$\omega_{zo}=\frac{\dfrac{R_1(R_2+R_3)}{R_3}}{L_4}=\frac{R_1(R_2+R_3)}{L_4R_3} \tag{2.167}$$

and

$$\omega_p=\frac{R_1\|(R_2+R_3)}{L_4} \tag{2.168}$$

To check our results, we have captured data in Mathcad® and in a SPICE simulator. Curves from the solver appear in Figure 2.44. Please note the architecture of the sheet in which time constants and gains clearly appear in the transfer function H. This way, should you identify a mistake at some point, you can simply change the term that is involved without affecting the whole derivation. This is the power of this approach which will become obvious as we increase the number of energy-storing elements.

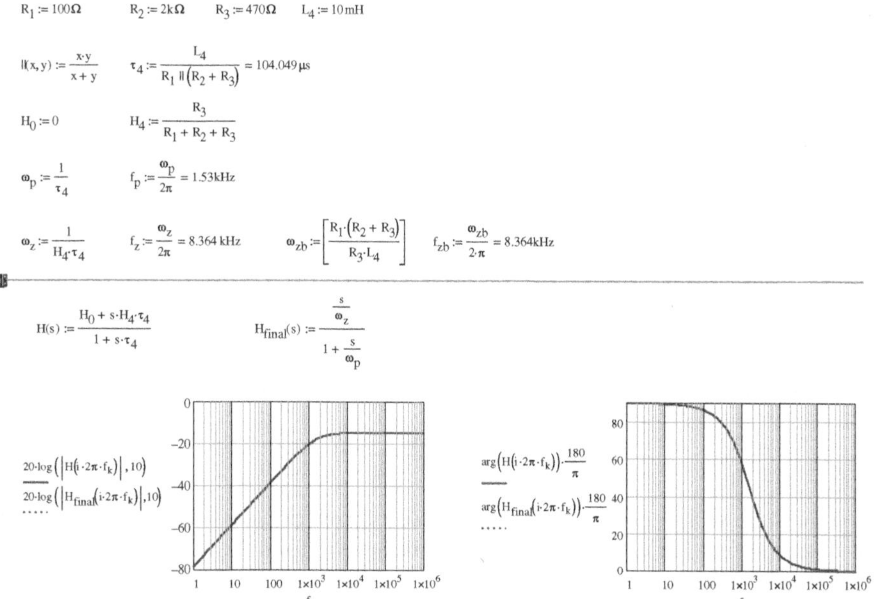

Figure 2.44 Mathcad® offers an easy and quick means to graph transfer functions and verify that final and low-entropy expressions deliver similar results.

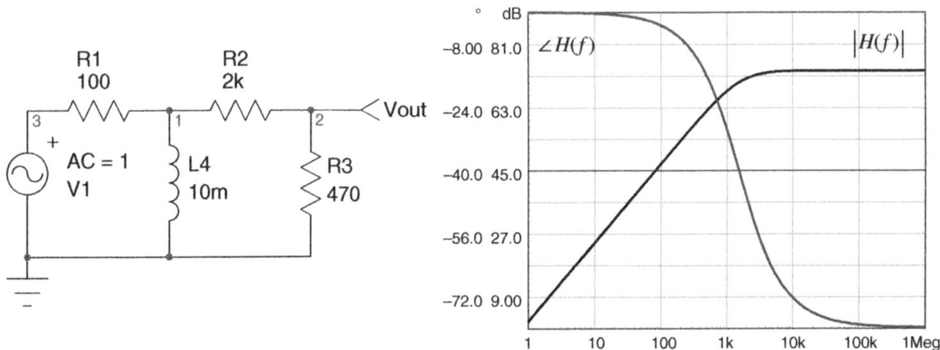

Figure 2.45 A SPICE simulation is also a possibility and offers a quick means to confirm the mathematical derivation.

The SPICE simulation circuit is shown in Figure 2.45. Simulation results appear in the right side of the sketch and show no difference with those of Mathcad®.

The last example is shown in Figure 2.46 and is slightly more complex than the ones we have seen so far. Nevertheless, we will stick to the 3-point methodology given above. First, we assess the circuit time constant by looking into the capacitor ports while the excitation is set to 0 V (Figure 2.46b). We can see that R_2 comes in parallel with the capacitor terminals while R_1 becomes paralleled with R_4. If necessary, you can go through another sketch where rearranging these resistors in a more convenient way can help see the solution. The resistance R is simply:

$$R = R_2 \| (R_3 + R_1 \| R_4) \tag{2.169}$$

The time constant τ_5 associated with capacitor C_5 is thus:

$$\tau_5 = [R_2 \| (R_3 + R_1 \| R_4)] C_5 \tag{2.170}$$

Next step, find the static gain for s equals 0. In this configuration, capacitor C_5 disappears as indicated in Figure 2.46c. The transfer function is obtained applying a voltage divider involving R_4 and the series-parallel arrangement of R_2 R_3 and R_1:

$$H_0 = \frac{R_4}{R_4 + (R_2 + R_3) \| R_1} \tag{2.171}$$

The final step is to calculate H^5 implying that capacitor C_5 value is set to infinity or replaced by a strap (Figure 2.46d). As R_2 right terminal is grounded, it comes in parallel with the source and plays no role in the gain. R_3 is now paralleled with R_4 and forms a resistive divider with R_1:

$$H^5 = \frac{R_4 \| R_3}{R_4 \| R_3 + R_1} \tag{2.172}$$

It was the final step, we have everything needed to build the transfer function using (2.146). This time, H_0 is not equal to zero and can be advantageously factored to form the leading term:

$$H(s) = H_0 \frac{1 + s \frac{H^5}{H_0} \tau_5}{1 + s\tau_5} \tag{2.173}$$

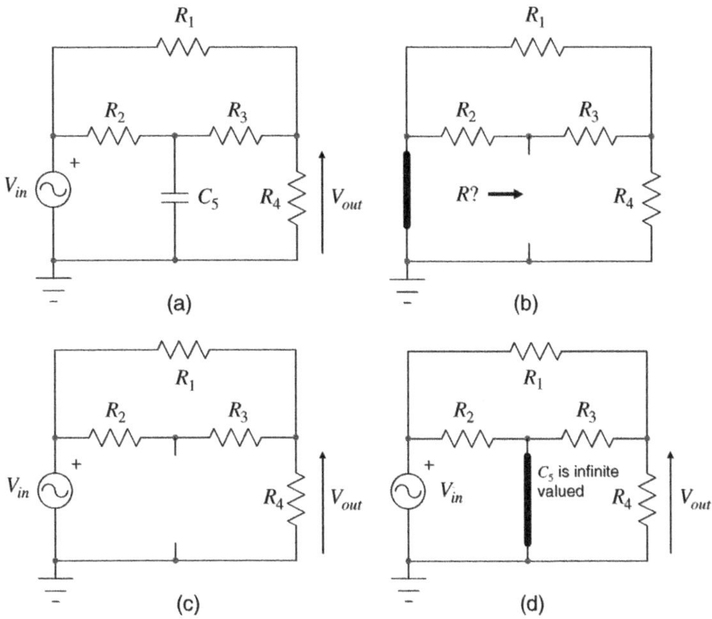

Figure 2.46 In this circuit, an element links the input and output, R_1. How can you quickly obtain its transfer function?

The pole occurs at:

$$\omega_p = \frac{1}{\left[R_2 \| (R_3 + R_1 \| R_4)\right] C_5} \tag{2.174}$$

and the zero is located at:

$$\omega_z = \frac{\dfrac{R_4}{R_4 + (R_2 + R_3) \| R_1}}{\dfrac{R_4 \| R_3}{R_4 \| R_3 + R_1}\left[R_2 \| (R_3 + R_1 \| R_4)\right] C_5} \tag{2.175}$$

We can now capture these terms in Mathcad® and check the delivered magnitude and phase curves. The sheet appears in Figure 2.47.

We have also captured this simple schematic into a SPICE circuit as shown in Figure 2.48. The simulated results exactly match those of Mathcad®.

In these examples, by applying a three-step analysis that we illustrated each time with a separate drawing, we have been able to derive 1st-order transfer functions in a very short time without writing a single equation. This is the strength of this technique which offers another interesting insight into the circuit. In the final step, we place the energy-storing element into its high-frequency state: the capacitor is replaced by a short circuit and the inductor is removed from the circuit. This gives us the gain H^i which, further combined with the network time constant, delivers the zero position. The numerator $N(s)$ follows the form:

$$N(s) = H_0 + H^i \tau_i s \tag{2.176}$$

$R_1 := 4.7\text{k}\Omega \qquad R_2 := 680\Omega \qquad R_3 := 2\text{k}\Omega \qquad R_4 := 1\text{k}\Omega \qquad C_5 := 0.22\mu\text{F}$

$\|(x,y) := \frac{x \cdot y}{x+y} \qquad \tau_5 := \left[R_2 \| \left[R_3 + \left(R_1 \| R_4\right)\right]\right] \cdot C_5 = 120.573\mu\text{s}$

$H_0 := \frac{R_4}{R_4 + \left[\left(R_2 + R_3\right) \| R_1\right]} \qquad H_5 := \frac{R_4 \| R_3}{R_1 + R_4 \| R_3}$

$\omega_p := \frac{1}{\tau_5} \qquad f_p := \frac{\omega_p}{2\pi} = 1.32\text{kHz}$

$\omega_z := \frac{H_0}{H_5 \cdot \tau_5} \qquad f_z := \frac{\omega_z}{2\pi} = 3.926\,\text{kHz} \qquad f_{za} := \frac{\frac{R_4}{R_4 + \left[\left(R_2 + R_3\right) \| R_1\right]}}{\frac{R_4 \| R_3}{R_1 + R_4 \| R_3} \cdot \left[R_2 \| \left[R_3 + \left(R_1 \| R_4\right)\right]\right] \cdot C_5} \cdot \frac{1}{2\pi} = 3.926\,\text{kHz}$

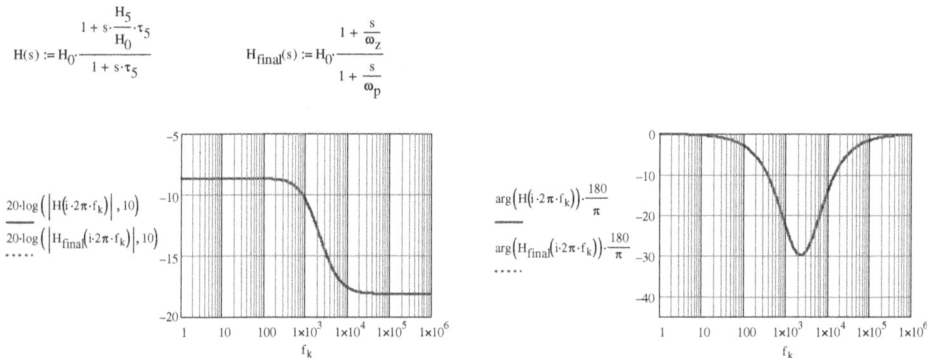

Figure 2.47 Using Mathcad® helps immediately check the dynamic response.

In case this gain H^i is equal to 0, the right term in (2.176) disappears and there is no zero in the circuit you are analyzing. In other terms, by inspection, when you short the capacitor or open the inductor, if the excitation waveform does not propagate to the output, the capacitor or the inductor do not contribute a zero in the transfer function. On the other hand, if you short the capacitor or remove the inductor and you can see that nothing blocks the input signal propagation to the output, then yes, there is a zero associated with the capacitor or the inductor.

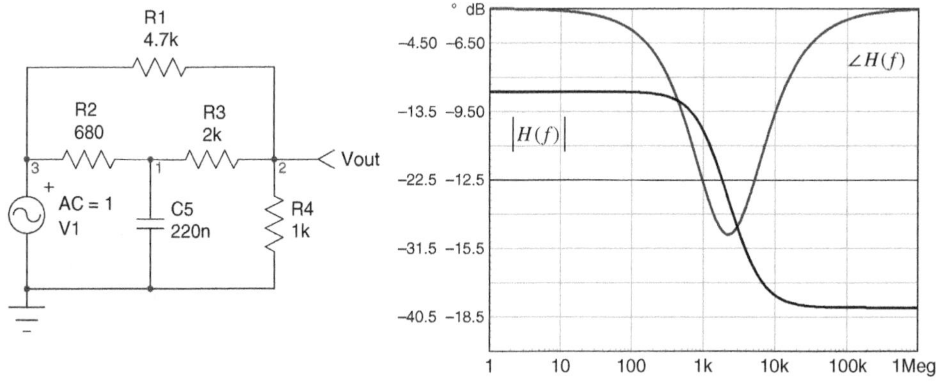

Figure 2.48 A SPICE simulation is also a possibility and offers a quick means to confirm the mathematical derivation.

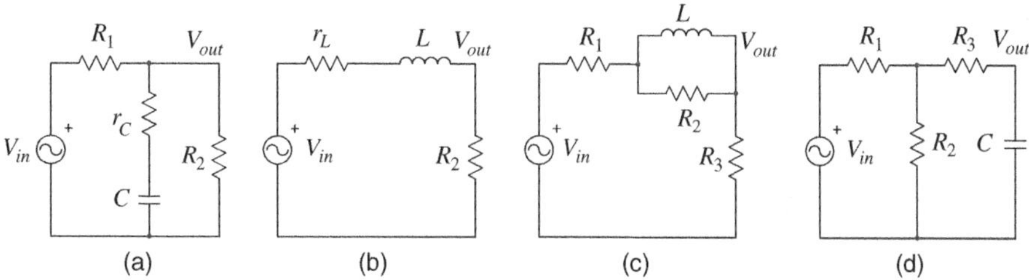

Figure 2.49 Short the capacitor or open the inductor and see if a response still exists. If yes, there is a zero associated with the considered energy-storing element.

This is a very important and useful rule to check for the presence of a zero in a 1st-order transfer function. Figure 2.49 gathers a few examples to see how we can exercise this fact. In sketch (a), if we short C, the response still exists: there is a zero with C. In example (b), if we remove L, there is no response: no zero in this circuit. In (c), removing L keeps R_2 in place and ensures a response: there is a zero with L. In (d), shorting C simply cancels the response and there is no zero in this circuit.

Back to Figure 2.34, if you short C_2, the response is 0 so C_2 does not contribute a zero. But if you short C_1, you still have a response, implying that C_1 is involved in a zero definition.

2.4.2 *Obtaining the Zero with the Null Double Injection*

When calculating the denominator pole, we evaluated the circuit time constant with the excitation source set to 0. We then looked into the capacitor or inductor terminals and measured or inferred (by inspection) the equivalent resistance driving the energy-storing element. Associated with the concerned component, it forms the inverse of the circuit pole. The pole is the denominator root which brings the transfer function to infinity when $s = s_p$.

The zero, on the other hand, appears in the transfer function numerator and zeroes the function for $s = s_z$. When a transformed circuit featuring a zero in its transfer function is examined at $s = s_z$, the excitation waveform does not propagate to the output and the response is a null noted $V_{out}(s) = 0$ or $\hat{v}_{out} = 0$ for a voltage output. We say a zero *nulls* the response at $s = s_z$. When the transformed circuit is analyzed at $s = s_z$, we can also look into the energy-storing element port and check what resistance R is offered in this condition.

As we will see later in Chapter 3, the newly-formed time constant involving R obtained in this mode further associated with the involved energy-storing component represents the inverse of the zero position. In this configuration, the excitation is kept in place while a test current source I_T is installed across the energy-storing component terminals to measure the resistance seen from its port (see Figure 1.4 for instance). As we run the analysis with two excitation sources (V_{in} and I_T) at $s = s_z$, the ac output voltage must be nulled: this approach is called the *Null Double Injection*, abbreviated NDI. In a circuit hosting zero(s), for a certain combination of V_{in} and I_T the output signal disappears and $V_{out}(s) = 0$. We will see how to demonstrate this result in Chapter 3 but assume it is true for now. It is important to note the difference between an output nulled at $s = s_z$ and a short circuit. Figure 2.50 illustrates the difference. When the output is nulled at $s = s_z$, the small-signal voltage is certainly 0 V but the small-signal current flowing into the load is also a null. If you install a short circuit on the output, you certainly null $\hat{v}_{out}$ but now have a short circuit current that is obviously different than zero. Therefore, nulling the output is different from shorting it.

This concept is probably one of the most difficult parts to grasp in the Fast Analytical Circuit Techniques and the Extra-Element Theorem. Similarity exists with the operational amplifier virtual

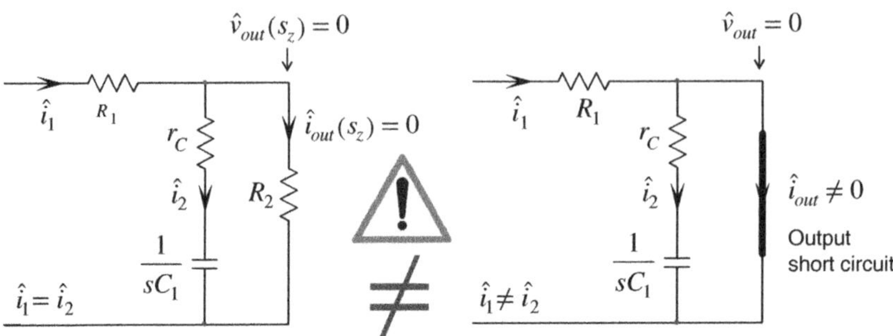

Figure 2.50 Nulling the output voltage is different than installing a short circuit across the output port.

ground as it appears in Figure 2.51. When the loop is closed with R_f, meaning the op amp output operates within its linear range (the output is not stuck to ground or V_{cc}), the op amp strives to maintain $V_{(-)} = V_{(+)}$. This is true in dc, meaning that the voltage at the inverting node is also V_{ref}(considering an infinite open-loop gain) but also in ac. As V_{ref} is perfectly stable, if V_{in} modulates the circuit, $\hat{v}_{ref} = 0$. Since both inverting and non-inverting pins are at the same potential, the ac level on the inverting pin is also zero: $\hat{v}_{(-)} = 0$. Then, as the inverting input of an ideal op amp features an infinite impedance, the current entering the inverting pin is also zero: $\hat{i}_{(-)} = 0$. The closed-loop op amp creates a null at node (−), this the so-called virtual ground.

Let's apply this technique to several examples. Consider the 1st-order circuit shown in Figure 2.52a. We have seen that the time constant is evaluated by setting the excitation source to 0. By inspection, when replacing V_{in} by a short circuit and looking into the inductor port in this condition, we immediately see that:

$$\tau_1 = \frac{L}{r_L + R_1 \| R_2} \tag{2.177}$$

Looking again at this circuit, the dc gain H_0 is child's play and obtained when L_1 is shorted:

$$H_0 = \frac{R_2 \| r_L}{R_2 \| r_L + R_1} \tag{2.178}$$

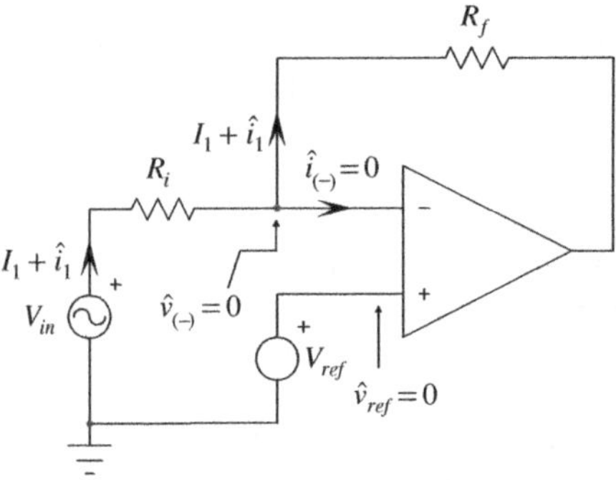

Figure 2.51 The op amp virtual ground also creates a null at the inverting input.

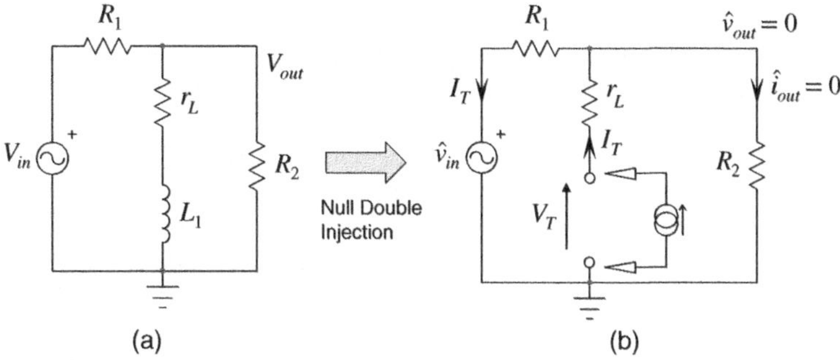

Figure 2.52 This high-pass filter involves a single energy-storing element, it is a 1st-order circuit. The zero can be observed by looking into the inductor port while $\hat{v}_{out} = 0$.

Now, is there a zero in this circuit? When L_1 is set to an infinite value (physically removed from the circuit), can the input excitation propagate to the output? Yes, the gain H^1 involves R_2 and R_1 as a simple voltage divider. A zero thus exists in this network and it involves L_1. To apply the NDI technique, look at Figure 2.52b where the circuit is observed at $s = s_z$: the source V_{in} is kept in place but as the transfer function is equal to zero at this particular frequency, $\hat{v}_{out} = 0$. A null in the output voltage implies a zero current in R_2. As a result, all the source current I_T flows in R_1 and the source V_{in}. Because the upper terminal of r_L is biased at a 0-V level, ($\hat{v}_{out} = 0$) then the current flowing into this element is I_T and defined as:

$$I_T = \frac{V_T}{r_L} \tag{2.179}$$

Otherwise stated:

$$\frac{V_T}{I_T} = r_L \tag{2.180}$$

and the time constant involving L_1 while $\hat{v}_{out} = 0$ is L/r_L. The zero is thus located at:

$$\omega_z = \frac{r_L}{L_1} \tag{2.181}$$

The complete transfer function can be expressed as:

$$H(s) = \frac{R_2 \| r_L}{R_2 \| r_L + R_1} \frac{1 + s\dfrac{L_1}{r_L}}{1 + s\dfrac{L_1}{r_L + R_1 \| R_2}} \tag{2.182}$$

and the zero has been found using the NDI technique. Should I have used the 3-step technique instead, the zero raw position would have been given as:

$$\omega_z = \frac{\dfrac{R_2 \| r_L}{R_2 \| r_L + R_1}}{\dfrac{R_2}{R_1 + R_2} \dfrac{L_1}{r_L + R_1 \| R_2}} \tag{2.183}$$

Believe it or not, (2.183) is the same as (2.181). As a preliminary conclusion, a zero obtained with the NDI is often expressed in a simpler manner than with the three-step technique described with (2.146).

Sometimes simplifications are obvious as shown in (2.165) but I doubt you can see it immediately in (2.183). However, and this is important, both expressions give the exact same result and end up with well-ordered terms. Depending on your taste and skill, you can choose the NDI or the three-step technique.

Let's exercise the NDI in Figure 2.46a example. The zero given by (2.175) is quite complex but is correct as confirmed by the SPICE simulation. Applying NDI to this example implies the addition of a test current generator I_T while the input source is reinstated (Figure 2.53b): this is a double injection. The newly-drawn circuit is observed at $s=s_z$. In this particular condition, $\hat{v}_{out}=0$. No current circulates in R_4. Therefore, the same current i_1 circulates in R_1 and R_3. R_4 is naturally eliminated from the calculation.

From Figure 2.53b, we can define the current circulating in R_3. As its right terminal is biased to $\hat{v}_{out}$ which is 0 V, then V_T is applied across R_3:

$$i_1=\frac{V_T}{R_3} \tag{2.184}$$

This current also circulates in R_1 as no current flows in R_4. Therefore the voltage across R_1 is defined as:

$$V_{R_1}=i_1R_1=\frac{V_T}{R_3}R_1 \tag{2.185}$$

The voltage across resistor R_2 is minus the sum of the voltage across R_1 plus that across R_3:

$$V_{R_2}=-\left(\frac{V_T}{R_3}R_1+V_T\right)=-V_T\left(\frac{R_1}{R_3}+1\right) \tag{2.186}$$

From KCL, current i_1 is the sum of i_2 and I_T:

$$i_1=i_2+I_T=\frac{V_{R_2}}{R_2}+I_T \tag{2.187}$$

Substituting (2.186) and (2.184) in (2.187) then rearranging, we have:

$$\frac{V_T}{R_3}=-\frac{V_T\left(\frac{R_1}{R_3}+1\right)}{R_2}+I_T \tag{2.188}$$

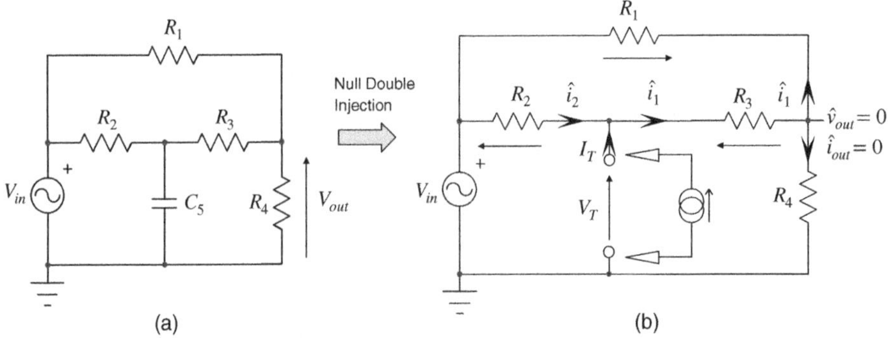

Figure 2.53 The zero can be observed by looking into the capacitor port while $\hat{v}_{out}=0$.

Factoring V_T in the left side gives:

$$V_T\left(\frac{1}{R_3}+\frac{\frac{R_1}{R_3}+1}{R_2}\right)=I_T \tag{2.189}$$

Finally,

$$\frac{V_T}{I_T}=\frac{1}{\frac{1}{R_3}+\frac{\frac{R_1}{R_3}+1}{R_2}}=\frac{R_2R_3}{R_1+R_2+R_3} \tag{2.190}$$

This expression is the resistance seen from the capacitor terminals during the NDI. The zero is thus defined as:

$$\omega_z=\frac{R_1+R_2+R_3}{R_2R_3C_5} \tag{2.191}$$

This expression is equal to that given in (2.175) despite the weak resemblance. To make sure this is correct, below are the Mathcad® results of equivalent resistances seen from C_5 terminals when applying NDI. There is the first expression we derived using the three-step technique (see (2.175)), the one obtained from the NDI and finally, which [2] gives on page 66.

$$\frac{\frac{R_4\|R_3}{R_1+R_4\|R_3}\cdot\{R_2\|[R_3+(R_1\|R_4)]\}}{\frac{R_4}{R_4+[(R_2+R_3)\|R_1]}}=184.282\,\Omega \qquad \frac{R_2\cdot R_3}{R_1+R_2+R_3}=184.282\,\Omega \qquad R_3\|\left(\frac{R_2}{1+\frac{R_1}{R_3}}\right)=184.282\,\Omega$$

These expressions are identical. Again, as we have already mentioned, the NDI approach gives the simplest expression but requires careful KVL/KCL analysis involving the nulled output. On the other hand, the 3-step approach did not. Both lead to identical results but the 3-step technique may need more work to simplify the zero, if this is possible. For instance, from the NDI analysis, one realizes that R_4 does not play a role. If R_4 is approaching infinity in the above left-side equation, then R simplifies to $\frac{R_3}{R_1+R_3}[R_2\|(R_3+R_1)]$. A new different expression but still giving 184.282 Ω!

A few remarks now: despite its presence in the schematic, the input source V_{in} plays no role in the expression of the zero position. This makes sense: why would changing the modulation amplitude or the bias level would affect the zero position after all? Unless, of course, if there is one or several controlled-sources linked to V_{in} (which is not the case in the example). Apart from this particular case, V_{in} has no role in determining the capacitor/inductor resistance during a NDI.

It can also be shown that the resistance offered by a capacitor or an inductor port while having the excitation source set to 0 – to calculate the circuit natural time constants – is unique. Should you want to derive different transfer functions on a circuit, meaning that the excitation source V_{in} is kept at the same position but you probe different nodes for V_{out}, the time constants are unchanged and $D(s)$ keeps the same for all derived transfer functions pertaining to the circuit under study (see the *network excitation* section below also). It is no longer true for the coefficients in the numerator. Depending on where you probe V_{out} for the same V_{in} position, the numerator will change, involving different zeros, if any. As a result, the resistance offered by a capacitor or inductor port in a NDI configuration – to calculate the numerator time constants – is not unique. In Figure 2.53a, should we probe V_{out} as the voltage across R_1 terminals, then $D(s)$ would remain unchanged but the resistance derived in (2.190) would be different.

In the above examples, we have concentrated our efforts on voltage transfer functions in which the null was applied across an impedance (R_4 in Figure 2.53). Figure 2.50 taught us that nulling the voltage across an impedance is different than shorting its terminals. If we now study impedances, the excitation source becomes a current source and the response is the voltage across its terminals. When we apply NDI to the circuit under study, the null is the voltage at the current source connecting points. If we null that voltage, there is 0 V across the source connecting terminals and the circulating current is I. If we replace it with a short circuit, the same current I circulates and there is still 0 V across the source. Figure 2.54 details an example when assessing an impedance.

This is a degenerate case useful when determining the zeros time constants during impedance analysis.

2.4.3 Checking Zeros Obtained in Null Double Injection with SPICE

In Chapter 1, we have seen how SPICE could help us verify that our analytical results were correct. By biasing the energy-storing component terminals with a dc current source (1 A for practical reasons), measuring the voltage at the injection node while the excitation source was set to 0 gave us the resistance seen by the capacitor or the inductor in this particular condition. This is extremely useful, in particular for high-order transfer functions involving controlled sources and complex architectures. I have been able several times to track an error I made in a particular time constant derivation by comparing the value Mathcad® gave me and what the SPICE dc analysis delivered.

For an NDI, it is a little bit more complicated. You must recreate the condition observed for $s = s_z$ in the transformed circuit – nulling the output response – while measuring the resistance offered by the energy-storing component in this condition. Practically speaking, you must adjust the current source I_T so that it brings V_{out} to zero while the circuit input excitation is still present. A possible test structure is presented in Figure 2.55 and is actually quite simple. The example is that of Figure 2.53.

The transconductance amplifier G_1 compares its inverting input to its grounded non-inverting input. This is a closed-loop system in which G_1 adjusts the delivered current so that both inputs are matched. As the transconductance gain is enormous, node 4, which is the output, is almost zero. The injected current is measured by V_2 (a dummy 0-V source) and B_1 computes the resistance seen at node 1, where the capacitor was connected. Make sure the right polarity is respected when considering the measurement nodes. It exactly measures 184.282 Ω as we have calculated via equations. This is the correct result. If V_{in} is changed to different values, G_1 adjusts its output to always

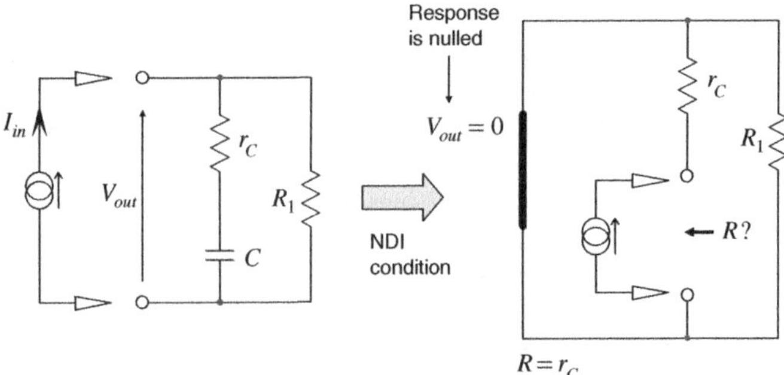

Figure 2.54 In this simple impedance calculation, the NDI implies that the output voltage across the current source (the response) is nulled. Simply replace the current source by a short circuit in this case: it is a degenerate case. The zero here is classically imposed by r_C.

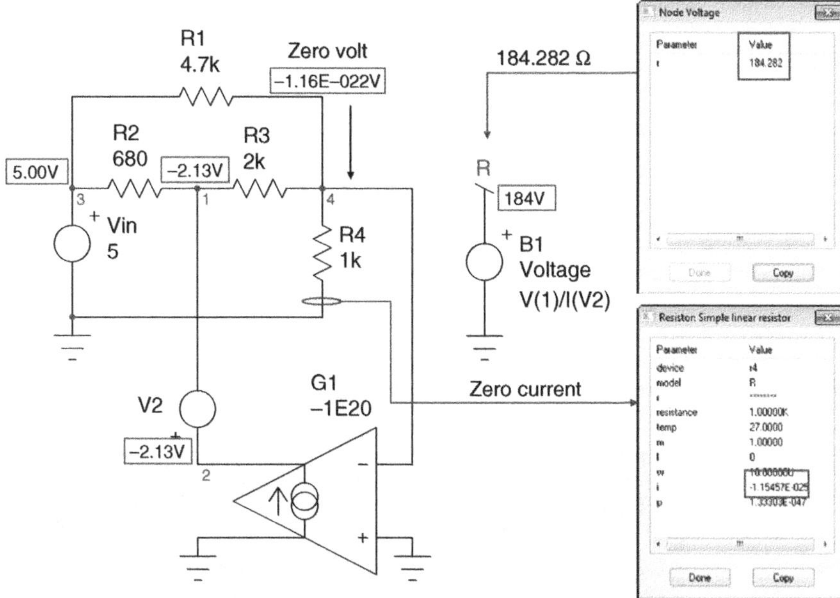

Figure 2.55 The nulling circuit lets you measure the resistance offered by the energy-storing elements while the output is brought to zero.

maintain the null at node 4 but the result computed by B_1 remains unchanged: V_{in} does not play a role in the resistance offered by node 4 during NDI.

In Figure 2.56, we have compiled all three measurement stages necessary to evaluate a circuit transfer function. In (a), the dc gain is evaluated and computed by B_2: V_4 injects 1 V and 322 mV appear across R_{15}. The gain is 0.322 or −9.84 dB. In (b), the 1-A test generator biases the capacitor node to measure the resistance offered at its terminals while the excitation is 0 V. A resistance of 408 Ω is measured and used to compute the pole position with B_4 (1.77 kHz). In (c), the NDI technique is applied to obtain the resistor offered by the capacitor terminals while the output is nulled. Source G_1 injects current to null the voltage across R_{11}. B_1 computes the resistance measured in this condition which amounts to 81.6 Ω. B_5 computes the corresponding zero evaluated at 8.86 kHz. Finally, subcircuit X_1 combines the computed gain, pole and zero into a Laplace expression block whose output is compared to the original one (V_{out1}). As shown beneath the schematic, all responses are similar.

The fact that we have a possibility to check what our individual port resistance calculations gave – in different analysis configurations – is great and constitutes an excellent option to ensure final results integrity. The transconductance amplifier can sense any nodes and inject current into floating ports without problems.

2.4.4 Network Excitation

Before we close this chapter, it is important to realize the role of the excitation signal and how we apply it to an existing circuit. We have seen in numerous examples how the excitation or drive waveform represents the electrical stimulus that propagates through the network to produce the response signal. This stimulus can be a voltage source or a current source but it needs to be injected or applied in a way that the network structure remains unaffected: a voltage source stimulus can only be

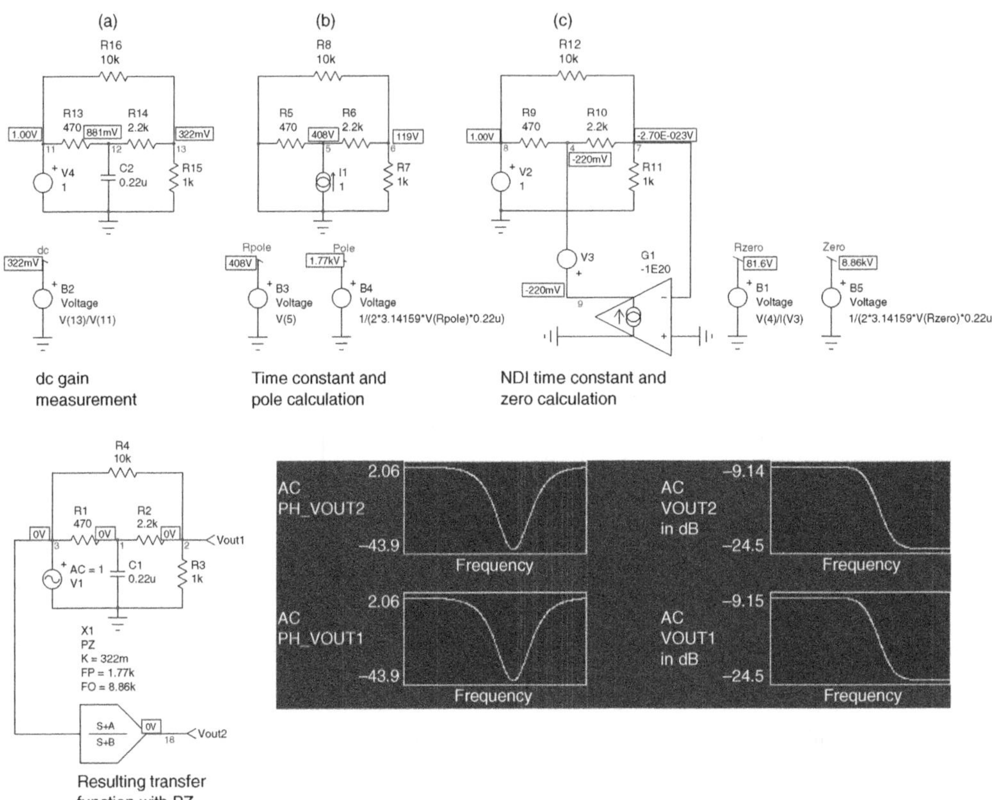

Figure 2.56 In this example, the pole, zero and gain are all separately measured and computed using the techniques we have described.

introduced inside a branch, *in series* with an existing element. Same for a current source that can only be applied across a branch, *in parallel* with a given element. This is because when you turn the excitation signal off, the circuit must return to its original structure. If not, no harm done of course, but you have a different network as we will see shortly. For instance, a voltage source turned off and replaced by a short circuit does not affect the network if it is placed somewhere in series inside a branch. Should you apply it in parallel with an existing resistor and later turn the source off, you short the resistor and change the circuit structure. Similarly, a current source applied across an existing element and later turned off to 0 A leaves the circuit intact as the current source disappears from the network when turned off. Should you insert the current source in series with an existing component instead, then turning the source off would leave the component with an unconnected terminal, again changing the network structure.

Figure 2.57 shows a simple 1st-order circuit whose single energy-storing element is capacitor C_1. There is no excitation and the output is not specified. It is what we consider the original form, the *structure*, of this network.

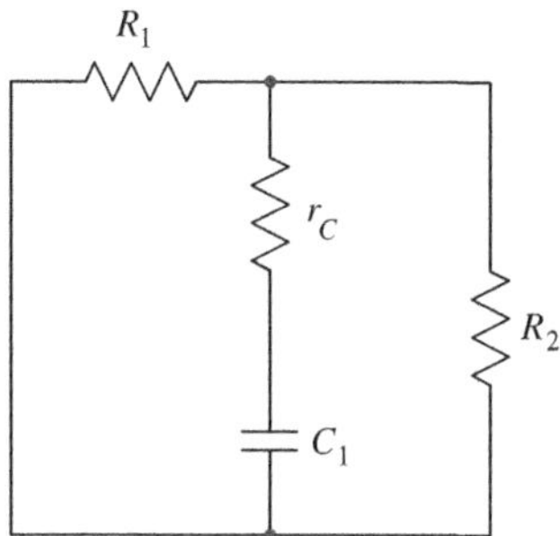

Figure 2.57 The natural time constants of a circuit depend on its structure, not on its stimulus value.

Assume we want to excite this network to obtain a transfer function linking the voltage across R_2, the response, to an excitation voltage source. To comply with the statement we have given before, a place to insert a voltage stimulus could be in series with R_1, C_1 or R_2 for instance. Figure 2.58 shows all options. If you now turn the excitation signal off, that is, replace the voltage source by a short circuit, you revert the circuit to its original form as shown in the right side of the figure.

If the excitation signal is a current source, you can apply it across several elements as drawn in Figure 2.59. When turned off, the source is removed and the circuit reverts to its original structure. One remarkable property affects natural time constants: because the circuit structure remains unaffected when sources are turned off in all of these configurations involving voltage or current

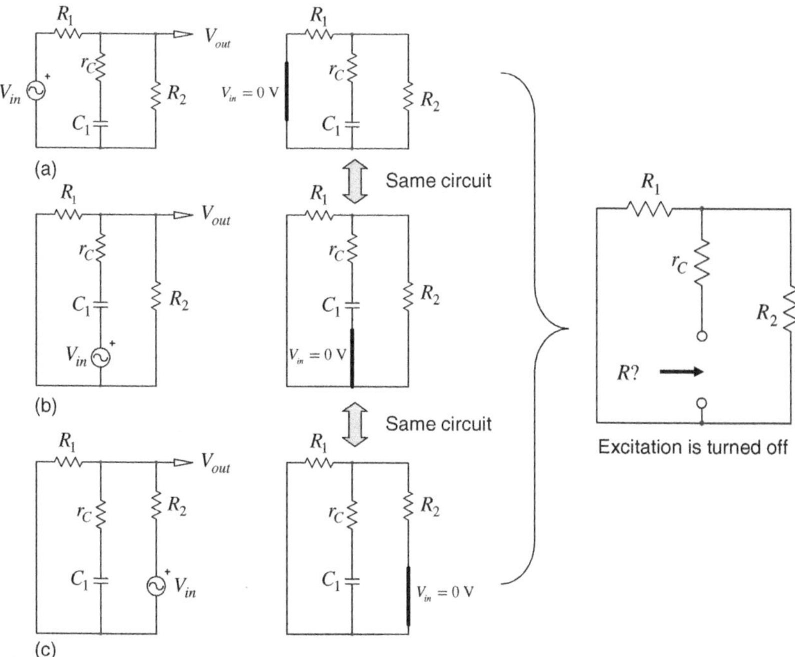

Figure 2.58 The place to insert a voltage stimulus is in series with R_1, with C_1 or with R_2. When the excitation is turned off, the structure is unchanged as the voltage source is replaced by a short circuit.

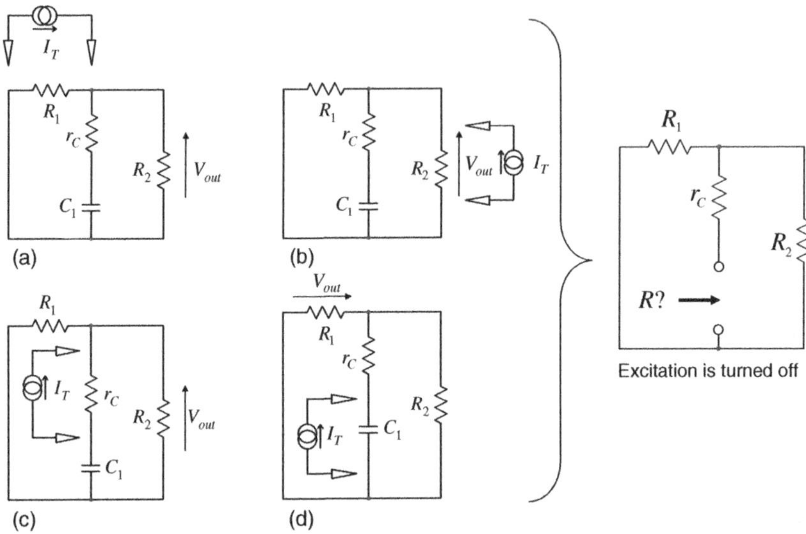

Figure 2.59 The place to insert a current source stimulus is in parallel with a component. When the excitation is turned off, the structure is unchanged as the current source disappears.

sources stimuli, the time constant affecting C_1 remains the same. As a result, a denominator expression $D(s)$ is common to all of these transfer functions. What will change though, are the zeros which depend on where you probe the response. The time constant involving C_1 is immediate and equal to:

$$\tau_2 = C_1\left(r_C + R_1 \| R_2\right) \tag{2.192}$$

leading to

$$D(s) = 1 + sC_1\left(r_C + R_1 \| R_2\right) \tag{2.193}$$

This property is interesting when you need to calculate several transfer functions pertinent to a network like gain or output impedance. Once you have the denominator, you can keep it for other calculations as long as the excitation source installed for the next transfer function determination does not change the structure when turned off. In Figure 2.58a, the dc gain H_0 linking V_{out} to V_{in} is:

$$H_0 = \frac{R_2}{R_1 + R_2} \tag{2.194}$$

and by inspection, r_C and C_1 form a zero: you have the transfer function in three lines:

$$H(s) = \frac{R_2}{R_1 + R_2} \frac{1 + sr_C C_1}{1 + sC_1\left(r_C + R_1 \| R_2\right)} = H_0 \frac{1 + \frac{s}{\omega_{z_1}}}{1 + \frac{s}{\omega_{p_1}}} \tag{2.195}$$

If you now want the output impedance instead, look at Figure 2.59b. The current source is applied across R_2 and when set to 0 A, the circuit reverts to its original form: time constants are unchanged, $D(s)$ in (2.193) is common to both transfer functions. In Figure 2.59b the dc resistance is:

$$R_0 = R_1 \| R_2 \tag{2.196}$$

and by inspection, the zero is the same as that of Figure 2.58a. The output impedance is thus defined as:

$$Z_{out}(s) = \left(R_1 \| R_2\right) \frac{1 + s r_C C_1}{1 + s C_1 \left(r_C + R_1 \| R_2\right)} = R_0 \frac{1 + \frac{s}{\omega_{z_1}}}{1 + \frac{s}{\omega_{p_1}}} \tag{2.197}$$

Now assume you want the input impedance of Figure 2.58a. The excitation signal becomes a current source and the response is across that current source, this is V_T in Figure 2.60. Inserting the current source in series with R_1 changes the structure of our circuit and the denominator we have calculated before no longer applies. When the current source is turned off, the time constant involving C_1 becomes:

$$\tau_2 = C_1(r_C + R_2) \tag{2.198}$$

and the denominator changes to:

$$D(s) = 1 + sC_1(r_C + R_2) \tag{2.199}$$

For $s = 0$, the input resistance is:

$$R_0 = R_1 + R_2 \tag{2.200}$$

The zero is found by nulling the response across the current source. The degenerate case of a current source having 0 V across its terminal is a short circuit as explained in Figure 2.54. If you replace the current source by a short circuit and determine the resistance looking into C_1's terminals in this condition, you find:

$$\tau_2 = C_1\left(r_C + R_1 \| R_2\right) \tag{2.201}$$

and

$$N(s) = 1 + sC_1\left(r_C + R_1 \| R_2\right) \tag{2.202}$$

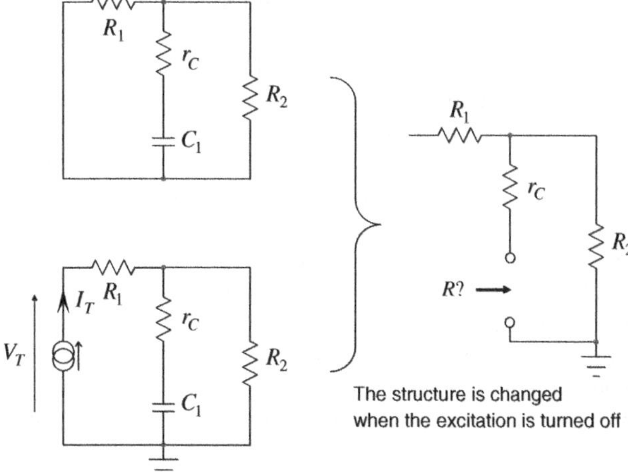

Figure 2.60 When calculating the input impedance of this network, the current source insertion changes the circuit structure and our previously-determined denominator no longer applies.

The circuit input impedance is thus:

$$Z_{in}(s) = (R_1 + R_2)\frac{1 + sC_1\left(r_C + R_1 \| R_2\right)}{1 + sC_1(r_C + R_2)} = R_0\frac{1 + \dfrac{s}{\omega_{z_2}}}{1 + \dfrac{s}{\omega_{p_2}}} \tag{2.203}$$

In this expression, the denominator is no longer the one we have calculated in (2.193). This is because the insertion of the current source has changed the circuit structure, affecting the natural time constants values. When you want to reuse a denominator expression determined for a given circuit transfer function and want to apply it in a second transfer function, ask yourself if the way you drive the circuit for this second exercise changes the network structure. If no, then you can reuse the denominator you have already on hand. If yes, you must re-derive it. Now, rather than determining the input impedance in Figure 2.60, we could have derived the input admittance Y_{in} with a voltage source V_T as the excitation and the input current I_T as a response. Since the network structure reverts to its original form when setting V_T to 0 V, we could have immediately reused $D(s)$ from (2.193), later obtaining the zero using NDI for instance (to null I_T in this case, r_C in series with C_1 and paralleled with R_2 form an infinite impedance for $s = s_z$). Then taking the inverse of Y_{in} would have lead to Z_{in}.

2.5 What Should I Retain from this Chapter?

In this second chapter, we have entered the world of the Fast Analytical Techniques and already obtained interesting results with 1st-order circuits. Below is a summary of what we learned in this chapter:

1. A system is said to be linear if it satisfies the superposition principle. When dealing with a nonlinear circuit, it must be linearized before applying the Laplace transform and other useful theorems such as Thévenin's and superposition.
2. A nonlinear circuit can be linearized by analyzing its response to a small-amplitude modulating ac waveform. After having sorted out ac and dc terms in the total response, you only keep the 1st-order ac component which is the small-signal or linear response of the considered circuit.
3. Time constants which are of the forms RC or L/R affect the natural response of the network. As these time constants are linked to the circuit structure only (how R, L and C are arranged in the electrical schematic), they are not affected by the excitation signal.
4. To determine the denominator terms or the circuit natural time constants, the excitation generator is turned off. Express the resistance R driving the energy-storing elements in this particular condition to obtain the circuit natural time constants. The resistance combined with the given element – RC or L/R – forms the inverse of the pole.
5. If the driving signal is a voltage source, setting it to 0 V means that you short its terminals: the source is replaced by a short circuit. If the excitation source is a current generator, setting it to 0 A implies that the generator is replaced by an open circuit.
6. A transfer function is said to be *low entropy* when it is expressed in a clear and ordered form so that you can immediately distinguish gain, poles and zeros without having to reveal them through extra work.
7. Polynomial forms of 2nd-order, when the quality factor is much smaller than 1, can be factored as well-separated poles (denominator) or zeros (numerator). High-order expressions can also be arranged as factorized poles or zeros when these poles or zeros are well spread.

8. Zeros are the numerator roots and can be identified by looking for conditions in the transformed circuit examined at $s = s_z$ which could prevent the modulating waveform from propagating to form an output response. If a transformed series impedance becomes infinite at $s = s_z$, there is a zero associated with this network. If a transformed network can shunt the signal path to ground at $s = s_z$ and also nulls the output, there is a zero associated with that impedance.
9. A generalized transfer function for a 1st-order system has been derived and requires three simple calculation steps to obtain the gain, the pole and the zero, if any.
10. The Null Double Injection (NDI) is a means to calculate the resistance associated with an energy-storing element while the network output is nulled. The resistance combined with the given element forms the inverse of the zero. Zeros in NDI are obtained by looking at the resistance driving the energy-storing element when the response is nulled.
11. Finally, the exercise to obtain poles or zeros is similar – look into the energy-storing element's terminals to determine the resistance driving the reactance – but carried in two different conditions: when the excitation is turned off for the poles, and when the response is nulled for the zeros.
12. A zero obtained through a NDI is usually of simpler form than that obtained with the three-step approach. NDI requires KVL and KCL analysis while the output is nulled whereas the three-step approach only requires simple gain calculations at $s = 0$ and when the energy-storing element is infinite-valued.
13. Inspection, when possible, always leads to the simplest numerator form.
14. SPICE can also be used in a very simple manner to verify the integrity of all calculations, including the zero obtained when the network is analyzed in a NDI.
15. Several transfer functions characterizing a given network share a common denominator $D(s)$ if the excitation source applied in each transfer function determination does not change the circuit structure when turned off. If it does, then denominators will be different.

References

1. Signals and Systems/Time Domain Analysis. http://en.wikibooks.org/wiki/Signals_and_Systems/Time_Domain_Analysis (accessed 12/12/2015).
2. Vorpérian, V. *Fast Analytical Techniques for Electrical and Electronic Circuits*. Cambridge University Press. 2002.
3. DiStefano, J., Stubberud, A., Williams, I. *Feedback and Control Systems*. Schaum's Outlines, McGraw-Hill. 1990.
4. C. Basso. *Designing Control Loops for Linear and Switching Power Supplies: a Tutorial Guide*. Artech House, Boston. 2012.
5. http://www.rdmiddlebrook.com/ (accessed 12/12/2015).
6. A. Hajimiri. Generalized time- and transfer-constant circuit analysis. *IEEE Transactions on Circuits and Systems*, **57** (6), 1105–1121. 2009.

2.6 Appendix 2A – Problems

Below are several problems based on the information delivered in this chapter.

Problem 1 – rearrange the transfer function of a compensation circuit into an expression where a leading term G_0 is factored. As a hint, you can factor s/ω_z in the numerator. The transfer function of a type 2 compensator is as follows:

$$G(s) = -\frac{1+\frac{s}{\omega_z}}{\frac{s}{\omega_{po}}\left(1+\frac{s}{\omega_p}\right)} \tag{2.204}$$

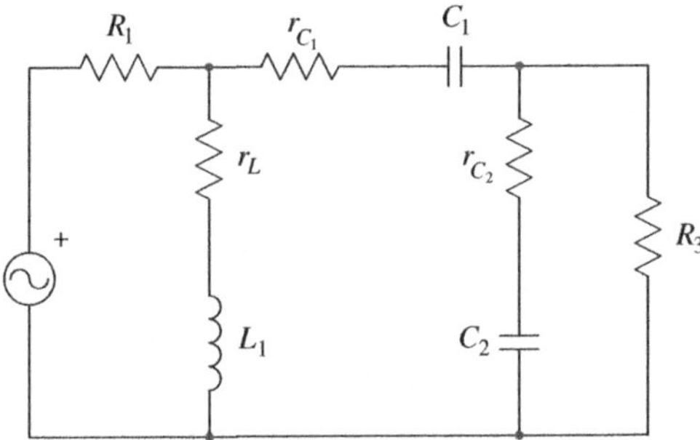

Figure 2.61 – Problem 2: What is the order of this network? What is the degree of its denominator $D(s)$?

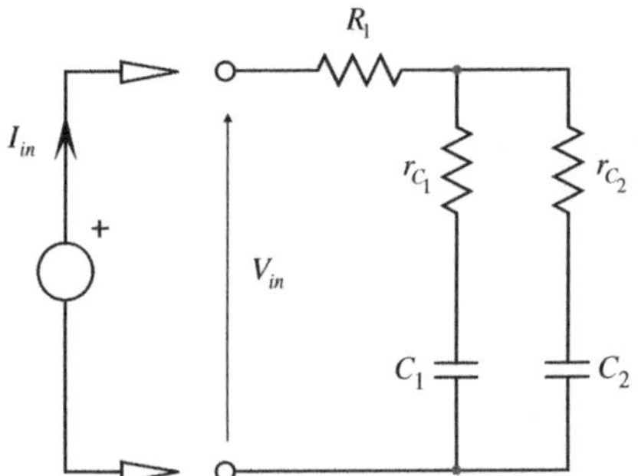

Figure 2.62 – Problem 3: What transfer function do we want to determine in the above circuit? Identify the stimulus and the response. What is the degree of its denominator $D(s)$?

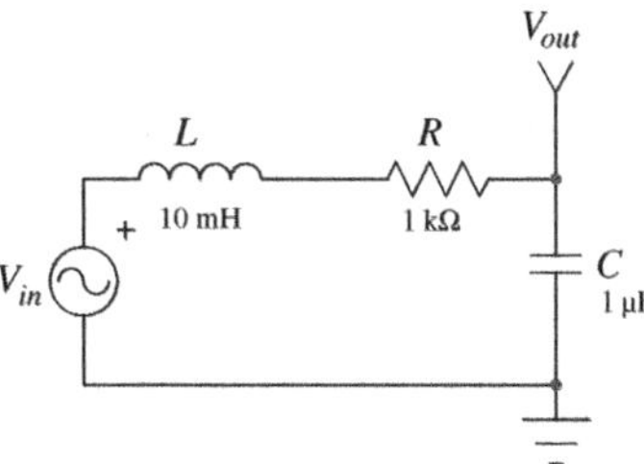

Figure 2.63 – Problem 4: With a Q factor defined as $\frac{1}{R}\sqrt{\frac{L}{C}}$ and a resonant frequency ω_0 expressed as $\frac{1}{\sqrt{LC}}$, calculate the two-pole equivalent expression describing this low-Q 2nd-order circuit.

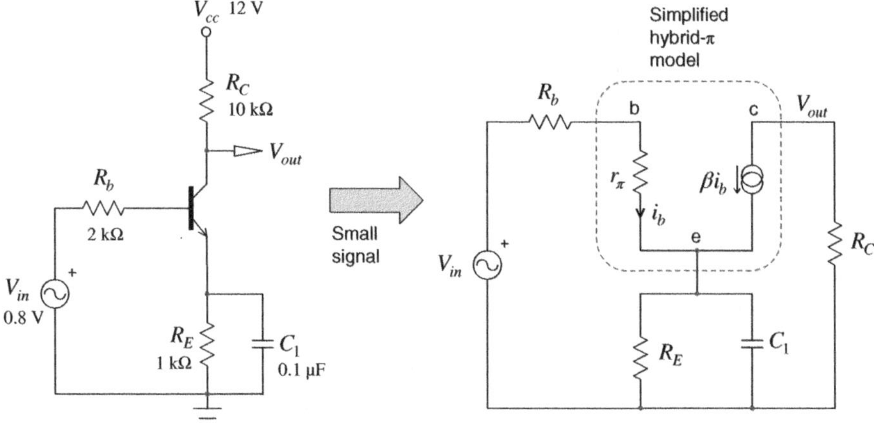

Figure 2.64 – Problem 5: In this circuit, check if C_1 contributes a zero and define its location using NDI. Use SPICE to verify your result.

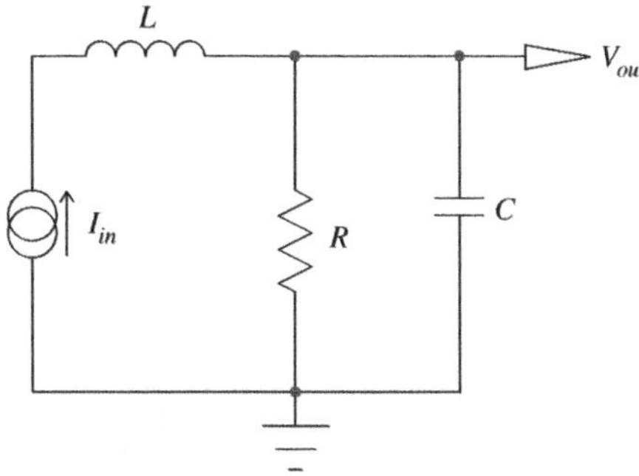

Figure 2.65 – Problem 6: What is the order of the above circuit linking V_{out} to I_{in}?

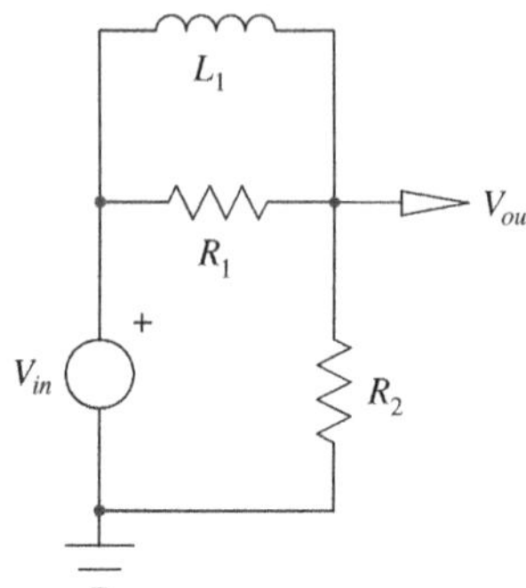

Figure 2.66 – Problem 7: Calculate the transfer function using the generalized 1st-order expression.

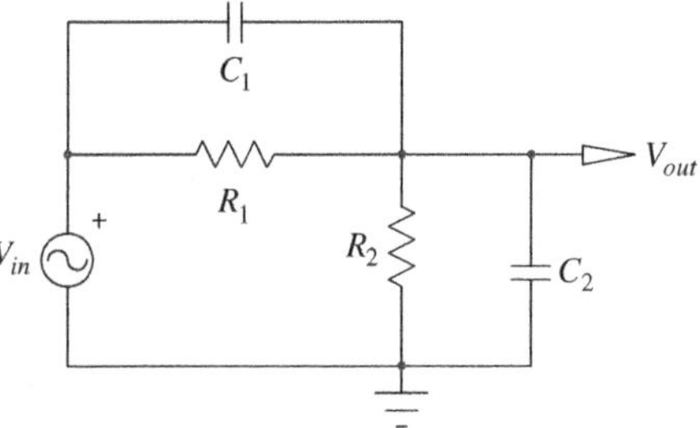

Figure 2.67 – Problem 8: Derive the transfer function of this circuit featuring two capacitors. Observe the capacitors configuration and infer the circuit order.

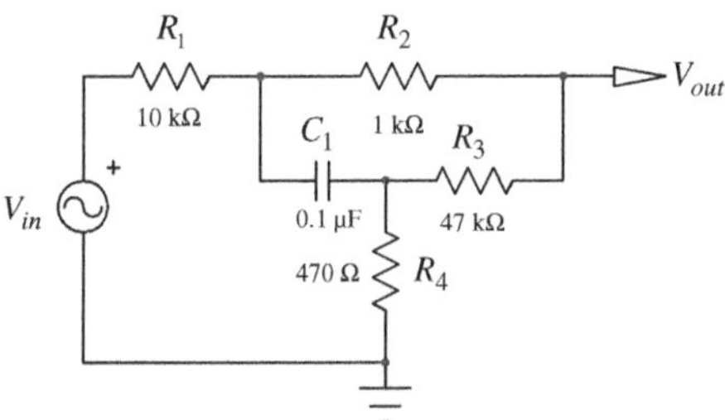

Figure 2.68 – Problem 9: Derive the transfer function of this circuit using NDI for the zero. Use SPICE to confirm your calculations.

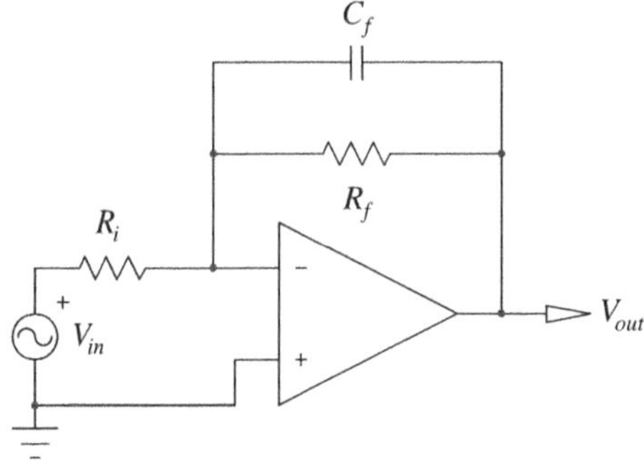

Figure 2.69 – Problem 10: Derive the transfer function of this low-pass active filter using a simplified model for the op amp.

Answers

Problem 1:

To unveil a leading term G_0, simply factor s/ω_z in the numerator:

$$G(s) = -\frac{\frac{s}{\omega_z}\left(\frac{\omega_z}{s}+1\right)}{\frac{s}{\omega_{po}}\left(1+\frac{s}{\omega_p}\right)} \tag{2.205}$$

Then have s/ω_z and s/ω_{po} become the leading term. Simplify by s and you have the ratio of a zero and a pole; this is the G_0 term you want:

$$G(s) = -\frac{\omega_{po}}{\omega_z}\frac{\frac{\omega_z}{s}+1}{1+\frac{s}{\omega_p}} = -G_0\frac{1+\frac{\omega_z}{s}}{1+\frac{s}{\omega_p}} \tag{2.206}$$

with $G_0 = \frac{\omega_{po}}{\omega_z}$.

In (2.205), there is a pole at the origin. When $s = 0$, the transfer function magnitude approaches infinity. The coefficient ω_{po} is the 0-dB crossover pole. When $s = \omega_{po}$, the magnitude of the function s/ω_{po} is 1 or 0 dB. By adjusting the value of ω_{po}, you change the G_0 gain without affecting ω_z or ω_p.

(2.206) describes the transfer function of a what is called a *type 2 compensator*. It is part of a control system. You adjust the so-called mid-band gain G_0, the pole and the zero to ensure a fast and precise response once the loop is closed. The pole ω_p and the zero ω_z are placed to generate phase boost (see Figure 2.15) at the selected crossover frequency. Once ω_z has been fixed, you can freely adjust ω_{po} to select the desired gain G_0. With (2.204), the relationship linking the 0-dB crossover pole and the gain G_0 does not readily appear. (2.206) is of *lower entropy* compared to (2.204). Reference [4] from Chapter 2 covers this subject at length.

Problem 2:

In this figure, count the storage elements and verify that all their state variables are independent: no loop including capacitors and no all-inductor node configuration are the most obvious degenerate cases. Here, we have three storage elements then the degree of D is three and it follows the form $D(s) = 1 + b_1s + b_2s^2 + b_3s^3$.

Problem 3:

In this is figure, we want to evaluate the admittance seen at the left-side connection terminals. The voltage source represents the excitation while the response is the current flowing in the voltage source. Again, count the storage elements and verify that all their state variables are independent. This is the case, both capacitors are physically separated by their equivalent series resistance, the ESR. This is a 2nd-order network. The degree of D is 2 and it follows the form $D(s) = 1 + b_1s + b_2s^2$.

Problem 4:

We first evaluate the quality factor using the given formula and SPICE abbreviations:

$$Q = \frac{1}{R}\sqrt{\frac{L}{C}} = \frac{1}{1k}\sqrt{\frac{10m}{1u}} = 0.1 \tag{2.207}$$

For this value, the low-Q approximation holds. The two equivalent poles are located at:

$$\omega_{p_1} = \omega_0 Q = \frac{1}{\sqrt{LC}}\frac{1}{R}\sqrt{\frac{L}{C}} = \frac{1}{RC} = \frac{1}{1u \times 1k} = 1\ \text{krd/s or} \approx 159\ \text{Hz} \tag{2.208}$$

$$\omega_{p_2} = \frac{\omega_0}{Q} = \frac{1}{\sqrt{LC}}R\sqrt{\frac{C}{L}} = \frac{R}{L} = \frac{1k}{10m} = 100\ \text{krd/s or} \approx 15.9\ \text{kHz} \tag{2.209}$$

In the low frequency range, the inductive contribution is weak and L is almost a short circuit. The RC time constant dominates. At high frequencies, the L/R time constant dominates, this is ω_{p2}.

Figure 2.70 compares the original response to its low-Q equivalent network. Two poles are cascaded: R_2 and C_2 generate ω_{p_1} while L_2 and R_3 build ω_{p_2}. Please note the presence of the buffer X_1 to isolate both circuits (L_2 does not load C_2).

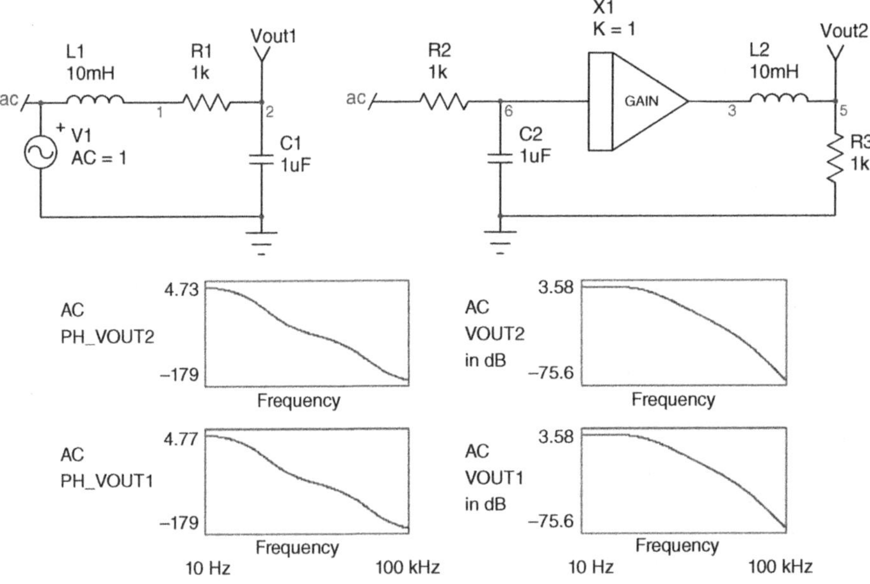

Figure 2.70 The low-Q approximation helps organizing a 2nd-order polynomial form into two spread poles.

Problem 5:

If we short C_1, the circuit transforms into a common-emitter configuration and we still have an output response: C_1 contributes a zero. To unveil the zero associated with C_1, redraw the small-signal circuit with collector, emitter and base currents a shown in Figure 2.71. If we are in an NDI situation, it means that the voltage response V_{out} is a null. The response appears across the collector resistor R_c. The reasoning can be done the following way:

A null in the response implies $\hat{v}_{out} = 0$

If $\hat{v}_{out} = 0$ then no current flows in the load thus $\hat{i}_c = 0$

The collector current is the base current multiplied by the transistor gain.

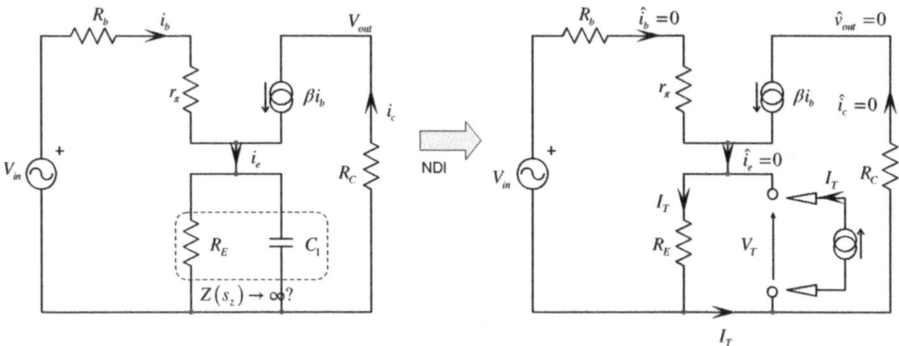

Figure 2.71 The NDI in this example helps solving the second zero position in a snapshot: the null condition propagates to the nodes and clears up the path for a simple solution.

If $\hat{i}_c = 0$ then $\beta\hat{i}_b = 0$

If $\beta\hat{i}_b = 0$ then the base current is also a null and no current circulate in r_π

If no current circulates in r_π and the collector current is a null, then the emitter current $(\beta + 1)\hat{i}_b$ is also a null: $\hat{i}_e = 0$

If $\hat{i}_e = 0$, then all the test current I_T circulates in the emitter resistor R_E. R_E is the resistance seen by the capacitor during NDI therefore the zero is positioned at

$$\omega_z = \frac{1}{R_E C_1}$$

We could also say that if the collector current is zero, then the emitter is not connected, implying that the impedance offered by R_E and C_1 is infinite at some frequency. Solving for the denominator root of $R_E \| \frac{1}{sC_1}$ leads to the above zero definition.

We can apply the technique learned in this chapter to check the resistance seen by C_1 in a NDI condition. The test circuit is shown in Figure 2.72. The bias points confirm the 0 V at the output and the resistance computed by the B_2-element is the emitter resistor, as expected. Please note that V_{cc} is set to 0 in this mode since in ac, its contribution would also be zero, hence node three (upper R_C terminal) grounded in the small-signal schematic of Figure 2.64. We have arbitrarily set V_{in} to 0.8 V for this exercise but any other value would work: G_1 will adjust the delivered current to maintain the output null and the value computed by B_2 won't change.

Problem 6:

In this circuit, the inductor lies in series with the current generator. Therefore, its state variable x_1 is entirely fixed by I_{in}. It is a degenerate case in which L has no role. It is a 1st-order response. In dc, the gain H_0 is R and the resistance seen by C while the excitation is 0 A (remove the current source) is also R. The transfer function is simply:

$$H(s) = H_0 \frac{1}{1 + s\tau_1} = R \frac{1}{1 + sRC} = R \frac{1}{1 + \frac{s}{\omega_p}} \tag{2.210}$$

with $\omega_p = \frac{1}{RC}$

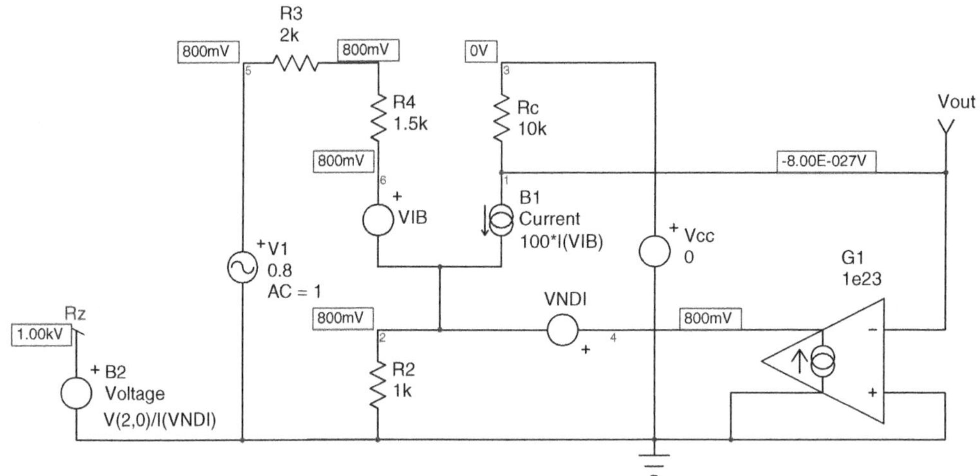

Figure 2.72 The added transconductance amplifier helps nulling the output and tells us what resistance R_z is seen across C_1 terminals when V_{out} is a null. Bias points confirm that the base current is also well nulled (0 V across R_4).

Problem 7:

To apply the 1st-order generalized transfer function expression, we start with the dc gain for which L_1 is a short. The gain in this case is:

$$H_0 = 1 \tag{2.211}$$

The gain when L_1 is infinite-valued (L_1 is physically removed from the circuit) is a simple resistive divider:

$$H^1 = \frac{R_2}{R_1 + R_2} \tag{2.212}$$

The time constant is found by setting the excitation to zero while looking into the inductor's terminals. The resistance seen in this mode is:

$$R = R_1 \| R_2 \tag{2.213}$$

The time constant τ_1 is thus:

$$\tau_1 = \frac{L_1}{R_1 \| R_2} \tag{2.214}$$

The transfer function comes easily then:

$$H(s) = \frac{H_0 + H^1 s\tau_1}{1 + s\tau_1} = \frac{1 + s\dfrac{R_2}{R_1 + R_2}\dfrac{L_1}{R_1 \| R_2}}{1 + s\dfrac{L_1}{R_1 \| R_2}} = \frac{1 + s\dfrac{L_1}{R_1}}{1 + s\dfrac{L}{R_1 \| R_2}} \tag{2.215}$$

Problem 8:

If you carefully look at the picture, you see that C_1 and C_2 are in a loop with V_{in} and their state variables x_1 and x_2 are linked:

$$V_{in} = x_1 + x_2 \tag{2.216}$$

This is another degenerate case and the circuit order is 1 despite the presence of two capacitors. Do we have zeros in this network? Let's see: if I short C_2, do I still have a response? No. If I now short C_1, is the response still existing? Yes, there is a zero associated with C_1. How can it block the driving voltage to deliver a response? If the parallel combination of R_1 and C_1 forms an infinite impedance:

$$Z(s) = R_1 \| C_1 \tag{2.217}$$

We know how to calculate this type of transfer function immediately: in dc, the impedance is R_1 and the time constant is R_1C_1:

$$Z(s) = R_1 \frac{1}{1 + sR_1C_1} \tag{2.218}$$

It goes to infinity if the denominator is 0. Otherwise stated, $1 + sR_1C_1 = 0$ implying a zero positioned at:

$$\omega_z = \frac{1}{R_1C_1} \tag{2.219}$$

What is the transfer function in dc, meaning all caps are removed from the circuit?

$$H_0 = \frac{R_2}{R_1 + R_2} \tag{2.220}$$

To obtain the circuit time constant, set V_{in} to 0 V. Without drawing anything, in your head, you see that you have all elements in parallel. C_1 and C_2 combined to form $C_1 + C_2$ while the resistors become $R_1 \| R_2$. The time constant is therefore:

$$\tau_1 = (R_1 \| R_2)(C_1 + C_2) \tag{2.221}$$

Combining the above results, the transfer function is:

$$H(s) = \frac{R_2}{R_1 + R_2} \frac{1 + sR_1C_1}{1 + s(R_1 \| R_2)(C_1 + C_2)} = H_0 \frac{1 + \frac{s}{\omega_z}}{1 + \frac{s}{\omega_p}} \tag{2.222}$$

In which ω_p is defined as $\omega_p = \frac{1}{(R_1 \| R_2)(C_1 + C_2)}$.

Problem 9:

If I short C_1, is the response still present? Yes, so a zero is associated with C_1. What is this zero position using the NDI approach? You need to redraw the schematic where the test generator I_T appears. Position currents and voltages then mark the node which is at 0 V. Let's start with the voltage V_T across the capacitor's terminals:

$$V_T = V_1 + V_3 \tag{2.223}$$

The two resistors R_2 and R_3 are crossed by the same current i_2, therefore:

$$V_T = i_2(R_2 + R_3) \tag{2.224}$$

The test generator current I_T splits into i_1 and i_2:

$$I_T = i_1 + i_2 \tag{2.225}$$

Current i_1 is extracted as:

$$i_1 = I_T - i_2 \tag{2.226}$$

The voltages across R_3 and R_4 are equal since the R_3's upper terminal is at 0 V:

$$R_4 i_1 = R_3 i_2 \tag{2.227}$$

From this expression, extract i_1, substitute it by (2.226). Solve for i_2:

$$i_2\left(1 + \frac{R_4}{R_3}\right) = \frac{R_4}{R_3} I_T \tag{2.228}$$

$$i_2 = \frac{R_4}{R_3} \frac{1}{1 + \frac{R_4}{R_3}} I_T = I_T \frac{R_4}{R_3 + R_4} \tag{2.229}$$

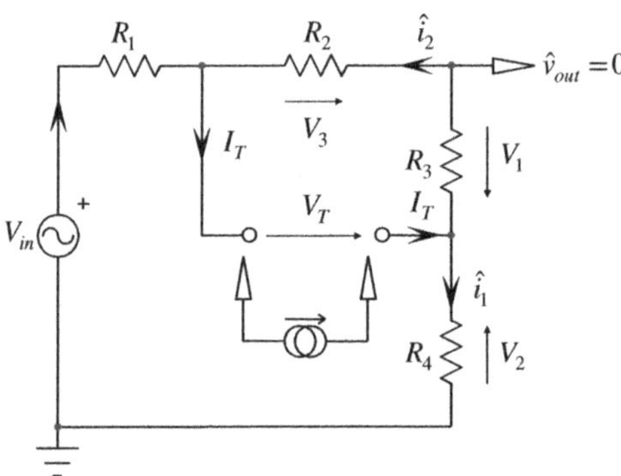

Figure 2.73 Null double injection requires the addition of a test generator to find the resistance seen across C_1's terminals when $\hat{v}_{out} = 0$.

If we now substitute (2.229) in (2.224), rearrange to unveil V_T/I_T, we have our resistance offered by the capacitor's terminals while the output is a null:

$$\frac{V_T}{I_T} = \frac{R_4}{R_3 + R_4}(R_2 + R_3) \tag{2.230}$$

The time constant τ associated with the capacitor is thus $\tau_1 = C_1 \frac{R_4}{R_3+R_4}(R_2 + R_3)$ and the zero is located at:

$$\omega_z = \frac{1}{\frac{R_4}{R_3 + R_4}(R_2 + R_3)C_1} \tag{2.231}$$

Where is the pole in this circuit? Set the excitation voltage to 0 V and look into the capacitor's terminals to see the resistance.

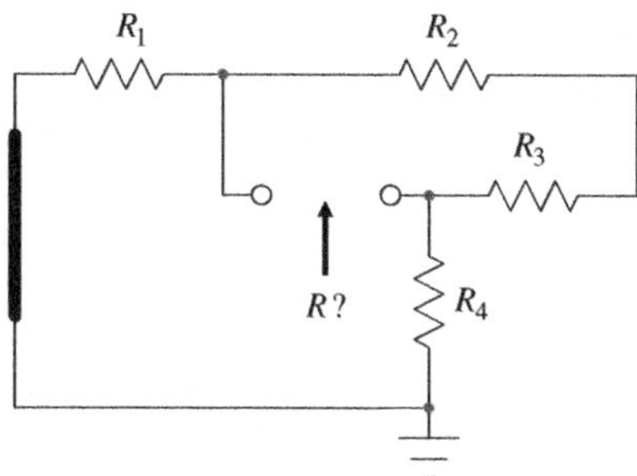

Figure 2.74 For the pole, the resistance driving the capacitor is obtained by setting the excitation to 0 V.

The series combination of R_2 and R_3 appears across the terminals together with the parallel combination of R_1 and R_4. Thus:

$$R = (R_4 + R_1) \| (R_2 + R_3) \tag{2.232}$$

The pole is located at:

$$\omega_p = \frac{1}{\tau_2} = \frac{1}{C_1\left[(R_4 + R_1)\|(R_2 + R_3)\right]} \tag{2.233}$$

What is the transfer function for $s=0$? By looking at Figure 2.68, when C_1 disappears, we immediately see a resistive divider loaded by R_3 and R_4:

$$H_0 = \frac{R_3 + R_4}{R_1 + R_2 + R_3 + R_4} \tag{2.234}$$

The final transfer function can be written as:

$$H(s) = \frac{R_3 + R_4}{R_1 + R_2 + R_3 + R_4} \frac{1 + s\dfrac{R_4}{R_3 + R_4}(R_2 + R_3)C_1}{1 + sC_1\left[(R_4 + R_1)\|(R_2 + R_3)\right]} = H_0 \frac{1 + \dfrac{s}{\omega_z}}{1 + \dfrac{s}{\omega_p}} \tag{2.235}$$

Now, should you consider the NDI approach a bit tricky, you could very well apply the three-step method we have described which uses the pole time constant calculation and two gain derivations. H_0 and τ_2 are already known so we omit 30% of the work. The second gain H^1 and is obtained when the component involved in time constant 1 (it is C_1) is infinite-valued. For a capacitor, we short it and install a strap across its terminals as in Figure 2.68. The gain is immediate as R_2 and R_3 are unloaded:

$$H^1 = \frac{R_4}{R_4 + R_1} \tag{2.236}$$

(2.235) could therefore be rewritten as:

$$H(s) = H_0 \frac{1 + \dfrac{H^1}{H_0} s\tau_2}{1 + s\tau_2} = \frac{R_3 + R_4}{R_1 + R_2 + R_3 + R_4} \frac{1 + s\dfrac{\dfrac{R_4}{R_1 + R_4}}{\dfrac{R_3 + R_4}{R_1 + R_2 + R_3 + R_4}} C_1\left[(R_4 + R_1)\|(R_2 + R_3)\right]}{1 + sC_1\left[(R_4 + R_1)\|(R_2 + R_3)\right]} \tag{2.237}$$

The numerator is more complex than in (2.235) but simplifications can be made: the paralleling of $R_4 + R_1$ and $R_2 + R_3$ ends up in a denominator summing all these resistors and as it already appears in H_0, the sum goes away. The remark we have formulated earlier still holds: Either you apply NDI and you 'suffer' with KVL and KCL while $V_{out}(s)$ is nulled (it is not obvious sometimes, I agree). The method results in a simple expression involving the energy-storing element straight away. Or you refuse to apply NDI and use the three-step technique but end-up with a more complex numerator. You then 'suffer' to simplify and rearrange it in a simpler shape. If you want, by the way, because the final result is already in a well-ordered form except that the numerator coefficient is more complex than in the NDI case. The choice is yours. Figure 2.75 compares responses obtained with the NDI or the 3-step approach. As expected, they are identical.

$$R_1 := 10k\Omega \quad R_2 := 1k\Omega \quad R_3 := 47k\Omega \quad R_4 := 470\Omega \quad C_1 := 0.1\mu F$$

$$\|(x,y) := \frac{x \cdot y}{x+y}$$

$$\tau_1 := (R_2 + R_3)\cdot\frac{R_4}{R_3 + R_4}\cdot C_1 = 47.525\mu s \qquad R_z := (R_2 + R_3)\cdot\frac{R_4}{R_3 + R_4} = 475.248\Omega$$

$$\tau_2 := (R_2 + R_3) \| (R_4 + R_1)\cdot C_1 = 859.518\mu s \qquad R_p := (R_2 + R_3) \| (R_4 + R_1) = 8.595k\Omega$$

$$H_0 := \frac{R_4 + R_3}{R_1 + R_2 + R_3 + R_4} = 0.812 \qquad H_1 := \frac{R_4}{R_1 + R_4}$$

$$\omega_p := \frac{1}{\tau_2} \qquad f_p := \frac{\omega_p}{2\pi} = 185.168\cdot Hz$$

$$\omega_z := \frac{1}{\tau_1} \qquad f_z := \frac{\omega_z}{2\pi} = 3.349\cdot kHz$$

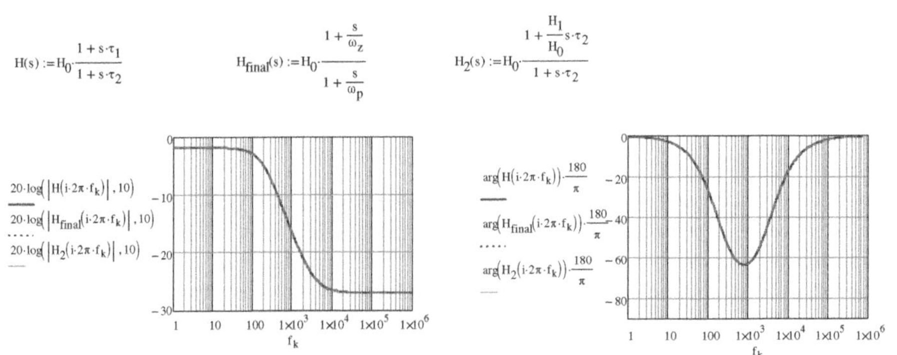

Figure 2.75 Mathcad® gives similar responses whether NDI or the 3-step technique is used.

The SPICE check involves the transconductance amplifier already described. We have pushed the analysis a bit farther by reproducing the test schematic from Figure 2.56. The configuration appears in Figure 2.76 and confirms our calculations. Please note that source V_2 value is not important in the NDI configuration, it is there to force a current from the transconductance amplifier to null the output. You can change it to any value, the resistance computed at the R_{zero} node will not vary.

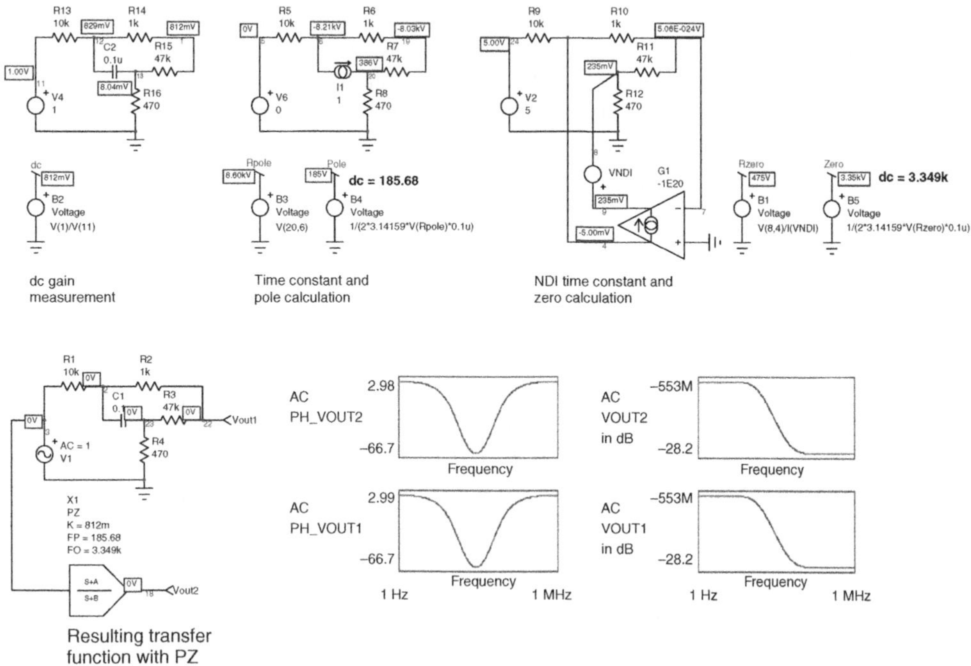

Figure 2.76 SPICE can test all the expressions in a convenient and simple arrangement.

Problem 10:

To derive the transfer function of this active filter, the op amp is replaced by its equivalent circuit as shown in Figure 2.77. ε (epsilon) represents the error voltage between both inputs while A_{OL} is the op amp open-loop gain. For $s = 0$, C_f disappears from the picture while R_f and R_i remain in place. The sketch updates to that of Figure 2.78a.

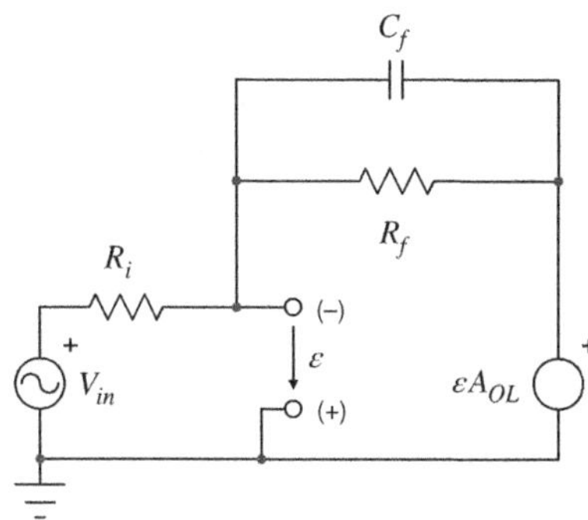

Figure 2.77 The equivalent simplified model of the op amp involves the error voltage ε.

In this sketch, you see a circuit involving two sources while the error voltage ε appears at the resistances junction. To solve this circuit dc transfer function swiftly, we can apply superposition. The theorem says that if a linear system involves different sources, the response of the circuit to the various excitations is the sum of individual responses obtained when one source is active while the other are set to 0. Mathematically, we can express this theorem the following way:

$$\begin{aligned} V_{out}(V_1, V_2, V_3 \ldots V_i) = {} & V_{out}(V_1)|_{V_2=V_3=V_i=0} + V_{out}(V_2)|_{V_1=V_3=V_i=0} + V_{out}(V_3)|_{V_1=V_2=V_i=0} \\ & + \ldots + V_{out}(V_i)|_{V_1=V_2=V_3=0} \end{aligned} \tag{2.238}$$

In our op amp circuit, we have two sources, V_{in} and V_{out}.

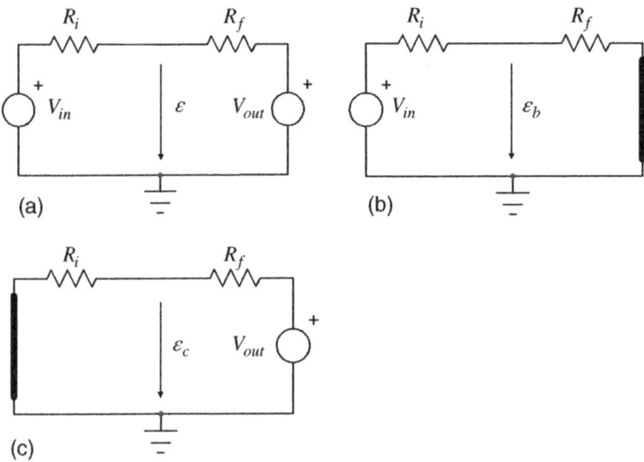

Figure 2.78 The superposition theorem is ideal when more than one source are involved.

To obtain the output voltage V_{out}, we will derive the error voltage ε in two steps where sources will alternatively be set to 0 while the output is computed in each case. In (b), the second source V_{out} is set to 0. The error voltage ε_b is equal to:

$$\varepsilon_b = -V_{in}\frac{R_f}{R_f + R_i} \tag{2.239}$$

In (c), the input voltage V_{in} is set to 0:

$$\varepsilon_c = -V_{out}\frac{R_i}{R_i + R_f} \tag{2.240}$$

Summing this two voltages gives us ε:

$$\varepsilon = \varepsilon_b + \varepsilon_c = -V_{in}\frac{R_f}{R_f + R_i} - V_{out}\frac{R_i}{R_f + R_i} \tag{2.241}$$

We know that ε is linked to V_{out} by the open-loop gain A_{OL}:

$$\varepsilon = \frac{V_{out}}{A_{OL}} \tag{2.242}$$

Equating (2.241) and (2.242) then collecting terms leads to:

$$V_{out}\left(\frac{1}{A_{OL}}+\frac{R_i}{R_f+R_i}\right)=-V_{in}\frac{R_f}{R_f+R_i} \tag{2.243}$$

Rearranging and factoring, we have the final dc transfer function G_0:

$$G_0=\frac{V_{out}}{V_{in}}=-\frac{R_f}{R_i}\frac{1}{\left(\frac{\frac{R_f}{R_i}+1}{A_{OL}}+1\right)} \tag{2.244}$$

For $A_{OL} \gg 1$, the formula simplifies to the well-known expression:

$$G_0 \approx -\frac{R_f}{R_i} \tag{2.245}$$

Now, regarding C_f, does it contribute a zero? In other terms, when infinite-valued (a short circuit), do we still have a response? No, when shorted, the R_f term in (2.244) is 0 and there is no response. No zero in this circuit.

For the pole, we set the input voltage to 0 and the circuit simplifies to that of Figure 2.79.

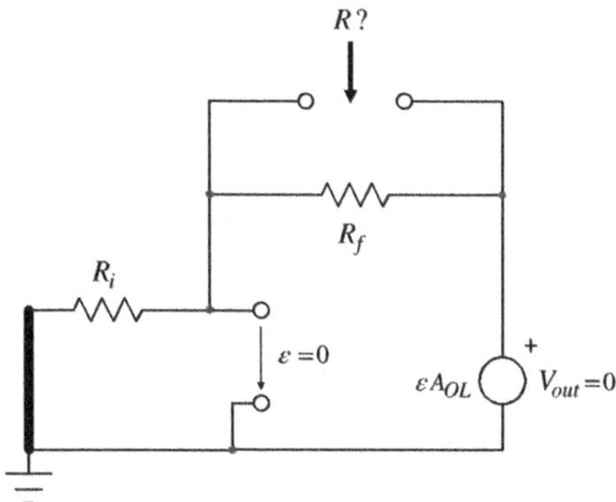

Figure 2.79 When the input voltage is set to 0, both R_f terminals are grounded.

According to (2.244), when V_{in} is 0, then V_{out} is also 0. R_f's right terminal is therefore grounded. If V_{out} is 0 V, then according to (2.242) ε is also equal to 0: R_f's left terminal is at a 0-V potential through the virtual ground. The resistance seen by capacitor C_f is thus resistor R_f. The pole is located at:

$$\omega_p=\frac{1}{R_f C_f} \tag{2.246}$$

The complete transfer function is expressed as:

$$G(s)=G_0\frac{1}{1+\frac{s}{\omega_p}} \tag{2.247}$$

With G_0 and ω_p given by (2.245) and (2.246).

3

Superposition and the Extra Element Theorem

In the first chapter, we reviewed some of the tools we often refer to when analyzing electrical networks. Among theorems we described, Norton's and Thévenin's are extremely useful and were also used at some point in Chapter 2. However, when dealing with more than one input, the superposition theorem occupies the top of the list as its extension naturally leads to the Extra Element Theorem (EET). By first introducing superposition theorem in a simple and graphical way, this third chapter paves the way towards the EET understanding in a hopefully simple and intelligible manner.

3.1 The Superposition Theorem

Figure 3.1 shows a box hosting a linear system fed by two inputs u_1 and u_2. These inputs can be current and/or voltage, it does not matter. These two stimuli simultaneously applied form a response y_1 defined as:

$$y_1 = f(u_1, u_2) \tag{3.1}$$

The superposition theorem states that output y_1 is the algebraic sum of the response obtained when u_1 is set to 0 and the response obtained when u_2 is set to 0.

You could write this statement in a generalized form such as:

$$y_1(u_1, u_2 \ldots u_i) = y_1(u_1)\big|_{u_2=u_i=0} + y_1(u_2)\big|_{u_1=u_i=0} + \ldots + y_1(u_n)\big|_{u_2=u_1=0} \tag{3.2}$$

Setting a source to 0 or turning it off was defined in the previous chapters: an independent voltage source is set to 0 V and replaced by a short circuit while an independent current source is set to 0 A by open-circuiting it or physically removing it from the circuit. Dependent sources are left untouched in the circuit unless the control source is the one that is set to 0 during an analysis step. Applying this theory to our simple diagram leads to an updated schematic shown in Figure 3.2.

From this diagram, we can write:

$$\left.\frac{y_1}{u_1}\right|_{u_2=0} \equiv A_1 \tag{3.3}$$

Linear Circuit Transfer Functions: An Introduction to Fast Analytical Techniques, First Edition. Christophe P. Basso.

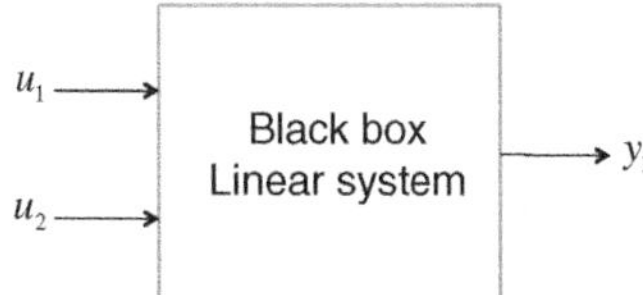

Figure 3.1 This simple system delivers an output y_1 formed by the two stimuli u_1 and u_2.

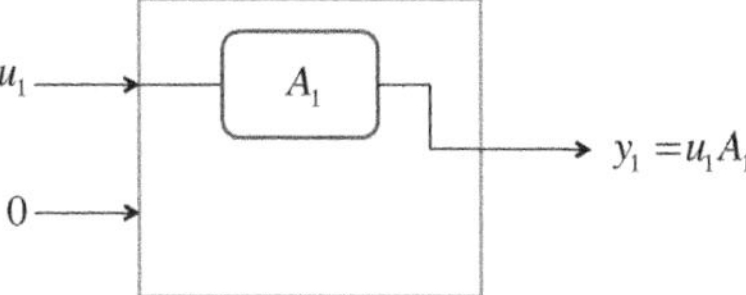

Figure 3.2 In this configuration, u_2 is set to 0 while u_1 alone contributes to the output.

Similarly, we can set u_1 to zero and keep u_2 active while observing y_1. This is what is shown in Figure 3.3.

Using the previous notation, we can write:

$$\left.\frac{y_1}{u_2}\right|_{u_1=0} \equiv A_2 \tag{3.4}$$

The combination of both inputs leads to adding (3.3) and (3.4). The illustrating sketch appears in Figure 3.4 which graphically illustrates:

$$y_1 = \left.\frac{y_1}{u_1}\right|_{u_2=0} u_1 + \left.\frac{y_1}{u_2}\right|_{u_1=0} u_2 = A_1 u_1 + A_2 u_2 \tag{3.5}$$

Let's check two simple examples to show how we can exercise the superposition theorem. Figure 3.5a represents a circuit combining two sources, voltage and current. The output voltage is measured across resistor R_3. We first set V_1 to 0 as shown in Figure 3.5b.

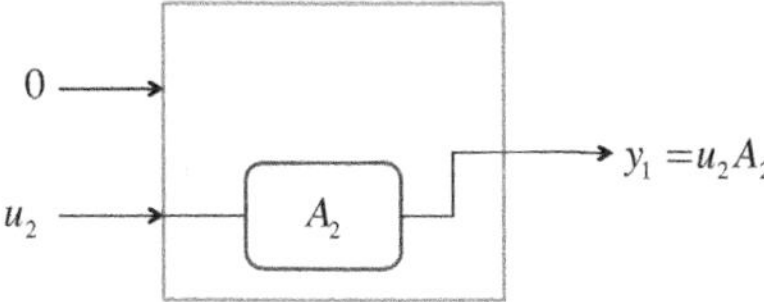

Figure 3.3 In this second configuration, u_1 is set to 0 while u_2 alone contributes to the output.

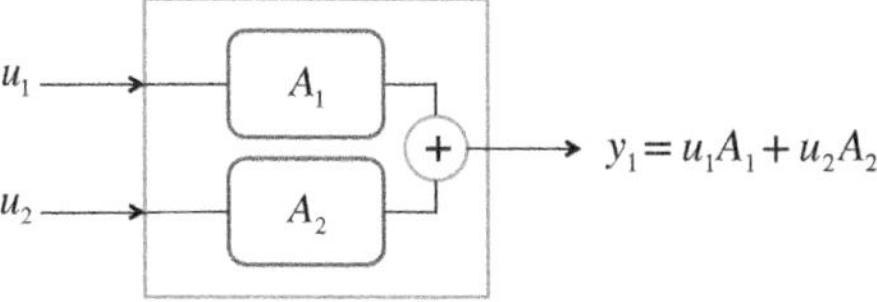

Figure 3.4 This output is the sum of the outputs independently contributed by u_1 and u_2 in the above conditions.

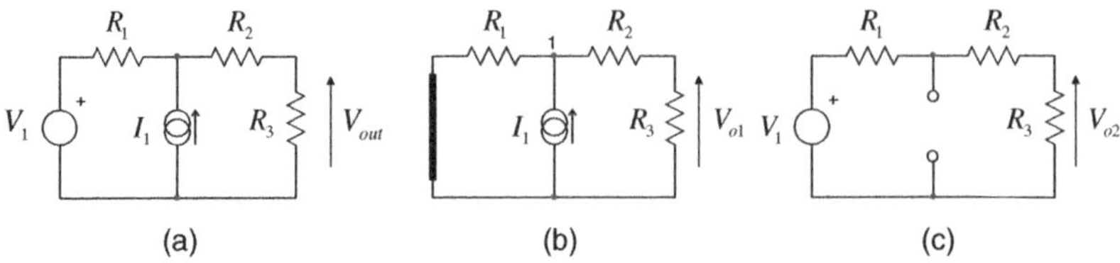

Figure 3.5 This simple circuit combines current and voltage sources.

R_1 appears in parallel with the series combination of R_2 and R_3. The voltage across the current source at node 1 is therefore:

$$V_{(1)} = I_1\left[R_1 \| (R_2 + R_3)\right] \tag{3.6}$$

The voltage across R_3 implies the resistive divider made of R_2 and R_3:

$$V_{o1} = V_{(1)}\frac{R_3}{R_2 + R_3} \tag{3.7}$$

Substituting (3.6) into (3.7) gives the first voltage V_{o1} when V_1 is equal to 0 V:

$$V_{o1} = I_1\left[R_1 \| (R_2 + R_3)\right]\frac{R_3}{R_2 + R_3} \tag{3.8}$$

The second voltage is obtained with diagram (c) in which the current source is turned off. The voltage V_{o2} is again obtained by using a resistive divider involving R_3 with the series arrangement of R_1 and R_2:

$$V_{o2} = V_1\frac{R_3}{R_1 + R_2 + R_3} \tag{3.9}$$

The final voltage across R_3 is thus the sum of V_{o1} and V_{o2} which gives:

$$V_{out} = V_{o1} + V_{o2} = I_1\left[R_1 \| (R_2 + R_3)\right]\frac{R_2}{R_2 + R_3} + V_1\frac{R_3}{R_1 + R_2 + R_3} \tag{3.10}$$

We can verify our calculation by capturing resistors and sources values in Mathcad® while running a dc operating point in SPICE. Figure 3.6 confirms our results.

A second example featuring a dependent source appears in Figure 3.7. A voltage source biases a resistive network and delivers a current I_1. Then a current source I_a biases the intermediate node 2 to which connects a current-controlled voltage source dependent on I_1 via resistor R_2. What is the value of I_1?

$R_1 := 15\Omega$ $R_2 := 38\Omega$ $R_3 := 50\Omega$ $V_1 := 12V$ $I_1 := 2A$ $\|(x,y) := \frac{x \cdot y}{x + y}$

$V_{o1} := I_1 \cdot \left[R_1 \| (R_2 + R_3)\right] \cdot \frac{R_3}{R_2 + R_3} = 14.563V$ $V_{o2} := V_1 \cdot \frac{R_3}{R_1 + R_2 + R_3} = 5.825V$

$V_{out} := V_{o1} + V_{o2} = 20.388V$

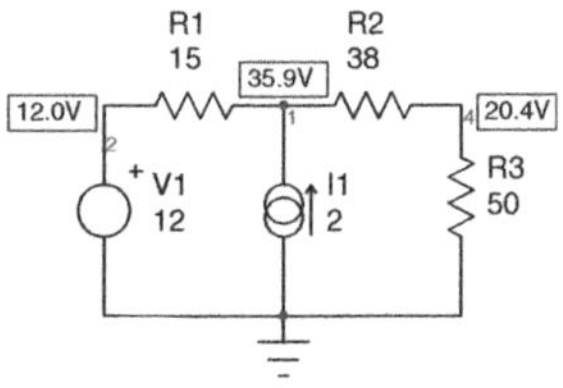

R1 15 | R2 38 | R3 50 | V1 12 | I1 2 | 12.0V | 35.9V | 20.4V

Figure 3.6 A Mathcad® sheet and a SPICE dc operating point tell us if our derivation is correct.

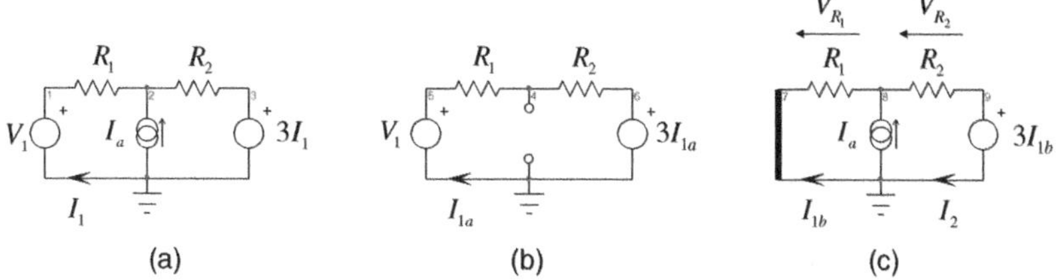

Figure 3.7 In this example, a dependent source delivering $3I_1$ is added to the right side of the sketch.

Applying superposition, we will alternately set the independent sources to 0 while the dependent source is left untouched. In Figure 3.7b, current I_{1a} is equal to the voltage difference across R_1 and R_2 in series divided by the sum of R_1 and R_2:

$$I_{1a} = \frac{V_1 - 3I_{1a}}{R_1 + R_2} \tag{3.11}$$

Developing and factoring, we obtain the first definition for I_{1a}:

$$I_{1a} = \frac{V_1}{R_1 + R_2 + 3} \tag{3.12}$$

Now, we set the voltage source V_1 to 0 (short its output terminals) as in Figure 3.7c and we solve a simple system of equations involving I_1 and I_2. The current in R_2 equals the voltage across its terminals, V_{R_2}, divided by R_2:

$$I_2 = \frac{V_{R_2}}{R_2} = \frac{-V_{R_1} - 3I_{1b}}{R_2} = \frac{-R_1 I_{1b} - 3I_{1b}}{R_2} = -\frac{I_{1b}(R_1 + 3)}{R_2} \tag{3.13}$$

Current I_2 is also equal to the sum of I_a and I_{1b}:

$$I_2 = I_a + I_{1b} \tag{3.14}$$

Equating (3.13) and (3.14), then factoring and rearranging I_1, we have our second current definition labeled I_{1b}:

$$I_{1b} = -\frac{I_a}{1 + \dfrac{R_1 + 3}{R_2}} = -\frac{R_2 I_a}{R_1 + R_2 + 3} \tag{3.15}$$

The final expression for the current I_1 is the sum of I_{1a} and I_{1b}:

$$I_1 = I_{1a} + I_{1b} = \frac{V_1}{R_1 + R_2 + 3} - \frac{R_2 I_a}{R_1 + R_2 + 3} = \frac{V_1 - R_2 I_a}{R_1 + R_2 + 3} \tag{3.16}$$

To verify this expression, we have captured resistors and sources values in a Mathcad® sheet and compared the delivered results with a SPICE dc point analysis. Please note that the factor three in the current-controlled source expression has the dimension of ohms. Figure 3.8 shows that our calculations are correct.

We won't go further with examples as the web abounds with tutorials and exercises on the subject (see [1] and [2] for instance). The link given in [2] is interesting. The second superposition example comes from this paper. The author revisits the theorem claiming that all sources in a circuit analyzed with superposition can be zeroed if certain conditions are met, dependent sources included. He exercises his findings on various examples which nicely support his theory.

$R_1 := 10\Omega \qquad R_2 := 25\Omega \qquad V_1 := 24V \qquad I_a := 7A \qquad \text{II}(x,y) := \dfrac{x \cdot y}{x+y}$

$I_{1a} := \dfrac{V_1}{R_1 + R_2 + 3\Omega} = 0.632A \qquad I_{1b} := -\dfrac{I_a}{1 + \dfrac{R_1 + 3\Omega}{R_2}} = -4.605A$

$I_1 := I_{1a} + I_{1b} = 3.974A$

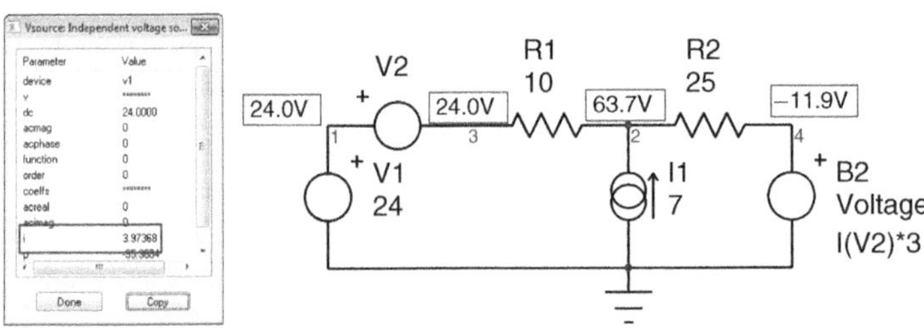

Figure 3.8 A Mathcad® sheet and a quick dc operating point confirm that our derivation is correct.

3.1.1 A Two-input/Two-output System

In Figure 3.1, our system features multiple inputs but only one output. Let's add a second output y_2 as shown in Figure 3.9. Responses formed at outputs y_1 and y_2 depend on stimuli u_1 and u_2. As we did with the 2-input/1-output system, we alternately set u_1 and u_2 to zero to unveil what y_1 and y_2 are made of. This is what Figure 3.10 shows.

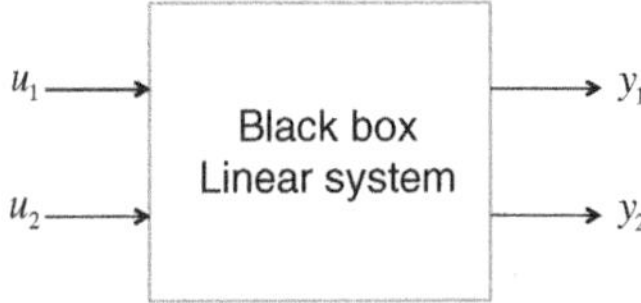

Figure 3.9 A second output is added to the box which makes it a 2-input/2-output network.

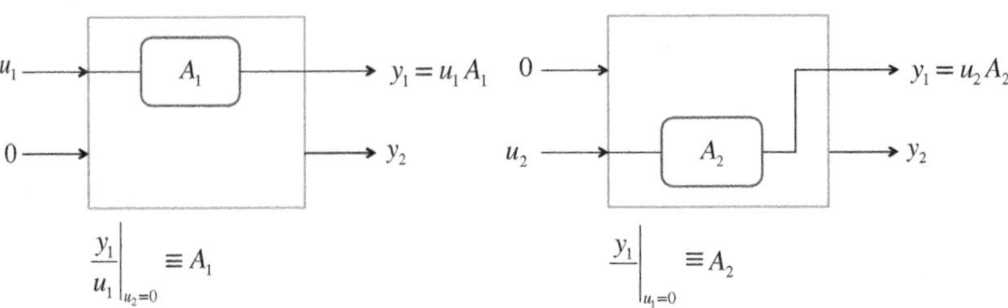

Figure 3.10 Output y_1 is first derived when u_1 and u_2 are alternately set to 0.

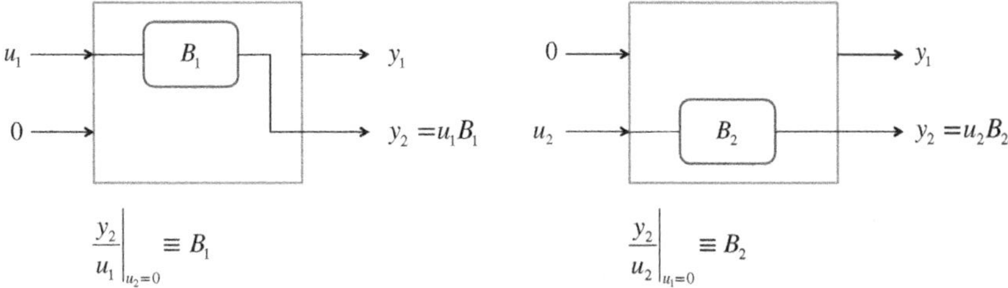

Figure 3.11 Output y_2 is then obtained for u_1 and u_2 alternately set to 0.

The first configuration gives the definition for A_1:

$$\left.\frac{y_1}{u_1}\right|_{u_2=0} \equiv A_1 \tag{3.17}$$

while the second test leads to the definition for A_2:

$$\left.\frac{y_1}{u_2}\right|_{u_1=0} \equiv A_2 \tag{3.18}$$

A similar exercise is carried out but this time looking at the second output, y_2. The configurations are shown in Figure 3.11.

From these drawings, we obtain the definitions for the gains B_1 and B_2:

$$\left.\frac{y_2}{u_1}\right|_{u_2=0} \equiv B_1 \tag{3.19}$$

$$\left.\frac{y_2}{u_2}\right|_{u_1=0} \equiv B_2 \tag{3.20}$$

By rearranging these results in a graphical form, we have Figure 3.12 in which all paths leading to y_1 and y_2 are well identified. In a 2-input system, the superposition theorem considers one input active at a time while the other is set to 0. We will now consider *two* inputs active at the same time, hence the term already encountered of *double injection*. Considering y_1 and y_2 linearly combining u_1 and u_2, they can take any value. The interesting situation, however, occurs when one of the two outputs is purposely driven to 0 by the combined action of u_1 and u_2. This is the *null double injection* or NDI

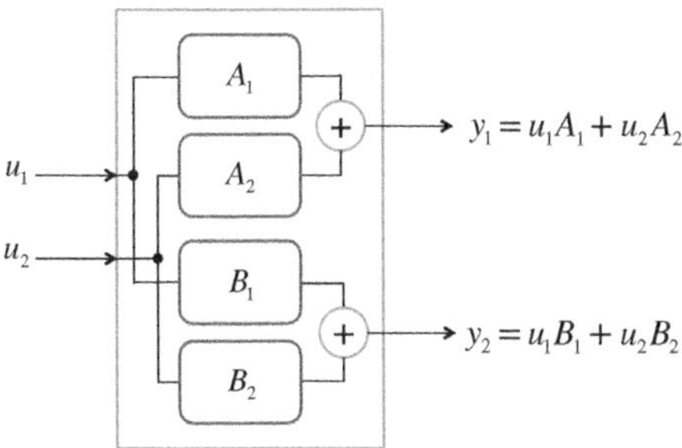

Figure 3.12 We can combine Figure 3.10 and Figure 3.11 results to finalize the content of our black box.

already seen when identifying zeros in the network. Let's see how we can null one of the two outputs, y_1, by adjusting u_1 and u_2. We have seen that:

$$y_1 = u_1A_1 + u_2A_2 \tag{3.21}$$

and

$$y_2 = u_1B_1 + u_2B_2 \tag{3.22}$$

If we null y_1, then the above equations can be re-written considering u_1 and u_2 nulling y_1:

$$0 = u_1|_{y_1=0}A_1 + u_2|_{y_1=0}A_2 \tag{3.23}$$

Output 2 also needs to be rewritten given u_1 and u_2 adjusted to null y_1:

$$y_2|_{y_1=0} = u_1|_{y_1=0}B_1 + u_2|_{y_1=0}B_2 \tag{3.24}$$

From (3.23), we can extract the value of u_1 which nulls y_1:

$$u_1|_{y_1=0} = -u_2|_{y_1=0}\frac{A_2}{A_1} \tag{3.25}$$

This expression can be substituted in (3.24) which now becomes:

$$y_2|_{y_1=0} = u_2|_{y_1=0}B_2 - u_2|_{y_1=0}\frac{A_2}{A_1}B_1 \tag{3.26}$$

Factoring $u_2|_{y_1=0}$, we have:

$$y_2|_{y_1=0} = u_2|_{y_1=0}\left[\frac{A_1B_2 - A_2B_1}{A_1}\right] \tag{3.27}$$

Rearranging in the form of a transfer function linking y_2 to u_2 when y_1 is at a null, we obtain:

$$\left.\frac{y_2}{u_2}\right|_{y_1=0} = \left[\frac{A_1B_2 - A_2B_1}{A_1}\right] \tag{3.28}$$

Please note that this definition is different than the one given by (3.20) which corresponds to input u_1 set to 0. Here, we talk about the same ratio, but obtained for a null in output y_1. Practically, for a 2-input/2-output system for which the internal gains A and B have been identified, I can obtain the gain linking y_2 to u_2 by nulling y_1 and applying (3.28) without actually measuring y_2/u_2.

At this point, let's conduct a simple experiment to see if these expressions are physically meaningful. Figure 3.13 shows what our original black box could look like with different values for A and B. Let' see if we can calculate bias values for u_1 and u_2 to null y_1 as indicated before. From (3.23) we extract the ratio of u_1 to u_2 when y_1 is nulled:

$$\left.\frac{u_1}{u_2}\right|_{y_1=0} = -\frac{A_2}{A_1} \tag{3.29}$$

leading to what we have found in (3.25):

$$u_1|_{y_1=0} = -\frac{A_2}{A_1}u_2|_{y_1=0} \tag{3.30}$$

Applying Figure 3.13 values

$$u_1|_{y_1=0} = -\frac{2}{5}u_2|_{y_1=0} = -0.4u_2|_{y_1=0} \tag{3.31}$$

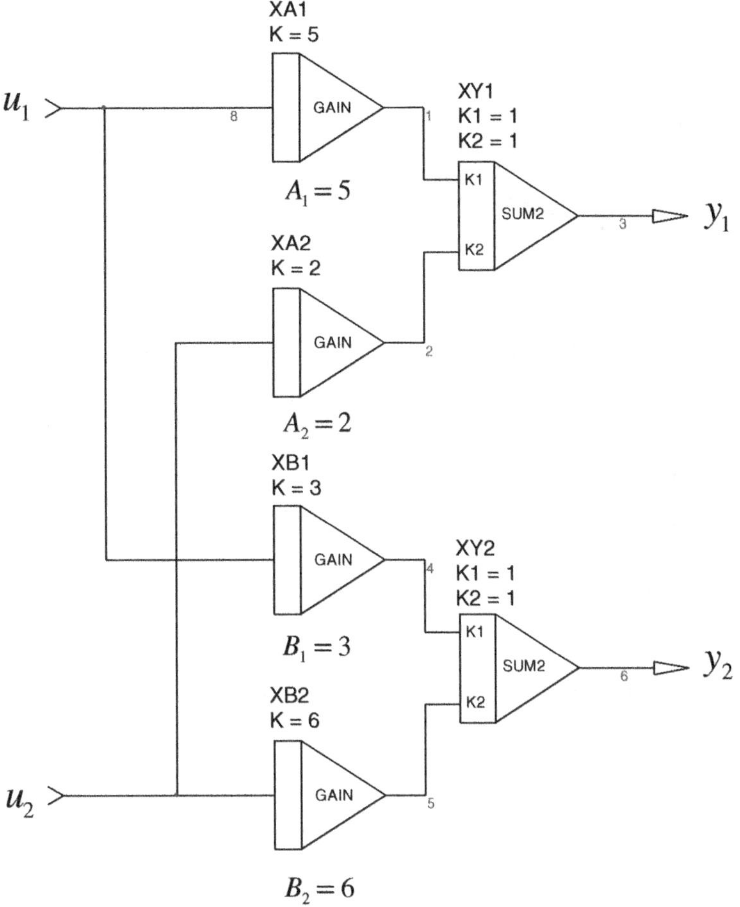

Figure 3.13 Our black box featuring various A and B gains.

Similarly, from (3.28), we can calculate the value of output 2 for u_1 and u_2 nulling y_1:

$$y_2|_{y_1=0} = \left[\frac{A_1B_2 - A_2B_1}{A_1}\right] u_2|_{y_1=0} = \frac{5 \times 6 - 2 \times 3}{5} u_2|_{y_1=0} = 4.8u_2|_{y_1=0} \tag{3.32}$$

Assume we arbitrarily bias u_2 to 5 V, what is the voltage applied to u_1 to impose a null on y_1? (3.30) gives us the value:

$$u_1|_{y_1=0} = -0.4u_2|_{y_1=0} = -0.4 \times 5 = -2\text{ V} \tag{3.33}$$

Then, when y_1 is at a null, what is the value of y_2? (3.32) helps us find:

$$y_2|_{y_1=0} = 4.8u_2|_{y_1=0} = 4.8 \times 5 = 24\text{ V} \tag{3.34}$$

If we sweep u_1 while u_2 is fixed to 5 V, we obtain the graph from Figure 3.14. As confirmed by the figure, for a −2-V bias applied to u_1, y_1 delivers 0 V and y_2 24 V. For this combination, the gain linking

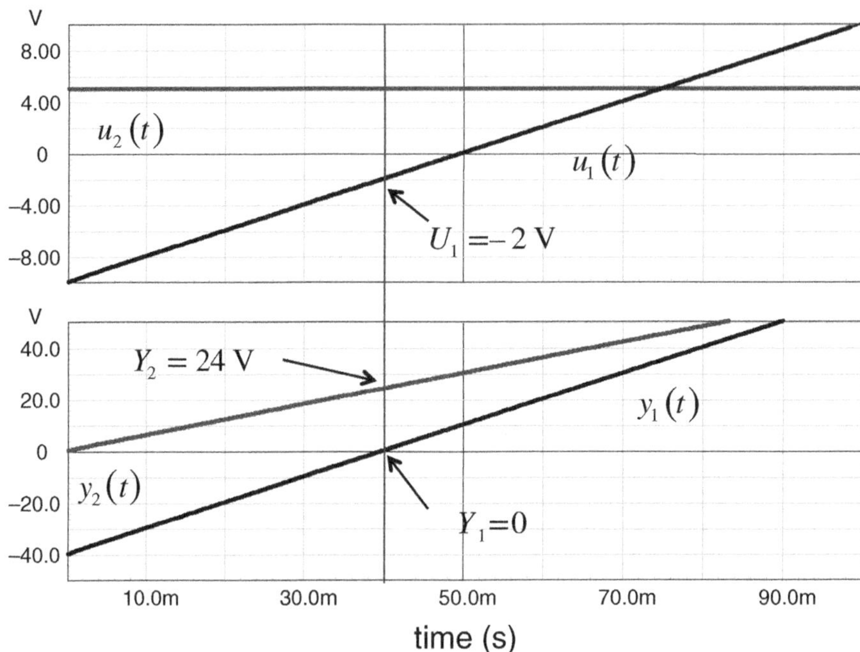

Figure 3.14 For a certain combination of u_1 and u_2, output y_1 delivers 0 V. In this example, I have swept dc sources to null output Y_1.

y_2 to u_2 is given by (3.28) and is equal to:

$$\left.\frac{y_2}{u_2}\right|_{y_1=0} = \frac{5\times 6 - 2\times 3}{5} = \frac{24}{5} = 4.8 \quad (3.35)$$

This is what Figure 3.14 gives for a 24-V output obtained from a 5-V bias.

This experiment also works in ac and it was part of the laboratory exercise Dr. Middlebrook described in his founding paper [3]. The simulation setup appears in Figure 3.15.

Node V_1 is a sinusoidal source featuring a 20-V peak amplitude. V_2 is the same as V_1 but phase-shifted by 180°. Then node *ramp* is a sawtooth ramping from 0 to 1 V over 100 ms. When V_1 is multiplied by V_{ramp} in source BU1, you obtain a sinusoidal waveform linearly ramping from 0 V to 20 V peak in 100 ms. When V_2 (constant amplitude) is applied as u_2 while u_1 is the ramping waveform, y_1 suddenly goes through a null as shown in Figure 3.16. The null in y_1 is obtained when U_1 reaches 8 V peak as indicated by (3.30): $20\times 0.4 = 8$ V. In this condition, (3.32) tells that y_2 reaches $4.8\times 20 = 96$ V. These two examples show how you can null one output in a two-input/two-output system when two signals are simultaneously injected.

Why are we interested in nulling an output? Because the null in the response is the mathematical manifestation of the zero when the transformed network is observed at $s = s_z$. We will see in a few lines how this interesting characteristic is exploited by the Extra Element Theorem.

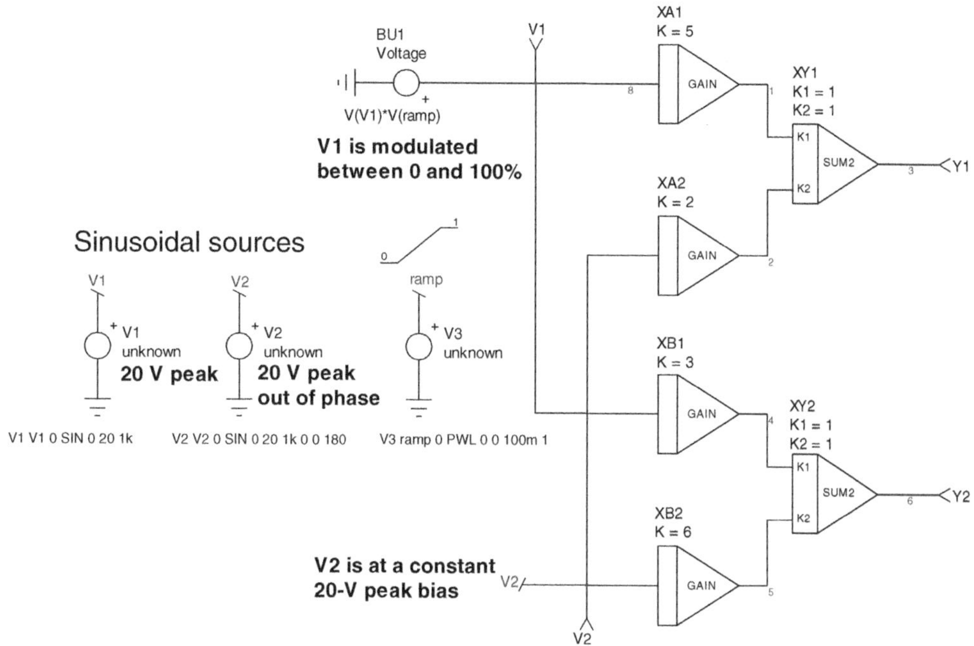

Figure 3.15 The black box structure inputs will now be ac-swept and outputs will be recorded.

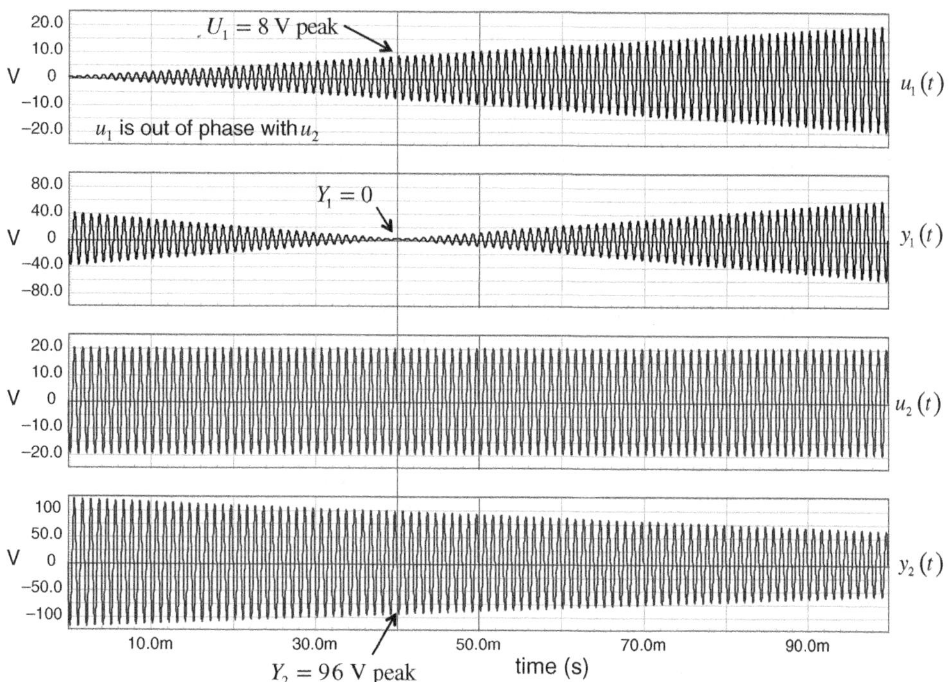

Figure 3.16 For a certain combination of input voltages, output y_1 goes to a null value.

3.2 The Extra Element Theorem

So far, u_1 and u_2 were stimuli but we did not define the type of excitation sources they were, voltage or current. Now assume source u_2 is a current generator i and the variable delivered at output y_2 is a voltage v. We can redraw Figure 3.12 into an updated figure appearing in Figure 3.17.

From that circuit, we can write the two output equations:

$$y_1 = u_1 A_1 + iA_2 \tag{3.36}$$

and

$$v = u_1 B_1 + iB_2 \tag{3.37}$$

If we consider that u_1 and i generate a null in y_1, then (3.36) returns 0 for this particular combination. If you extract u_1 in (3.36), substitute it into (3.37) then rearrange inthe form of a ratio v/i, you should obtain what we have derived in (3.28):

$$\left.\frac{v}{i}\right|_{y_1=0} = \frac{A_1 B_2 - A_2 B_1}{A_1} \tag{3.38}$$

This expression, a transfer function linking an excitation current i to a voltage response v, is actually a transimpedance as defined in Chapter 1. It is not an impedance because v and i are not measured at common port terminals but at two different places. A second expression can be derived from (3.37), it is when u_1, the excitation, is set to 0. In this case, we obtain another transimpedance definition for the ratio i/v following the form of (3.20):

$$\left.\frac{v}{i}\right|_{u_1=0} \equiv B_2 \tag{3.39}$$

In the above steps, we have considered an input and an output physically separated, hence the transimpedance definition. However, the technique we have used so far makes use of impedances or resistances seen from an observation port in which i and v are measured at the same location. How could we modify Figure 3.17 drawing to unveil impedances or resistances instead? Simply change the location of the second output y_2 and make it the voltage v generated across the current generator i. The updated drawing becomes that of Figure 3.18.

The above sketch is a theoretical representation of equations (3.38) and (3.39). In reality, the port at which the current generator is connected – it now becomes clearer – corresponds to an element's terminals, the *extra element*. This is what the updated drawing in Figure 3.19 illustrates.

In this configuration, as i and v are observed at the same physical location – the current source's connecting terminals – our definitions in (3.38) and (3.39) become values interpreted as *driving point*

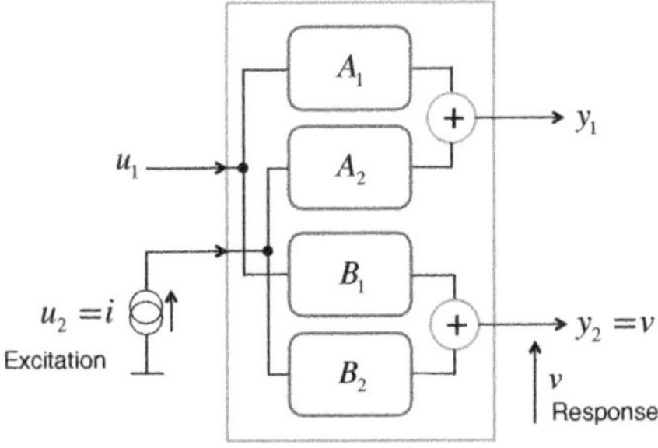

Figure 3.17 Input u_2 is now defined as a current source.

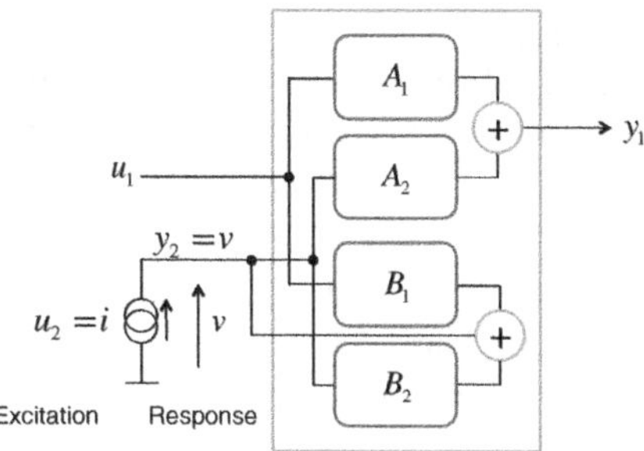

Figure 3.18 Output y_2 is now considered as the voltage generated across the current generator terminals: the ratio i/v becomes an impedance.

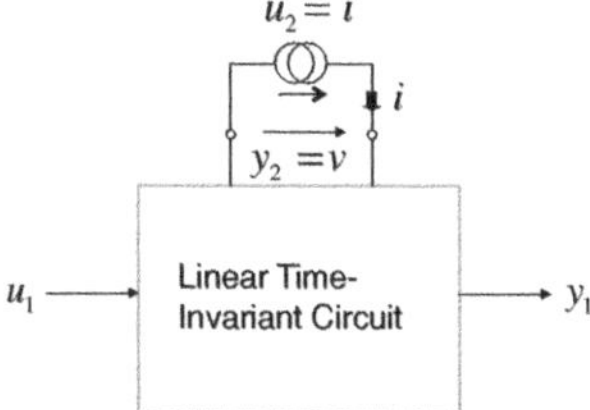

Figure 3.19 The current and voltage labeled as i and v are those observed at a component port.

impedances (DPI, labeled Z_{DP}). Because (3.38) is obtained when y_1 is nulled, it is called Z_n while (3.39) being obtained at a zeroed excitation, it is labeled Z_d. The 'n' stands for numerator – when the numerator is 0, the function returns a null, $y_1 = 0$ – and the 'd' stands for denominator whose coefficients are obtained for a zeroed excitation, $u_1 = 0$. Taking advantage of these notations, we can write:

$$\left.\frac{v}{i}\right|_{y_1=0} = Z_{DP}\big|_{y_1=0} \equiv Z_n = \frac{A_1B_2 - A_2B_1}{A_1} \tag{3.40}$$

$$\left.\frac{v}{i}\right|_{u_1=0} = Z_{DP}\big|_{u_1=0} \equiv Z_d = B_2 \tag{3.41}$$

Rather than considering a constant current generator, replace the source by an impedance Z while i and v are kept in the same direction for this exercise. The new circuit appears in Figure 3.20. In this

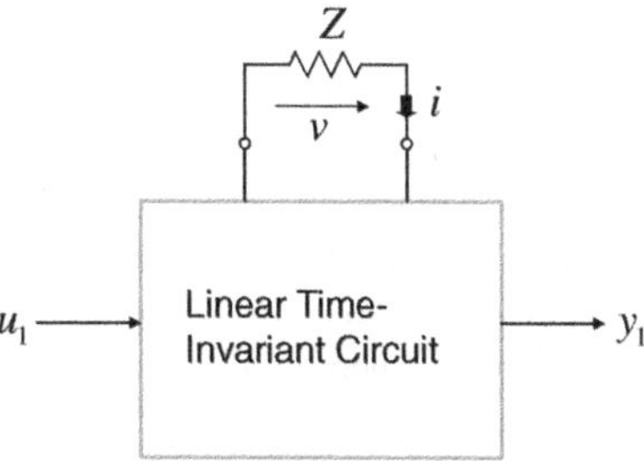

Figure 3.20 An impedance Z is installed in lieu of the current generator. Still i and v from Figure 3.19 remain present but are collinear in this representation.

representation, the current i is no longer forced by the external current generator but depends on the voltage v applied over the impedance Z. Mathematically, the relationships derived in (3.21) and (3.22) are still valid and the relationship between y_2 and u_2 now imposed by Z translates to:

$$i = -\frac{v}{Z} \tag{3.42}$$

The negative sign comes from the fact that i and v are taken in the same direction to agree with Figure 3.18 voltage and current orientations.

With this new definition on hand, (3.36) can be updated as:

$$y_1 = u_1 A_1 - \frac{A_2}{Z} v \tag{3.43}$$

From which the voltage v can be extracted:

$$v = \frac{u_1 A_1 - y_1}{A_2} Z \tag{3.44}$$

(3.44) and (3.42) are now substituted in (3.37) to form the following equation:

$$\frac{A_1 u_1 Z}{A_2} - \frac{y_1 Z}{A_2} = B_1 u_1 + \frac{B_2 y_1}{A_2} - \frac{A_1 B_2 u_1}{A_2} \tag{3.45}$$

Collect terms on both sides

$$u_1 \left(\frac{A_1}{A_2} Z + \frac{A_1 B_2}{A_2} - B_1 \right) = y_1 \left(\frac{Z}{A_2} + \frac{B_2}{A_2} \right) \tag{3.46}$$

and define the ratio y_1/u_1:

$$\frac{y_1}{u_1} = \frac{\frac{A_1}{A_2} Z + \frac{A_1 B_2 - B_1 A_2}{A_2}}{\frac{Z}{A_2} + \frac{B_2}{A_2}} \tag{3.47}$$

Factor $\frac{A_1 Z}{A_2}$ to obtain:

$$\frac{y_1}{u_1} = \frac{\frac{A_1}{A_2} Z \left(1 + \frac{A_2}{A_1 Z} \frac{A_1 B_2 - A_2 B_1}{A_2} \right)}{\frac{Z}{A_2} \left(1 + \frac{A_2}{Z} \frac{B_2}{A_2} \right)} \tag{3.48}$$

Simplify by $\frac{Z}{A_2}$ in the numerator and the denominator to obtain:

$$\frac{y_1}{u_1} = A_1 \frac{1 + \frac{1}{Z} \frac{A_1 B_2 - A_2 B_1}{A_1}}{1 + \frac{1}{Z} B_2} \tag{3.49}$$

In this expression, the term $\frac{A_1 B_2 - A_2 B_1}{A_1}$ is defined by (3.40): it is Z_n, the impedance seen from the injection port when the output y_1 is a null. In the denominator, the term B_2 is simply the impedance seen from the injection port when input u_1 is set to 0: it is Z_d, defined by (3.41). Rewriting (3.49) with these definitions leads to the Extra Element Theorem definition:

$$\frac{y_1}{u_1} = A_1 \frac{1 + \frac{Z_n}{Z}}{1 + \frac{Z_d}{Z}} \tag{3.50}$$

When Z is physically removed from the circuit, it is set to infinity in (3.50). In this particular condition, the transfer function becomes:

$$\left.\frac{y_1}{u_1}\right|_{Z\to\infty} = A_1 \frac{1+\frac{Z_n}{\infty}}{1+\frac{Z_d}{\infty}} = A_1 \tag{3.51}$$

If we designate the transfer function y_1/u_1 by the label A whose value depends on the *extra element* Z, then using (3.51), we can reformulate (3.50) in a more formal way:

$$A|_Z = A|_{Z=\infty} \frac{1+\frac{Z_n}{Z}}{1+\frac{Z_d}{Z}} \tag{3.52}$$

This theorem tells us that the gain of a linear system considering an extra element Z is equal to the gain of the system with the element Z physically disconnected further multiplied by a correction factor involving the extra element Z and two impedances seen from the extra element port when the output is nulled (Z_n) and the excitation is zeroed (Z_d).

The expression in (3.52) can be rearranged differently if we factor $\frac{Z_n}{Z}$ and $\frac{Z_d}{Z}$ respectively in the numerator and the denominator:

$$A|_Z = A|_{Z=\infty} \frac{\frac{Z_n}{Z}\left(\frac{Z}{Z_n}+1\right)}{\frac{Z_d}{Z}\left(\frac{Z}{Z_d}+1\right)} = \left(A|_{Z=\infty}\frac{Z_n}{Z_d}\right)\frac{1+\frac{Z}{Z_n}}{1+\frac{Z}{Z_d}} \tag{3.53}$$

In this new formula, Z lies in numerator and can be set to 0 instead of infinity as in (3.51):

$$A|_{Z=0} = \left(A|_{Z=\infty}\frac{Z_n}{Z_d}\right)\frac{\frac{0}{Z_n}+1}{\frac{0}{Z_d}+1} = A|_{Z=\infty}\frac{Z_n}{Z_d} \tag{3.54}$$

Capitalizing on this expression, $A|_{Z=\infty}\frac{Z_n}{Z_d}$ in (3.53) can simply be replaced by $A|_{Z=0}$ to form the second definition of the Extra Element Theorem:

$$A|_Z = A|_{Z=0} \frac{1+\frac{Z}{Z_n}}{1+\frac{Z}{Z_d}} \tag{3.55}$$

This second form tells us that the gain of a linear system considering an extra element Z is equal to the gain of the system with the element Z physically shorted further multiplied by a correction factor involving the extra element Z and two impedances seen from the extra element port when the output is nulled (Z_n) and the excitation is zeroed (Z_d).

Applying the EET to a 1st-order circuit implies the following steps:

1. Identify the extra element Z. It can be an energy-storage element L or C but also a resistor R. The EET also works for dependent sources but this option will not be explored here. The extra element is usually selected as the element that 'annoys' you, with which, the transfer function becomes more complex to solve.

2. Decide whether you can short or remove the extra element. In some cases, if you remove the element, the transfer function may go to zero and (3.52) cannot be applied. It is the case for circuits having a zero at the origin for instance. Short it instead and use (3.55). Once the energy-storing element is put in its *reference state* (open or shorted), calculate the leading term $A|_{z=0}$ or $A|_{z=\infty}$. This term is named the *reference gain.*
3. Apply the techniques we have seen in Chapters 1 and 2. Set the excitation source to 0 and evaluate the resistance seen from the port created when the extra element is removed. You have Z_d.
4. Use the Null Double Injection (NDI) to determine the resistance offered by the port created when the extra element is removed and the response is nulled. You have Z_n.
5. If the reference circuit is purely resistive, $Z_d = R_d$ and $Z_n = R_n$ are resistances. The correction factor directly gives the corner frequencies of the circuit under study.

We now have an operating manual for the EET. Let's exercise our new knowledge with a few examples.

3.2.1 The EET at Work on Simple Circuits

Now that we have derived the extra element theorem, we will exercise our skills with a series of 1st-order circuits. The first one appears in Figure 3.21a and represents a resistance bridge. This is a fully resistive circuit, without energy-storing element. The output voltage is observed across resistor R_4. What is the transfer function linking V_{out} to V_{in}? At first glance, the element that complicates the circuit is R_5. A different observer would perhaps pick R_4 (or any other resistance) and make it the extra element. There is no problem, the flow would remain the same. Assume R_5 is selected as the extra element. We set it to infinity and check if the gain exists. The drawing is in Figure 3.21b. String R_1-R_3 plays no role and V_{out} is linked to V_{in} via R_4-R_2 forming a simple voltage divider. The first step to find our reference gain is immediate:

$$\left.\frac{V_{out}}{V_{in}}\right|_{R_5\to\infty} = \frac{R_4}{R_4 + R_2} \tag{3.56}$$

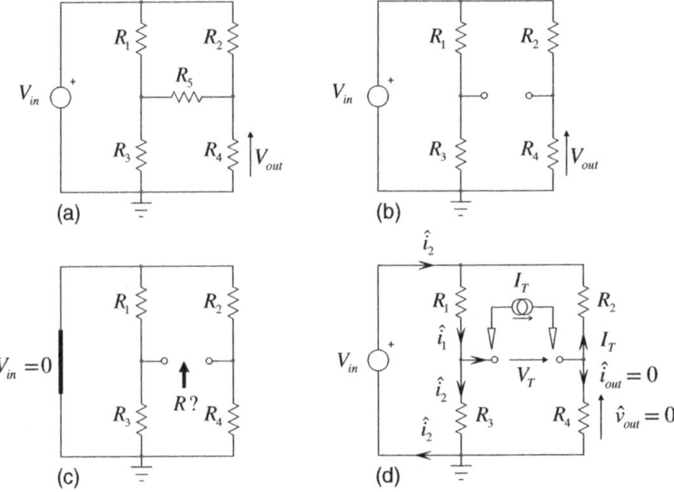

Figure 3.21 This circuit combines resistors to form a divider linking V_{out} to V_{in}. What is the transfer function?

Then, the excitation is set to 0, meaning the V_{in} source is replaced by a short circuit as shown in Figure 3.21c. Folding back R_1/R_2 upper terminals to ground, the resistance seen from R_5 terminals is simply the series arrangement of R_1/R_3 paralleled and R_2/R_4 paralleled:

$$R_d = R_1 \| R_3 + R_2 \| R_4 \tag{3.57}$$

For the final step, we consider that $\hat{v}_{out}$ is nulled while looking into R_5 terminals, determining what resistance we see. The updated schematic appears in Figure 3.21d, V_{in} is back in place for the double injection. As previously noted, associated with the current source I_T, the input source ensures the null in the output. It is never necessary to know V_{in} in order to determine the resistance seen from the considered energy-storing (or resistive) element's terminals. As $\hat{v}_{out}$ is a null, $\hat{i}_{out}$ is also zero. Therefore all the test current I_T flows into R_2 whose lower terminal is at 0 V. The voltage across R_3 is thus minus V_T and current i_2 is simply:

$$\hat{i}_2 = -\frac{V_T}{R_3} \tag{3.58}$$

Current i_1 is the sum of I_T and i_2:

$$\hat{i}_1 = I_T + \hat{i}_2 \tag{3.59}$$

Voltage V_T appears across R_2 and R_1, therefore:

$$V_T = I_T R_2 + \hat{i}_1 R_1 \tag{3.60}$$

We can extract $\hat{i}_1$:

$$\hat{i}_1 = \frac{V_T - R_2 I_T}{R_1} \tag{3.61}$$

Substituting (3.58) and (3.61) into (3.59) gives us:

$$\frac{V_T - R_2 I_T}{R_1} = -\frac{V_T}{R_3} + I_T \tag{3.62}$$

Factoring V_T and I_T leads to:

$$V_T\left(1 + \frac{R_1}{R_3}\right) = I_T(R_1 + R_2) \tag{3.63}$$

The resistance seen from R_5's terminals when V_{out} is a null is thus:

$$R_n = \frac{V_T}{I_T} = \frac{R_1 + R_2}{1 + \frac{R_1}{R_3}} \tag{3.64}$$

The final transfer function is found by assembling (3.56), (3.57) and (3.64):

$$\frac{V_{out}}{V_{in}} = \frac{R_4}{R_4 + R_2} \frac{1 + \frac{\frac{R_1 + R_2}{1 + R_1/R_3}}{R_5}}{1 + \frac{R_1 \| R_3 + R_2 \| R_4}{R_5}} \tag{3.65}$$

This expression has been captured in a Mathcad® sheet and a screenshot appears in Figure 3.22.

$R_1 := 250\Omega$ $R_2 := 12k\Omega$ $R_3 := 18k\Omega$ $R_4 := 150\Omega$ $R_5 := 470\Omega$

$||(x,y) := \frac{x \cdot y}{x+y}$ $V_1 := 3V$

$$H_1 := \frac{R_4}{R_4+R_2} \frac{1+\frac{\frac{R_1+R_2}{1+\frac{R_1}{R_3}}}{R_5}}{1+\frac{R_1 \,||\, R_3 + R_2 \,||\, R_4}{R_5}} = 0.179$$

Figure 3.22 Mathcad® efficiently computes the transfer function involving paralleled elements.

How do we know this is correct? Call SPICE for help, as we did before. Figure 3.23 shows the whole circuit in which all the above steps are simulated and finally assembled to form the last equation (source B_7). Node *TF* displays the original transfer function calculation with all elements in place. It indicates 0.179, what (3.65) returns in Mathcad®. Then the three calculation steps are performed: step 1 is when R_5 is removed, step 2 is when V_{in} is set to 0 and finally, step 3 uses the transconductance amplifier to obtain R_n. The combination of all these steps is given by node *TFEET* which displays 0.179, confirming the path we took. Should you modify source V_4 (V_{in}) in the NDI setup, the injected current through V_5 would adjust to maintain $\hat{v}_{out} = 0$ but the resistance computed by B_4 would remain constant. Make sure the voltage and current measurements around the transconductance amplifier respect the polarity of Figure 3.19. In SPICE, a current flowing in an element (resistor, source and so on) is considered positive when it leaves the considered element by its negative (−) terminal. To obtain the correct resistance sign, please note the polarity of source V_5 in series with G_1 and nodes 14–15 used by B_4 to calculate Z_n.

3.2.2 The EET at Work – Example 2

For the second example, consider an inductor in lieu of R_5. The new circuit appears in Figure 3.24. With this element in place, we have the choice to apply (3.52) or (3.55). If we apply the first definition, we have nothing to do; all the work has already been done! The only thing we need is replace R_5 by the inductor impedance which is sL and the following formula appears:

$$\frac{V_{out}(s)}{V_{in}(s)} = \frac{R_4}{R_4+R_2} \frac{1+\frac{\frac{R_1+R_2}{1+R_1/R_3}}{sL}}{1+\frac{R_1 \| R_3 + R_2 \| R_4}{sL}} \tag{3.66}$$

This is the new transfer function linking V_{out} to V_{in} in Figure 3.24. However, in (3.66), you see that s lies in the denominator and it does not very conveniently match our traditional low-entropy expression:

$$H(s) = H_0 \frac{1+\frac{s}{\omega_z}}{1+\frac{s}{\omega_p}} \tag{3.67}$$

Rather than applying (3.52), let's use (3.55) instead. In this expression, the impedance Z is brought to 0, meaning that L is replaced by a short circuit. Yes, this is the dc transfer function H_0 we have

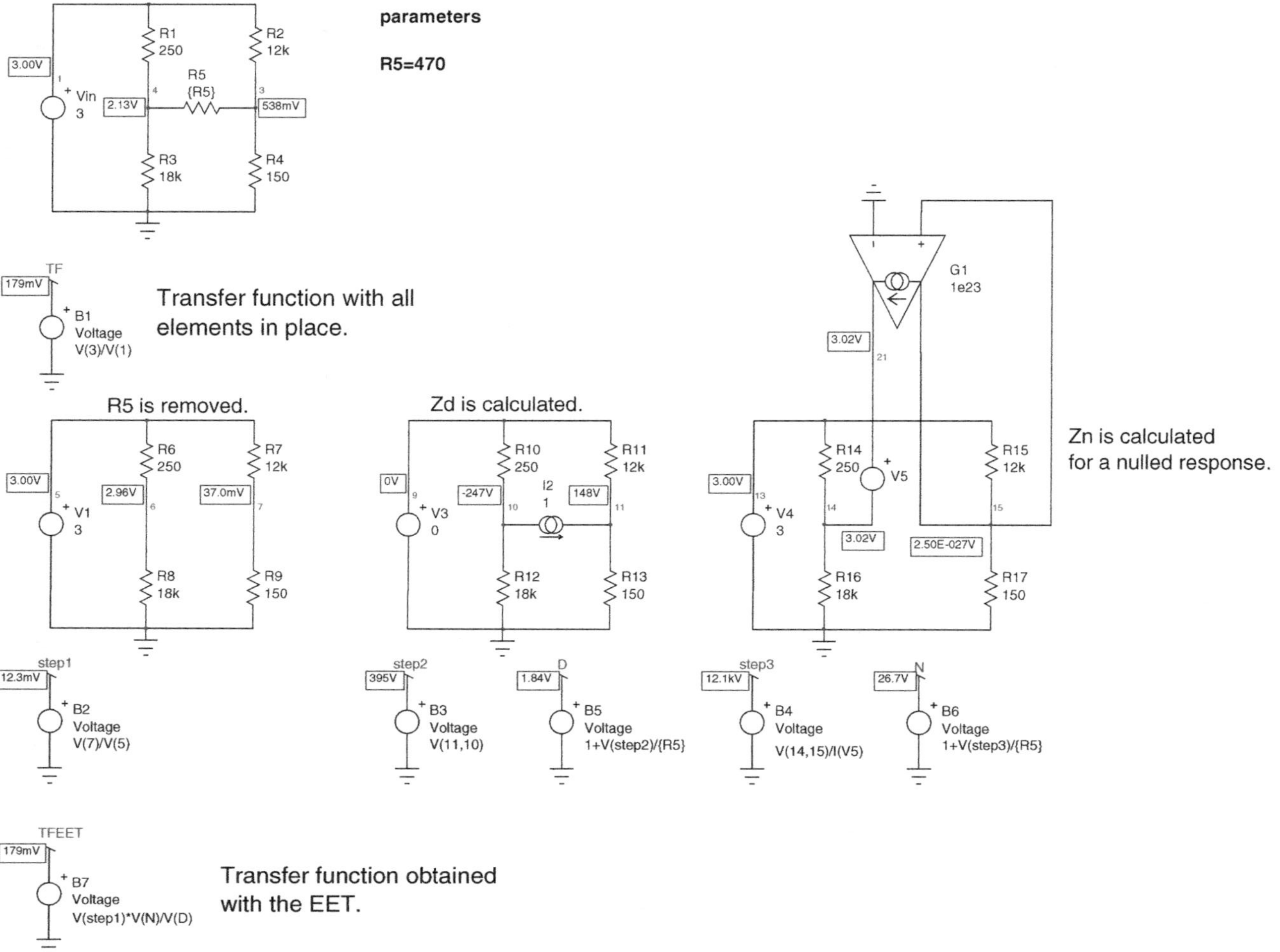

Figure 3.23 A SPICE simulation helps to quickly check your results.

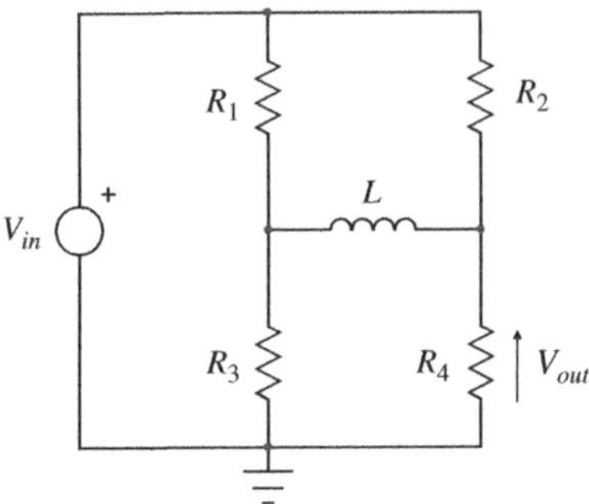

Figure 3.24 An inductor is now replacing resistor R_5.

calculated several times in Chapters 1 and 2. Looking at Figure 3.24, there is no need to write a single line of algebra as we see that, when L is replaced by a short, R_1/R_2 are paralleled as well as R_3 and R_4. The set forms a resistive divider equal to:

$$\left.\frac{V_{out}}{V_{in}}\right|_{L=0} = \frac{R_3 \| R_4}{R_1 \| R_2 + R_3 \| R_4} \tag{3.68}$$

This is it, we already calculated the resistances driving L when the excitation is set to 0 and when the output is a null. All we need is to assemble the pieces as indicated by (3.55) in which Z is simply sL:

$$H(s) = \frac{R_3 \| R_4}{R_1 \| R_2 + R_3 \| R_4} \frac{1 + s\dfrac{\dfrac{L}{R_1 + R_2}}{1 + R_1/R_3}}{1 + s\dfrac{L}{R_1 \| R_3 + R_2 \| R_4}} \tag{3.69}$$

This is the low-entropy form we are familiar with and it fits the format given in (3.67) with:

$$H_0 = \frac{R_3 \| R_4}{R_1 \| R_2 + R_3 \| R_4} \tag{3.70}$$

$$\omega_z = \frac{R_1 + R_2}{(1 + R_1/R_3)L} \tag{3.71}$$

and

$$\omega_p = \frac{R_1 \| R_3 + R_2 \| R_4}{L} \tag{3.72}$$

The two formulas expressed in (3.66) and (3.69) are identical and again, Mathcad® and SPICE agree very well as shown in Figure 3.25 and Figure 3.26. The pole and zero calculated by SPICE respectively through sources B_5/B_6 match the value found in Figure 3.25. V_4 in Figure 3.26 is arbitrarily set to 4 V but any other value would not make a difference as current in V_5 would adjust to maintain the output null.

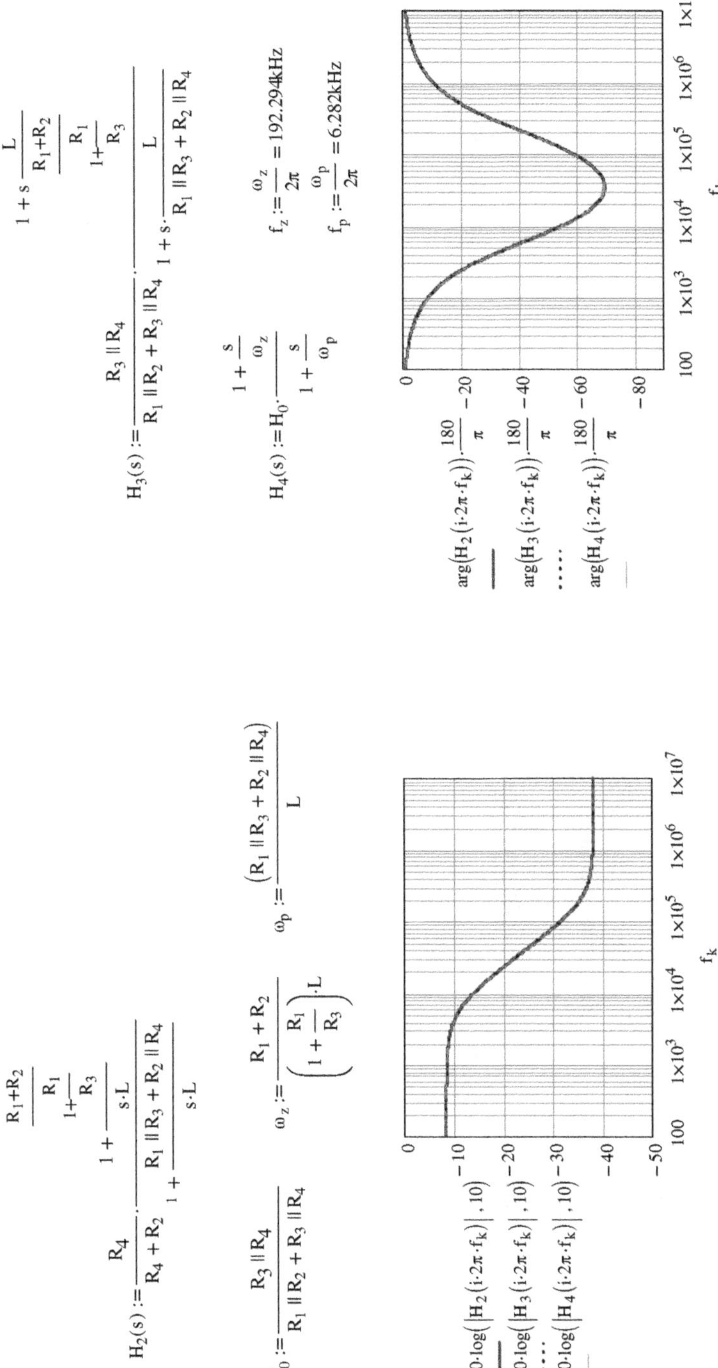

Figure 3.25 Mathcad® confirms that (3.66) and (3.69) are identical.

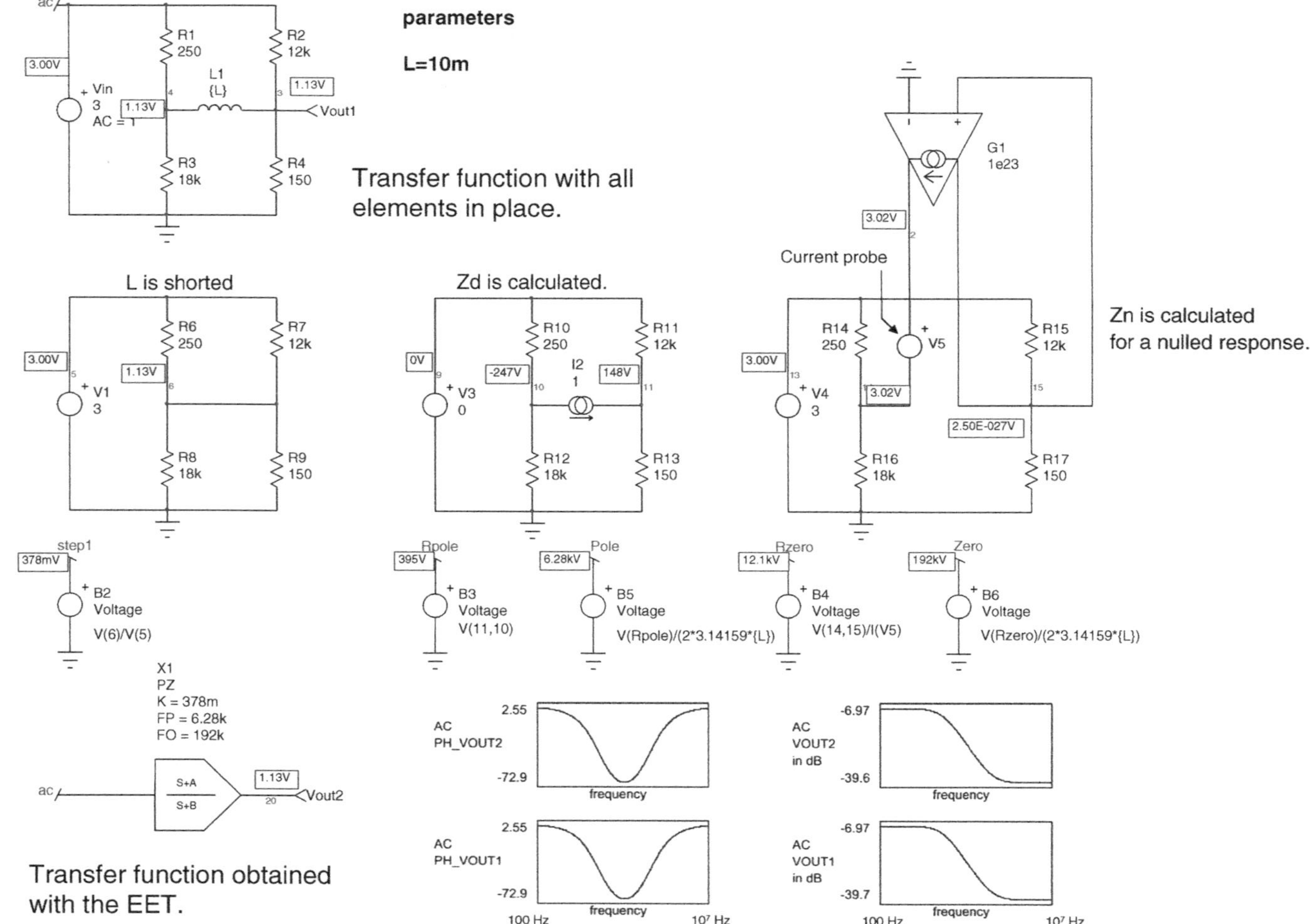

Figure 3.26 SPICE automates the pole and zero calculations to form a transfer function whose response matches the original one.

3.2.3 The EET at Work – Example 3

A third example appears in Figure 3.27a. It is a 1st-order system built around a capacitor. If capacitor C is infinite valued, meaning it is replaced by a short circuit, does the response linking V_{out} to V_{in} still exists? Yes, thus there is a zero associated with C. We can first remove the capacitor and obtain the first transfer function, the reference gain. It is a dc gain, immediately found to be:

$$H|_{Z\to\infty} = H_0 = \frac{R_3}{R_3 + R_1} \tag{3.73}$$

The second step is setting V_{in} to 0 as in Figure 3.27c. Given the circuit simplicity, the resistance seen from the capacitor's terminals is:

$$R_d = R_2 + R_1 \| R_3 \tag{3.74}$$

Finally, setting $\hat{v}_{out}$ to 0 is represented in Figure 3.27d. As no current circulates in R_3, all the I_T current flows in R_1 and returns through R_2. The resistance seen by the capacitor in this mode is simply:

$$R_n = R_1 + R_2 \tag{3.75}$$

The final transfer function $H(s)$ is given by assembling the above equations following (3.52):

$$H(s) = \frac{R_3}{R_3 + R_1}\frac{1 + \dfrac{R_1 + R_2}{1/sC}}{1 + \dfrac{R_2 + R_1\|R_3}{1/sC}} = \frac{R_3}{R_3 + R_1}\frac{1 + s(R_1 + R_2)C}{1 + sC\left[R_2 + R_1\|R_3\right]} \tag{3.76}$$

We can rearrange this transfer function in the classical form:

$$H(s) = H_0\frac{1 + \dfrac{s}{\omega_z}}{1 + \dfrac{s}{\omega_p}} \tag{3.77}$$

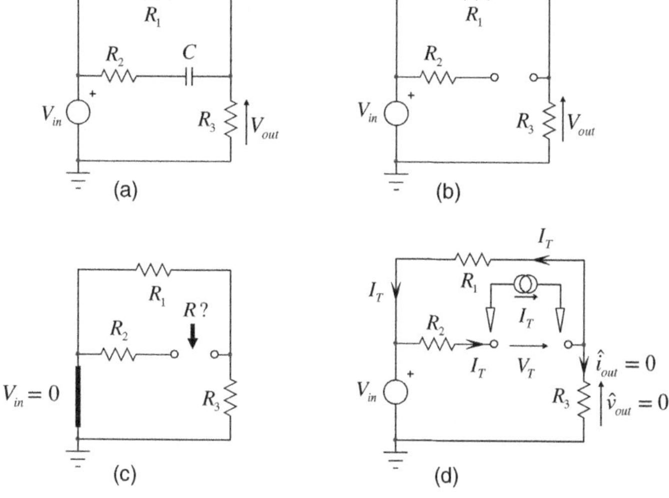

Figure 3.27 The third example involves a capacitor and three resistors.

In which H_0 is defined by (3.73) while the pole and zero are expressed as:

$$\omega_z = \frac{1}{(R_1 + R_2)C} \tag{3.78}$$

$$\omega_p = \frac{1}{C\left[R_2 + R_1 \| R_3\right]} \tag{3.79}$$

3.2.4 The EET at Work – Example 4

The next example is given in Figure 3.28. This is a simple bipolar amplifier having a local feedback brought by resistor R_f. The equivalent small-signal model appears in Figure 3.29a. We will first assume that capacitor C_i is a dc block capacitor and can be considered a short circuit for the analysis. In this configuration, the element that annoys us is resistance R_f. Let's set it to infinity (remove it from the circuit) and calculate the transfer function in this condition shown in Figure 3.29b. The output voltage is observed across R_C in which the collector current flows, βi_b:

$$V_{out} = -\beta i_b R_C \tag{3.80}$$

The input voltage splits between the dynamic input resistance r_π and the emitter resistor:

$$V_{in} = r_\pi i_b + (\beta + 1) i_b R_E = i_b[r_\pi + (\beta + 1)R_E] \tag{3.81}$$

We can extract the base current from this equation:

$$i_b = \frac{V_{in}}{r_\pi + (\beta + 1)R_E} \tag{3.82}$$

and substitute it into (3.80). Factoring V_{out} and V_{in}, we have:

$$\left.\frac{V_{out}}{V_{in}}\right|_{R_f \to \infty} = -\frac{\beta R_C}{r_\pi + (\beta + 1)R_E} \tag{3.83}$$

Now that we have one part of the answer, let's have a look at the resistance R_d seen from R_f's terminals when the input source V_{in} is a short circuit. As indicated by Figure 3.29c, resistor r_π and R_E are shorted to ground. In this condition, there is no base current, $i_b = 0$. The current generator βi_b is thus zero (it disappears from the sketch) and the only resistor that is seen from R_f's terminals in this condition is the collector resistor, R_C:

$$R_d = R_C \tag{3.84}$$

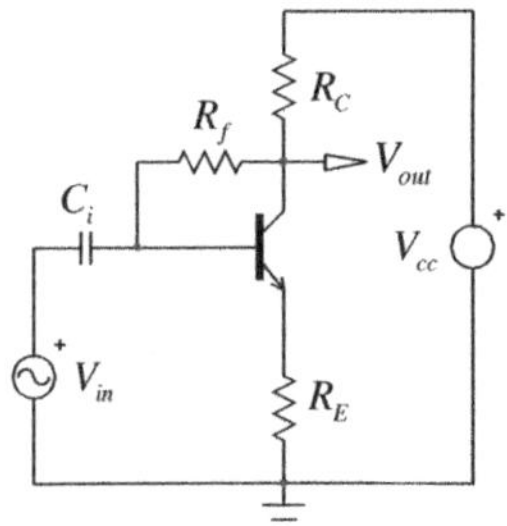

Figure 3.28 A bipolar transistor has a local feedback through resistor R_f. What is the transfer function of this design?

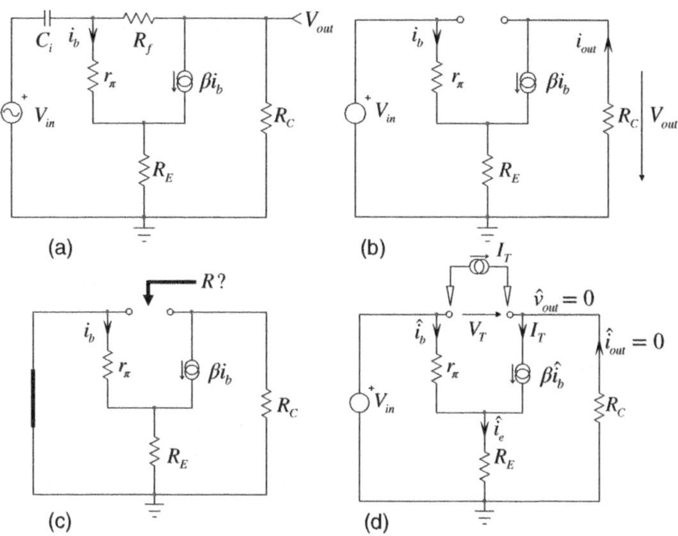

Figure 3.29 The transfer function can be found by applying the Extra Element Theorem in three distinct steps.

The final step is to find the resistance R_n offered by R_f's terminals when the output voltage is a null. The schematic updates to that of Figure 3.29d. If the output voltage is a null, then the output current is also a null. Therefore, all the test current I_T is absorbed by the current source $\beta \hat{i}_b$:

$$I_T = \beta \hat{i}_b \tag{3.85}$$

The base current is therefore:

$$\hat{i}_b = \frac{I_T}{\beta} \tag{3.86}$$

In the emitter resistor, the circulating current is thus the sum of the base current and the test current:

$$\hat{i}_e = \frac{I_T}{\beta} + I_T = I_T \left(\frac{1}{\beta} + 1 \right) \tag{3.87}$$

Since the right terminal of the test generator is at 0 V, then its left terminal is at minus V_T. That voltage is the drop across r_π and R_E:

$$-V_T = r_\pi \frac{I_T}{\beta} + R_E \left(1 + \frac{1}{\beta} \right) I_T \tag{3.88}$$

If we rearrange this expression, we obtain the resistance offered by R_f's terminals while $\hat{v}_{out} = 0$:

$$R_n = \frac{V_T}{I_T} = -\left[\frac{r_\pi}{\beta} + R_E \left(1 + \frac{1}{\beta} \right) \right] \tag{3.89}$$

This is a negative resistance. If a capacitor was connected in place of R_f, then the resulting negative time constant would give a nice Right Half-Plane Zero.

We now have all the pieces and we can compute the transfer function we need using (3.52):

$$\frac{V_{out}}{V_{in}}=-\frac{\beta R_C}{r_\pi+(\beta+1)R_E}\frac{1-\frac{\frac{r_\pi}{\beta}+R_E\left(1+\frac{1}{\beta}\right)}{R_f}}{1+\frac{R_C}{R_f}}=-\frac{\beta}{\beta+1}\frac{R_C}{\frac{r_\pi}{\beta+1}+R_E}\frac{1-\frac{\frac{r_\pi}{\beta}+R_E\left(1+\frac{1}{\beta}\right)}{R_f}}{1+\frac{R_C}{R_f}} \tag{3.90}$$

As usual, we have tested all these individual expressions in SPICE and compared the results with a Mathcad® spreadsheet. The results appear in Figure 3.30 and Figure 3.31, and they perfectly match.

3.2.5 The EET at Work – Example 5

The exercise now consists of incorporating capacitor C_i back into the circuit, as shown by Figure 3.28. After all, considering its impedance a short circuit for the beginning of the exercise, it was already a form of the EET – see (3.55) – and we already derived part of it with (3.90). The full expression including C_i is:

$$\left.\frac{V_{out}(s)}{V_{in}(s)}\right|_{Z_{C_i}}=\left.\frac{V_{out}(s)}{V_{in}(s)}\right|_{Z_{C_i}\to 0}\frac{1+\frac{Z_{C_i}}{R_n}}{1+\frac{Z_{C_i}}{R_d}} \tag{3.91}$$

The time constant involving R_d is found by setting V_{in} to 0 V, leading to Figure 3.32 circuit. With this type of arrangement, unfortunately, finding the resistance which drives capacitor C_i by inspection is not an easy task simply because of the current-controlled source. The solution is found by adding a test generator I_T and finding the voltage across its terminals, V_T. The ratio V_T/I_T is the resistance we look for.

From KCL, we have

$$I_T=\hat{i}_1+\hat{i}_b \tag{3.92}$$

but also

$$I_T+\hat{i}_c=(\beta+1)\hat{i}_b \tag{3.93}$$

The current flowing in R_f is the voltage across its terminals divided by R_f:

$$\hat{i}_1=\frac{V_T+R_C\hat{i}_c}{R_f} \tag{3.94}$$

The base current $\hat{i}_b$ is derived realizing that V_T is applied across the bridge made of r_π and the emitter resistor:

$$V_T=\hat{i}_b r_\pi+(\beta+1)\hat{i}_b R_E \tag{3.95}$$

Extracting the base current gives:

$$\hat{i}_b=\frac{V_T}{r_\pi+(\beta+1)R_E} \tag{3.96}$$

Substituting (3.94) into (3.92) gives:

$$I_T=\frac{V_T+R_C\hat{i}_c}{R_f}+\hat{i}_b \tag{3.97}$$

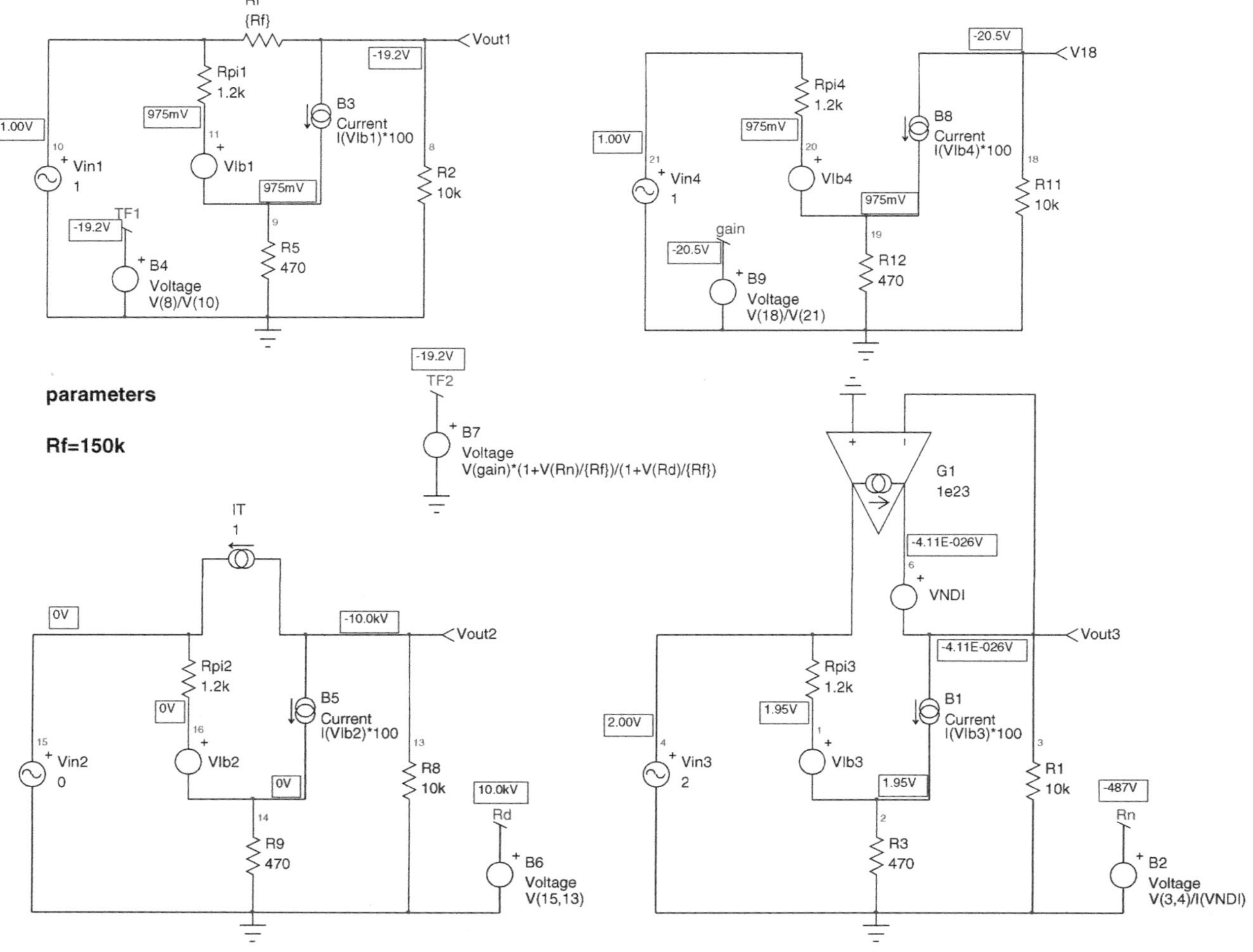

Figure 3.30 The SPICE dc point analysis gives a gain of −19.2 for the complete circuit (node TF1), similar to the EET result in node TF2.

$$\beta := 100 \qquad R_C := 10\text{k}\Omega \qquad r_\pi := 1.2\text{k}\Omega \qquad R_E := 470\Omega \qquad R_f := 150\text{k}\Omega$$

$$R_n := -\left[\frac{r_\pi}{\beta} + R_E \cdot \left(1 + \frac{1}{\beta}\right)\right] = -486.7\Omega \qquad R_d := R_C \qquad H_0 := -\frac{\beta \cdot R_C}{r_\pi + (\beta + 1) \cdot R_E} = -20.547$$

$$G_1 := -\frac{\beta \cdot R_C}{r_\pi + (\beta + 1) \cdot R_E} \cdot \frac{1 - \frac{\frac{r_\pi}{\beta} + R_E \cdot \left(1 + \frac{1}{\beta}\right)}{R_f}}{1 + \frac{R_C}{R_f}} = -19.2$$

Figure 3.31 The Mathcad® results match those obtained by the SPICE dc operating point calculation in Figure 3.30.

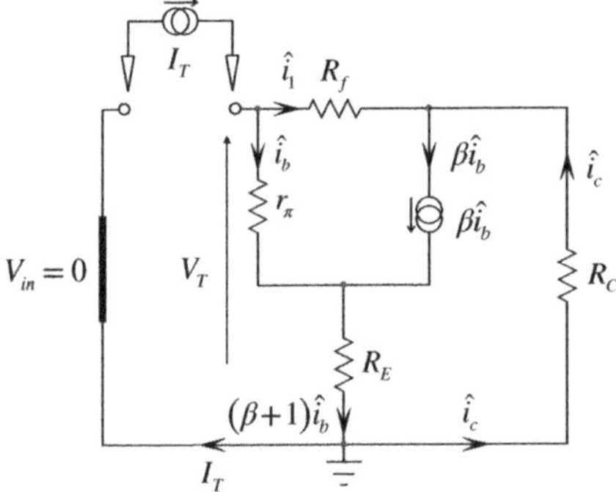

Figure 3.32 The resistance driving the capacitor when the excitation is set to 0 requires KCL and KVL at work because of the current-controlled current source.

Now extracting $\hat{i}_c$ from (3.93) and substituting it into (3.97) leads to:

$$I_T = \frac{V_T + R_C\left[(\beta + 1)\hat{i}_b - I_T\right]}{R_f} + \hat{i}_b \tag{3.98}$$

Substitute in this equation the base current from (3.96), factor, rearrange and you have the resistance R_d seen from the capacitor terminals when V_{in} is set to 0:

$$\frac{V_T}{I_T} = R_d = \frac{(R_C + R_f)(R_E + r_\pi + \beta R_E)}{R_C + R_E + R_f + r_\pi + \beta(R_C + R_E)} \tag{3.99}$$

The second resistance value R_n is obtained when the output is nulled. The new circuit appears in Figure 3.33.

In this circuit, as the output is a null, so is the collector current $\hat{i}_c$:

$$\hat{v}_{out} = 0 \tag{3.100}$$

$$\hat{i}_c = 0 \tag{3.101}$$

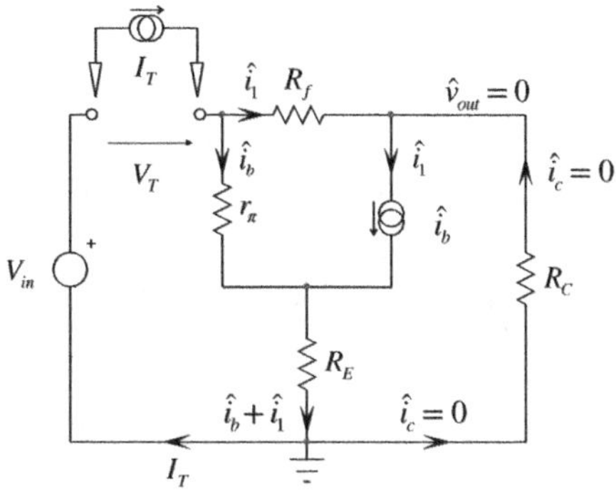

Figure 3.33 In this configuration, the output is nulled, meaning there is no current circulating in the collector resistor R_C.

Given a nulled collector current, the current $\hat{i}_1$ in R_f is $\beta\hat{i}_b$. Given the null in the output, the voltage across R_f is thus equal to the voltage across r_π and the emitter resistor R_E:

$$\beta\hat{i}_b R_f = \hat{i}_b[r_\pi + (\beta + 1)R_E] \tag{3.102}$$

The only practical way to satisfy this equation is to have $\hat{i}_b$ equal to 0. If $\hat{i}_b$ is 0, then $\hat{i}_1$ is also a null. As the test current I_T is the sum of $\hat{i}_b$ and $\hat{i}_1$, then I_T is also equal to zero. As a result, the resistance seen from the capacitor terminals in a null-double injection is:

$$R_n = \left.\frac{V_T}{I_T}\right|_{I_T=0} \to \infty \tag{3.103}$$

This result confirms the zero placed at the origin immediately observed in dc: when C_i is removed from the circuit, there is no output, the gain is 0. We can now assemble all these results to obtain a transfer function following (3.91):

$$\left.\frac{V_{out}}{V_{in}}\right|_{C_i} = -\frac{\beta R_C}{r_\pi + (\beta + 1)R_E}\frac{1 - \dfrac{\dfrac{r_\pi}{\beta} + R_E\left(1 + \dfrac{1}{\beta}\right)}{R_f}}{1 + \dfrac{R_C}{R_f}}\frac{1 + \dfrac{1/sC}{\infty}}{1 + \dfrac{1/sC}{\dfrac{(R_C + R_f)(R_E + r_\pi + \beta R_E)}{R_C + R_E + R_f + r_\pi + \beta(R_C + R_E)}}} \tag{3.104}$$

This expression can be rearranged to fit a friendlier format:

$$\frac{V_{out}}{V_{in}} = H_\infty \frac{1}{1 + \dfrac{\omega_p}{s}} \tag{3.105}$$

in which

$$H_\infty = -\frac{\beta R_C}{r_\pi + (\beta + 1)R_E}\frac{1 - \dfrac{\dfrac{r_\pi}{\beta} + R_E\left(1 + \dfrac{1}{\beta}\right)}{R_f}}{1 + \dfrac{R_C}{R_f}} \tag{3.106}$$

and

$$\omega_p = \frac{1}{C_1 \dfrac{(R_C + R_f)(R_E + r_\pi + \beta R_E)}{R_C + R_E + R_f + r_\pi + \beta(R_C + R_E)}} \tag{3.107}$$

(3.105) combines a zero at the origin plus a pole. Please note its unusual writing form, which differs from (3.77) in which s appears in a different place. This what is called an *inverted pole* and you can jump to the next section for more information on this writing notation.

To check our calculations, we have simulated Figure 3.28 circuit and compared results to what Mathcad® returns. The SPICE circuit is in Figure 3.34. The gain at 1 Hz is −22.189 dB while the 3-dB pole is evaluated at 247 Hz. The high-frequency asymptote is the gain computed by eqn. (3.90). The automated sheet is in Figure 3.35 and formulas return values and graph similar to those given by SPICE.

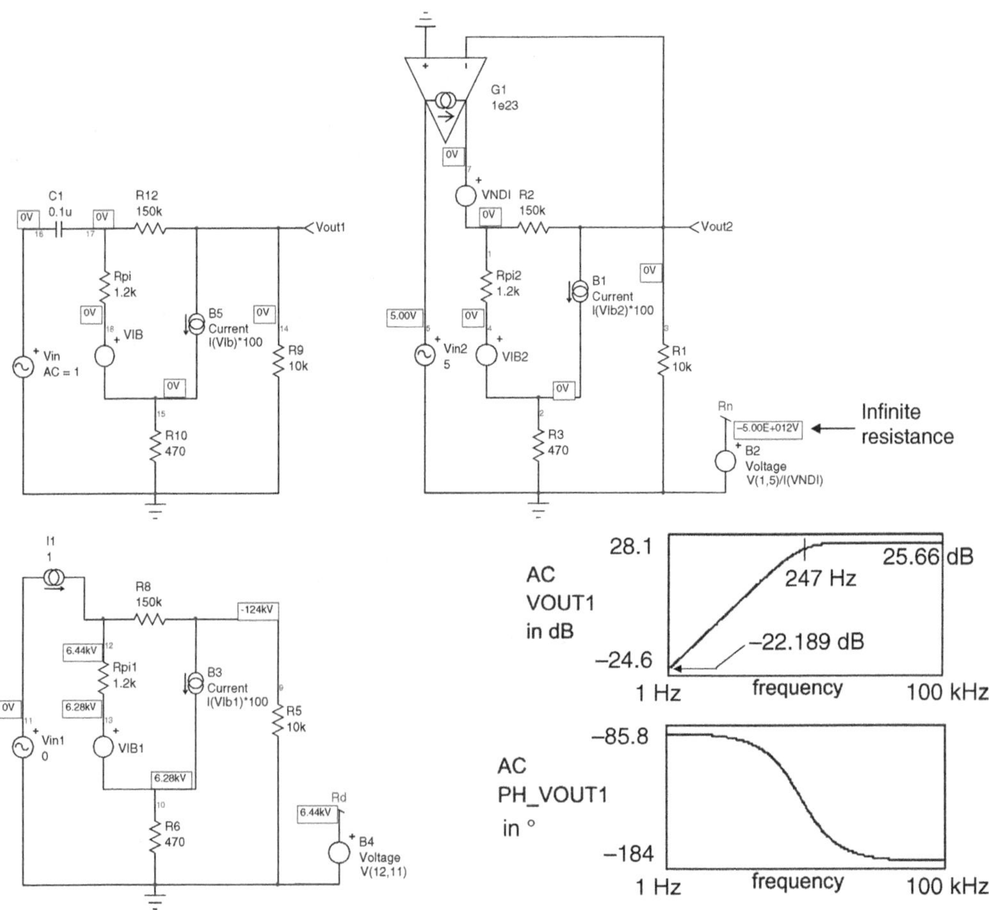

Figure 3.34 Our simple test fixtures help us confirm calculations regarding the various resistances driving the series capacitor. For a zero at the origin, the input impedance is infinite in the nulled-V_{out} configuration.

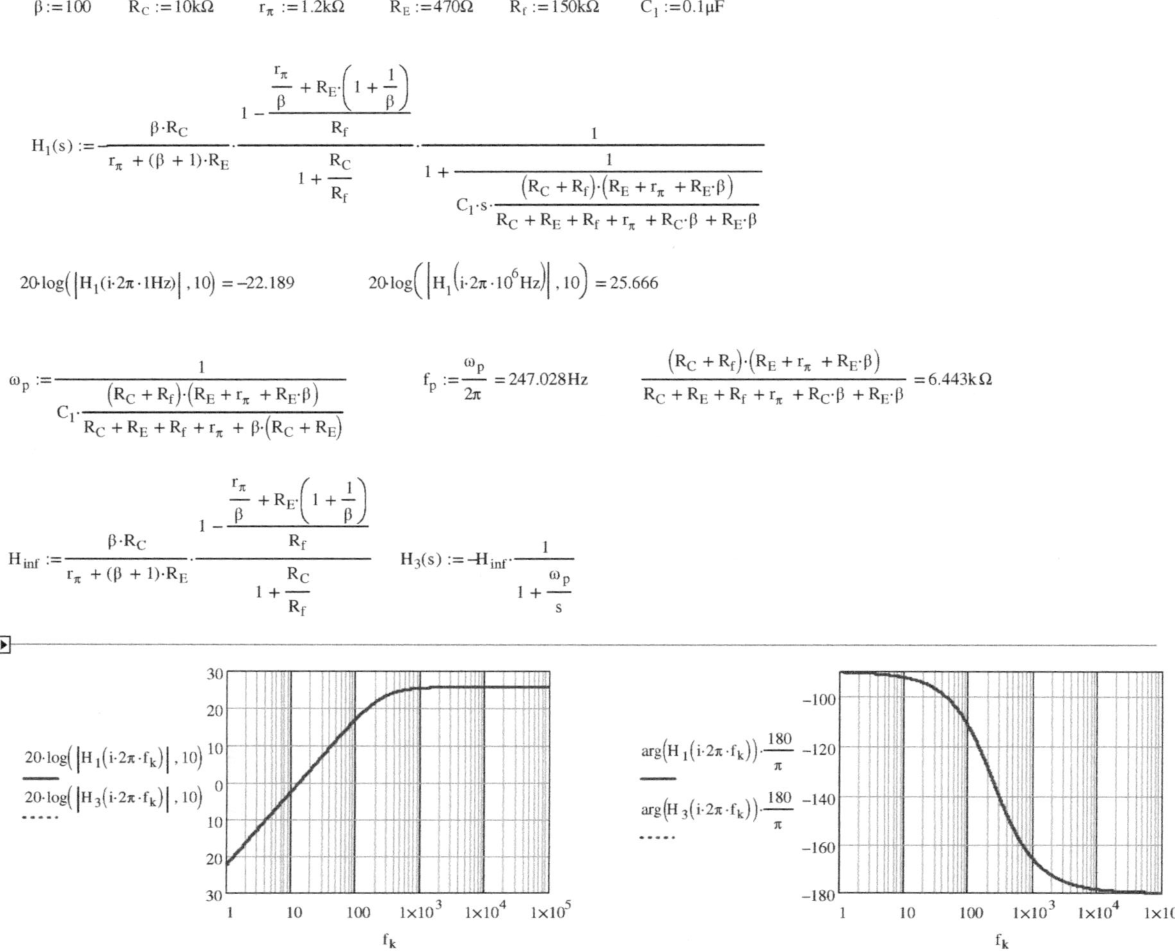

Figure 3.35 Mathcad® results confirm what we have found. Salient points perfectly match SPICE simulation results.

3.2.6 The EET at Work – Example 6

In Chapter 1 problems, we proposed a 1st-order circuit featuring an inductor. You were asked to find the circuit time constant and later, the dc input resistance expression. The circuit appears in Figure 3.36. If we consider L_1 as the extra element, we can apply (3.52) or (3.55). Here, if we consider L_1 as a short circuit, we end up with the circuit shown in Figure 3.37. This is a bridge configuration in which you cannot derive the resistance R? seen from the connecting terminals by inspection. You have to apply the EET with R_3 as the extra element before you carry on. It is not insurmountable but requires an extra effort. On the other hand, if L_1 is made infinite, therefore removing it from the circuit, then you can derive the input resistance R? by inspection, as confirmed by Figure 3.38. If we choose the EET form from (3.52), then we have:

$$Z|_{L_1\to\infty} = R_2 + R_4 \| (R_3 + R_5) \tag{3.108}$$

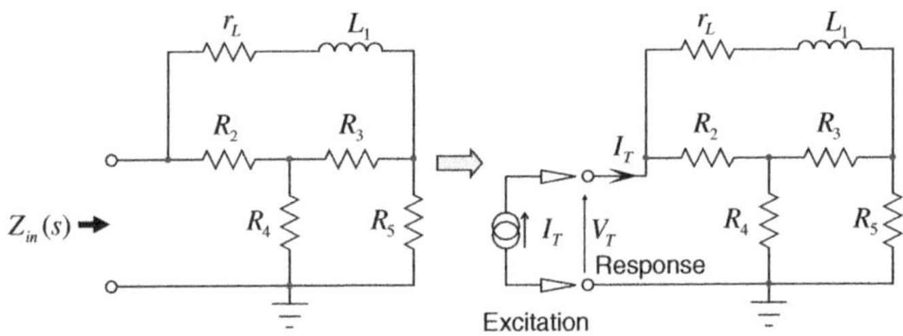

Figure 3.36 To derive this 1st-order circuit input impedance, a current source I_T is applied and the response voltage V_T across its terminals is calculated.

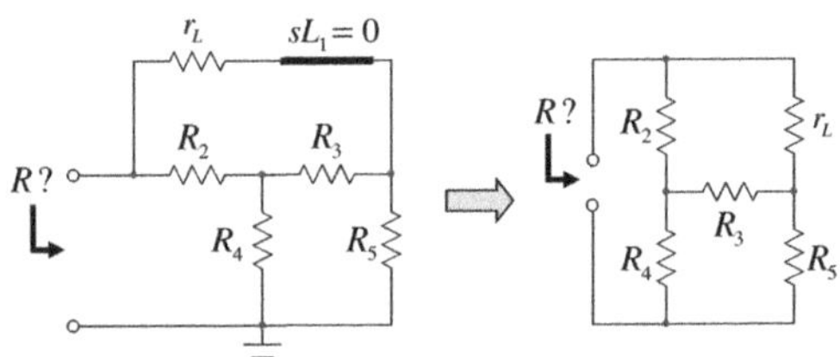

Figure 3.37 If L_1 is set to 0, then the bridge input resistance is quite complex to calculate and requires application of the EET with R_3.

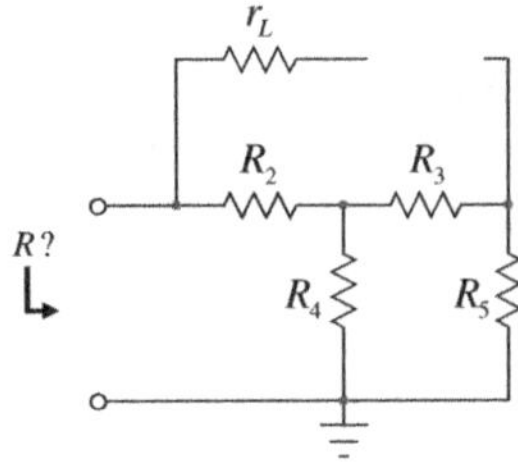

Figure 3.38 Setting L_1 to an infinite value makes it disappear from the circuit, offering an obvious value for R? by inspection.

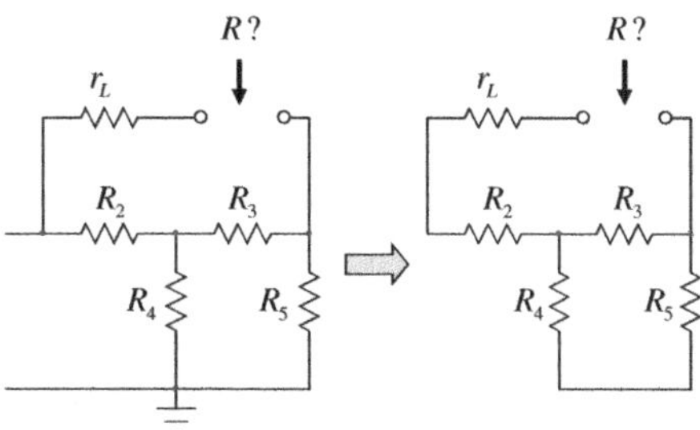

Figure 3.39 Setting the excitation signal to 0 is similar to removing the current source.

Now, let's find the resistance seen from the inductor terminals once the excitation is set to 0. As we derive an impedance, the excitation signal is a current source. Setting it to 0 is similar as removing it from the circuit for the time constant calculation. The circuit now updates to that of Figure 3.39 that we redraw in a more convenient form. The result is that given in Chapter 1 solution where R_4 and R_5 are in series and paralleled with R_3. This network appears in series with R_2 and r_L to form a total resistance R_d equal to:

$$R_d = r_L + R_2 + (R_5 + R_4) \| R_3 \tag{3.109}$$

For the R_n term, we must find the resistance seen from the inductor terminals when the response is nulled. If you recall what we explained in Chapter 2, a null across a current generator is similar to replacing the current generator by a strap. This is what we did in Figure 3.40 and after a few rearrangements, the resistance $R?$ is obvious:

$$R_n = r_L + R_5 \| \left[(R_2 \| R_4) + R_3 \right] \tag{3.110}$$

The final transfer function is thus given by applying (3.52):

$$Z_{in}(s) = R_2 + R_4 \| (R_3 + R_5) \frac{1 + \dfrac{r_L + R_5 \| \left[(R_2 \| R_4) + R_3 \right]}{sL_1}}{1 + \dfrac{r_L + R_2 + (R_5 + R_4) \| R_3}{sL_1}} \tag{3.111}$$

A numerical application is given in Figure 3.41 and shows two asymptotes at dc and high frequency. We selected the EET form where the high-frequency asymptote is represented by the leading term. If the other form would have been chosen, the leading term would be the gain at dc. To check our calculation, we have built a SPICE circuit in which we computed all individual terms. Results appear in Figure 3.42 and agree with the results delivered by Mathcad®.

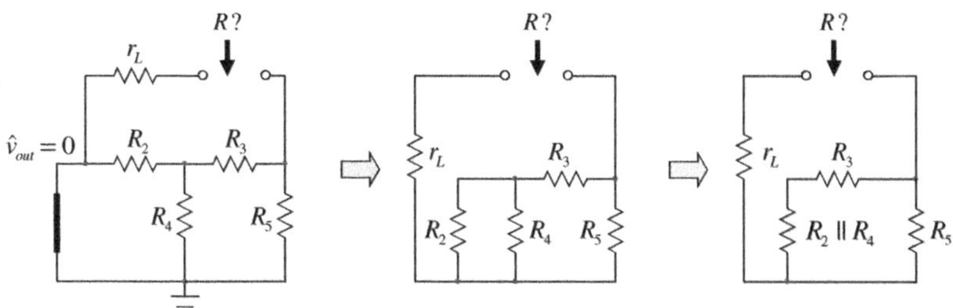

Figure 3.40 A null across a current source is the same as shorting the current generator.

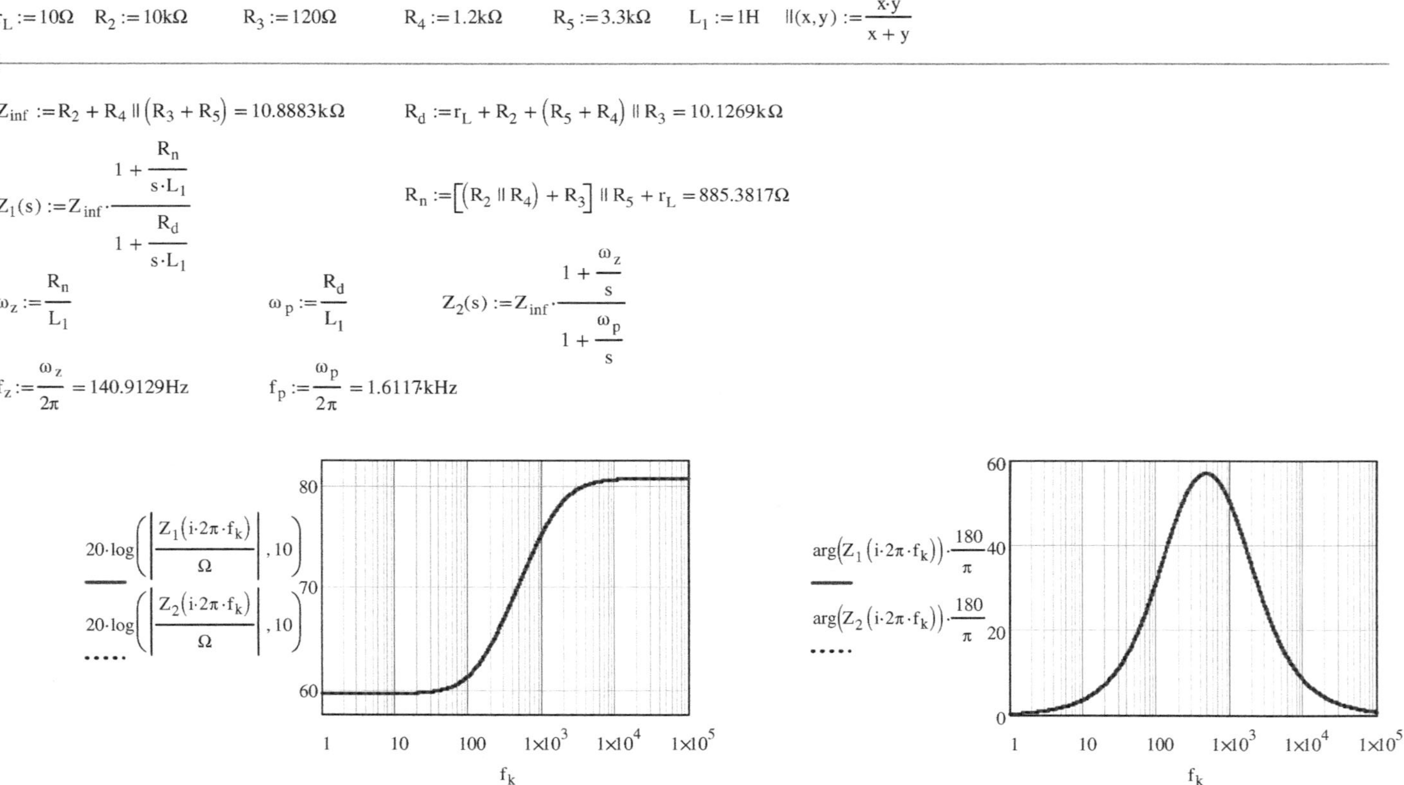

Figure 3.41 The final formula is captured into a Mathcad® sheet to run a numerical application.

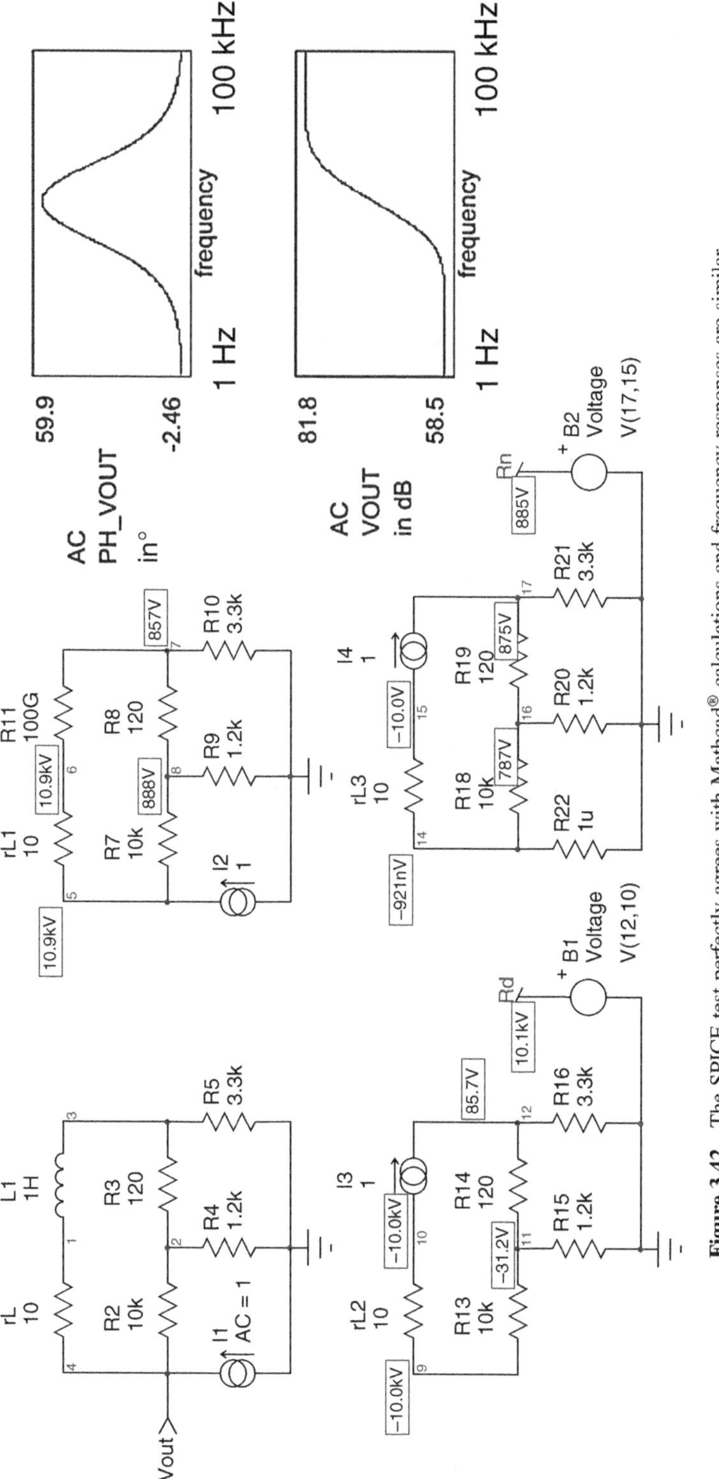

Figure 3.42 The SPICE test perfectly agrees with Mathcad® calculations and frequency responses are similar.

3.2.7 Inverted Pole and Zero Notation

In (3.67), we see a zero and a pole affecting the frequency-response of an expression whose static gain is H_0. When s equals 0, the gain is well identified. Now assume a transfer function which features a zero at the origin and a high-frequency asymptote H_∞ occurring after a pole takes effect. Such a transfer function could be of the following form:

$$H(s) = H_\infty \frac{\frac{s}{\omega_1}}{1 + \frac{s}{\omega_1}} \tag{3.112}$$

and its magnitude response could be that of Figure 3.35. However, this expression could be written in a different way by factoring s/ω_1 in the numerator and the denominator. The updated expression looks like:

$$H(s) = H_\infty \frac{\frac{s}{\omega_1}}{\frac{s}{\omega_1}} \cdot \frac{1}{\frac{\omega_1}{s} + 1} = H_\infty \frac{1}{1 + \frac{\omega_1}{s}} \tag{3.113}$$

Here, when s approaches infinity, we immediately see that the magnitude of H reaches the asymptote H_∞. The expression $1 + \omega_1/s$ in the denominator is called an *inverted pole*. This notation lets you rewrite (3.112) in a more compact and readable form, satisfying the concept of *low entropy*.

Sometimes, it is required that you factor terms in the raw expression to match (3.113) simplicity and obtain the insight of a *low-entropy* formula. What is interesting with the EET is that it directly leads to a result following the form of (3.113). Example 5 showed you how but let's apply it one more time to Figure 3.43 and see the expression it gives.

In sketch (a), we see a capacitor in series with the excitation signal. This is our extra element. Among the two forms of the EET, (3.52) and (3.55), we chose the definition which gives a value for a low or high-frequency asymptote. Here, should we decide to remove C_1, hence (3.52), there would be no static gain to factor. Let's pick (3.55) instead and calculate the gain for Z set to zero with the help of Figure 3.43b:

$$H_\infty|_{Z\to 0} = \frac{R_2}{R_1 + R_2} \tag{3.114}$$

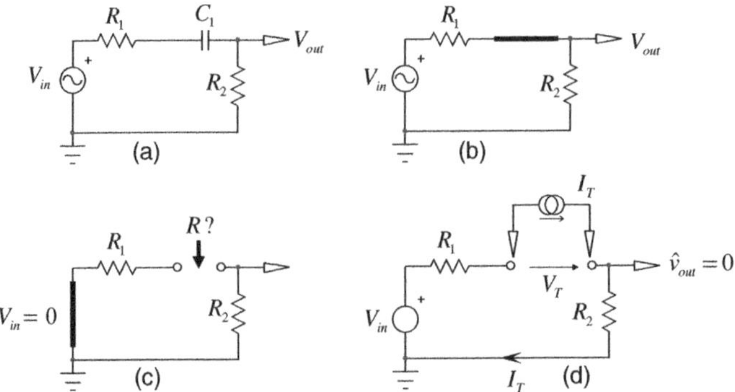

Figure 3.43 In this expression, there is a zero at the origin brought by capacitor C_1 blocking dc.

The resistance R_d driving the capacitor when the excitation is set to 0 is obvious by looking at Figure 3.43c:

$$R_d = R_1 + R_2 \tag{3.115}$$

The resistance R_n driving the capacitor for a null in the output is immediate (Figure 3.43d). The output is a null if the current in R_2 is also 0. As this current is the test generator I_T, it can only be null if the resistance offered by the capacitor terminals is infinite, hence:

$$R_n = \left.\frac{V_T}{I_T}\right|_{I_T=0} \to \infty \tag{3.116}$$

According to (3.55), the transfer function is thus:

$$H(s) = \frac{R_2}{R_1+R_2}\frac{1+\frac{1/sC_1}{\infty}}{1+\frac{1/sC_1}{R_1+R_2}} = \frac{R_2}{R_1+R_2}\frac{1}{1+\frac{1}{sC_1(R_1+R_2)}} = H_\infty\frac{1}{1+\frac{\omega_p}{s}} \tag{3.117}$$

in which

$$\omega_p = \frac{1}{C_1(R_1+R_2)} \tag{3.118}$$

(3.117) is expressed in a low-entropy form with an *inverted pole* in the denominator.

Let's check now how an *inverted zero* looks like. Assume we want the transfer function of the filter shown in Figure 3.44a. We can replace the op amp by a voltage-controlled voltage source amplifying the error voltage between the two inputs by a gain A_{OL}.

In this representation, we designate C_1 as the extra element and one option is to consider the gain when C_1 is a short and thus use (3.55). Figure 3.44b shows the configuration in this mode and the gain is obtained after a few lines of algebra involving the superposition theorem:

$$\varepsilon|_{V_{in}=0} = -V_{out}\frac{R_1}{R_1+R_2} \tag{3.119}$$

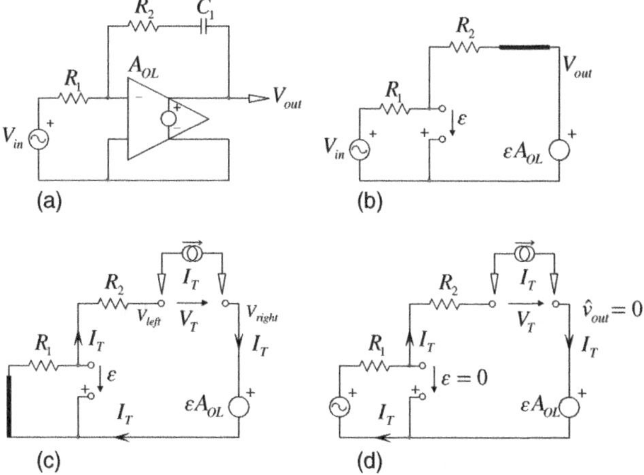

Figure 3.44 The EET can be applied to op amp-based circuits without problems.

$$\varepsilon|_{V_{out}=0} = -V_{in}\frac{R_2}{R_1+R_2} \tag{3.120}$$

The total error voltage is the sum of (3.119) and (3.120):

$$\varepsilon = \varepsilon|_{V_{in}=0} + \varepsilon|_{V_{out}=0} = -V_{out}\frac{R_1}{R_1+R_2} - V_{in}\frac{R_2}{R_1+R_2} \tag{3.121}$$

The error voltage ε is V_{out} divided by the op amp open-loop gain A_{OL}, thus:

$$\frac{V_{out}}{A_{OL}} = -V_{out}\frac{R_1}{R_1+R_2} - V_{in}\frac{R_2}{R_1+R_2} \tag{3.122}$$

Rearranging and factoring give us the gain definition we want:

$$\left.\frac{V_{out}}{V_{in}}\right|_{Z\to 0} = -\frac{R_2}{R_1+\dfrac{R_1+R_2}{A_{OL}}} \tag{3.123}$$

The time constant is obtained by forming the circuit presented in Figure 3.44c where the excitation is set to 0 V. The voltage at the inverting pin is easy to find and is:

$$V_{(-)} = -I_T R_1 \tag{3.124}$$

The current generator left terminal is thus biased at the sum of the negative voltage generator over R_1 and R_2:

$$V_{left} = -I_T R_1 - I_T R_2 = -I_T(R_1+R_2) \tag{3.125}$$

The voltage at the right terminal is:

$$V_{right} = \varepsilon A_{OL} = \left(V_{(+)} - V_{(-)}\right)A_{OL} = I_T R_1 A_{OL} \tag{3.126}$$

therefore

$$V_T = V_{right} - V_{left} = I_T R_1 A_{OL} + I_T(R_1+R_2) \tag{3.127}$$

The resistance seen by the capacitor when the excitation is set to 0 is therefore:

$$\frac{V_T}{I_T} = R_d = R_1(A_{OL}+1) + R_2 \tag{3.128}$$

The zero is found by a null double injection as shown in Figure 3.44d. As we have a null on the output, it clearly implies that ε is also a null: R_2's left terminal is grounded and the current generator only 'sees' this resistor. Thus, we have:

$$\frac{V_T}{I_T} = R_n = R_2 \tag{3.129}$$

This is it, we have all our pieces and we can assemble them according to (3.55). The final transfer function is made of (3.123), (3.128) and (3.129):

$$\frac{V_{out}(s)}{V_{in}(s)} = -\frac{R_2}{R_1+\dfrac{R_1+R_2}{A_{OL}}}\,\frac{1+\dfrac{1/sC_1}{R_2}}{1+\dfrac{1/sC_1}{R_1(A_{OL}+1)+R_2}} = -\frac{R_2}{R_1+\dfrac{R_1+R_2}{A_{OL}}}\,\frac{1+\dfrac{1}{sR_2C_1}}{1+\dfrac{1}{sC_1[R_1(A_{OL}+1)+R_2]}} \tag{3.130}$$

In the above expression, if the op amp open-loop gain approaches infinity, the denominator expression simplifies to 1 and (3.130) becomes:

$$H(s) = H_{\infty}\left(1 + \frac{\omega_z}{s}\right) \tag{3.131}$$

in which H_{∞} is defined by

$$H|_{\infty} = -\frac{R_2}{R_1} \tag{3.132}$$

and the zero by

$$\omega_z = \frac{1}{R_2 C_1} \tag{3.133}$$

The expression in (3.131) uses an *inverted zero* notation.

Figure 3.45 shows the different responses for classical pole/zero notations and their inverted versions. The zero ac response is seen as the magnitude and phase responses of the pole but inverted with respect to their vertical axes. An inverted pole (or zero) can be seen as the magnitude and phase responses of the pole (or zero) but inverted with respect to the logarithmic horizontal axis. This notation offers a way to unveil a high-frequency asymptote (H_{∞}) in a *low-entropy* form.

3.3 A Generalized Transfer Function for 1st-order Systems

The EET can be applied in two different ways, whether the extra element is considered a short circuit or an open circuit. Both formulas in (3.52) and (3.55) thus express the exact same transfer function H:

$$H|_{Z=\infty}\frac{1+\dfrac{Z_n}{Z}}{1+\dfrac{Z_d}{Z}} = H|_{Z=0}\frac{1+\dfrac{Z}{Z_n}}{1+\dfrac{Z}{Z_d}} \tag{3.134}$$

This expression can be rearranged the following way:

$$\frac{1+\dfrac{Z_n}{Z}}{1+\dfrac{Z_d}{Z}} = \frac{H|_{Z=0}}{H|_{Z=\infty}}\frac{1+\dfrac{Z}{Z_n}}{1+\dfrac{Z}{Z_d}} \tag{3.135}$$

Isolating the gain ratios on the right side leads to:

$$\frac{\left(1+\dfrac{Z_n}{Z}\right)\left(1+\dfrac{Z}{Z_d}\right)}{\left(1+\dfrac{Z_d}{Z}\right)\left(1+\dfrac{Z}{Z_n}\right)} = \frac{H|_{Z=0}}{H|_{Z=\infty}} \tag{3.136}$$

If you develop and simplify the left term, you obtain:

$$\frac{Z_n}{Z_d} = \frac{H|_{Z=0}}{H|_{Z=\infty}} \tag{3.137}$$

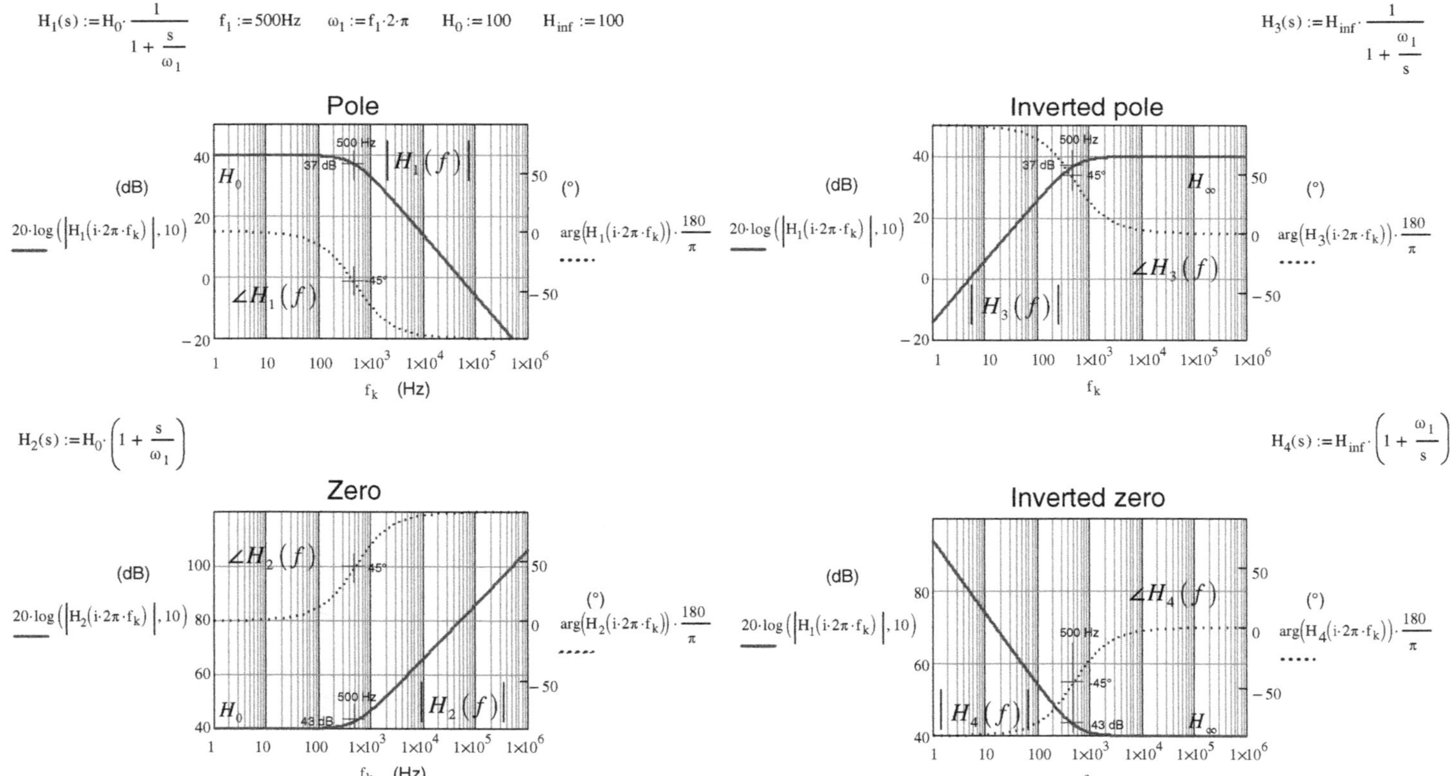

Figure 3.45 Classical pole/zero responses and their inverted counterpart responses.

Otherwise stated:

$$Z_n = \frac{H|_{Z=0}}{H|_{Z=\infty}} Z_d \tag{3.138}$$

or

$$Z_d = \frac{H|_{Z=\infty}}{H|_{Z=0}} Z_n \tag{3.139}$$

In the above expressions:

$Z=0$ means that the extra-element impedance is a short circuit: C and L are replaced by a short.

$Z=\infty$ means that the extra element is open: C and L are removed from the circuit.

Using these definitions, we can rework the founding EET equations by replacing Z_n by (3.138). If we substitute (3.138) in (3.55), we have:

$$H|_Z = H|_{Z=0} \frac{1 + \dfrac{Z}{\dfrac{H|_{Z=0}}{H|_{Z=\infty}} Z_d}}{1 + \dfrac{Z}{Z_d}} = \frac{H|_{Z=0} + H|_{Z=0} \dfrac{Z}{\dfrac{H|_{Z=0}}{H|_{Z=\infty}} Z_d}}{1 + \dfrac{Z}{Z_d}} = \frac{H|_{Z=0} + H|_{Z=\infty} \dfrac{Z}{Z_d}}{1 + \dfrac{Z}{Z_d}} = H|_{Z=0} \frac{1 + \dfrac{H|_{Z=\infty}}{H|_{Z=0}} \dfrac{Z}{Z_d}}{1 + \dfrac{Z}{Z_d}} \tag{3.140}$$

Similarly, if we substitute (3.138) in (3.52), we obtain:

$$H|_Z = H|_{Z=\infty} \frac{1 + \dfrac{\dfrac{H|_{Z=0}}{H|_{Z=\infty}} Z_d}{Z}}{1 + \dfrac{Z_d}{Z}} = \frac{H|_{Z=\infty} + H|_{Z=\infty} \dfrac{\dfrac{H|_{Z=0}}{H|_{Z=\infty}} Z_d}{Z}}{1 + \dfrac{Z_d}{Z}} = \frac{H|_{Z=\infty} + H|_{Z=0} \dfrac{Z_d}{Z}}{1 + \dfrac{Z_d}{Z}} = H|_{Z=\infty} \frac{1 + \dfrac{H|_{Z=0}}{H|_{Z=\infty}} \dfrac{Z_d}{Z}}{1 + \dfrac{Z_d}{Z}} \tag{3.141}$$

Expressions (3.140) and (3.141) are equivalent. Now, in (3.140), assume the impedance Z characterizes an inductance L. We can update this equation as:

$$H(s) = \frac{H|_{Z=0} + H|_{Z=\infty} \dfrac{sL}{Z_d}}{1 + \dfrac{sL}{Z_d}} \tag{3.142}$$

If we consider that the first term is a dc gain ($Z=0$ implies that L is shorted) while the second one is a high-frequency gain (L is removed from the circuit), the notation we introduced in Chapter 2 for the generalized transfer function is used:

$$H_0 = H|_{Z=0} \tag{3.143}$$

$$H^1 = H|_{Z\to\infty} \tag{3.144}$$

Following these steps, (3.142) can be rewritten as:

$$H(s) = \frac{H_0 + H^1 s \dfrac{L}{Z_d}}{1 + s \dfrac{L}{Z_d}} \tag{3.145}$$

The term L/Z_d is the circuit time constant τ_1 and appears in the numerator and the denominator. Finally, we have:

$$H(s) = \frac{H_0 + H^1 s\tau_1}{1 + s\tau_1} \tag{3.146}$$

If Z is now a capacitor C, (3.141) becomes:

$$H(s) = \frac{H|_{Z=\infty} + H|_{Z=0}\dfrac{Z_d}{1/sC}}{1 + \dfrac{Z_d}{1/sC}} \tag{3.147}$$

If we consider that the first term is a dc gain ($Z = \infty$ implies that C is removed from the circuit or zero valued) while the second one is a high-frequency gain (C is replaced by a short circuit or infinite valued), the following notation is adopted:

$$H_0 = H|_{Z=\infty} \tag{3.148}$$

$$H^1 = H|_{Z\to 0} \tag{3.149}$$

Owing to these notations, (3.147) can be rewritten as

$$H(s) = \frac{H_0 + H^1 sZ_dC}{1 + sZ_dC} \tag{3.150}$$

Z_dC is the time constant τ_1 of the circuit and appears in the denominator and the numerator. (3.150) can thus be rewritten as:

$$H(s) = \frac{H_0 + H^1 s\tau_1}{1 + s\tau_1} \tag{3.151}$$

(3.151) and (3.146) are identical expressions and work for all 1st-order circuits with an inductor or a capacitor. This generalized 1st-order expression is the same as the one we derived in Chapter 2 using a different path but also given in [5]. The term H_0 can take on different values: it can be 0 if you have a zero at the origin. It can be of course a gain or an attenuation determined for $s = 0$. But it could also be an infinite value in case your circuit features a pole at the origin. However, in a practical implementation, this gain would be bounded by the op amp (or any other type of amplifier) open-loop gain A_{OL} as shown in example 5.

3.3.1 Generalized Transfer Function – Example 1

In Figure 3.37, setting L_1 to zero transforms the circuit into a resistive bridge whose resistance must be found. We could apply the EET but an NDI exercise would be necessary. Let's apply the generalized transfer function definition from (3.140) which perfectly applies to a resistive bridge since there is no energy-storage element. We start by setting R_3 to 0. The new circuit is given by Figure 3.46 and leads to an immediate resistance obtained by inspection:

$$R|_{Z=0} = r_L \| R_2 + R_5 \| R_4 \tag{3.152}$$

The second element we must find is the resistance offered by the circuit when R_3 is physically removed (R_3 set to an infinite value). The new network appears in Figure 3.47.

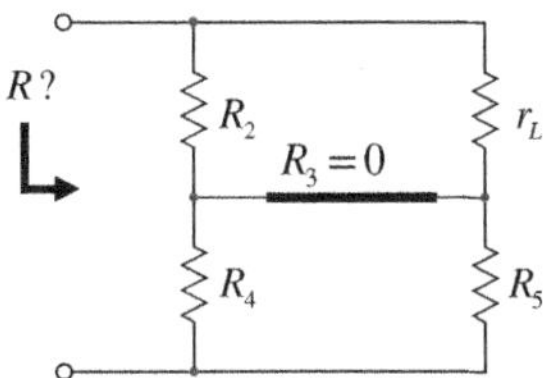

Figure 3.46 Setting R_3 to 0 unveils a simple series-parallel arrangement.

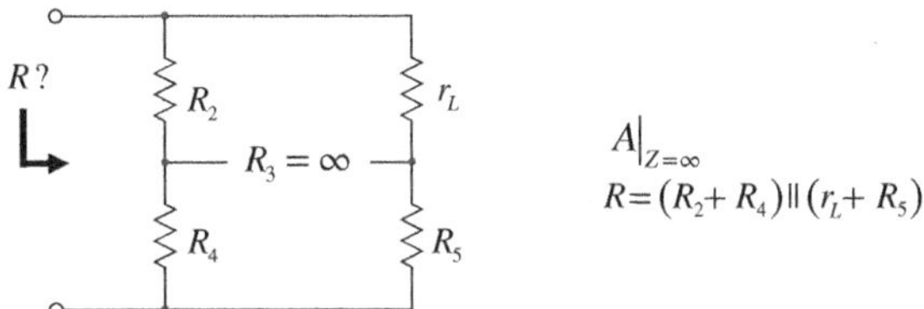

Figure 3.47 Setting R_3 to infinity shows another series-parallel arrangement.

Again, no special trick; we can find the input resistance by inspecting the diagram. The resistance is equal to:

$$R|_{Z=\infty} = (R_2 + R_4)\|(r_L + R_5) \tag{3.153}$$

The final step is obtained by looking into R_3 connecting terminals while the excitation is set to 0. For a current source, setting it to 0 A is similar to removing it from the circuit. This is what Figure 3.48 shows.

The resistance seen from R_3 terminals is immediate:

$$Z_d = (r_L + R_2)\|(R_5 + R_4) \tag{3.154}$$

We have all the needed elements to form the input resistance from Figure 3.37 arrangement. From (3.140), we have:

$$R_{in} = R|_{Z=0}\frac{1+\dfrac{R|_{Z=\infty}}{R|_{Z=0}}\dfrac{Z}{Z_d}}{1+\dfrac{Z}{Z_d}} = \left(r_L\|R_2 + R_5\|R_4\right)\frac{1+\dfrac{(R_2+R_4)\|(r_L+R_5)}{r_L\|R_2+R_5\|R_4}\dfrac{R_3}{(r_L+R_2)\|(R_5+R_4)}}{1+\dfrac{R_3}{(r_L+R_2)\|(R_5+R_4)}} \tag{3.155}$$

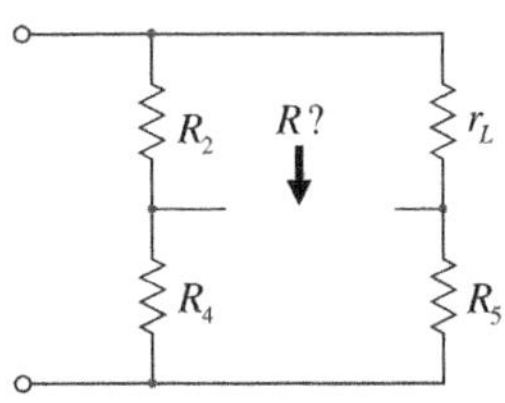

Figure 3.48 Looking into R_3's terminals is an easy exercise in this case.

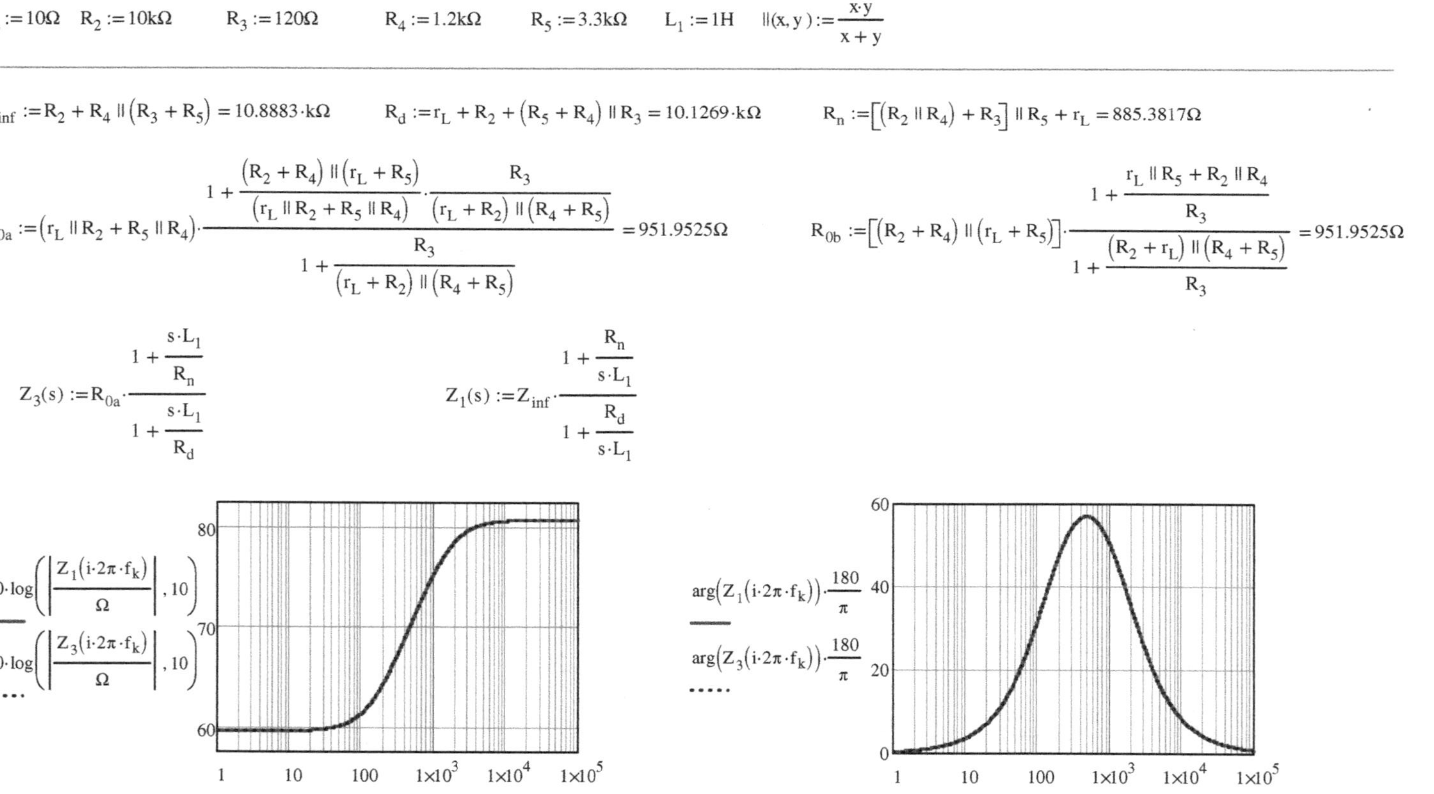

Figure 3.49 Using the extra element theorem and the generalized 1st-order transfer function lead to a similar result.

Reference [4] offers a different formula however obtained with the EET and the NDI configuration:

$$R_{in} = (R_2 + R_4)\|(r_L + R_5)\frac{1 + \dfrac{R_2\|R_4 + r_L\|R_5}{R_3}}{1 + \dfrac{(R_2 + r_L)\|(R_4 + R_5)}{R_3}} \tag{3.156}$$

The two expressions are equivalent. The remark we formulated in Chapter 2, comparing what the generalized transfer function delivers and what you obtain with the EET, makes sense in this example: the generalized form does not need to solve a null double injection configuration and goes through three simple steps in this particular case. The result is in an ordered form but the numerator coefficients are complicated. On the other hand, the EET requires an NDI – it is simple here because the null output is a short across the test generator – but it leads to a simpler formula.

Now that we have our dc expression – when L_1 is set to 0 – we can reuse (3.109) and (3.110) together with the second form of the EET given in (3.55) to obtain another definition for Figure 3.36 input impedance:

$$Z_{in}(s) = R_{in}\frac{1 + s\dfrac{L_1}{r_L + R_5\|\left[(R_2\|R_4) + R_3\right]}}{1 + s\dfrac{L_1}{r_L + R_2 + (R_5 + R_4)\|R_3}} \tag{3.157}$$

Expressions (3.111) and (3.157) are identical but written differently. The Mathcad® sheet from Figure 3.49 compares these results and plots the ac impedance responses.

3.3.2 Generalized Transfer Function – Example 2

Figure 3.50 shows a bipolar transistor whose collector delivers a voltage labeled V_{out}. The circuit monitors node V_{in} and brings V_{out} down as soon as the Zener diode starts conducting and biases Q_1. This is a very classical structure found in primary-regulated switching power supplies in which V_{out}

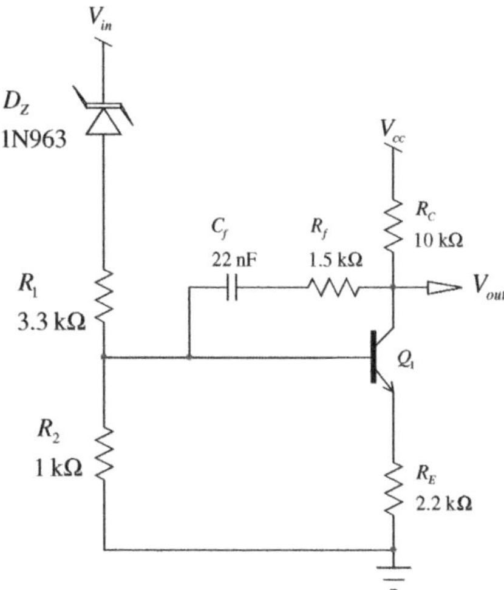

Figure 3.50 This simple bipolar amplifier is often used in primary-regulated switching power supplies.

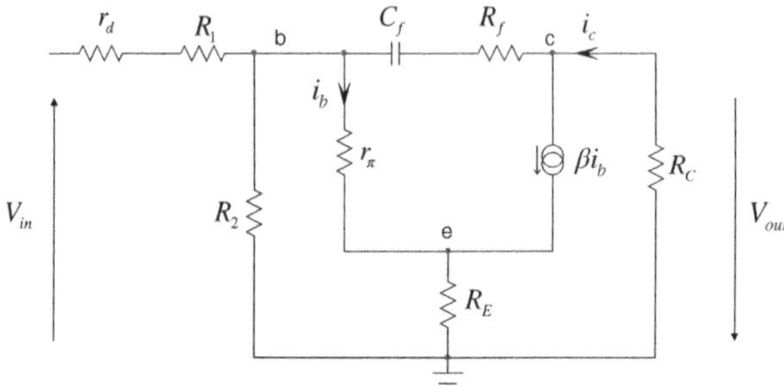

Figure 3.51 The small-signal model of Figure 3.50 is very similar to that which we have already studied in previous examples.

connects to the feedback pin of the selected integrated controller. V_{in} is the rectified auxiliary V_{cc} from the primary side. We start with the equivalent small-signal circuit given in Figure 3.51.

For this example, we will generalize the EET definition for a capacitor C (or an inductor L). As we have a single energy-storing element, the electrical circuit describes a 1st-order circuit. In Figure 3.51, if we short C_f, do we still have a response in V_{out}? Yes, the circuit reduces to something close to what we already studied in Figure 3.29a: there is a zero associated with C_f. A 1st-order transfer function featuring a pole, a zero and a low-frequency asymptote can be put in the following form:

$$H(s) = H_0 \frac{1 + s\tau_1}{1 + s\tau_2} \tag{3.158}$$

It is actually expression (3.52) in which Z_n and Z_d are the driving resistances R_n and R_d. R_n is determined in an NDI condition while R_d is obtained with the excitation set to zero. Z is simply the capacitor impedance $1/sC$ (it could be sL with an inductor in place). In this definition, τ_1 is the time constant involving C_f and the driving resistance R_n while τ_2 is the circuit time constant involving R_d obtained for the excitation set to 0. Let's start with the dc gain H_0 and a modified circuit shown in Figure 3.52. The output voltage is classically defined as:

$$V_{out} = -\beta i_b R_C \tag{3.159}$$

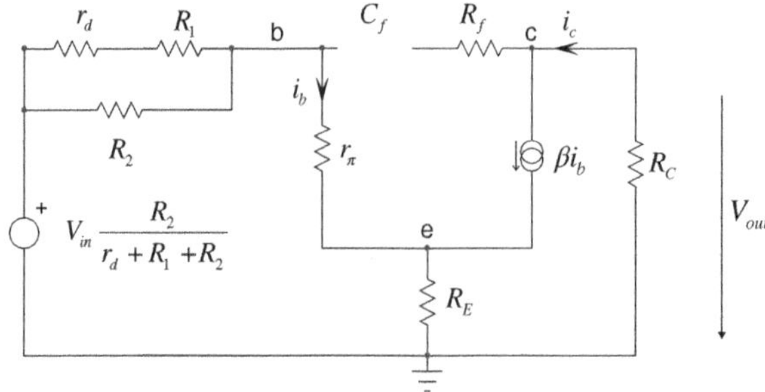

Figure 3.52 In dc conditions, capacitor C_f disappears.

The base current is flowing through the input source. It is defined as:

$$i_b = \frac{V_{in}\dfrac{R_2}{r_d + R_1 + R_2} - V_b}{(r_d + R_1) \| R_2} \tag{3.160}$$

The base voltage is equal to the drop across r_π plus the emitter voltage $v_{(e)}$:

$$V_b = r_\pi i_b + v_{(e)} = i_b(r_\pi + (\beta + 1)R_E) \tag{3.161}$$

Substituting (3.161) in (3.160) and solving for i_b gives:

$$i_b = \frac{V_{in}}{(R_1 + r_d)\left(\dfrac{[r_\pi + R_E(\beta + 1)](R_1 + R_2 + r_d)}{R_2(R_1 + r_d)} + 1\right)} \tag{3.162}$$

In (3.159), we replace i_b by its definition from (3.162) and we have the dc transfer function we need:

$$H_0 = -\frac{\beta R_C}{(R_1 + r_d)\left(\dfrac{[r_\pi + R_E(\beta + 1)](R_1 + R_2 + r_d)}{R_2(R_1 + r_d)} + 1\right)} \tag{3.163}$$

To obtain R_n, let's draw the schematic for an NDI condition. It appears in Figure 3.53. Fortunately, we have already calculated this resistance with the schematic from Figure 3.29d. In this circuit, you see that V_T, the test voltage is directly present at the base connection. This voltage does not involve the V_{in} mesh and regardless of the added resistors r_d, R_1 and R_2, the NDI analysis for Figure 3.53 results in an expression similar to that which was given in (3.89) except that we now have R_f in series:

$$R_n = \frac{V_T}{I_T} = -\left[\frac{r_\pi}{\beta} + R_E\left(1 + \frac{1}{\beta}\right)\right] + R_f \tag{3.164}$$

Please note that R_n is negative without R_f but becomes positive for R_f greater than:

$$R_f > \frac{r_\pi}{\beta} + R_E\left(1 + \frac{1}{\beta}\right) \tag{3.165}$$

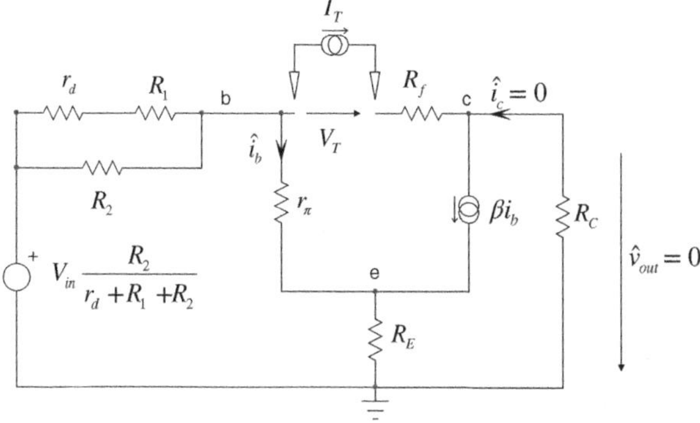

Figure 3.53 The NDI condition requires an output null.

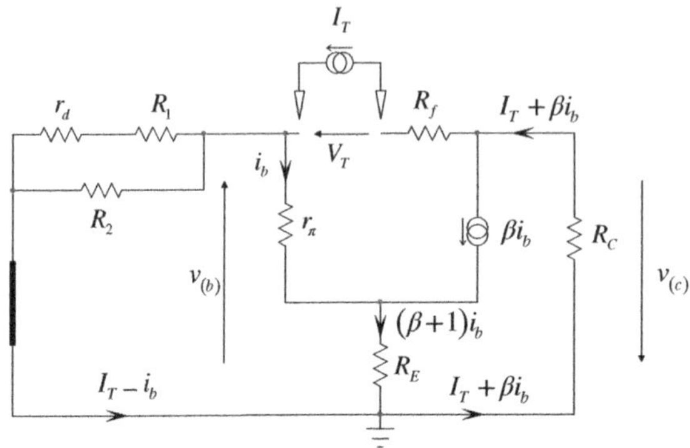

Figure 3.54 The time constant is found by setting the excitation to 0.

The numerator time constant τ_1 is defined as:

$$\tau_1 = C_f\left[R_f - \frac{r_\pi}{\beta} - R_E\left(1+\frac{1}{\beta}\right)\right] \tag{3.166}$$

The circuit time constant is found by setting the excitation to 0. This is what is shown in Figure 3.54.

We will run the analysis without R_f that we will add at the end to the result. This is a classical trick of fast analytical techniques that also works for paralleled elements. Here, despite the absence of input voltage, a current circulates across the input network made of r_d, R_1 and R_2 and produces a voltage drop $v_{(b)}$. This voltage is equal to:

$$v_{(b)} = R_{eq}(I_T - i_b) \tag{3.167}$$

where

$$R_{eq} = (R_1 + r_d)\|R_2 \tag{3.168}$$

The same voltage splits between r_π and R_E. Therefore:

$$v_{(b)} = i_b r_\pi + (\beta+1)i_b R_E = i_b[r_\pi + (\beta+1)R_E] \tag{3.169}$$

Equating (3.167) and (3.169), solving for i_b and rearranging gives:

$$i_b = kI_T \tag{3.170}$$

in which

$$k = \frac{1}{\dfrac{r_\pi + (\beta+1)R_E}{(R_1 + r_d)\|R_2} + 1} \tag{3.171}$$

The voltage at node c is the current flowing out of R_C multiplied by R_C:

$$v_{(c)} = -R_C(I_T + \beta i_b) \tag{3.172}$$

The resistance we want is thus defined as:

$$R = \frac{V_T}{I_T} = \frac{v_{(b)} - v_{(c)}}{I_T} = \frac{R_{eq}(I_T - k\cdot I_T) + R_C(I_T + k\cdot I_T\beta)}{I_T} = R_{eq}(1-k) + R_C(1+k\beta) \tag{3.173}$$

The final value is then:

$$R_d = R + R_f \tag{3.174}$$

Replacing R_{eq} and k by their values, the second time constant τ_2 is defined as:

$$\tau_2 = C_f\left[[(R_1+r_d)\|R_2]\left(1-\frac{1}{\dfrac{r_\pi+(\beta+1)R_E}{(R_1+r_d)\|R_2}+1}\right)+R_C\left(1+\frac{1}{\dfrac{r_\pi+(\beta+1)R_E}{(R_1+r_d)\|R_2}+1}\beta\right)+R_f\right] \tag{3.175}$$

The final transfer function is assembled by combining (3.163), (3.166) and (3.175):

$$H(s) = H_0\frac{1+\dfrac{s}{\omega_z}}{1+\dfrac{s}{\omega_p}} \tag{3.176}$$

in which the dc gain is defined as:

$$H_0 = -\frac{\beta R_C}{(R_1+r_d)\left(\dfrac{[r_\pi+R_E(\beta+1)](R_1+R_2+r_d)}{R_2(R_1+r_d)}+1\right)} \tag{3.177}$$

the zero by

$$\omega_z = \frac{1}{C_f\left[R_f-\dfrac{r_\pi}{\beta}-R_E\left(1+\dfrac{1}{\beta}\right)\right]} \tag{3.178}$$

and the pole equals

$$\omega_p = \frac{1}{C_f\left[[(R_1+r_d)\|R_2]\left(1-\dfrac{1}{\dfrac{r_\pi+(\beta+1)R_E}{(R_1+r_d)\|R_2}+1}\right)+R_C\left(1+\dfrac{1}{\dfrac{r_\pi+(\beta+1)R_E}{(R_1+r_d)\|R_2}+1}\beta\right)+R_f\right]} \tag{3.179}$$

To check these complex results, we have captured data into a Mathcad® sheet (Figure 3.55) and compared results with a SPICE simulation (Figure 3.56). They perfectly agree with each other.

Expression (3.165) tells us that the sign of R_n can change depending on the series resistance value, R_f. If R_f is strapped, then the zero in (3.178) jumps into the right half-plane of the s-chart. Rather than having a phase going up as the zero kicks in, we will add an extra phase lag as any right half-plane zero would do. Then, if R_f exactly cancels the zero (R_f= 1.2147 kΩ), the transfer function turns into a single-pole response who maximum lag is 90°. Finally, when R_f is above 1.2147 kΩ, you obtain the pole/zero response of Figure 3.56. Figure 3.57 summarizes these three scenarios.

3.3.3 Generalized Transfer Function – Example 3

Figure 3.58 shows an op amp wired in an inverting way where a capacitor connects the middle point of the feedback resistors to ground. To derive the transfer function, we will use the generalized 1st-order transfer function. First, we calculate the dc gain, removing C_1 from the picture as indicated in Figure 3.59. The op amp offers an open-loop gain A_{OL} therefore the error voltage ε is not equal to zero.

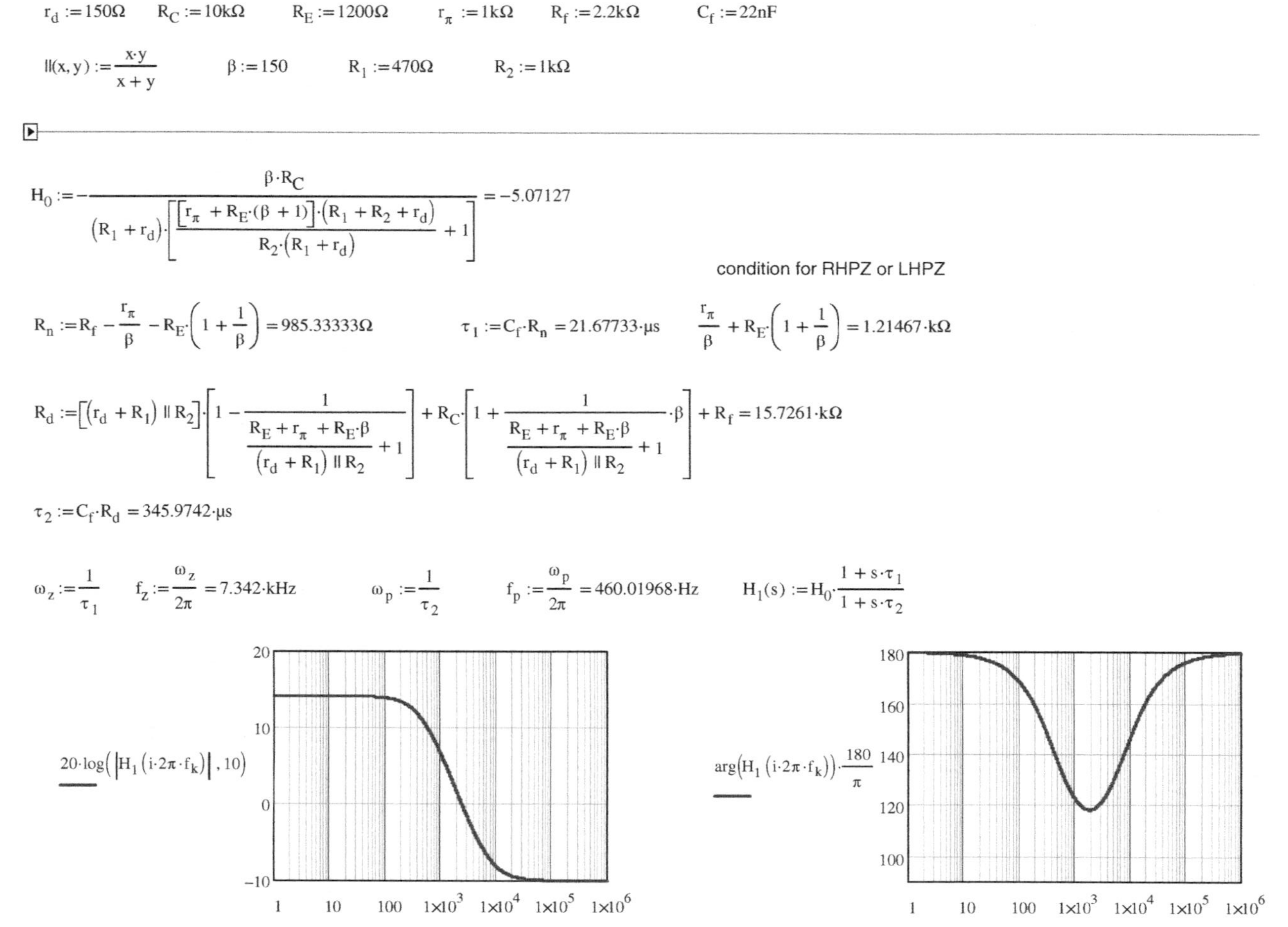

Figure 3.55 The sheet shows a flat response at dc followed by a pole and a higher frequency zero.

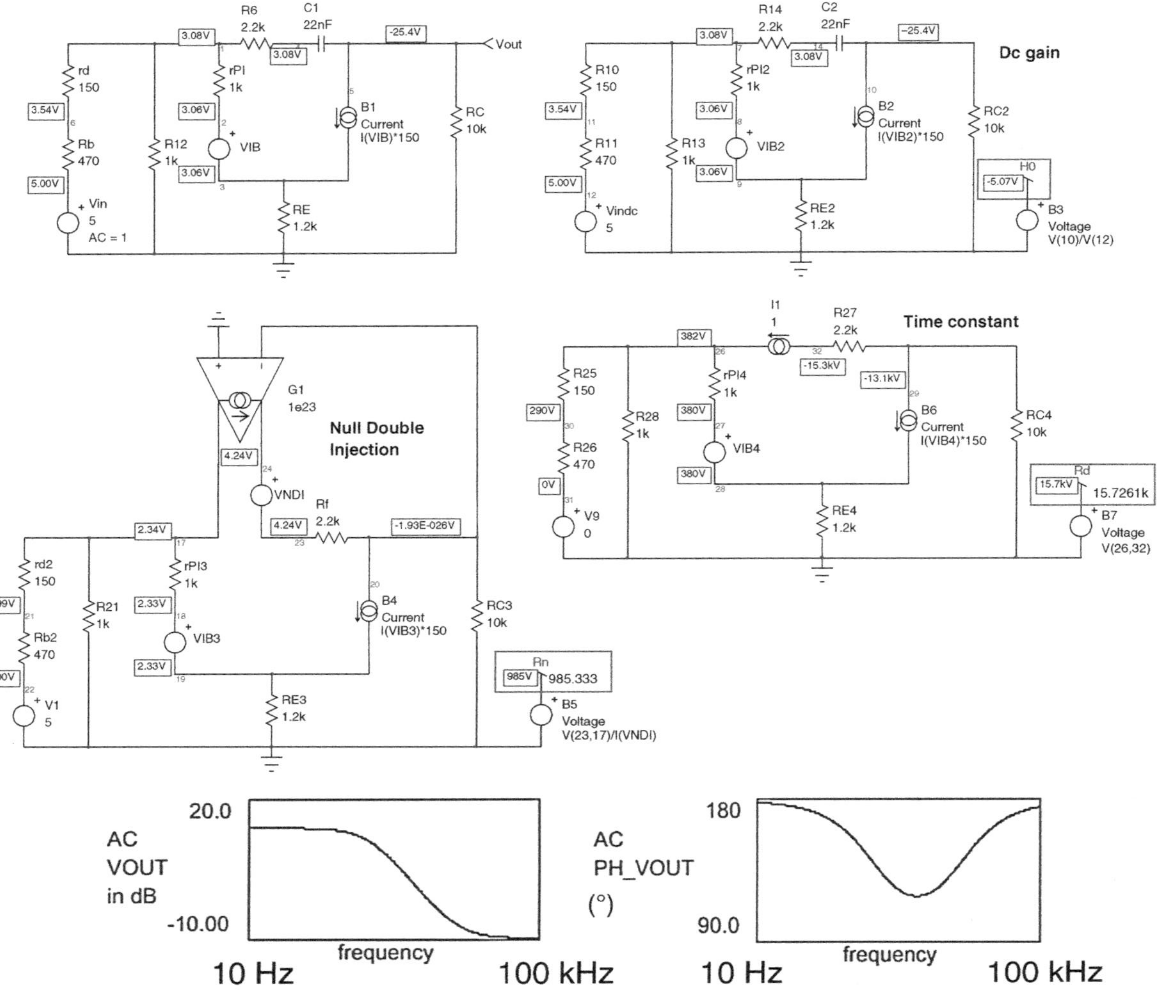

Figure 3.56 SPICE confirms numerical values and ac responses.

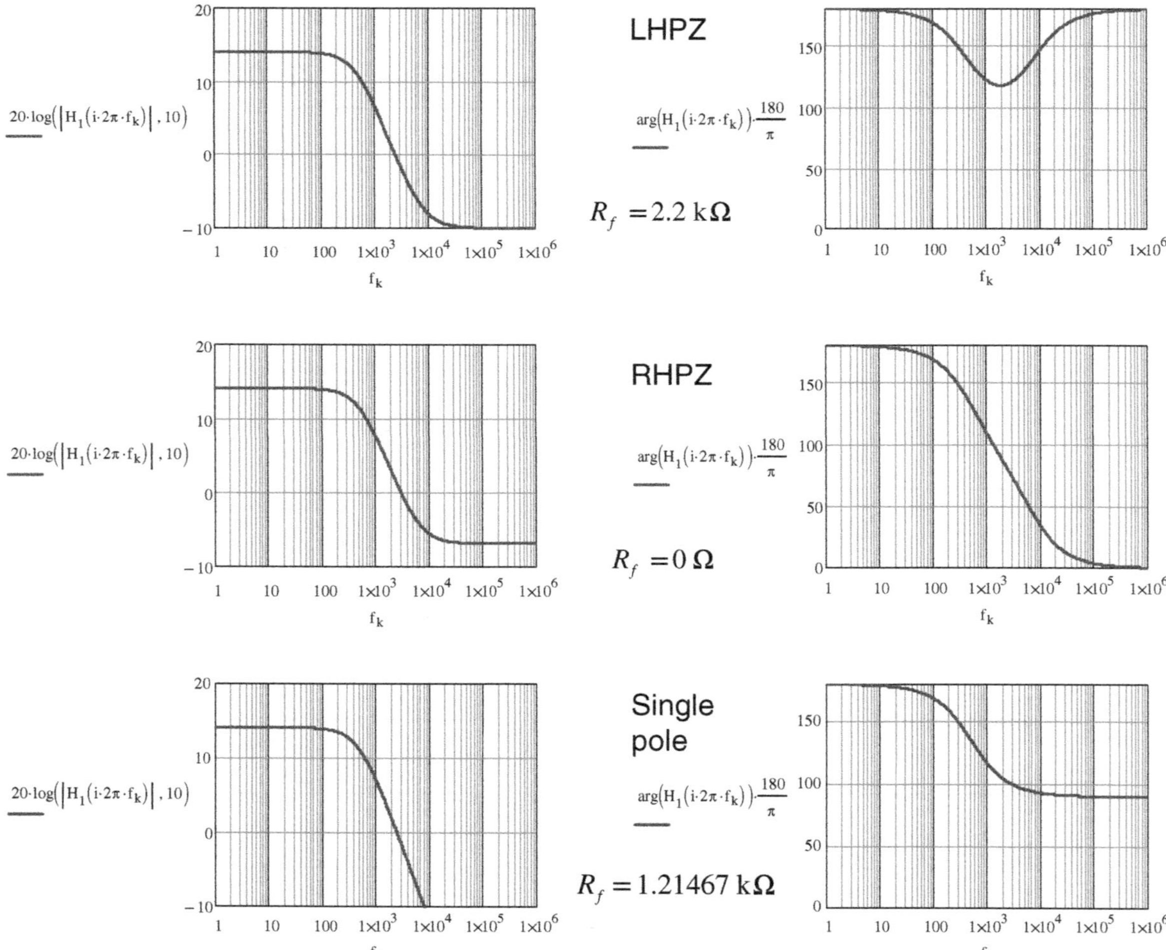

Figure 3.57 Different responses can be obtained by changing the value of R_f.

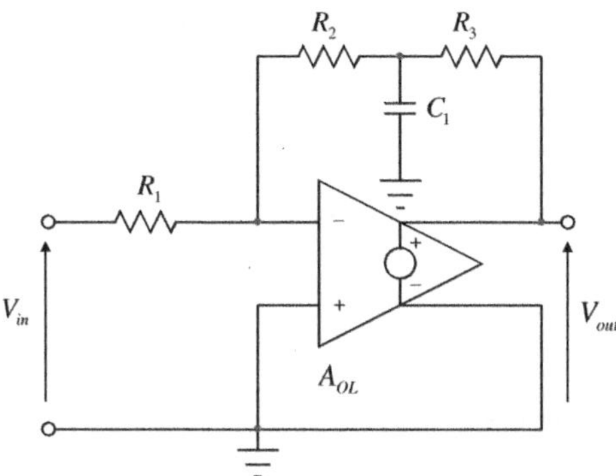

Figure 3.58 The op amp combines resistors and a capacitor to form a 1st-order active filter.

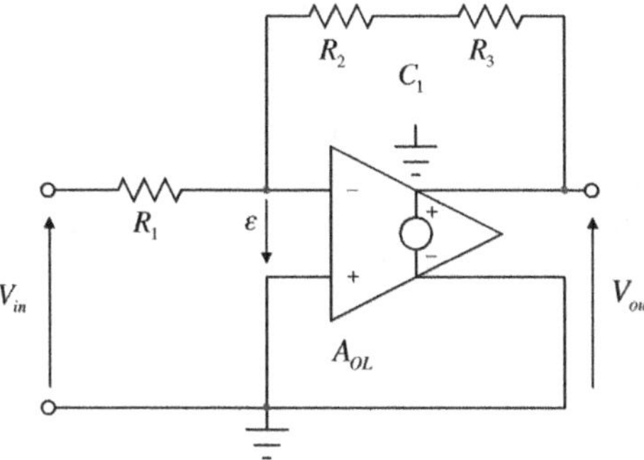

Figure 3.59 Removing C_1 shows an op amp in a classical inverter configuration.

Fortunately, we have already calculated this transfer function in problem 10 of Chapter 2. The gain is given by:

$$H_0 = -\frac{R_2 + R_3}{R_1} \frac{1}{\left(\frac{\frac{R_2 + R_3}{R_1} + 1}{A_{OL}} + 1 \right)} \tag{3.180}$$

The second gain we need is when C_1 is a short circuit. The updated circuit appears in Figure 3.60. R_3 disappears from the picture as it loads the op amp output (perfect op amp with 0-Ω output impedance). We have:

$$V_{out} = A_{OL}\varepsilon \tag{3.181}$$

The op amp is no longer in closed-loop condition and the virtual ground is lost. The voltage between the input pins is then given by a simple resistive divider configuration:

$$\varepsilon = -\frac{R_2}{R_1 + R_2} V_{in} \tag{3.182}$$

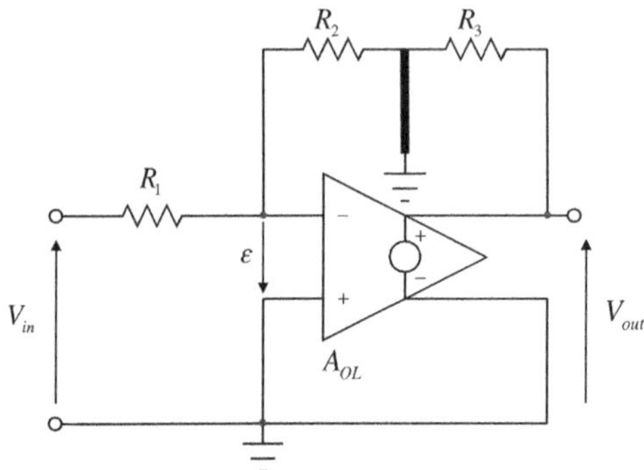

Figure 3.60 Shorting C_1 lets us calculate the high-frequency gain H^1.

Substituting (3.182) into (3.181) gives us the high-frequency gain:

$$H^1 = -\frac{R_2}{R_1 + R_2} A_{OL} \tag{3.183}$$

We now need to determine the time constant of this circuit. To do so, we set the excitation to 0 V and we have the circuit shown in Figure 3.61. Here, it is not possible to obtain the resistance by inspection, we need to go through the current source configuration and determine the voltage across its terminals. This is what Figure 3.62 illustrates. Please note that I purposely reversed ε for easier current conventions and thus also reversed εA_{OL}.

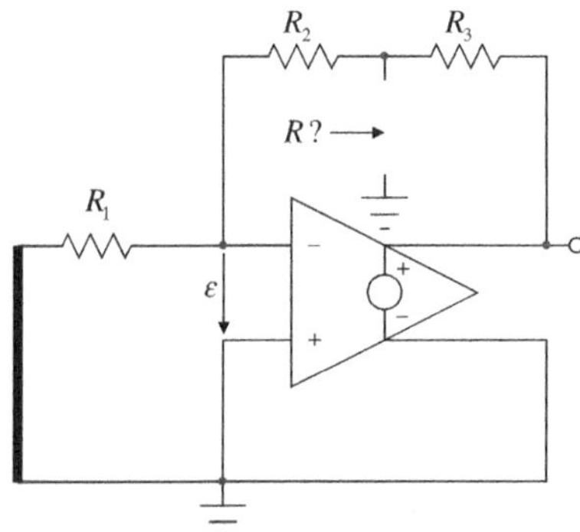

Figure 3.61 Setting the excitation voltage to 0 V is the way to determine the circuit time constant.

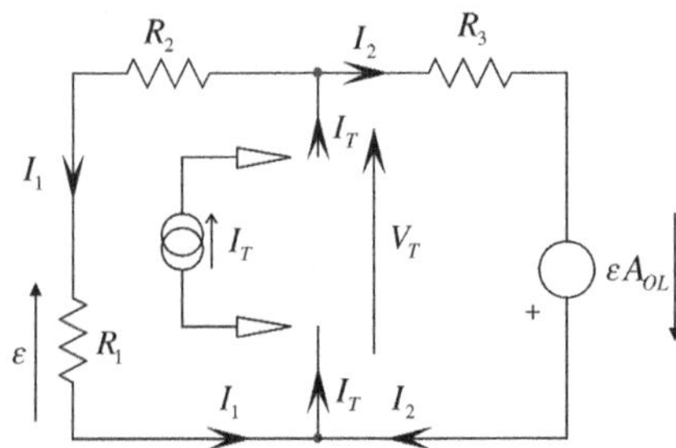

Figure 3.62 In this arrangement, we need to find the expression of V_T, the voltage across the test generator I_T.

Let's start by defining current I_1:

$$I_1 = \frac{\varepsilon}{R_1} \tag{3.184}$$

but it is also equal to

$$I_1 = \frac{V_T - \varepsilon}{R_2} \tag{3.185}$$

Equating both expressions and solving for ε leads to:

$$\varepsilon = V_T \frac{R_1}{R_1 + R_2} \tag{3.186}$$

The test current I_T is the sum of I_1 and I_2:

$$I_T = I_1 + I_2 = \frac{\varepsilon}{R_1} + \frac{V_T + \varepsilon A_{OL}}{R_3} \tag{3.187}$$

Substituting (3.186) in (3.187) gives an expression for the test current I_T as a function of V_T:

$$I_T = V_T \left(\frac{1}{R_1 + R_2} + \frac{1 + \frac{R_1}{R_1 + R_2} A_{OL}}{R_3} \right) \tag{3.188}$$

which ends up in the following resistance definition:

$$R = \frac{V_T}{I_T} = \frac{1}{\left(\frac{1}{R_1 + R_2} + \frac{1 + \frac{R_1}{R_1 + R_2} A_{OL}}{R_3} \right)} \tag{3.189}$$

Rearranging the above expression, the circuit time constant involving C_1 is thus:

$$\tau_1 = C_1 \frac{R_3(R_1 + R_2)}{R_1 + R_2 + R_3 + A_{OL} R_1} \tag{3.190}$$

We have all the needed expressions to assemble the final transfer function. It is defined by

$$\frac{V_{out}(s)}{V_{in}(s)} = H_0 \frac{1 + \frac{H^1}{H_0} s\tau_1}{1 + s\tau_1} = H_0 \frac{1 + \frac{s}{\omega_z}}{1 + \frac{s}{\omega_p}} \tag{3.191}$$

in which

$$H_0 = -\frac{R_2 + R_3}{R_1} \frac{1}{\left(\frac{\frac{R_2 + R_3}{R_1} + 1}{A_{OL}} + 1 \right)} \tag{3.192}$$

$$\omega_z = \frac{H_0}{H_1\tau_1} = \frac{-\frac{R_2+R_3}{R_1}\frac{1}{\left(\frac{\frac{R_2+R_3}{R_1}+1}{A_{OL}}+1\right)}}{-\frac{R_2}{R_1+R_2}A_{OL}C_1\frac{R_3(R_1+R_2)}{R_1+R_2+R_3+A_{OL}R_1}} \tag{3.193}$$

Simplifying the above equation leads to:

$$\omega_z = \frac{1}{C_1\left(R_2 \| R_3\right)} \tag{3.194}$$

while the pole is given by

$$\omega_p = \frac{1}{C_1\frac{R_3(R_1+R_2)}{R_1+R_2+R_3+A_{OL}R_1}} \tag{3.195}$$

We have captured all these data in a Mathcad® sheet and calculation results are shown in Figure 3.63. Once this is done, a simple SPICE simulation will confirm or show a calculation error in case discrepancies are found. The SPICE circuit is given in Figure 3.64 and dc operating points show that our results are correct. We purposely kept the op amp open-loop gain low (100 or 40 dB) to see its impact on calculations.

Subcircuit X_1 includes a pole, a zero and gain. The passed parameters are those computed by B_4 and B_5 then its ac response (V_{out2}) is compared to that of the original circuit built around E_2, V_{out}. As the bottom graphs show, responses are identical.

3.3.4 Generalized Transfer Function – Example 4

A fourth example circuit is shown in Figure 3.65. It is again built around an op amp this time affected by an infinite gain. Let's calculate the dc gain when the capacitor is removed from the circuit. The corresponding schematic is given in Figure 3.66. The gain can be obtained by applying superposition as shown in the right side of the figure. The first output, when R_1's left terminal is grounded, places the circuit in a non-inverting configuration affected by a gain H_a equal to:

$$H_a = \frac{R_2}{R_1} + 1 \tag{3.196}$$

While the second gain H_b is obtained when R_3's left terminal is grounded and places the circuit in an inverting structure. The gain is:

$$H_b = -\frac{R_2}{R_1} \tag{3.197}$$

The final gain is thus the sum of H_a and H_b which gives:

$$H_0 = -\frac{R_2}{R_1} + \left(\frac{R_2}{R_1} + 1\right) = 1 \tag{3.198}$$

The next step is to calculate the gain when C_1 is infinite valued as shown in Figure 3.67.

$R_1 := 12\text{k}\Omega \qquad R_3 := 10\text{k}\Omega \qquad R_2 := 22\text{k}\Omega \qquad C_1 := 22\text{nF} \qquad A_{OL} := 100$

$$\|(x,y) := \frac{x \cdot y}{x + y}$$

$$H_0 := -\frac{R_2 + R_3}{R_1} \cdot \frac{1}{\dfrac{\dfrac{R_2 + R_3}{R_1} + 1}{A_{OL}} + 1} = -2.57235 \qquad H_1 := -\left(\frac{R_2}{R_1 + R_2}\right) \cdot A_{OL} = -64.70588$$

$$R_d := \frac{R_3 \cdot (R_1 + R_2)}{R_1 + R_2 + R_3 + A_{OL} \cdot R_1} = 273.3119\ \Omega \qquad \tau_1 := C_1 \cdot R_d \qquad H_2(s) := H_0 \cdot \frac{1 + \dfrac{H_1}{H_0} \cdot s \cdot \tau_1}{1 + s \cdot \tau_1}$$

$$\omega_z := \frac{H_0}{H_1 \cdot \tau_1} = 6.61157 \times 10^3\ \frac{1}{s} \qquad f_z := \frac{\omega_z}{2\pi} = 1.05226\ \text{kHz} \qquad \omega_p := \frac{1}{\tau_1} = 1.6631 \times 10^5\ \frac{1}{s} \qquad f_p := \frac{\omega_p}{2\pi} = 26.46908\ \text{kHz}$$

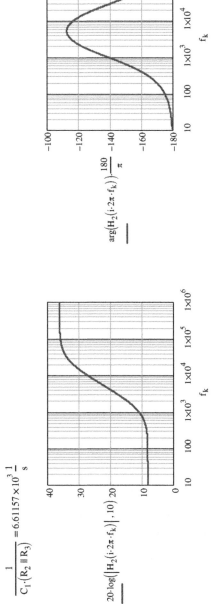

$$\frac{1}{C_1 \cdot (R_2 \| R_3)} = 6.61157 \times 10^3\ \frac{1}{s}$$

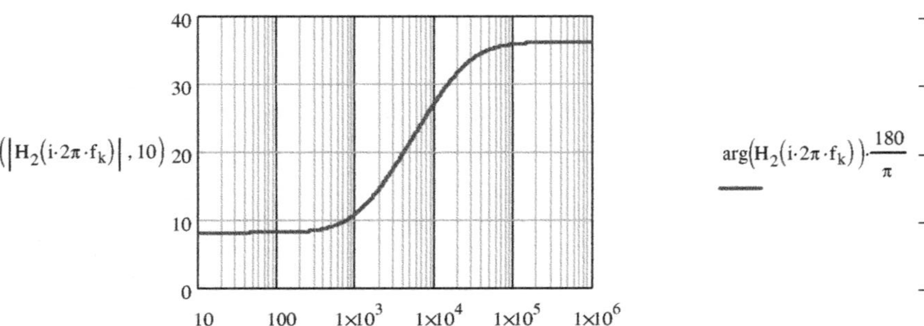

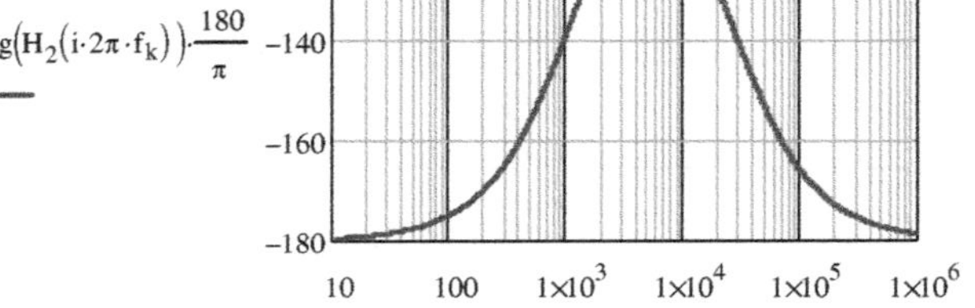

Figure 3.63 Mathcad® presents the results in a clear or ordered form that can be compared to what SPICE gives.

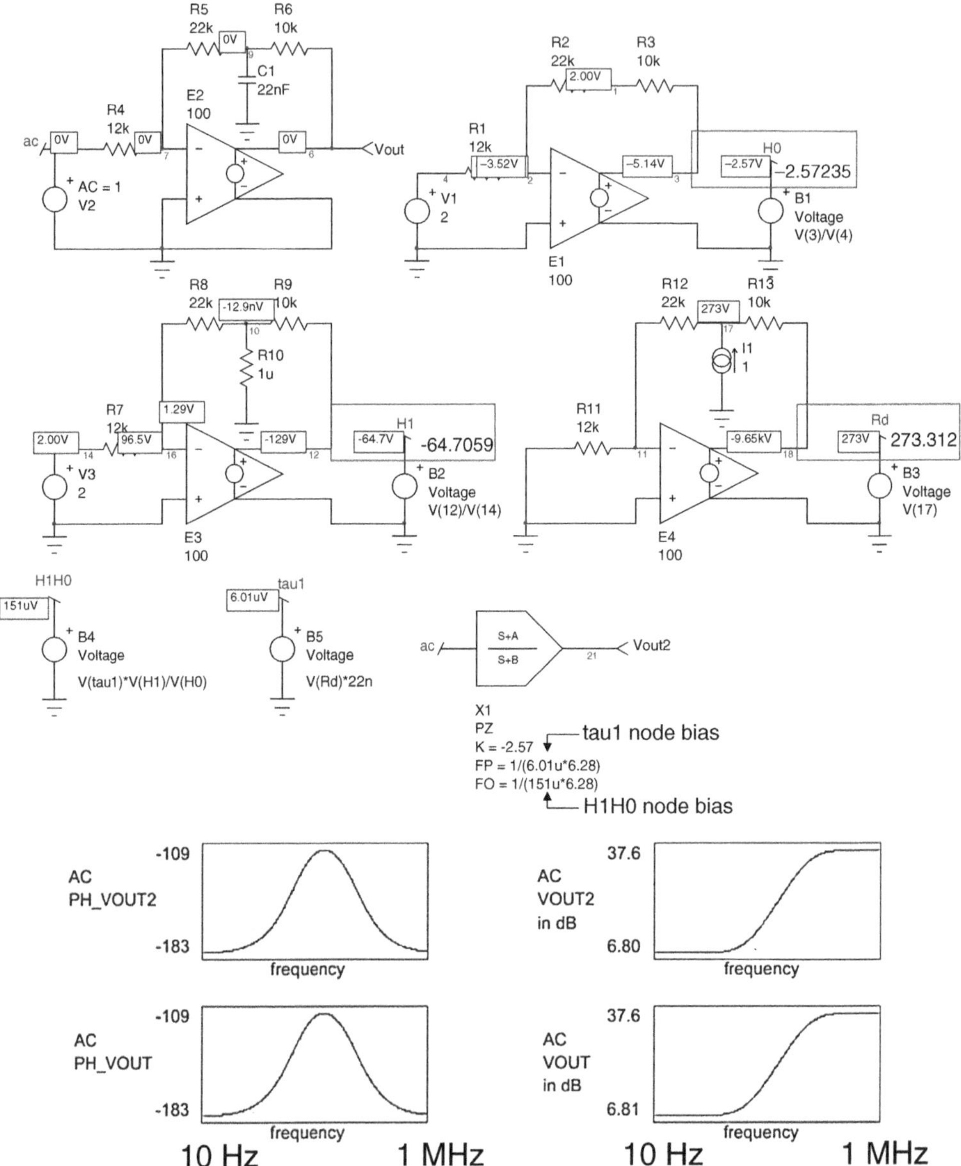

Figure 3.64 SPICE is really helpful to check that the derived equations are correct. The 1 μΩ resistance in H^1 calculation is a short circuit as SPICE would not accept a 0 Ω value.

We have a simple inverter whose gain is similar to that of (3.196):

$$H^1 = -\frac{R_2}{R_1} \tag{3.199}$$

The circuit time constant is obtained while setting the excitation voltage to 0 V as illustrated in Figure 3.68. In this mode, the only resistance seen from the connecting terminals is R_3 since the

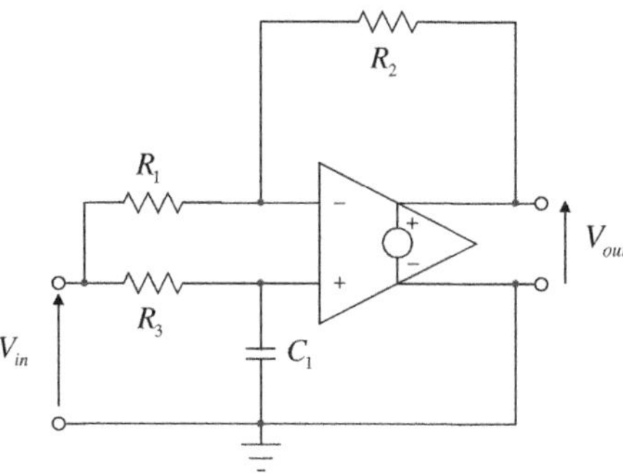

Figure 3.65 This op amp filter is built with three resistors and a capacitor. This is a 1st-order system.

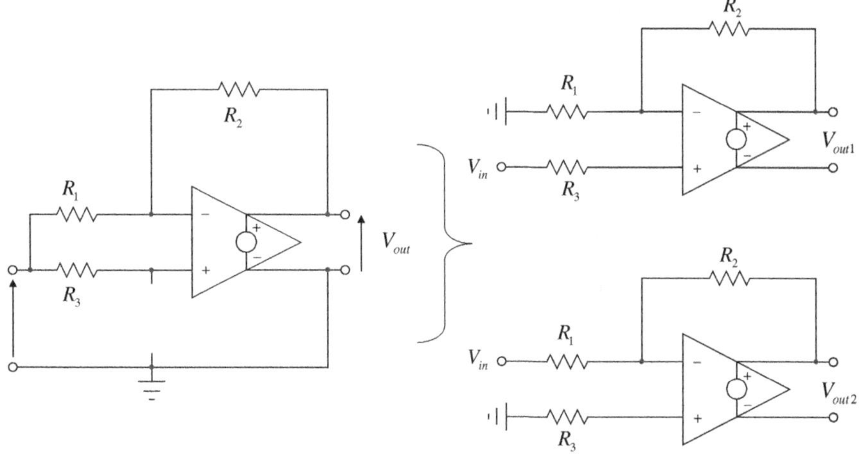

Figure 3.66 Using superposition is a quick means to derive this dc gain.

non-inverting pin offers an infinite input resistance. The time constant of this circuit is therefore:

$$\tau_1 = R_3C_1 \tag{3.200}$$

We now have all the ingredients to assemble the final transfer function. It is defined by:

$$H(s) = H_0\frac{1+\frac{H^1}{H_0}s\tau_1}{1+s\tau_1} = \frac{1-s\frac{R_2}{R_1}R_3C_1}{1+sR_3C_1} = \frac{1-\frac{s}{\omega_z}}{1+\frac{s}{\omega_p}} \tag{3.201}$$

in which

$$\omega_z = \frac{R_1}{R_2R_3C_1} \tag{3.202}$$

and

$$\omega_p = \frac{1}{R_3C_1} \tag{3.203}$$

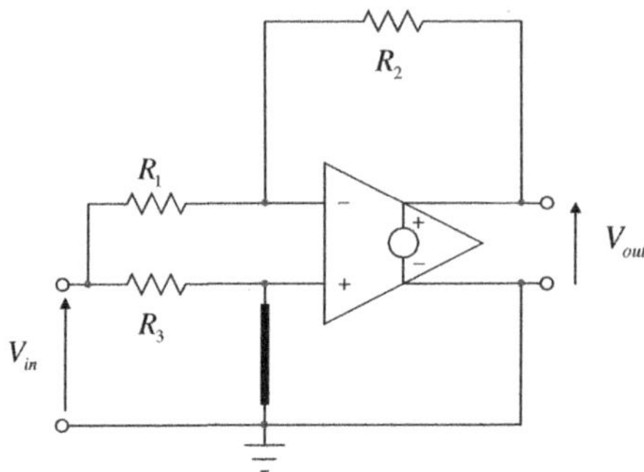

Figure 3.67 When C_1 is replaced by a strap, the gain becomes a simple expression.

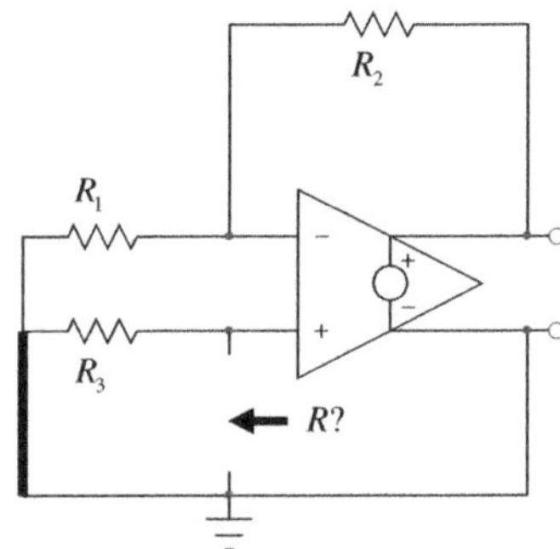

Figure 3.68 The resistance driving the capacitor is easy to determine, it is R_3.

As you can see with (3.201), the circuit features a right half-plane zero. Associated with the 90° pole lag, it will bring the total lag to 180°. Figure 3.69 and Figure 3.70 confirm these results through Mathcad® and SPICE analysis.

3.3.5 Generalized Transfer Function – Example 5

The last example will apply the generalized transfer function formula to the simple integrator shown in Figure 3.71. The dc gain H_0 is obtained when capacitor C_1 is removed from the circuit as illustrated in Figure 3.72.

In this case, the gain is simply

$$H_0 = -A_{OL} \tag{3.204}$$

Then the excitation is set to 0 V and we install a current generator I_T across C_1 as indicated in Figure 3.73. The voltage across the generator is

$$V_T = R_1 I_T - V_{out} \tag{3.205}$$

The op amp output voltage is given by:

$$V_{out} = \varepsilon \cdot A_{OL} \tag{3.206}$$

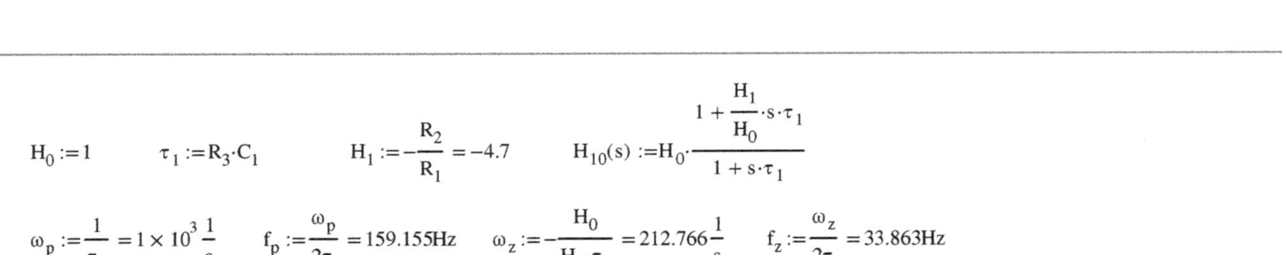

$R_1 := 1k\Omega \qquad R_2 := 4.7k\Omega \qquad R_3 := 10k\Omega \qquad C_1 := 0.1\mu F$

$H_0 := 1 \qquad \tau_1 := R_3 \cdot C_1 \qquad H_1 := -\frac{R_2}{R_1} = -4.7 \qquad H_{10}(s) := H_0 \cdot \frac{1 + \frac{H_1}{H_0} \cdot s \cdot \tau_1}{1 + s \cdot \tau_1}$

$\omega_p := \frac{1}{\tau_1} = 1 \times 10^3 \frac{1}{s} \qquad f_p := \frac{\omega_p}{2\pi} = 159.155Hz \qquad \omega_z := -\frac{H_0}{H_1 \cdot \tau_1} = 212.766 \frac{1}{s} \qquad f_z := \frac{\omega_z}{2\pi} = 33.863Hz$

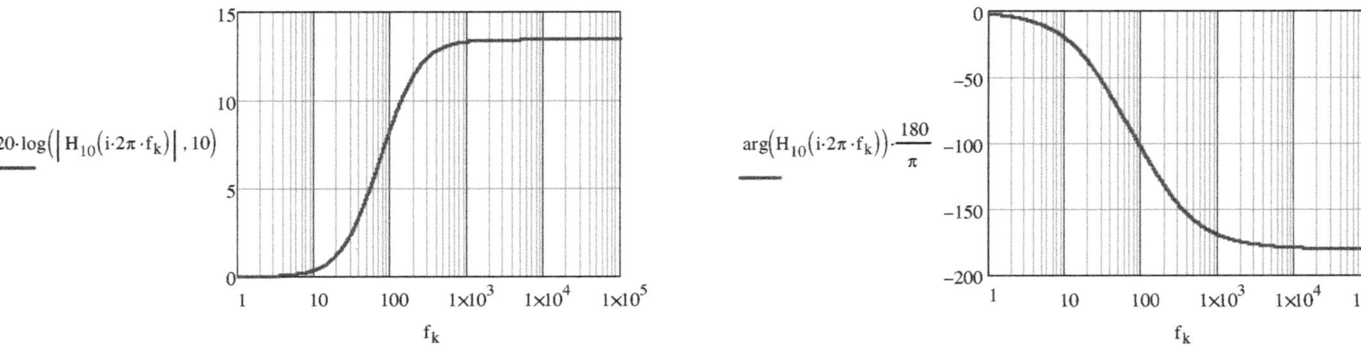

Figure 3.69 Mathcad® plots the response that we can compare with those of SPICE.

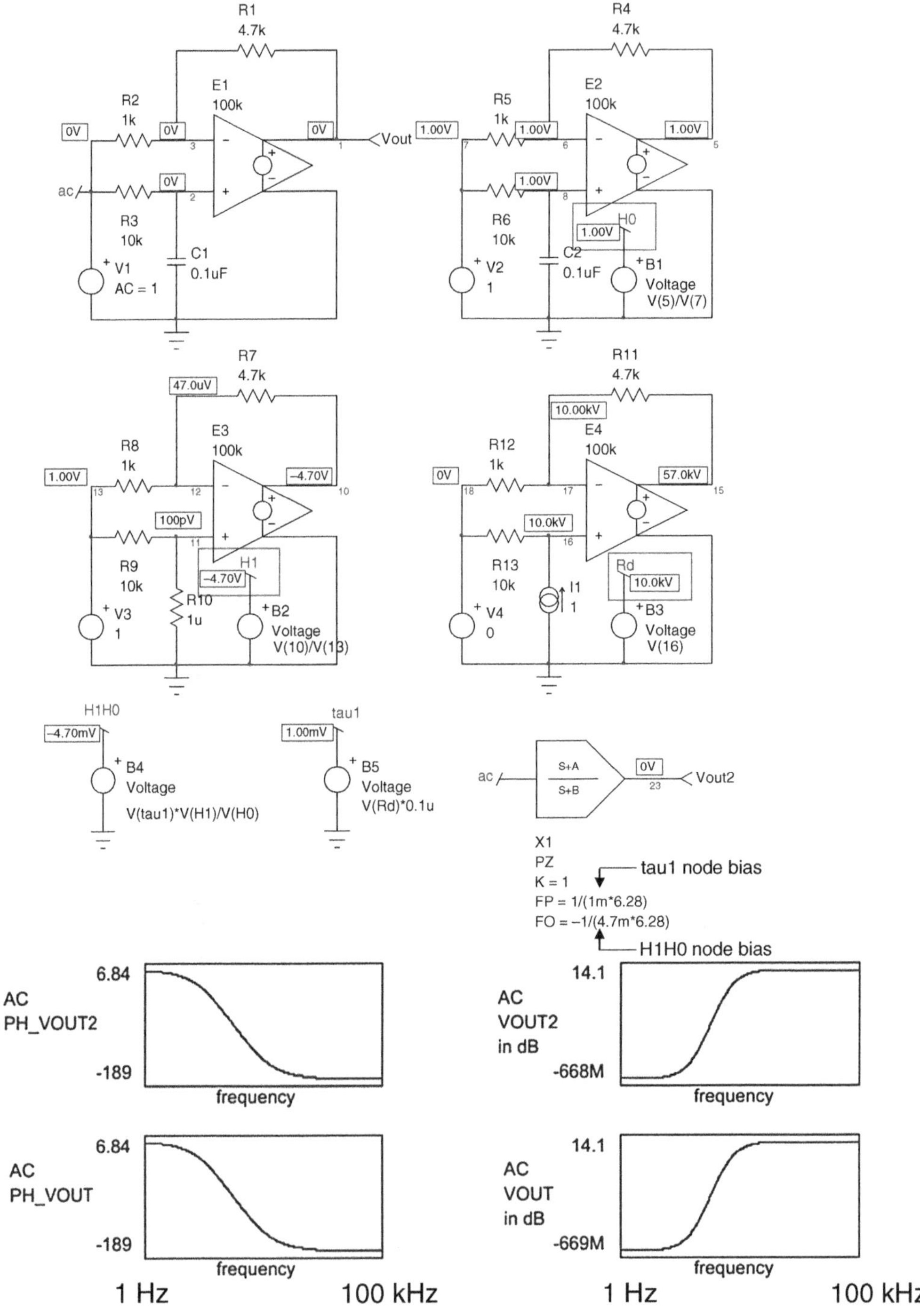

Figure 3.70 SPICE confirms our calculations for the independent calculation steps and the final transfer function.

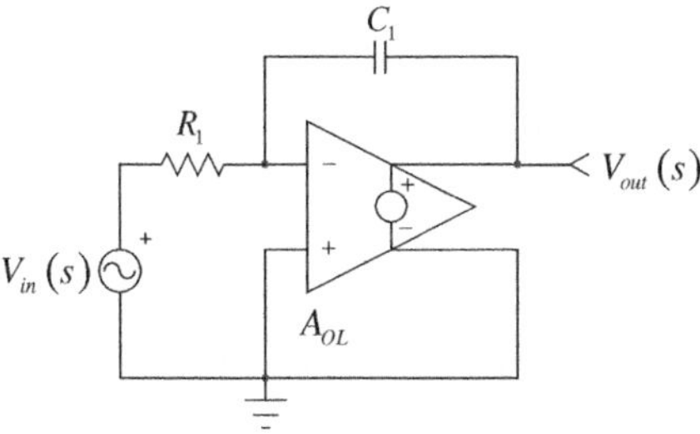

Figure 3.71 A simple integrator is formed with an op amp featuring an open-loop gain A_{OL}.

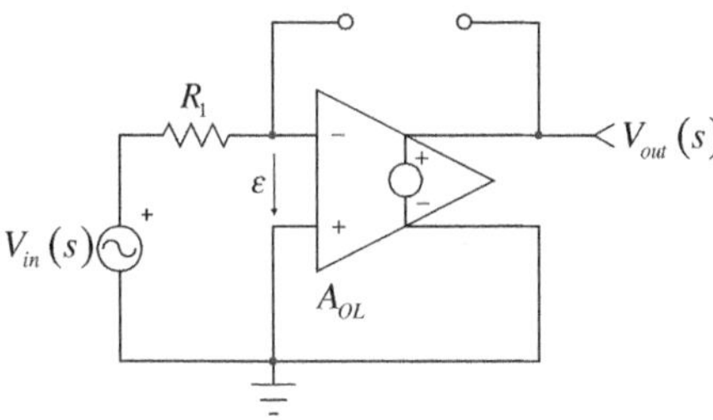

Figure 3.72 When the circuit is studied for $s = 0$, the capacitor is physically removed.

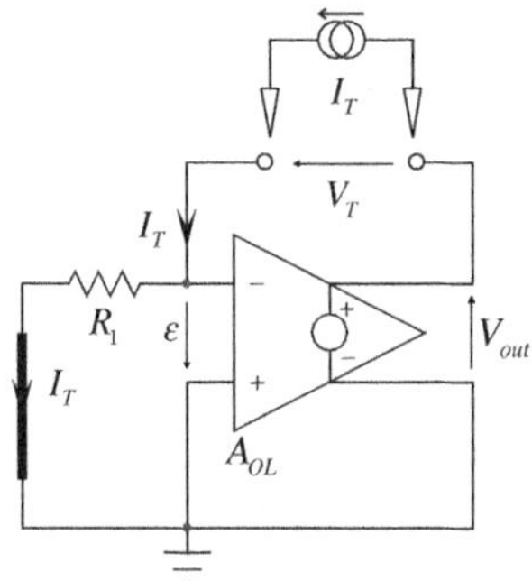

Figure 3.73 The current generator lets us determine the resistance driving C_1 at zero excitation.

where

$$\varepsilon = -R_1 I_T \tag{3.207}$$

Substituting (3.207) in (3.206) gives:

$$V_{out} = -R_1 I_T A_{OL} \tag{3.208}$$

which substituted in (3.205) leads to

$$V_T = I_T R_1 + R_1 I_T A_{OL} = I_T R_1 (1 + A_{OL}) \tag{3.209}$$

The resistance driving C_1 is then derived as:

$$R = \frac{V_T}{I_T} = R_1(1 + A_{OL}) \tag{3.210}$$

The time constant τ_1 is thus:

$$\tau_1 = C_1R_1(1 + A_{OL}) \tag{3.211}$$

When setting C_1 in its high-frequency state (a short circuit) as in Figure 3.74, the response is 0 V. Therefore:

$$H^1 = 0 \tag{3.212}$$

Applying the generalized transfer function expression, we find:

$$H(s) = \frac{H_0 + H^1 s\tau_1}{1 + s\tau_1} = -\frac{A_{OL}}{1 + sR_1C_1(1 + A_{OL})} = -\frac{A_{OL}}{1 + \dfrac{s}{\omega_p}} \tag{3.213}$$

in which the low-frequency pole is expressed as:

$$\omega_p = \frac{1}{R_1C_1(1 + A_{OL})} \tag{3.214}$$

We can now factor A_{OL} to obtain:

$$H(s) = -\frac{A_{OL}}{A_{OL}}\frac{1}{\left(\dfrac{1}{A_{OL}} + sR_1C_1\dfrac{1 + A_{OL}}{A_{OL}}\right)} = -\frac{1}{\dfrac{1}{A_{OL}} + sR_1C_1\dfrac{1 + A_{OL}}{A_{OL}}} \tag{3.215}$$

When A_{OL} is very large, this expression simplifies to:

$$H(s) = -\frac{1}{sR_1C_1} = -\frac{1}{\dfrac{s}{\omega_{po}}} \tag{3.216}$$

in which

$$\omega_{po} = \frac{1}{R_1C_1} \tag{3.217}$$

is the 0-dB crossover pole, the frequency at which the gain is 0 dB. Figure 3.75 represents the dynamic responses of both transfer functions. H_2 represents (3.213) and confirms the low-frequency pole whose value depends on A_{OL} while H_3 confirms a 0-dB crossover frequency of 1.6 kHz when A_{OL} is considered infinite.

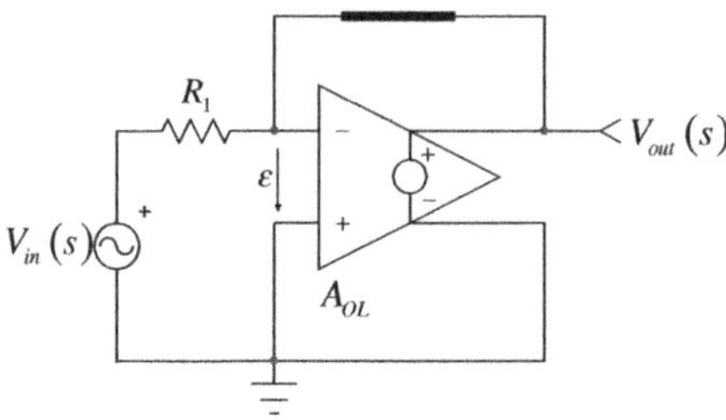

Figure 3.74 The transfer function when C_1 is a short circuit is 0.

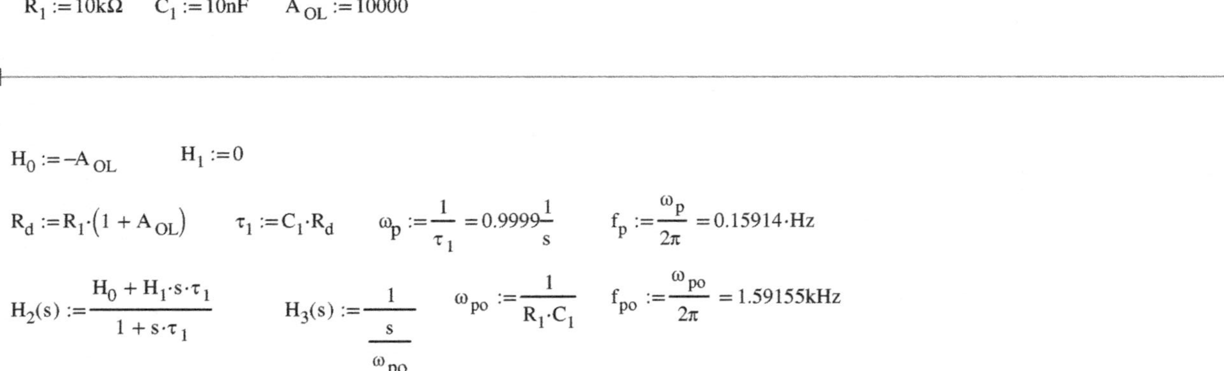

$R_1 := 10k\Omega \quad C_1 := 10nF \quad A_{OL} := 10000$

$H_0 := -A_{OL} \quad H_1 := 0$

$R_d := R_1 \cdot (1 + A_{OL}) \quad \tau_1 := C_1 \cdot R_d \quad \omega_p := \frac{1}{\tau_1} = 0.9999\frac{1}{s} \quad f_p := \frac{\omega_p}{2\pi} = 0.15914 \cdot Hz$

$H_2(s) := \frac{H_0 + H_1 \cdot s \cdot \tau_1}{1 + s \cdot \tau_1} \quad H_3(s) := \frac{1}{\frac{s}{\omega_{po}}} \quad \omega_{po} := \frac{1}{R_1 \cdot C_1} \quad f_{po} := \frac{\omega_{po}}{2\pi} = 1.59155kHz$

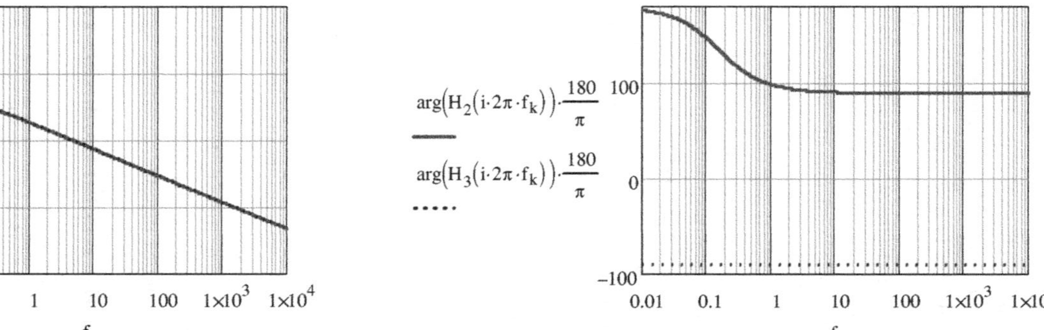

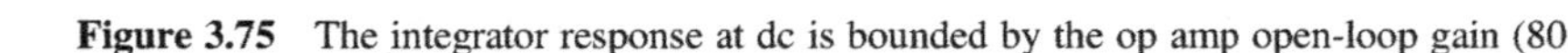

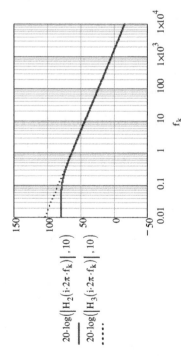

Figure 3.75 The integrator response at dc is bounded by the op amp open-loop gain (80 dB).

3.4 Further Reading

This fifth example ends our chapter on the Extra Element Theorem and its applications. I encourage you to further dig into the subject by reading other papers and documents available on the web or in dedicated books. Besides the founding paper written by the late Dr. Middlebrook in [3], you can visit the website address given in [6] which contains most of the presentations and seminars taught by Dr. Middlebrook about Design-Oriented Analysis. The course tackles the EET and abounds with examples and interesting information on design techniques. Reference [4] cannot be ignored and rigorously describes the EET with a lot of examples ranging from the simple 1st order circuit to more complex networks including many uncommon structures. Reference [5] is interesting as it presents a different approach in which NDI is not performed. Again, it is a matter of choice, whether you accept NDI as a necessary step to get the simplest numerator expression or you prefer to go through the generalized transfer function expression and 'suffer' later to simplify the numerator. Both expressions are similar in terms of results but it is true that with its manipulation of sometimes complex H^1 gain expressions, the generalized transfer function can lead to quite large formulas. Reference [7] offers an interesting introduction to the technique with historical details on how the EET and Fast Analytical Circuit Techniques emerged as a solution for solving circuit equations. Reference [8] links to a university website in Belgium where the authors have implemented various theorems from Dr. Middlebrook (EET but also the General Feedback Theorem, the GFT) in a simulation environment like Cadence®. Plenty of links and documents are available from this address. Reference [9] points to a Yahoo! users group where designers discuss FACTs and interact with experts in this field. Finally, [10] does not directly discuss EET but covers techniques for loop analysis using the GFT and it contains a lot of useful information.

Despite all these documents and examples, nothing replaces practical exercises: you must invest time in solving circuits using the EET and its different forms to become skilled in the art. The nice thing is that SPICE can immediately tell you if you are wrong or in which parts currents circulate and in what direction when you null the output or look for a particular time constant. This is a tremendous help to track down any mistake you could have missed.

Now that we described how to solve 1st-order circuits, Chapter 4 will teach us how to tackle second order networks. Before reading the next chapter, be sure to exercise yourself with the problem examples given at the end of this one.

3.5 What Should I Retain from this Chapter?

In this third chapter, we explored the EET, the heart of the Fast Analytical Circuits Techniques that we applied in several examples. Below is a summary of what we learned in this chapter:

1. In a linear circuit involving more than one individual source, superposition theorem states that the response (a current or a voltage) in a branch is the algebraic sum of the responses obtained from each of the individual sources while the other sources are turned off. A voltage source that is turned off or set to 0 V is replaced by a short circuit while a current source set to 0 A is open circuited.
2. In the above definition, one output (the response) and several inputs (the stimuli) are considered. The total output is calculated by combining responses obtained while only one source is active at a time. The Extra Element Theorem (EET) derives from the superposition theorem by considering two outputs and two inputs. Unlike superposition, two inputs can be active at the same time, engendering two output signals. For a certain inputs combination, one of the two outputs can be nulled, meaning it delivers 0 (V or A). This condition is called the Null Double Injection or NDI.
3. To apply the EET, identify one element in the circuit that complicates the circuit architecture: this is the extra element. Depending on the EET form, this extra element can be physically removed or

replaced by a strap. One usually selects the form that leads to a simpler circuit once the extra element is removed or shorted. The transfer function is then evaluated in either condition according to the selected EET form.

4. The zero, if present, is located in the numerator and combines the energy-storing element with a resistance R_n. This variable is the resistance seen from the energy-storing element port while an NDI is performed: the input signal is kept alive and one must determine the port resistance while the output is nulled.
5. The pole is obtained by zeroing the excitation signal and looking into the energy-storing element terminals to determine the driving resistance R_d. That resistance combined with the energy-storing element determines the circuit time constant τ.
6. The EET can be rearranged in a different manner to form a 1st-order generalized transfer function. In this expression, there is no need to obtain R_n via an NDI but rather calculate a dc gain H_0 and a high-frequency gain H^1. Combining these two elements with the circuit time constant τ forms the numerator. R_d is determined the same way as before. This technique shields you from performing an NDI but usually leads to slightly more complex coefficients compared to an EET derivation. The choice is yours to apply either techniques.
7. The EET can be applied to a variety of circuits and examples show it in action with passive circuits, bipolar transistors or operational amplifiers. The EET applies equally well for active circuits but determining R_n and R_d complicates the exercise with dependent sources.

The Extra Element Theorem (EET)
A is a transfer function linking a response to an excitation
Z is the extra element
Z_n is the impedance seen from the extra element port for a null in the response with excitation left in place
Z_d is the impedance seen from the extra element port for a zeroed excitation

$$A\big|_Z = A\big|_{Z=\infty}\frac{1+\dfrac{Z_n}{Z}}{1+\dfrac{Z_d}{Z}} \qquad A\big|_Z = A\big|_{Z=0}\frac{1+\dfrac{Z}{Z_n}}{1+\dfrac{Z}{Z_d}}$$

1. Calculate A when Z is removed
2. Calculate Z_n for a nulled response
3. Calculate Z_d for a zeroed excitation

1. Calculate A when Z is shorted
2. Calculate Z_n for a nulled response
3. Calculate Z_d for a zeroed excitation

If Z is a capacitor C or an inductor L
R_n is the resistance seen from C or L port for a null in the response with excitation left in place
R_d is the resistance seen from C or L port for a zeroed excitation

$$Z = \frac{1}{sC} \quad \tau_1 = R_n C \quad \tau_2 = R_d C$$

If a dc gain H_0 exists:

$$H(s) = H_0\frac{1+s\tau_1}{1+s\tau_2} = H_0\frac{1+\dfrac{s}{\omega_{z_1}}}{1+\dfrac{s}{\omega_{p_1}}}$$

$$Z = sL \quad \tau_1 = \frac{L}{R_n} \quad \tau_2 = \frac{L}{R_d}$$

If a gain H_∞ exists as s approaches ∞:

$$H(s) = H_\infty\frac{1+\dfrac{1}{s\tau_1}}{1+\dfrac{1}{s\tau_2}} = H_\infty\frac{1+\dfrac{\omega_{z_1}}{s}}{1+\dfrac{\omega_{p_1}}{s}}$$

Figure 3.76 Summary figure for the EETs and 1st-order transfer functions.

The generalized 1st-order transfer function
A is a transfer function linking a response to an excitation
A_0 is the transfer function when the circuit is observed at $s = 0$ (dc)
A^1 is the transfer function when the circuit is observed at $s \to \infty$
τ_1 is the time constant involving the energy-storing element

$$A(s) = \frac{A_0 + A^1 s\tau_1}{1 + s\tau_1}$$

$A^{\boxed{1}}$ ← The element involved in τ_1 is put in its high frequency state.

1. Calculate A_0 when C is removed or L is shorted
2. Determine R looking into the energy-storing terminals for a zeroed excitation
3. Compute $\tau_1 = RC$ or $\tau_1 = L/R$
4. Calculate A^1 when C is shorted or L is removed
5. If A_0 exists (different than 0), factor it in the numerator

$$A(s) = A_0 \frac{1 + \frac{A^1}{A_0} s\tau_1}{1 + s\tau_1}$$

A_0 ← The energy-storage element is put in its dc state.

Figure 3.77 Summary figure for the generalized 1st-order transfer function.

8. SPICE is an excellent resource to verify calculations either obtained by hand or from a mathematical solver. SPICE lets you find R_d with a simple dc current source driving the energy-storing element connecting terminals at a zeroed excitation but also provides an easy means to determine R_n via an NDI involving a transconductance amplifier biasing the energy-storing element terminals while ensuring an output null.
9. Figure 3.76 summarizes the different forms of EET and how you can combine time constants to form the numerator (zero) and the denominator (pole) of the transfer function under study. Figure 3.77 gives the definition of the generalized 1st-order transfer function form which avoids the NDI.

References

1. http://www.solved-problems.com/circuits/circuits-articles/839/turning-sources-off/ (last accessed 12/12/2015).
2. http://users.ece.gatech.edu/mleach/papers/superpos.pdf (last accessed 12/12/2015).
3. R. D. Middlebrook. Null Double Injection and the Extra Element Theorem. *IEEE Transactions on Education*, **32** (3), 167–180.
4. V. Vorpérian. *Fast Analytical Techniques for Electrical and Electronic Circuits*. Cambridge University Press, 2002, p 4.
5. A. Hajimiri. Generalized Time- and Transfer-Constant Circuit Analysis. *Transactions on Circuits and Systems*, **57** (6), 1105–1121. 2009. http://www.ece.ucsb.edu/Faculty/rodwell/Classes/mixed_signal/Hajimiri_MOTC.pdf (last accessed 12/12/2015).
6. http://www.rdmiddlebrook.com/ (last accessed 12/12/2015).
7. http://www.edn.com/electronics-blogs/outside-the-box-/4404226/Design-oriented-circuit-dynamics (last accessed 12/12/2015).
8. http://www.analogdesign.be/ (last accessed 12/12/2015).
9. https://groups.yahoo.com/neo/groups/Design-Oriented_Analysis_D-OA/info (last accessed 12/12/2015).
10. https://sites.google.com/site/frankwiedmann/loopgain (last accessed 12/12/2015).

3.6 Appendix 3A – Problems

Below are several problems based on the information delivered in this chapter.

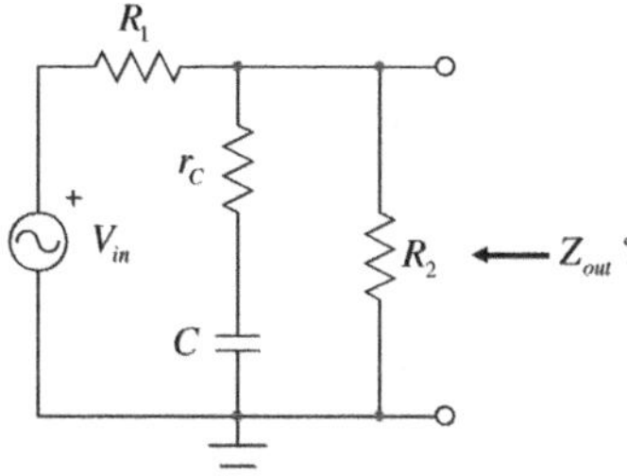

Figure 3.78 Problem 1 – Find the output impedance of this network.

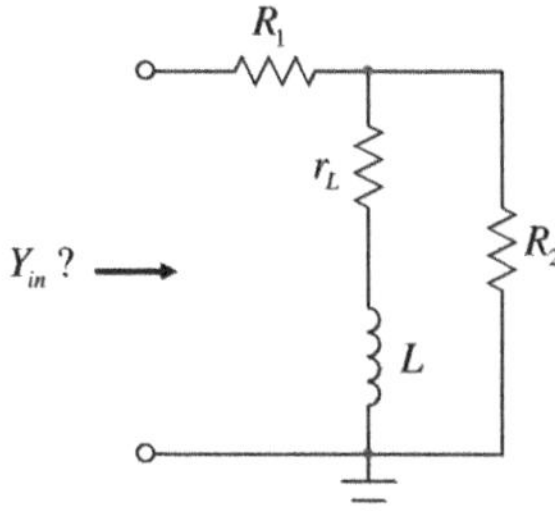

Figure 3.79 Problem 2 – Find the input admittance of this network.

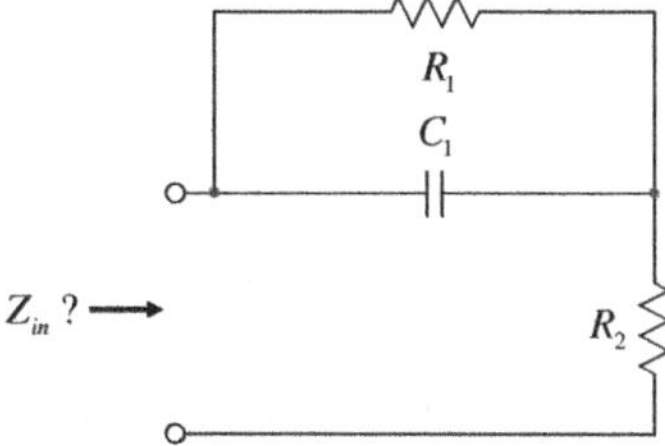

Figure 3.80 Problem 3 – Find the input impedance of this network.

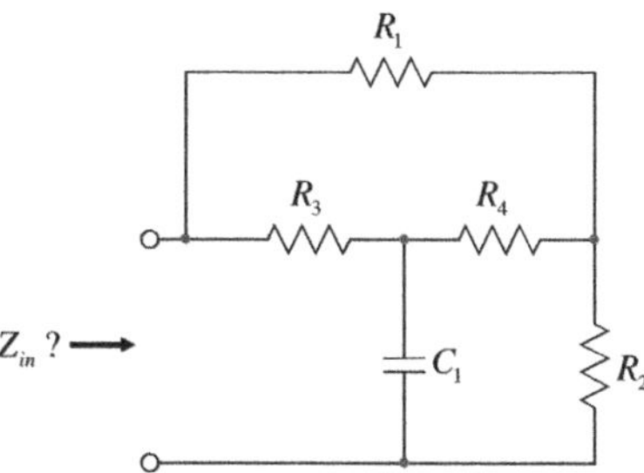

Figure 3.81 Problem 4 – Find the input impedance of this network using the 1st-order generalized transfer function without NDI and compare results obtained with NDI.

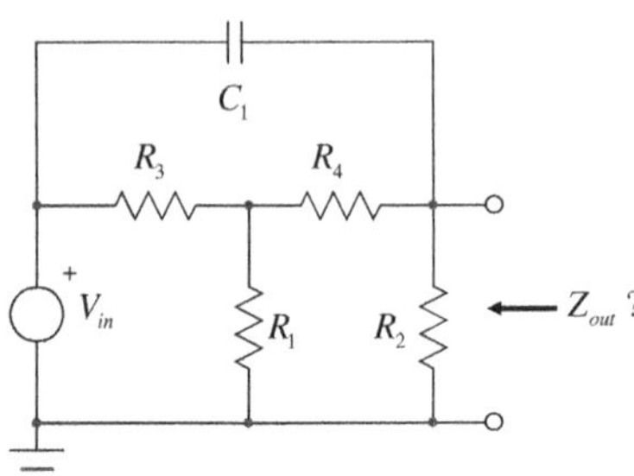

Figure 3.82 Problem 5 – Find the output impedance of this network.

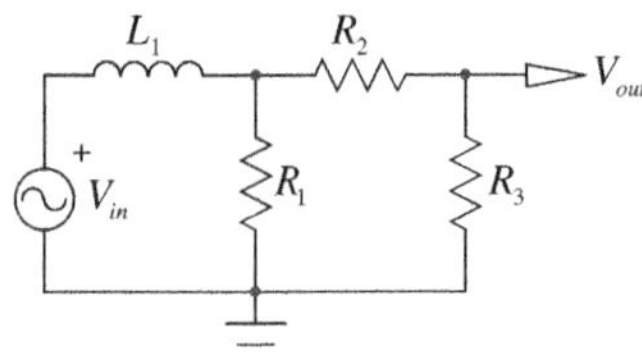

Figure 3.83 Problem 6 – Find the transfer function of this network.

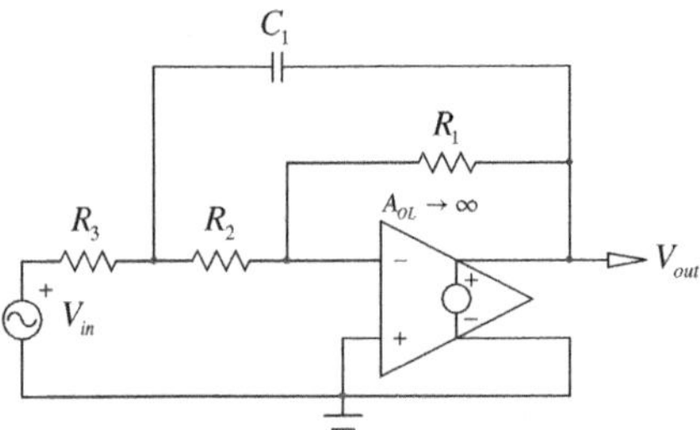

Figure 3.84 Problem 7 – Find the transfer function of this circuit featuring an op amp having an infinite open-loop gain.

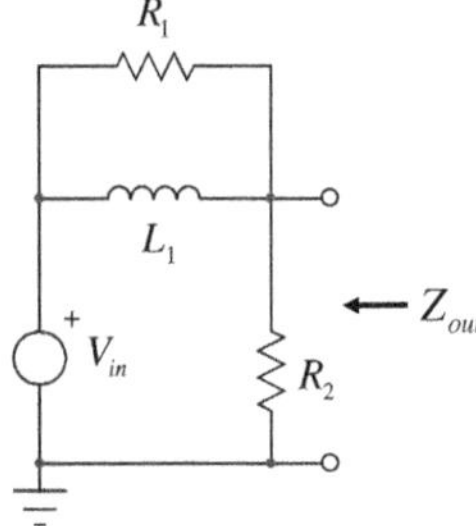

Figure 3.85 Problem 8 – Find the output impedance of this network.

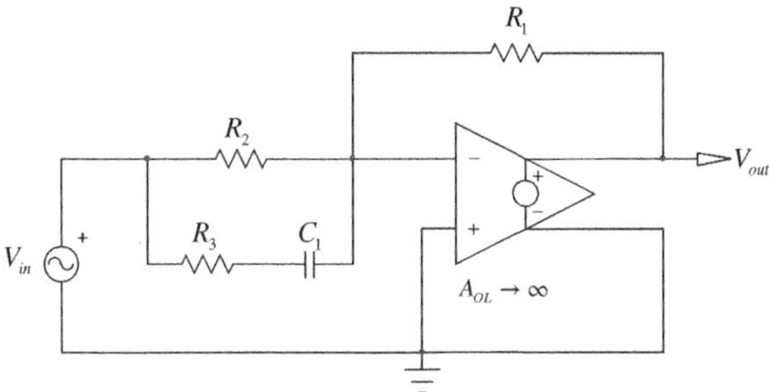

Figure 3.86 Problem 9 – Find the transfer function of this circuit.

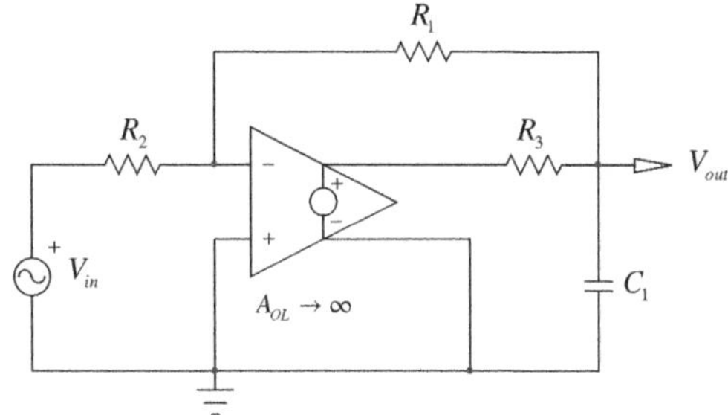

Figure 3.87 Problem 10 – Find the transfer function of this circuit.

Answers

Problem 1:

This network can be rearranged as shown in Figure 3.88. The driving signal is the current generator I_T while the response is the voltage V_T. There is one storage element, this is a 1st-order system. Is there a zero? If C is infinite valued or shorted, do we still have a voltage V_T? Yes, there is a zero associated with C. The transfer function follows the form:

$$Z_{out}(s) = R_0 \frac{1 + s\tau_1}{1 + s\tau_2} \tag{3.218}$$

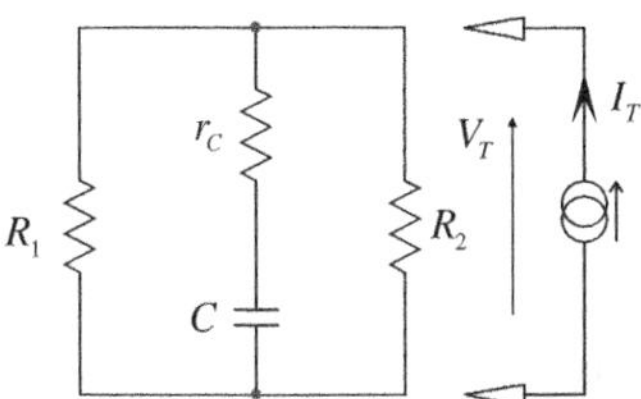

Figure 3.88 The excitation signal is I_T while the response is V_T.

R_0 is obtained in dc, when C is removed from the diagram. We have:

$$R_0 = R_2 \| R_1 \tag{3.219}$$

In Figure 3.88, what impedance combination could prevent the response voltage V_T to exist when I_T drives the circuit? If a short circuit exists across R_2, there is no response. This short circuit can be obtained if the impedance made of r_C in series with C becomes zero at an angular frequency s_z. In other terms:

$$r_C + \frac{1}{sC} = \frac{1 + sr_C C}{sC} = 0 \tag{3.220}$$

The solution is:

$$s_z = -\frac{1}{r_C C} \tag{3.221}$$

or

$$\omega_z = \frac{1}{r_C C} \tag{3.222}$$

If we prefer to calculate a null instead, look at the circuit in Figure 3.89. Because a null across a current source is similar to replacing the source by a short circuit, then we see that the upper side of r_C is grounded. Therefore, the resistance seen from the capacitor terminals is r_C and the zero time constant is simply:

$$\tau_1 = r_C C \tag{3.223}$$

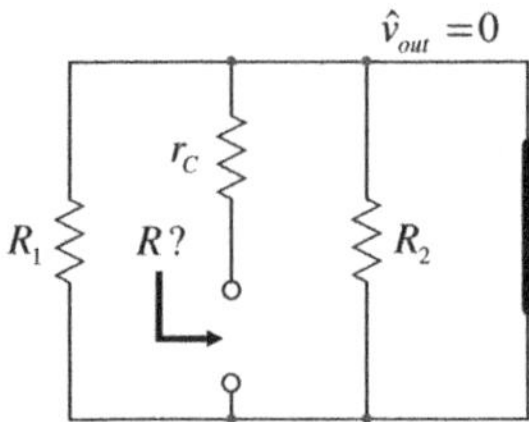

Figure 3.89 A null across a current source is equivalent to a strapped generator.

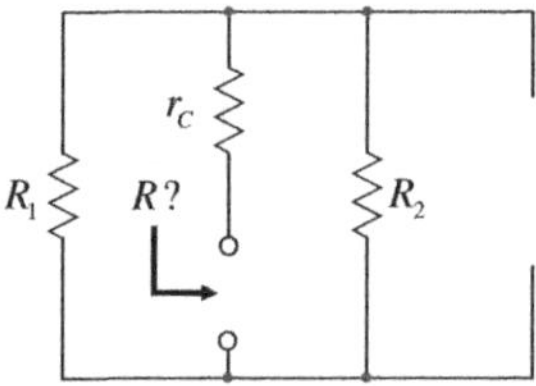

Figure 3.90 Setting the excitation to zero means removing the current generator.

The circuit time constant is identified by setting the excitation to zero. As we have a current generator, setting it to zero means open circuiting it. This is what we have in Figure 3.90. In this configuration, the resistance seen from the capacitor terminals is simply:

$$R = r_C + R_1 \| R_2 \tag{3.224}$$

leading to a time constant τ_2 equal to:

$$\tau_2 = \left(r_C + R_1 \| R_2\right) C \tag{3.225}$$

The output impedance expression is thus:

$$Z_{out}(s) = R_0 \frac{1 + \frac{s}{\omega_z}}{1 + \frac{s}{\omega_p}} \tag{3.226}$$

in which

$$R_0 = R_2 \| R_1 \tag{3.227}$$

$$\omega_z = \frac{1}{r_C C} \tag{3.228}$$

$$\omega_p = \frac{1}{(r_C + R_1 \| R_2)C} \tag{3.229}$$

Problem 2:

The input admittance is derived by driving the network with a voltage source and considering the current as the response. Several options are available to find this expression. Let's adopt the EET and consider L as a short circuit first:

$$Y|_{L=0} = \frac{1}{R_1 + r_L \| R_2} \tag{3.230}$$

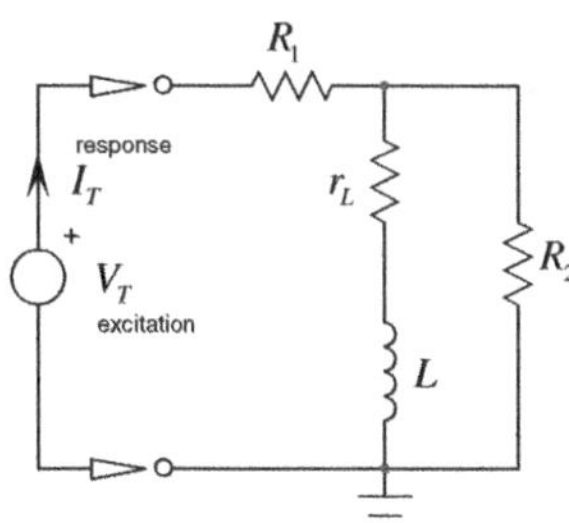

Figure 3.91 An admittance is obtained while driving the network with a voltage source.

What resistor R_n drives the inductor L when the response is a null? In other words, what is the resistance offered by the inductor terminals when current I_T is 0? The configuration appears in Figure 3.92 and the answer is immediate as R_1 is excluded from the picture:

$$R_n = r_L + R_2 \tag{3.231}$$

The zero could also be derived without going through NDI. If the current is 0 at an angular frequency s_k (our zero), the impedance defined by:

$$Z = R_1 + (r_L + s_k L) \| R_2 \tag{3.232}$$

is infinite. If you solve (3.232) for this condition, you find:

$$s_k L + r_L + R_2 = 0 \tag{3.233}$$

whose solution is

$$s_k = -\frac{r_L + R_2}{L} \tag{3.234}$$

The time constant involving the inductor is obtained when the excitation is set to 0 V. Figure 3.93 shows this configuration where the source is short circuited. The answer is immediate:

$$R_d = r_L + R_1 \| R_2 \tag{3.235}$$

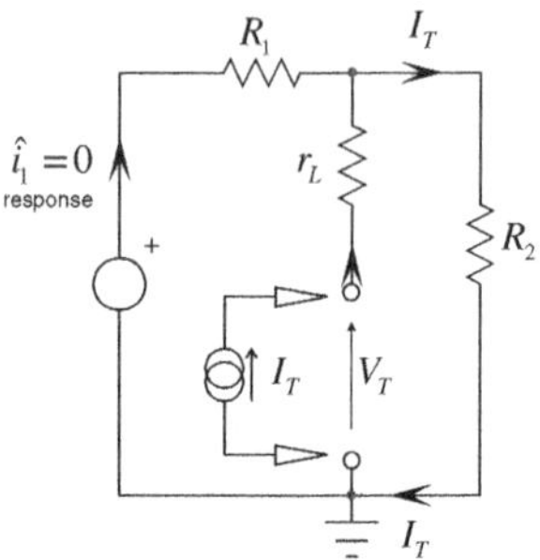

Figure 3.92 The response is the current i_1 in this case and is null: all the current I_T crosses r_L and R_2.

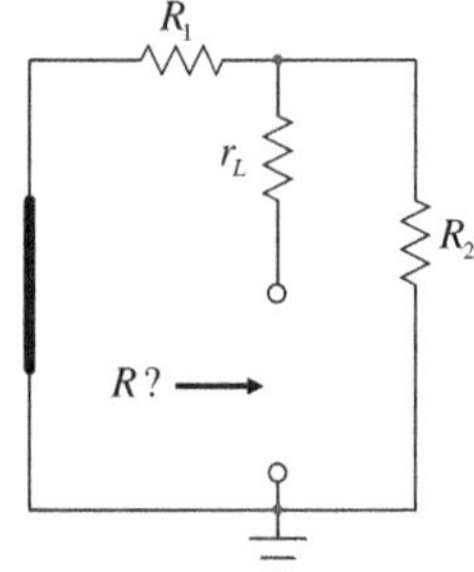

Figure 3.93 Setting the excitation to zero means strapping the voltage source.

We can assemble our transfer function applying (3.55):

$$Y|_Z = Y|_{Z=0} \frac{1 + \frac{sL}{R_n}}{1 + \frac{sL}{R_d}} = \frac{1}{R_1 + r_L \| R_2} \frac{1 + s\frac{L}{r_L + R_2}}{1 + s\frac{L}{r_L + R_1 \| R_2}} \tag{3.236}$$

This expression fits the form:

$$Y(s) = Y_0 \frac{1 + \frac{s}{\omega_z}}{1 + \frac{s}{\omega_p}} \tag{3.237}$$

in which

$$Y_0 = \frac{1}{R_1 + r_L \| R_2} \tag{3.238}$$

$$\omega_z = \frac{r_L + R_2}{L} \tag{3.239}$$

$$\omega_p = \frac{r_L + R_1 \| R_2}{L} \tag{3.240}$$

To verify these data, we have built a Mathcad® sheet and a simulation bench. They respectively appear in Figure 3.94 and Figure 3.95. They match each other very well.

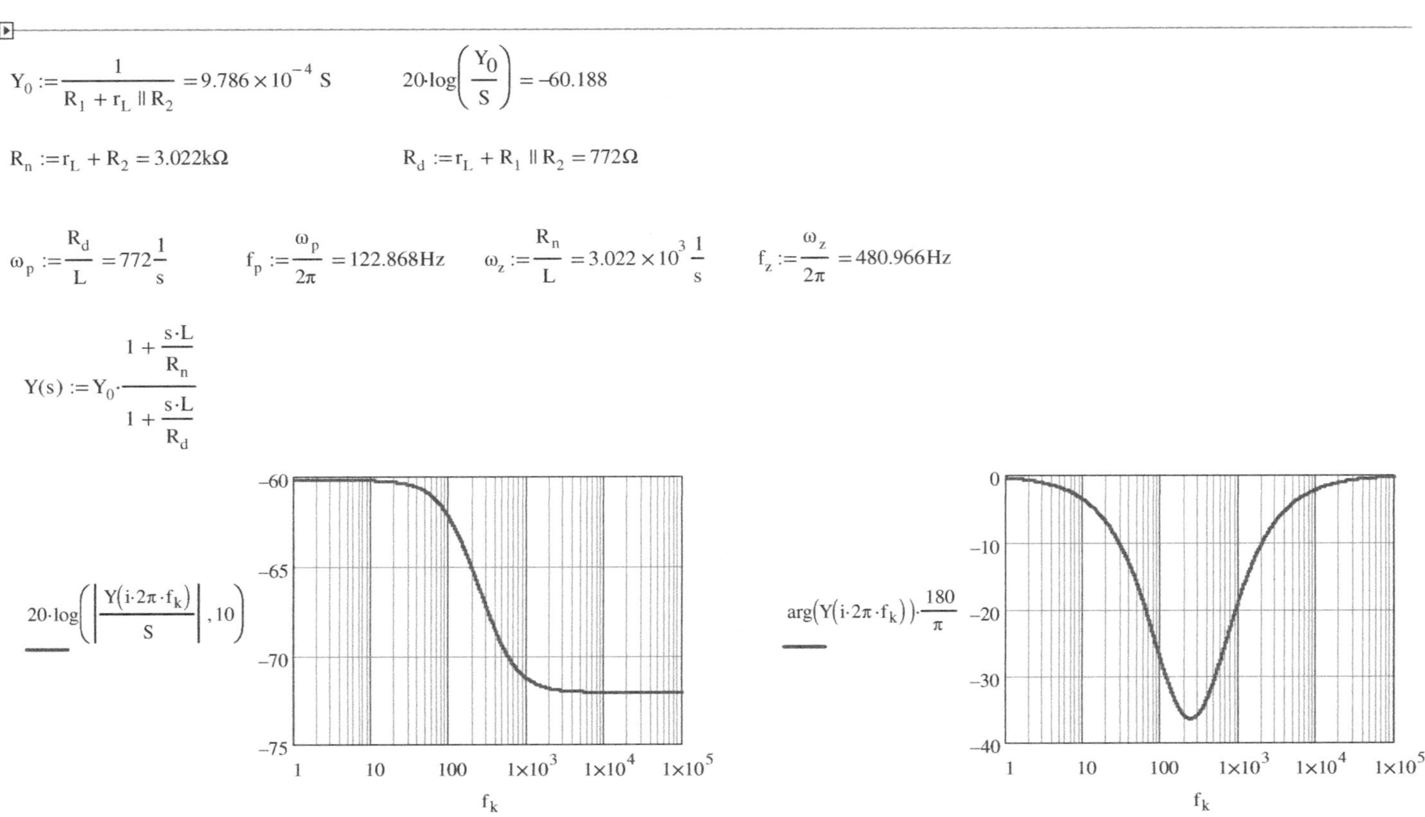

Figure 3.94 Mathcad® quickly plots the admittance magnitude and phase versus frequency.

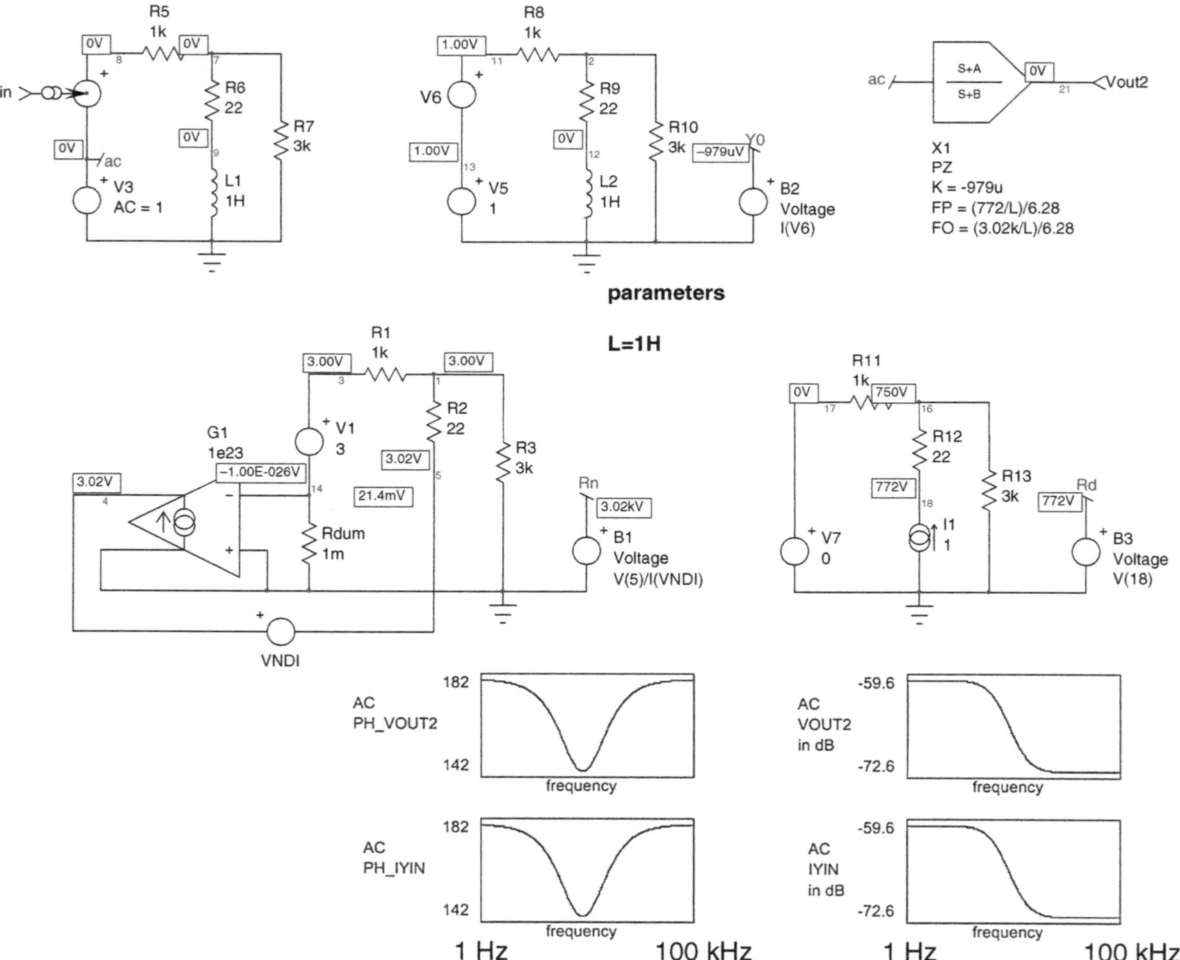

Figure 3.95 SPICE confirms all our calculations and shows an ac response similar to that of Mathcad®.

Problem 3:
The input impedance is found by driving the network connecting terminals by a current source as shown in Figure 3.96. Is there a zero in this transfer function? By shorting C_1, do we still have a response, a voltage across the current source? Yes, so there is a zero associated with C_1.

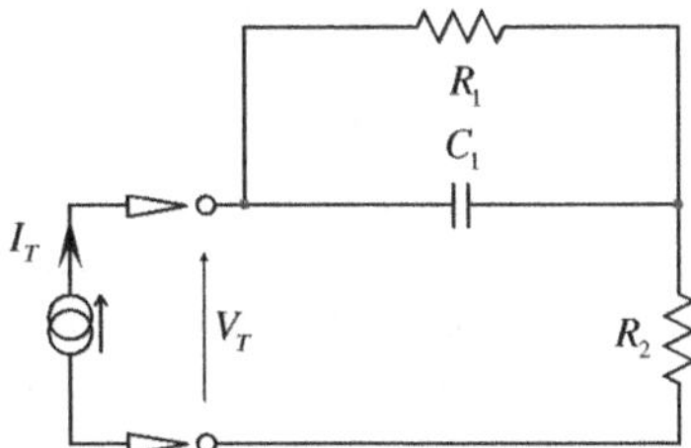

Figure 3.96 A current source is driving the network impedance. The response is the voltage across its terminals.

The transfer function follows the form:

$$Z_{in}(s) = R_0 \frac{1 + s\tau_1}{1 + s\tau_2} \tag{3.241}$$

The dc resistance R_0 is obtained by inspection when removing C_1:

$$R_0 = R_1 + R_2 \tag{3.242}$$

The first time constant τ_1 is obtained by evaluating the resistance seen from the capacitor terminals while the response is a null. The circuit appears in Figure 3.97. The nulled response ($V_T = 0$) implies that R_1 left terminal is at 0 V as a null across a current source is equivalent of having the source short circuited.

We have:

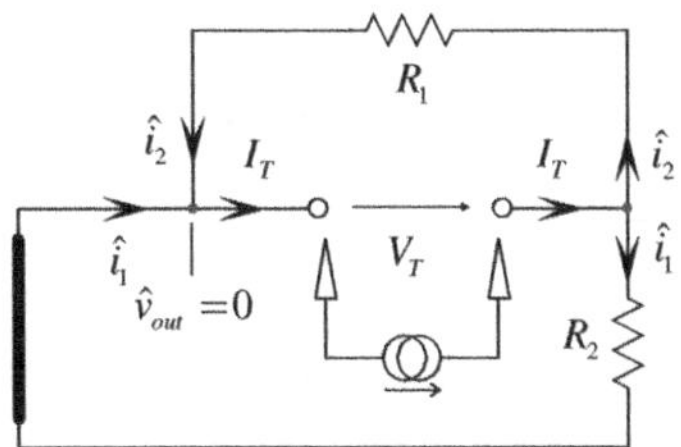

Figure 3.97 A null in the output implies that the voltage across the first current source is 0.

$$I_T = \hat{i}_1 + \hat{i}_2 \tag{3.243}$$

As V_T is present across R_2 but also R_1 given the null in the output, we can write:

$$\hat{i}_1 = \frac{V_T}{R_2} \tag{3.244}$$

$$\hat{i}_2 = \frac{V_T}{R_1} \tag{3.245}$$

Substituting these definition in (3.243) leads to:

$$I_T = V_T\left(\frac{1}{R_1} + \frac{1}{R_2}\right) \tag{3.246}$$

Rearranging, we can write:

$$R_n = \frac{1}{\frac{1}{R_1} + \frac{1}{R_2}} = R_1 \| R_2 \tag{3.247}$$

Without going through KCL, you could immediately see from Figure 3.97 that the left terminal of R_1 is electrically grounded and both resistors are in parallel. The NDI SPICE simulation appears in Figure 3.98 and shows that the value computed by B_1 is 909 Ω, the paralleling of the 10-kΩ resistor and 1 kΩ. I_1 is arbitrarily set to 1 A for this example but could take any value, B_1 would still deliver the same value.

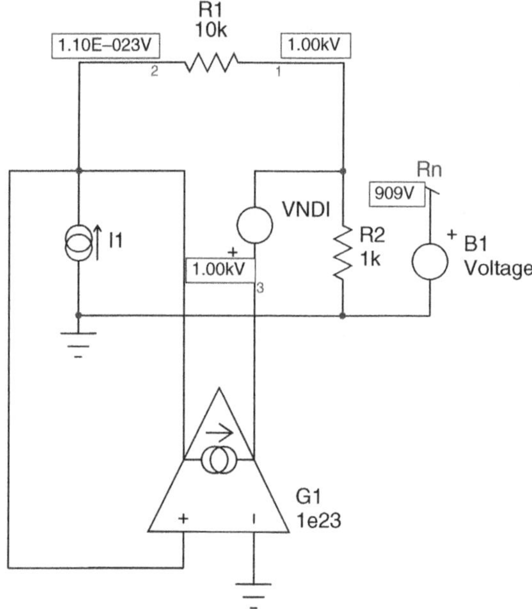

Figure 3.98 The SPICE simulation confirms the paralleling of R_1 and R_2.

The time constant τ_1 is thus defined as:

$$\tau_1 = C_1\left(R_1 \| R_2\right) \tag{3.248}$$

The second time constant is immediate: set the excitation to 0 – this is a current source so it disappears from the schematic – and look at the resistance seen from the capacitor. It is R_1 as R_2 is left unconnected. The second time constant is thus:

$$\tau_2 = R_1 C_1 \tag{3.249}$$

The final transfer function is obtained by assembling the variables according to (3.241):

$$Z_{in}(s) = (R_1 + R_2)\frac{1 + sC_1\left(R_1 \| R_2\right)}{1 + sR_1C_1} = R_0\frac{1 + \frac{s}{\omega_z}}{1 + \frac{s}{\omega_p}} \tag{3.250}$$

$R_1 := 10k\Omega \qquad R_2 := 1k\Omega \qquad C_1 := 100nF \qquad \|(x,y) := \frac{x \cdot y}{x+y}$

$R_0 := R_1 + R_2 = 11k\Omega \qquad 20 \cdot \log\left(\frac{R_0}{\Omega}\right) = 80.828\,\text{dBohm}$

$R_n := R_1 \| R_2 = 909.091 \cdot \Omega \qquad R_d := R_1 = 10 \cdot k\Omega$

$\omega_p := \frac{1}{R_d \cdot C_1} = 1 \times 10^3 \frac{1}{s} \quad f_p := \frac{\omega_p}{2\pi} = 159.155 \cdot Hz \qquad \omega_z := \frac{1}{R_n \cdot C_1} = 1.1 \times 10^4 \frac{1}{s} \qquad f_z := \frac{\omega_z}{2\pi} = 1.751 \times 10^3 \cdot Hz$

$Z(s) := R_0 \cdot \frac{1 + s \cdot C_1 \cdot R_n}{1 + s \cdot C_1 \cdot R_d}$

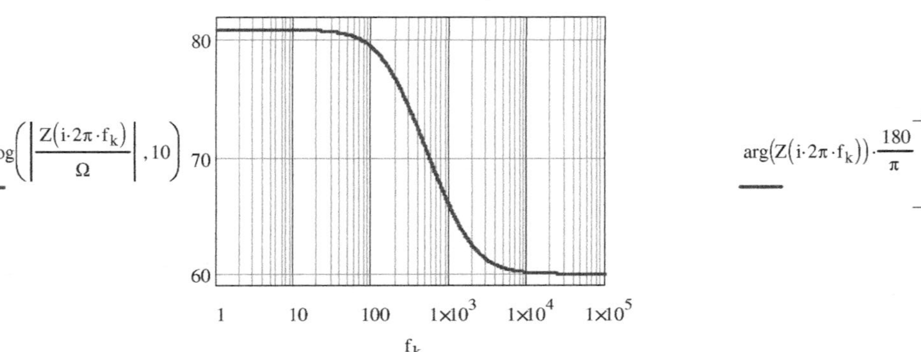

Figure 3.99 The ac response reveals the presence of a pole and a zero as predicted by (3.250).

in which

$$R_0 = R_1 + R_2 \tag{3.251}$$

$$\omega_z = \frac{1}{(R_1 \| R_2) C_1} \tag{3.252}$$

$$\omega_p = \frac{1}{R_1 C_1} \tag{3.253}$$

The ac response appears in Figure 3.99.

Problem 4:
The input impedance is found by driving the network connecting terminals by a current source as shown in Figure 3.100. Is there a zero in this transfer function? By shorting C_1, do we still have a response, a voltage across the current source? Yes, so there is a zero associated with C_1. We start with the dc input resistance when C_1 is removed as shown in Figure 3.101.

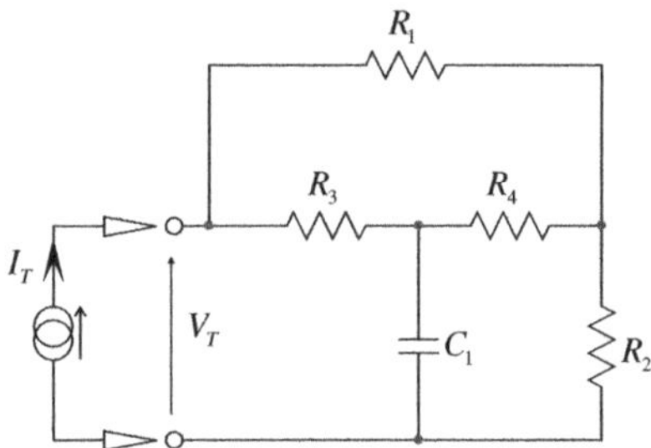

Figure 3.100 The input impedance is computed using a current source as the driving signal while the response is the voltage across its terminals.

In this configuration, the resistance R_0 is simply the paralleling of R_1 with R_3 plus R_4, in series with R_2:

$$R_0 = R_1 \| (R_3 + R_4) + R_2 \tag{3.254}$$

The time constant is found by setting the current generator to 0 A, and looking into the resistance offered by the capacitor terminals (see Figure 3.102). By inspection, you can see that R_3 and R_1 come in parallel with R_4, then in series with R_2:

$$R_d = (R_1 + R_3) \| R_4 + R_2 \tag{3.255}$$

The time constant τ_2 is thus:

$$\tau_2 = \left[(R_1 + R_3) \| R_4 + R_2\right] C_1 \tag{3.256}$$

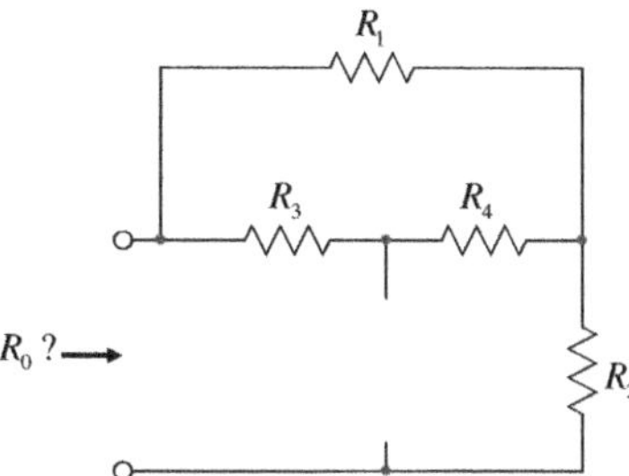

Figure 3.101 The dc resistance is easily found without calculations.

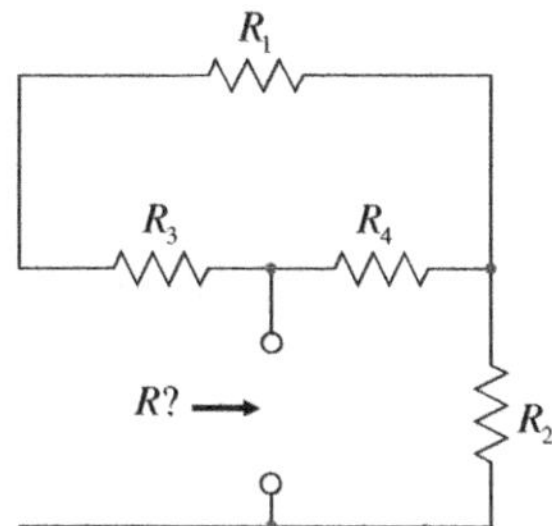

Figure 3.102 The time constant is evaluated when the capacitor is removed and the excitation current set to 0.

To obtain the numerator expression, we will calculate the input impedance while C_1 is infinite valued or replaced by a short circuit as shown in Figure 3.103a. It is perhaps easier to redraw the schematic to reveal the resistance in a quicker way as proposed in Figure 3.103b. In this case, the resistance is equal to:

$$R^1 = R_3 \| \left(R_1 + R_4 \| R_2\right) \tag{3.257}$$

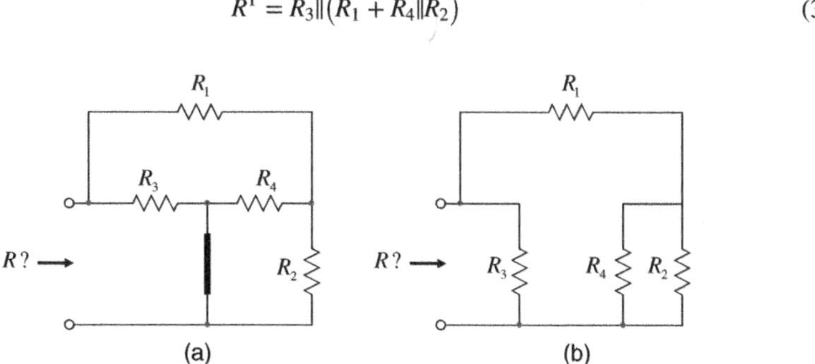

Figure 3.103 When infinite valued, capacitor C_1 shorts the middle point of R_3 and R_4.

By combining these results, the input impedance is found to be:

$$Z_{in}(s) = R_0 \frac{1 + s\dfrac{R^1}{R_0}\tau_2}{1 + s\tau_2} = \left(R_1 \| (R_3 + R_4) + R_2\right) \frac{1 + s\dfrac{R_3 \| \left(R_1 + R_4 \| R_2\right)}{R_1 \| (R_3 + R_4) + R_2}\left[(R_1 + R_3) \| R_4 + R_2\right]C_1}{1 + s\left[(R_1 + R_3) \| R_4 + R_2\right]C_1} \tag{3.258}$$

It can be rewritten in a more compact form:

$$Z_{in}(s) = R_0 \frac{1 + \frac{s}{\omega_z}}{1 + \frac{s}{\omega_p}} \tag{3.259}$$

in which you have

$$R_0 = R_1 \| (R_3 + R_4) + R_2 \tag{3.260}$$

then a zero

$$\omega_z = \frac{R_0}{\tau_2 R^1} = \frac{R_1 \| (R_3 + R_4) + R_2}{\left[R_3 \| \left(R_1 + R_4 \| R_2\right)\right] \left[(R_1 + R_3) \| R_4 + R_2\right] C_1} \tag{3.261}$$

and finally a pole

$$\omega_p = \frac{1}{\tau_2} = \frac{1}{\left[(R_1 + R_3) \| R_4 + R_2\right] C_1} \tag{3.262}$$

The expression for the zero is rather complex and is due to the approach we have adopted in which we did not perform NDI. Should we want to simplify it further, we need to go through NDI which, in an impedance calculation, is easier than in other applications because a null in the response implies a frank short circuit across the current generator. Figure 3.104a shows the update. We have redrawn the circuit to consider it from a different view. In this case, the resistance R_n is immediate and equal to:

$$R_n = R_3 \| \left(R_4 + R_2 \| R_1\right) \tag{3.263}$$

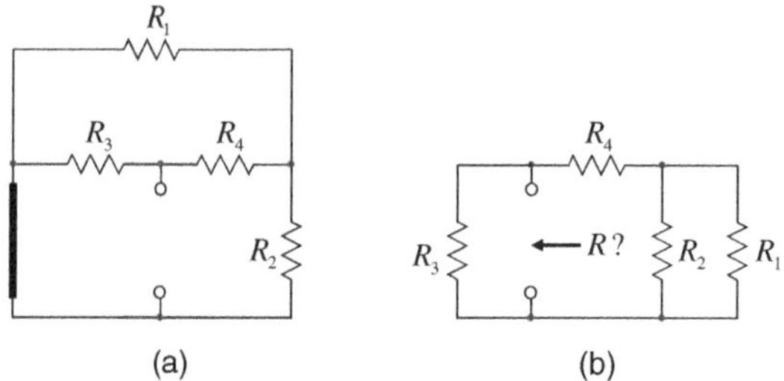

Figure 3.104 An NDI with a current generator as driving source implies that its terminals are shorted for the analysis.

and the new transfer function becomes

$$Z_{in}(s) = R_0 \frac{1 + s\tau_1}{1 + s\tau_2} = \left(R_1 \| (R_3 + R_4) + R_2\right) \frac{1 + s\left[R_3 \| \left(R_4 + R_2 \| R_1\right)\right] C_1}{1 + s\left[(R_1 + R_3) \| R_4 + R_2\right] C_1} = R_0 \frac{1 + \frac{s}{\omega_z}}{1 + \frac{s}{\omega_p}} \tag{3.264}$$

R_0 and ω_p are the same as in expressions defined in (3.260) and (3.262) but the zero becomes simpler:

$$\omega_z = \frac{1}{\left[R_3 \| \left(R_4 + R_2 \| R_1\right)\right] C_1} \tag{3.265}$$

Mathematically, (3.258) and (3.264) are identical except that we did not perform NDI in (3.258). For input impedance calculations, where the current generator is replaced by a short circuit, the NDI is simpler than with circuits where the response is a voltage. Again, you choose whichever is easier for you knowing the differences at the end.

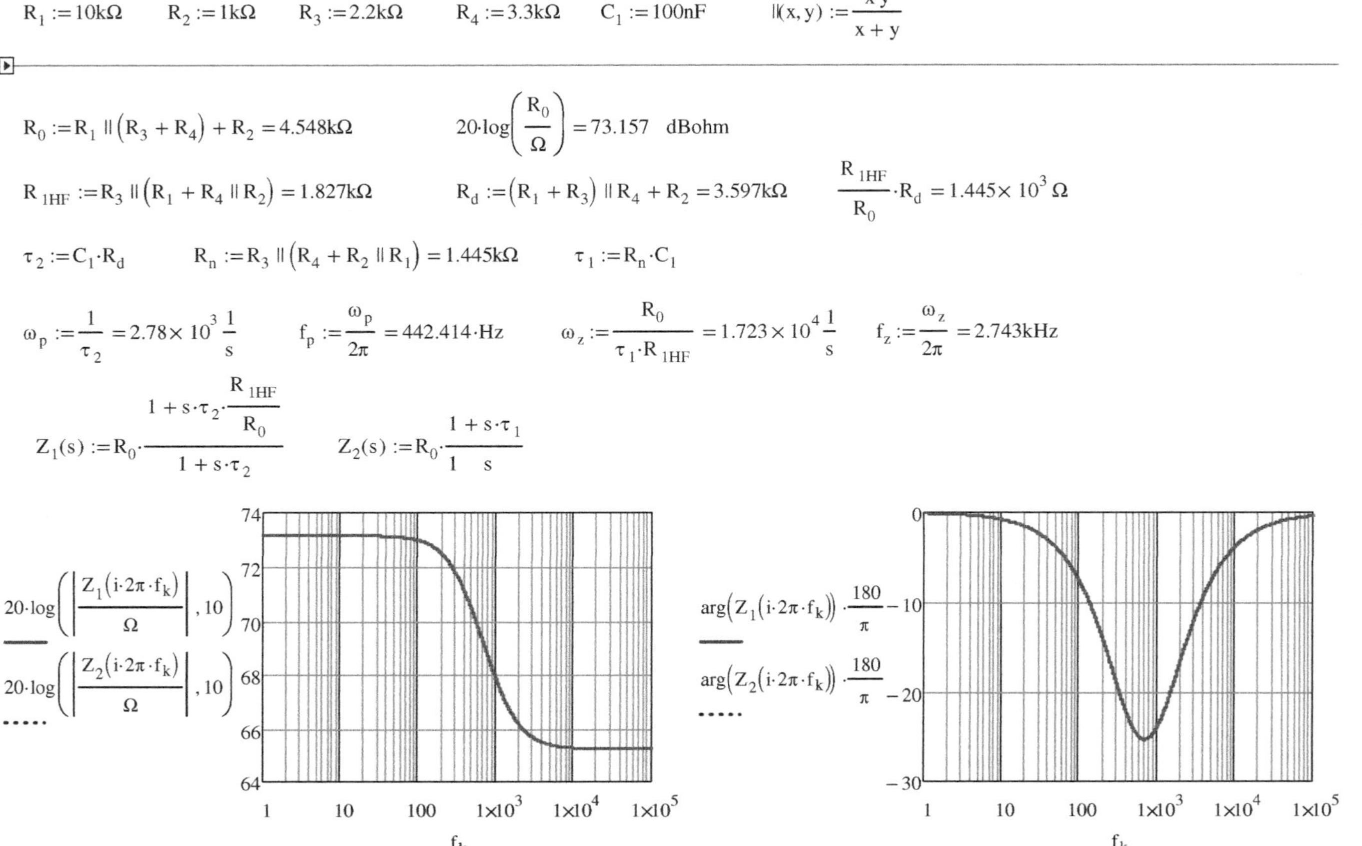

Figure 3.105 Mathcad® plots the various input impedance definitions and helps compare results with and without NDI: they are identical as expected.

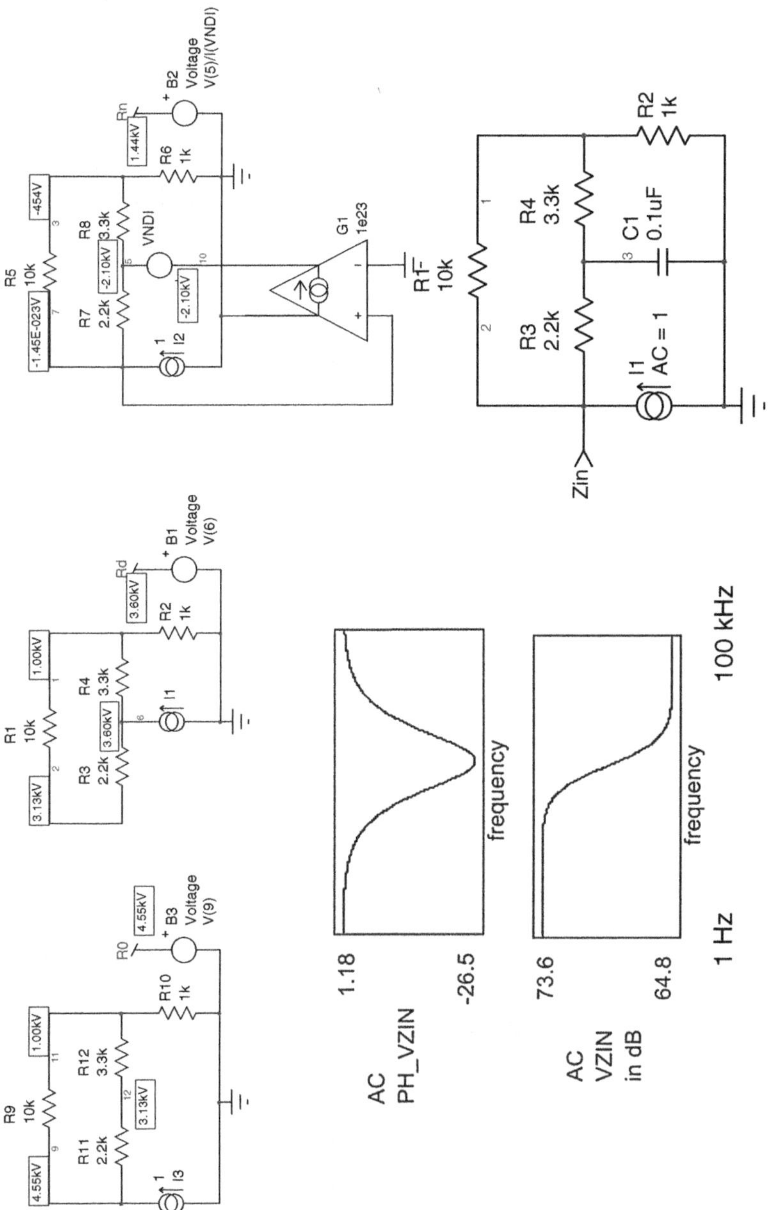

Figure 3.106 A simple SPICE simulation tells us if our calculations are correct.

Problem 5:

The output impedance is obtained by driving the output with a current generator. Again, as in the previous examples, the voltage V_T is the response and current I_T is the stimulus. In this example, you can immediately see that the left terminals of C_1 and R_3 are grounded. Without using Fast Analytical techniques, we could write:

$$Z_{out}(s) = R_2 \| \frac{1}{sC_1} \| (R_4 + R_1 \| R_3) \tag{3.266}$$

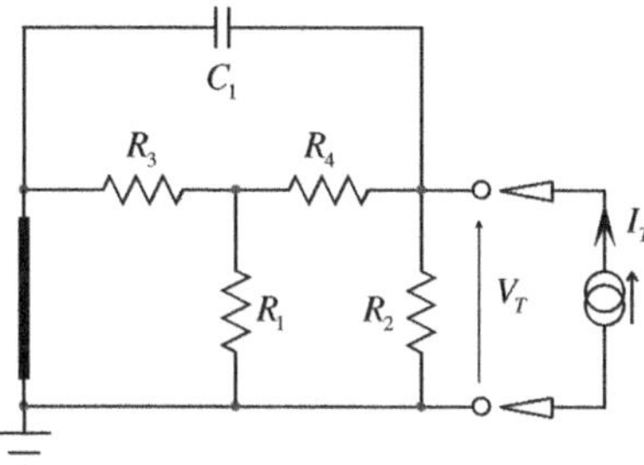

Figure 3.107 A current source sweeps the network while the dc input generator is short circuited.

Should you try to expand this expression, you would get a meaningless result in which ordering coefficients to show the pole and the zero would require extra energy. Is there a zero, by the way? If C_1 is shorted, do we still have a response V_T? No because C_1 being in parallel with R_2, shorting it cancels the response, so no zero in this structure. The transfer function thus follows a simple 1st-order transfer function:

$$Z_{out}(s) = R_0 \frac{1}{1 + s\tau_1} \tag{3.267}$$

The resistance offered in dc, when C_1 is removed, is quite easy as shown by Figure 3.108.

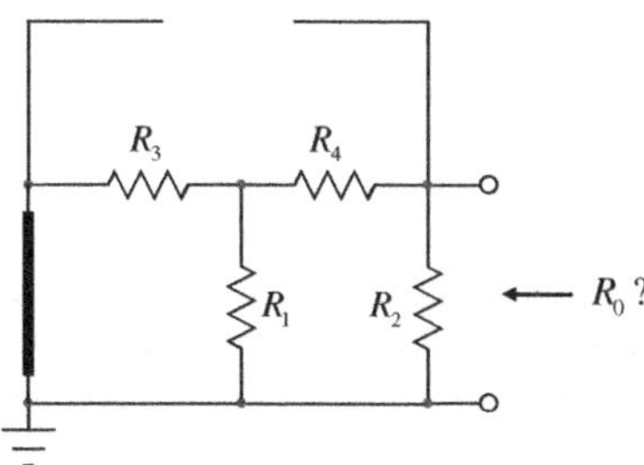

Figure 3.108 The dc resistance R_0 in this configuration is simple to find by inspection.

By inspection, we see that:

$$R_0 = R_2 \| (R_4 + R_1 \| R_3) \tag{3.268}$$

The time constant is obtained by setting the excitation current to 0, physically removing the generator as shown in Figure 3.109. The left terminal of the capacitor is grounded while its right one also goes to ground via R_2 with a series-parallel combination involving R_4, R_3 and R_1. We have:

$$R_d = R_2 \| (R_4 + R_3 \| R_1) \tag{3.269}$$

It is the same as the resistor we found in (3.268). The circuit time constant is therefore:

$$\tau_2 = C_1 \left[R_2 \| (R_4 + R_3 \| R_1) \right] \tag{3.270}$$

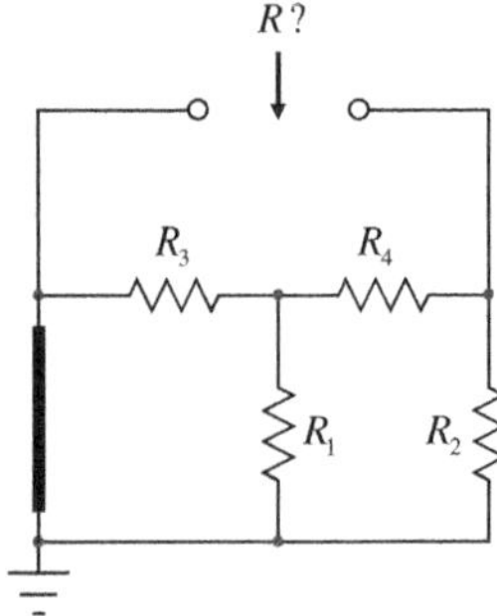

Figure 3.109 The resistance offered by the capacitor terminals is, again, quite easy to derive.

The final transfer function is given by:

$$Z_{out}(s) = \left[R_2 \| (R_4 + R_1 \| R_3) \right] \frac{1}{1 + sC_1 \left[R_2 \| (R_4 + R_3 \| R_1) \right]} = R_0 \frac{1}{1 + \frac{s}{\omega_p}} \tag{3.271}$$

in which we have

$$R_0 = R_2 \| (R_4 + R_1 \| R_3) \tag{3.272}$$

and

$$\omega_p = \frac{1}{C_1 \left[R_2 \| (R_4 + R_3 \| R_1) \right]} \tag{3.273}$$

Problem 6:
First, we can check if there is a zero in the transfer function by considering L_1 infinite valued. Do we still have a response in this case, when L_1 is physically removed? No, so we don't have a zero then. As this is a 1st-order network, the transfer function obeys the following form:

$$H(s) = H_0 \frac{1}{1 + s\tau_2} \tag{3.274}$$

The dc gain is obtained when L_1 is shorted as shown in Figure 3.111. We have a resistive divider involving R_2 and R_3:

$$H_0 = \frac{R_3}{R_2 + R_3} \tag{3.275}$$

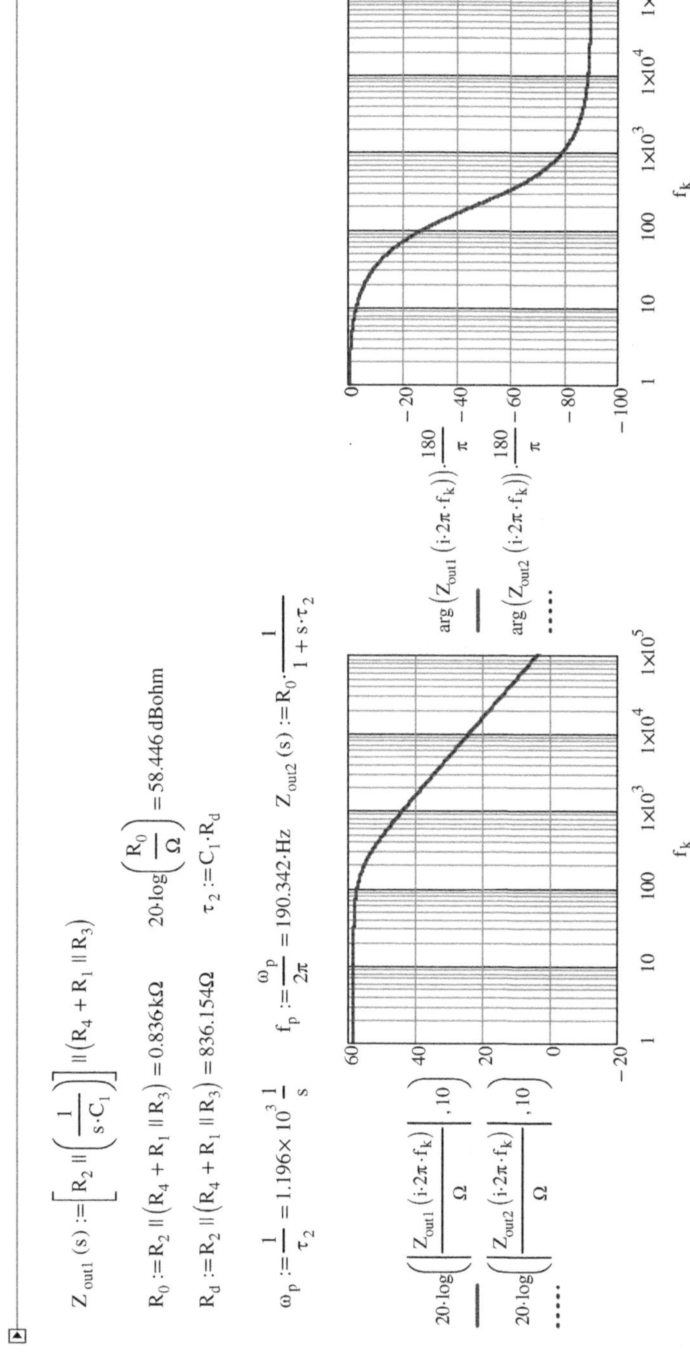

Figure 3.110 As Mathcad® shows, expressions in (3.266) and (3.271) deliver identical plots.

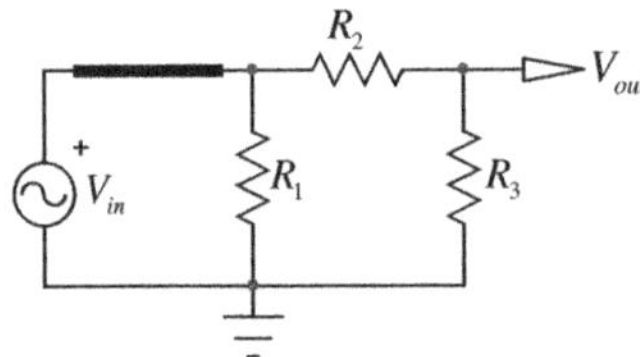

Figure 3.111 In dc the inductor is a short circuit and the transfer function H_0 is immediate.

The time constant is obtained by setting the excitation voltage to 0 as sketched in Figure 3.112. The resistance seen from the inductor terminals is thus:

$$R_d = R_1 \| (R_2 + R_3) \tag{3.276}$$

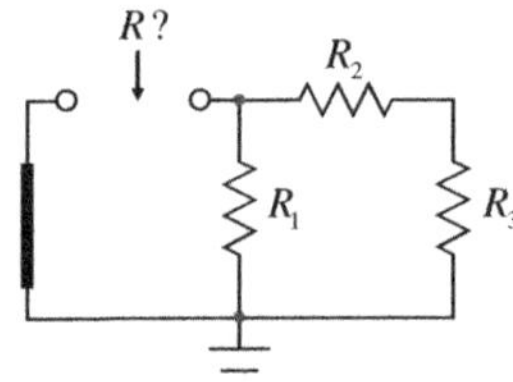

Figure 3.112 Setting the excitation voltage to 0 V leads to calculating the resistance driving the inductor.

The circuit time constant is:

$$\tau_2 = \frac{L_1}{R_1 \| (R_2 + R_3)} \tag{3.277}$$

and the final transfer function is

$$H(s) = \frac{R_3}{R_2 + R_3} \frac{1}{1 + s\frac{L_1}{R_1 \| (R_2 + R_3)}} = H_0 \frac{1}{1 + \frac{s}{\omega_p}} \tag{3.278}$$

In this expression, the dc gain is:

$$H_0 = \frac{R_3}{R_2 + R_3} \tag{3.279}$$

and the pole is expressed as

$$\omega_p = \frac{R_1 \| (R_2 + R_3)}{L_1} \tag{3.280}$$

Figure 3.113 shows the Mathcad® calculation sheet which plots the response of a simple 1st-order low-pass filter.

$R_1 := 1k\Omega \quad R_2 := 470\Omega \quad R_3 := 100\Omega \quad L_1 := 100mH \quad \|(x,y) := \frac{x \cdot y}{x+y}$

$H_0 := \frac{R_3}{R_2+R_3} = 0.175 \qquad 20 \cdot \log(H_0) = -15.117 \text{ dB}$

$R_d := R_1 \| (R_2+R_3) = 363.057\Omega \qquad \tau_2 := \frac{L_1}{R_d}$

$\omega_p := \frac{1}{\tau_2} = 3.631 \times 10^3 \frac{1}{s} \qquad f_p := \frac{\omega_p}{2\pi} = 577.824Hz \qquad H_1(s) := H_0 \cdot \frac{1}{1+s \cdot \tau_2}$

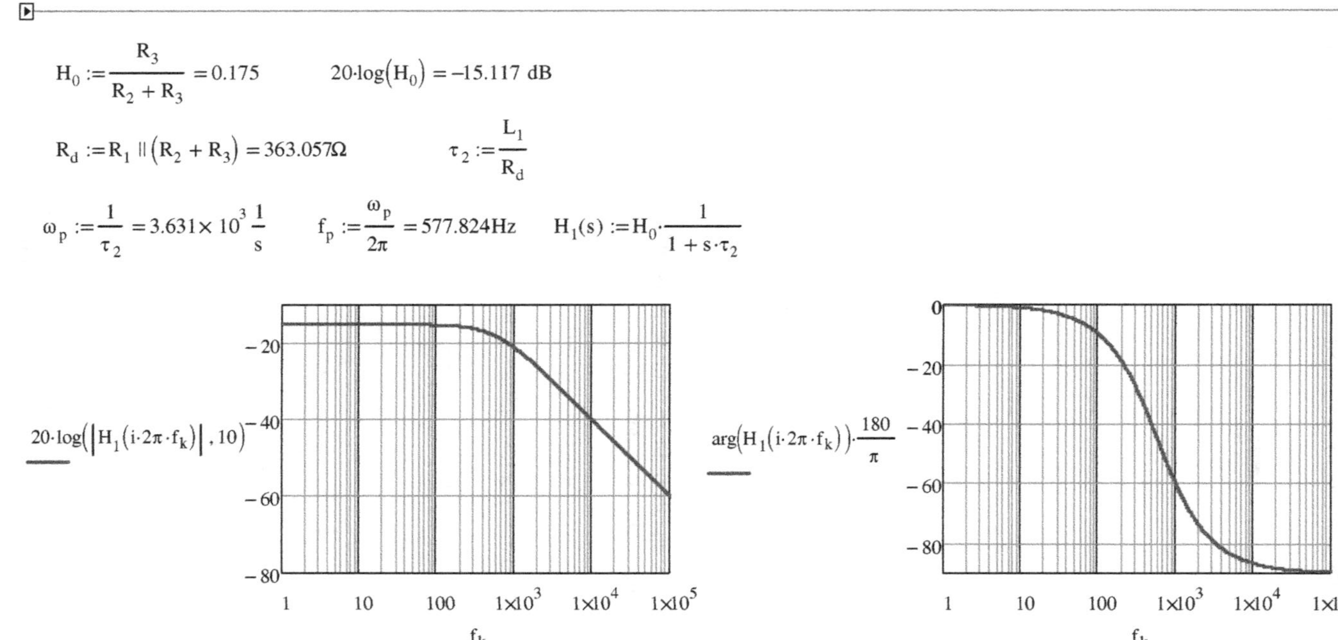

Figure 3.113 The response is that of a simple low-pass filter.

Problem 7:
This op amp-based circuit features one energy-storing element so it is 1st-order. As usual, does C_1 contribute a zero? To check it, we can replace it by a short and look at the output as proposed in Figure 3.114.

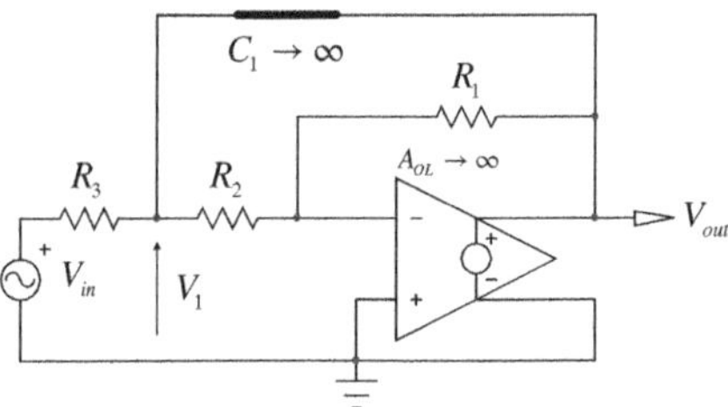

Figure 3.114 To check the presence of a zero, we look at the circuit gain when the capacitor is replaced by a short circuit.

Without the short circuit, the op amp realizes a buffer which links V_{out} to V_1 by:

$$V_{out} = -\frac{R_1}{R_2} V_1 \tag{3.281}$$

Now, if C_1 is a short circuit, V_1 becomes V_{out} and (3.281) updates to:

$$V_{out} = -\frac{R_1}{R_2} V_{out} \tag{3.282}$$

which can only be satisfied if V_{out} equals 0: there is no response when C_1 is shorted and thus, the transfer function does not feature a zero. It is defined as:

$$G(s) = G_0 \frac{1}{1 + s\tau_2} \tag{3.283}$$

In dc, the capacitor is removed and the circuit simplifies to that of Figure 3.115. The gain is simply:

$$G_0 = -\frac{R_1}{R_2 + R_3} \tag{3.284}$$

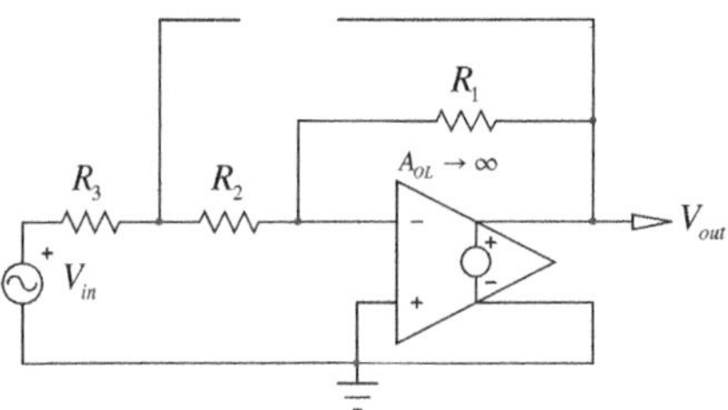

Figure 3.115 In dc, the circuit simplifies to a simple inverting circuit.

The time constant for this circuit is found by reducing the excitation voltage to 0 V while looking into the capacitor terminals to obtain the resistance R_d. In active circuits, such as here with an op amp, finding the resistance by inspection is a difficult exercise not to say impossible. We will thus install a current generator as we did before and the new circuit is that of Figure 3.116.

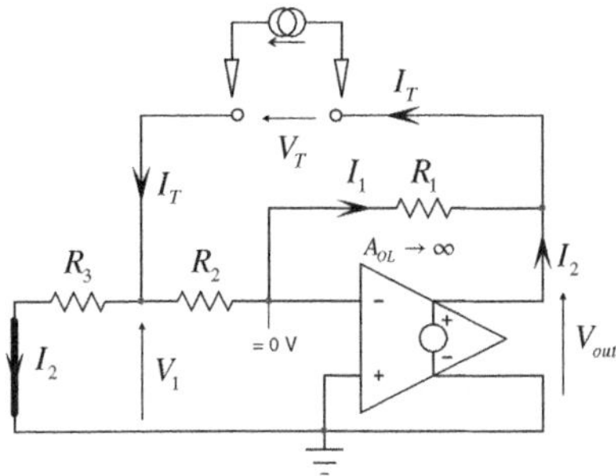

Figure 3.116 The current generator will help us find the resistance offered by the capacitor terminals.

Because the op amp gain approaches infinity, the voltage at the inverting pin is 0 V as indicated in the sketch. Therefore, the current I_1 is defined either by the voltage across R_2 or by V_{out} across R_1:

$$I_1 = \frac{V_1}{R_2} = -\frac{V_{out}}{R_1} \tag{3.285}$$

The voltage across R_3, labeled V_1, is defined by I_T flowing in the parallel combination of R_2 and R_3:

$$V_1 = I_T(R_2||R_3) \tag{3.286}$$

From (3.285), we can further define I_1 by:

$$I_1 = \frac{I_T(R_3||R_2)}{R_2} \tag{3.287}$$

The output voltage can be expressed as:

$$V_{out} = -\frac{I_T(R_3||R_2)}{R_2}R_1 \tag{3.288}$$

The voltage across the current generator, V_T, is given by V_1–V_{out}:

$$V_T = I_T(R_2||R_3) + I_T\left[(R_2||R_3)\frac{R_1}{R_2}\right] \tag{3.289}$$

After rearranging and factoring, we have:

$$R_d = \frac{V_T}{I_T} = (R_2||R_3)\left(1 + \frac{R_1}{R_2}\right) \tag{3.290}$$

The time constant τ_2 is thus:

$$\tau_2 = (R_2||R_3)\left(1 + \frac{R_1}{R_2}\right)C_1 \tag{3.291}$$

The final transfer function looks like:

$$G(s) = -\frac{R_1}{R_2 + R_3}\frac{1}{1 + s(R_2||R_3)\left(1 + \frac{R_1}{R_2}\right)C_1} = G_0\frac{1}{1 + \frac{s}{\omega_p}} \tag{3.292}$$

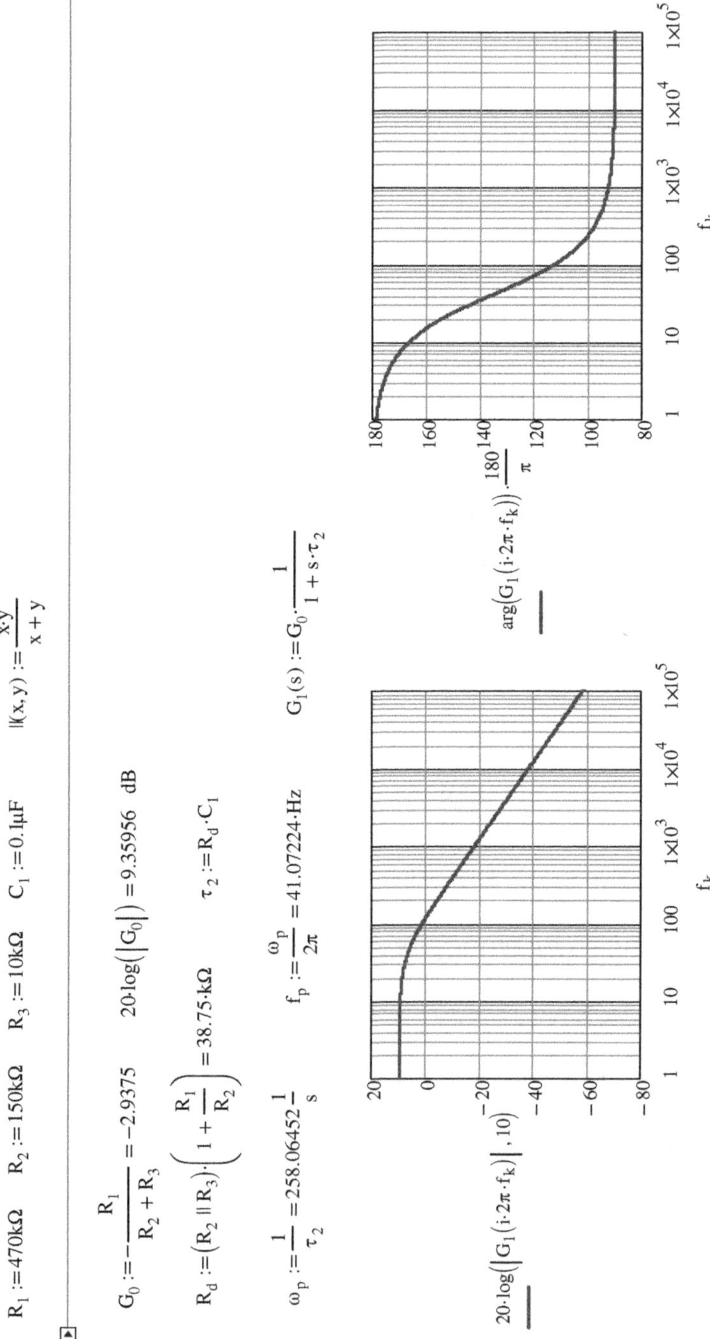

Figure 3.117 The solver delivers the answer quickly and let you compare results with SPICE.

where

$$G_0 = -\frac{R_1}{R_2 + R_3} \tag{3.293}$$

and the pole is given by

$$\omega_p = \frac{1}{(R_2 \| R_3)\left(1 + \frac{R_1}{R_2}\right)C_1} \tag{3.294}$$

The Mathcad® sheet is pasted in Figure 3.117 while SPICE simulation results are given in Figure 3.118. They agree well with each other.

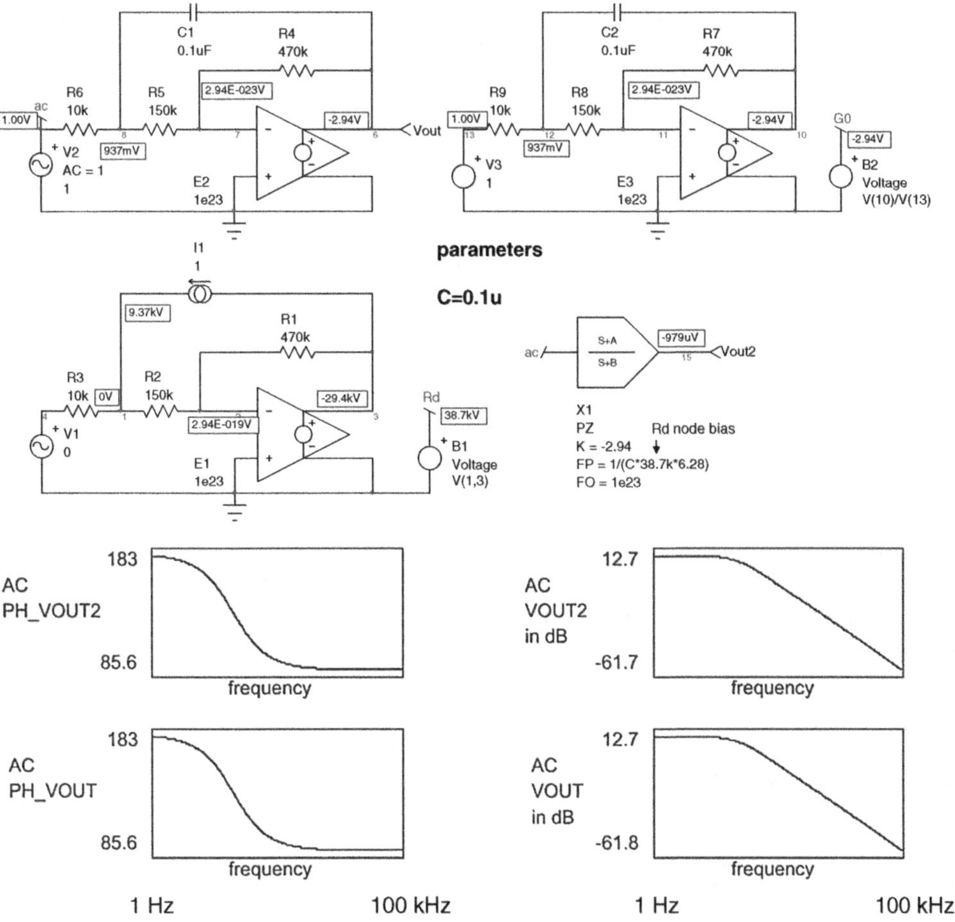

Figure 3.118 SPICE is of great help in these active circuits involving a virtual ground.

Problem 8:

The output impedance is computed by installing a current generator as illustrated in Figure 3.119:

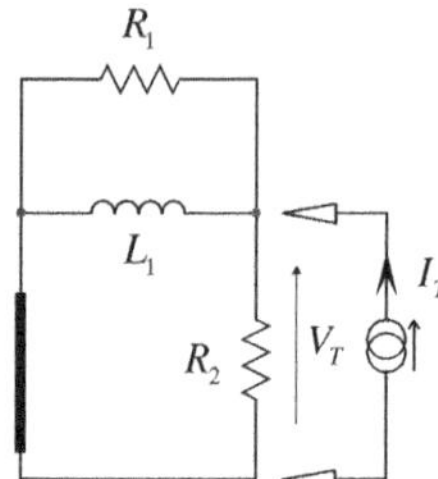

Figure 3.119 The output impedance is derived by using a current generator as a stimulus while the voltage across its terminals is the response.

The first thing is to check the presence of a zero: if L_1 is infinite valued (physically removed), do we still have a response while the current generator drives the network? Yes, so we have a zero associated with L_1 and the transfer function obeys the following expression:

$$Z_{out}(s) = R_0 \frac{1 + s\tau_1}{1 + s\tau_2} \tag{3.295}$$

The dc term R_0 is found by shorting L_1:

$$R_0 = 0 \tag{3.296}$$

This means that the expression selected in (3.295) is not appropriate since the leading term is null. Instead, let's use the alternate version described in Figure 3.76:

$$Z_{out}(s) = R_\infty \frac{1 + \dfrac{1}{s\tau_1}}{1 + \dfrac{1}{s\tau_2}} \tag{3.297}$$

In this case, when the inductance is infinite valued, R_1 comes in parallel with R_2 and the output resistance is:

$$R_\infty = R_1 \| R_2 \tag{3.298}$$

The first time constant τ_1 is found when the response is a null. In other words, with a current generator, a null across its terminals is similar to a short circuit. Figure 3.120 illustrates the configuration. Without hesitation, the resistance R_n seen from the connecting terminals is 0. Therefore, the time constant τ_1 is infinite:

$$\tau_1 = \frac{L_1}{0} \to \infty \tag{3.299}$$

The second circuit time constant τ_2 is found by setting the current to 0 A and looking into the inductor terminals (see Figure 3.121). The resistance R_d is defined as:

$$R_d = R_1 \| R_2 \tag{3.300}$$

and the time constant τ_2 as

$$\tau_2 = \frac{L_1}{R_1 \| R_2} \tag{3.301}$$

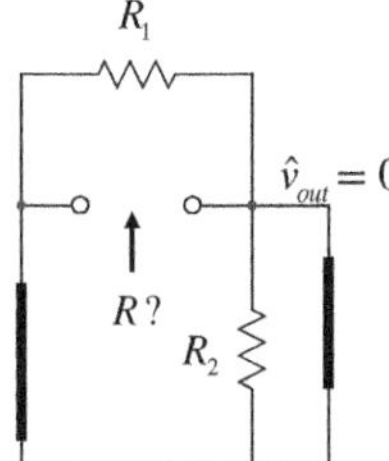

Figure 3.120 In this particular case, the resistance seen from the inductor terminals is 0.

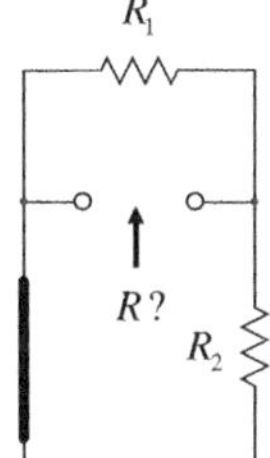

Figure 3.121 The time constant is obtained by setting the stimulus to 0 A.

The transfer function is then expressed the following way:

$$Z_{out}(s) = R_\infty \frac{1 + \dfrac{1}{s \cdot \infty}}{1 + \dfrac{1}{s\tau_2}} = R_\infty \frac{1}{1 + \dfrac{1}{s\tau_2}} = R_\infty \frac{1}{1 + \dfrac{\omega_p}{s}} \tag{3.302}$$

in which

$$R_\infty = R_1 \| R_2 \tag{3.303}$$

and

$$\omega_p = \frac{R_1 \| R_2}{L_1} \tag{3.304}$$

A quick Mathcad® sheet details the calculation results in Figure 3.122. Please note that (3.302) uses the inverted pole notation. In other words, next to a pole there is also a zero at the origin.

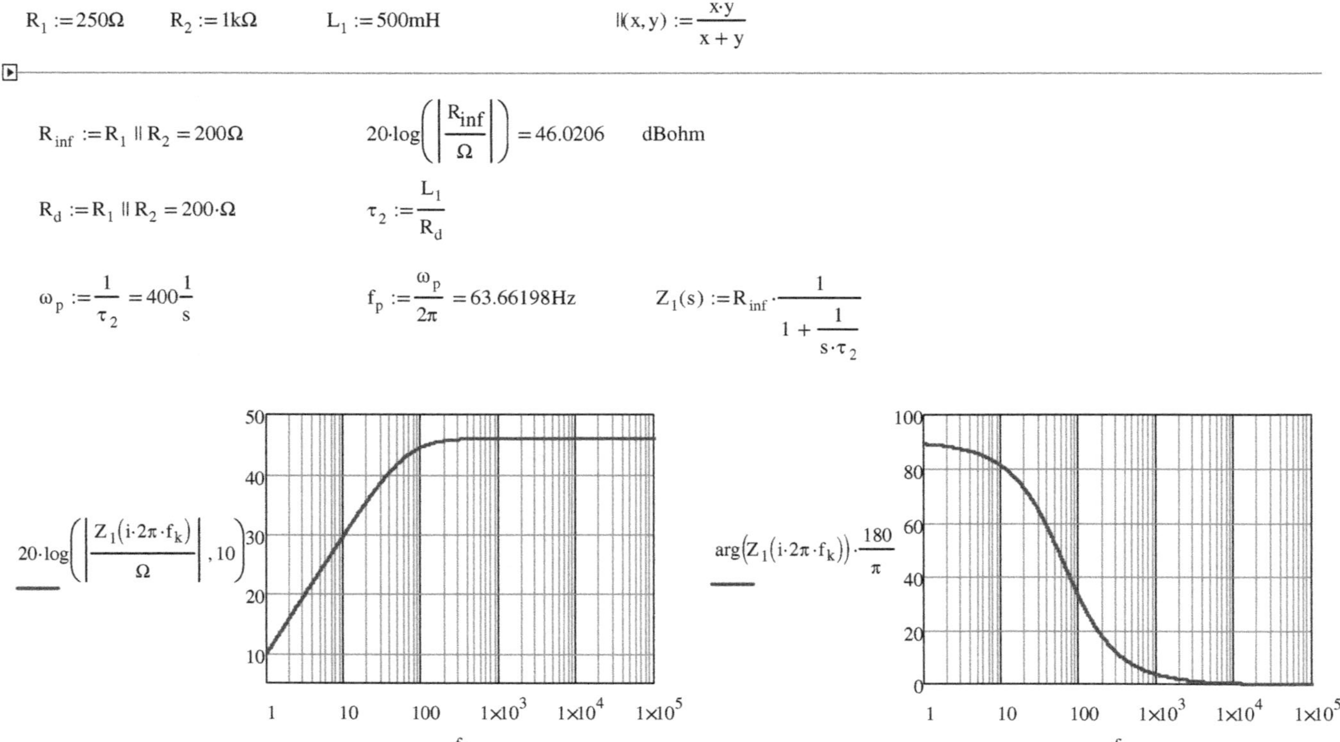

Figure 3.122 The circuit shows a zero at the origin and a higher-frequency pole.

Problem 9:

This is a 1st-order circuit having C_1 as an energy storing element. Is there a zero associated with this element? If C_1 is shorted, is there still a response on V_{out}? Sure, then we have a zero associated with C_1. The transfer function obeys the following expression:

$$G(s) = G_0 \frac{1 + s\tau_1}{1 + s\tau_2} \tag{3.305}$$

The dc gain is obtained by removing C_1 from the circuit as drawn in Figure 3.123.

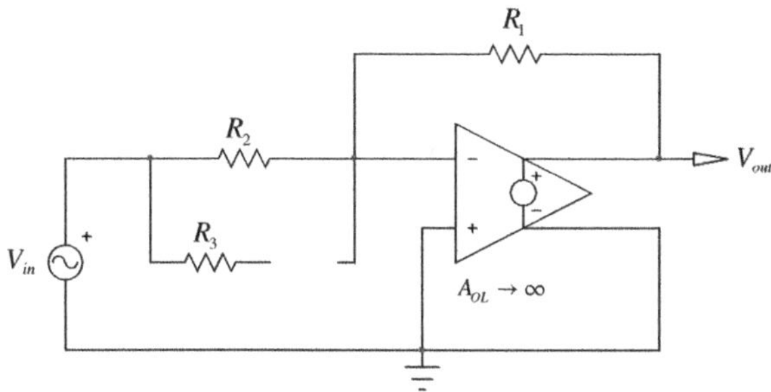

Figure 3.123 In dc, C_1 plays no role and this is a simple inverting amplifier.

The dc gain G_0 is immediate and equal to:

$$G_0 = -\frac{R_1}{R_2} \tag{3.306}$$

The zero is obtained by installing a test generator at the capacitor terminals to determine resistance R_n while in a NDI configuration. Figure 3.124 shows the adopted configuration.

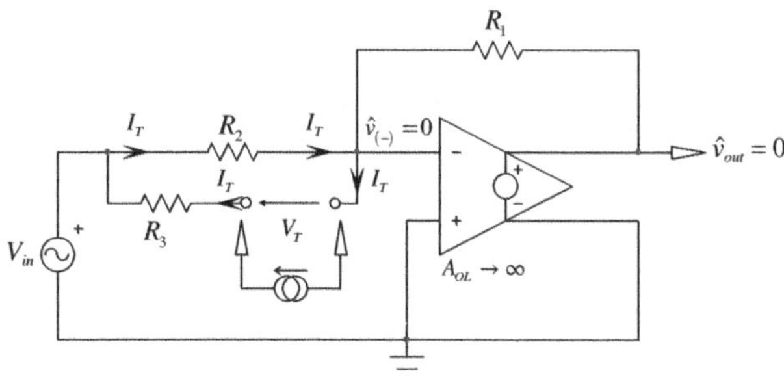

Figure 3.124 The zero is revealed by looking at R_n seen from the capacitor terminals while the response is a null.

Actually, inspection could work here, without the test generator. Since the output is equal to 0, the inverting pin is also at 0 V so no current flows in R_1. All the current I_T flows in R_2 and R_3 therefore the resistance R_n is equal to:

$$R_n = R_2 + R_3 \tag{3.307}$$

The time constant τ_1 is then expressed as:

$$\tau_1 = (R_2 + R_3)C_1 \tag{3.308}$$

For the circuit time constant τ_2, we set the stimulus voltage to 0 V and look at the resistance offered by the capacitor terminals in this mode. We could do that by inspection, but a test generator I_T is helpful in this case as depicted by Figure 3.125.

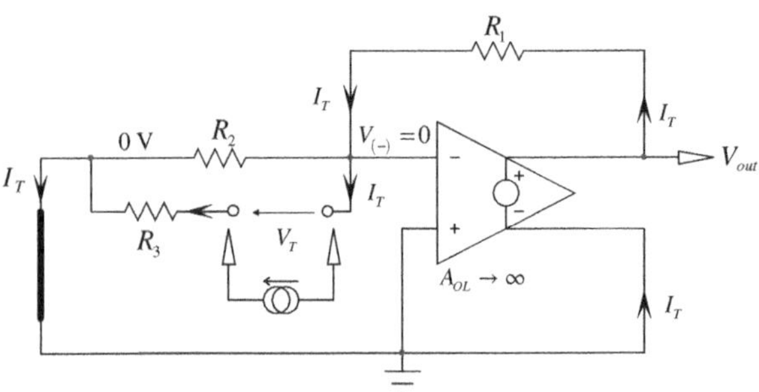

Figure 3.125 The time constant of the circuit is revealed while the stimulus voltage is set to 0 V.

In here, you see the current I_T flowing through the stimulus source set to 0 and coming back via R_1 to return to the generator. However, as the inverting pin is at 0 V (because of the virtual ground), the only resistance the current generator sees is R_3. Therefore:

$$R_d = R_3 \tag{3.309}$$

and

$$\tau_2 = R_3 C_1 \tag{3.310}$$

The complete transfer function is equal to:

$$G(s) = -\frac{R_1}{R_2}\frac{1 + s(R_2 + R_3)C_1}{1 + sR_3C_1} = G_0\frac{1 + \frac{s}{\omega_z}}{1 + \frac{s}{\omega_p}} \tag{3.311}$$

in which

$$G_0 = -\frac{R_1}{R_2} \tag{3.312}$$

$$\omega_z = \frac{1}{(R_2 + R_3)C_1} \tag{3.313}$$

and

$$\omega_p = \frac{1}{R_3C_1} \tag{3.314}$$

In Figure 3.126 and Figure 3.127, you respectively have the Mathcad® graphs and the SPICE test circuit. Again, results perfectly match. Should you see a difference, it indicates that something is wrong in the equations.

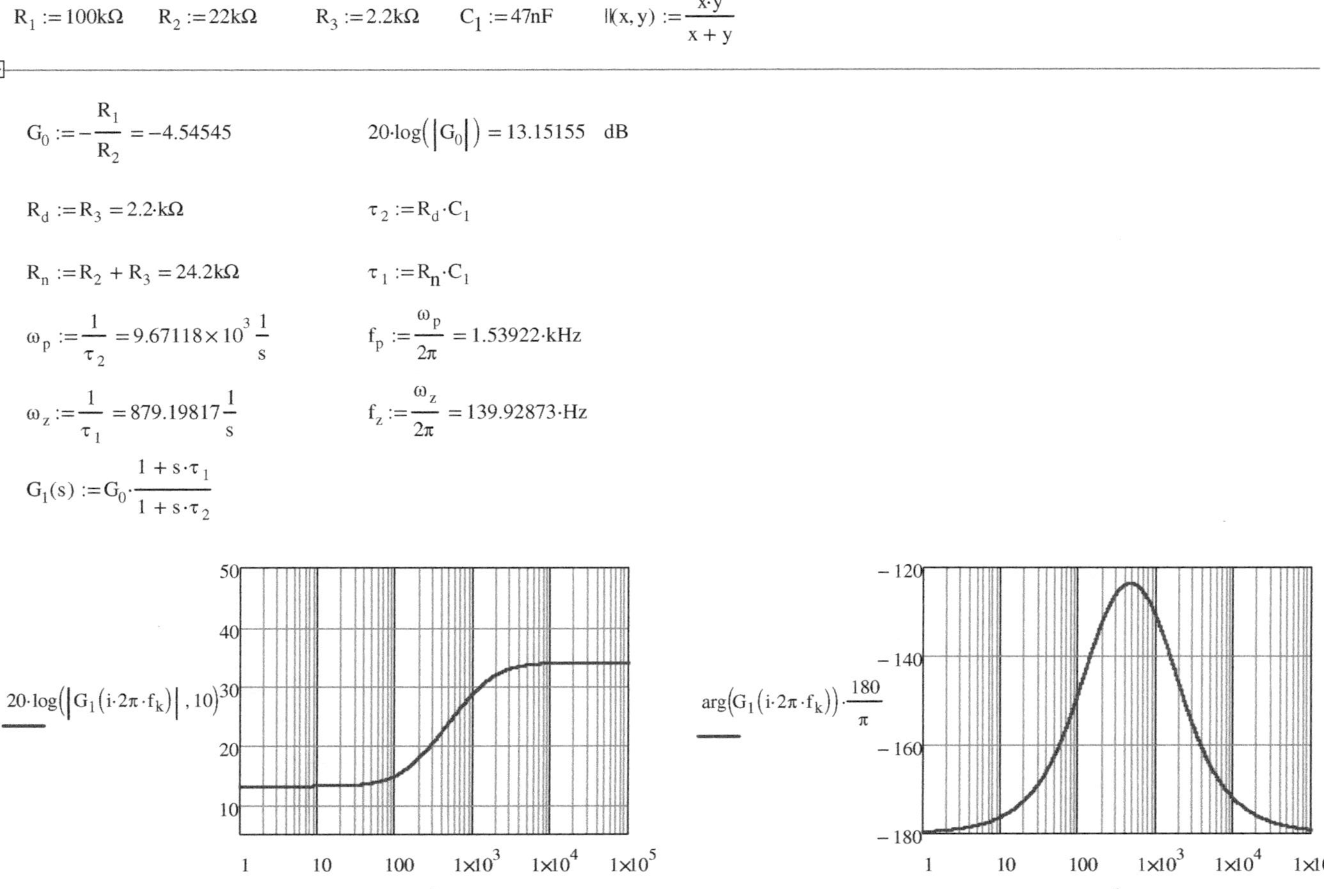

Figure 3.126 The Mathcad® sheet shows the response of the op amp-based circuit.

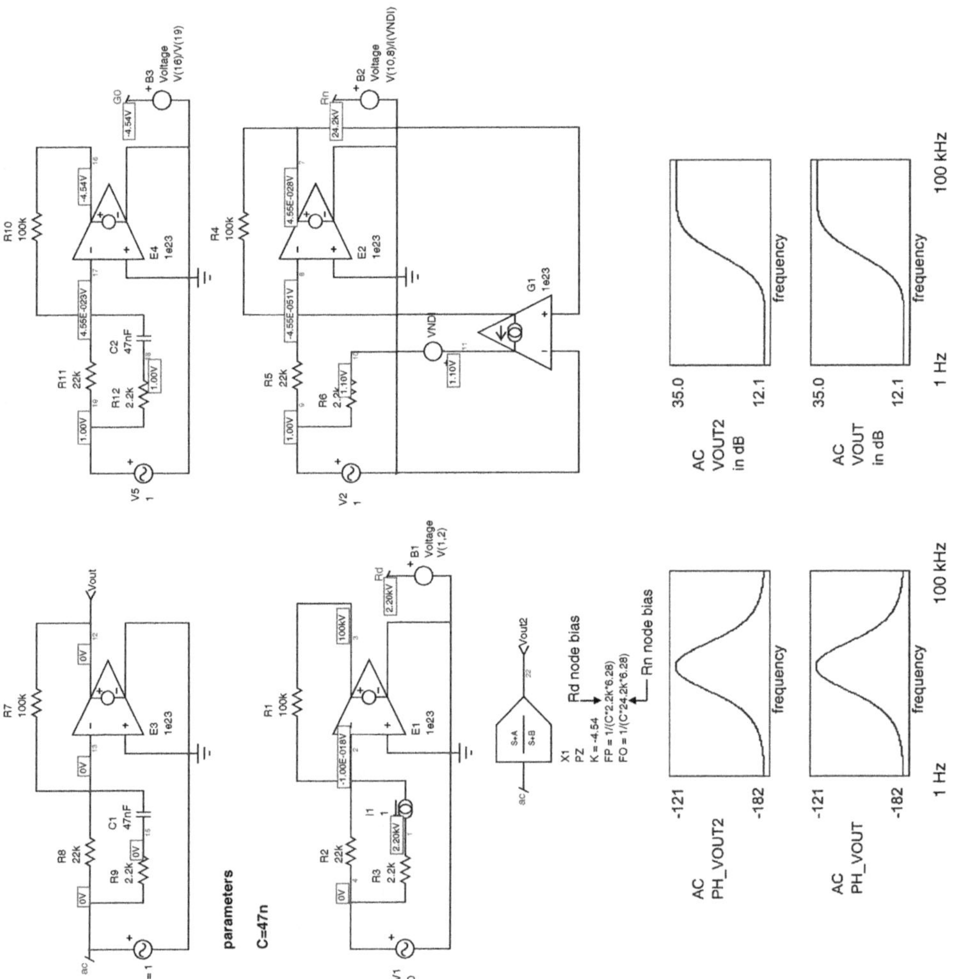

Figure 3.127 SPICE lends itself well to testing this circuit.

Problem 10:

The circuit includes a single energy-storing element and is thus a 1st-order network. If C_1 is infinite valued (replaced by a short circuit), the response disappears (V_{out} is short circuited) so there is no zero in this circuit. The transfer function obeys the following transfer function:

$$G(s) = G_0 \frac{1}{1 + s\tau_2} \tag{3.315}$$

The dc gain is found by removing C_1 as shown in Figure 3.128. Considering a perfect op amp, I_1 is the only circulating current. Given the virtual ground on the inverting pin, I_1 can be defined as:

$$I_1 = \frac{V_{in}}{R_2} \tag{3.316}$$

and

$$V_{out} = -I_1 R_1 \tag{3.317}$$

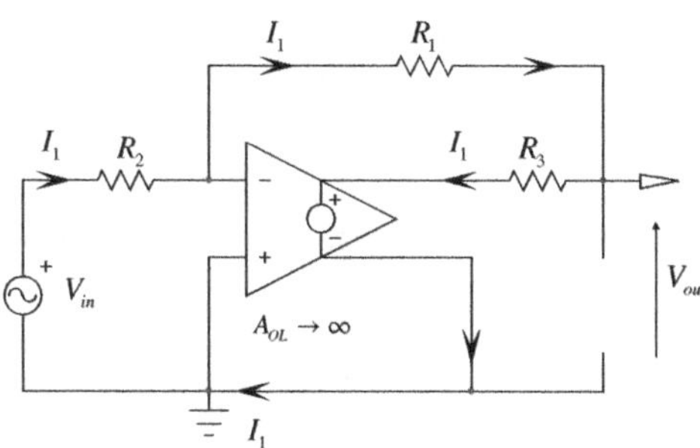

Figure 3.128 The gain of this inverter is unchanged despite the presence of R_3.

Combining both expressions leads to:

$$G_0 = -\frac{R_1}{R_2} \tag{3.318}$$

R_3 play no role in the dc gain expression. It slightly changes the op amp operating point though as I_1 circulates in R_3. The circuit time constant is found by setting the stimulus voltage to 0 and looking into the capacitor terminals. This is what Figure 3.129 describes.

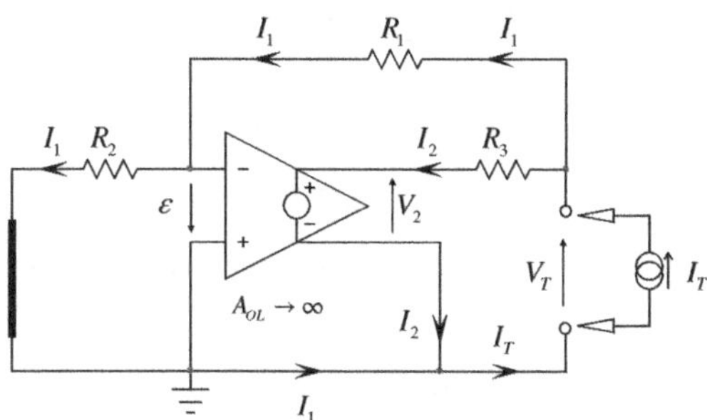

Figure 3.129 The time constant is obtained by looking into the capacitor terminals while the stimulus voltage is set to 0 V.

We can start by writing the relationship linking currents:

$$I_T = I_1 + I_2 \tag{3.319}$$

Let's assume the op amp has a finite open-loop gain noted A_{OL}. Then the error voltage ε is expressed as:

$$\varepsilon = -I_1 R_2 \tag{3.320}$$

The output voltage V_2 is given by:

$$V_2 = A_{OL} \cdot \varepsilon = -I_1 R_2 A_{OL} \tag{3.321}$$

The current flowing into resistor R_3 is defined as:

$$I_2 = \frac{V_T - V_2}{R_3} = \frac{V_T + I_1 R_2 A_{OL}}{R_3} \tag{3.322}$$

Current I_1 is defined as:

$$I_1 = \frac{V_T + \varepsilon}{R_1} \tag{3.323}$$

Substituting (3.320) in (3.323), we have:

$$I_1 = \frac{V_T - I_1 R_2}{R_1} \tag{3.324}$$

Rearranging this expression and factoring V_T, you have:

$$I_1 = \frac{V_T}{R_1 + R_2} \tag{3.325}$$

From (3.319), we extract I_2 and substitute it into (3.322):

$$I_T - I_1 = \frac{V_T + I_1 R_2 A_{OL}}{R_3} \tag{3.326}$$

Now replace I_1 by its definition in (3.325) and collect I_T and V_T terms. You should find:

$$I_T R_3 = V_T \left(1 + \frac{R_2 A_{OL}}{R_1 + R_2} + \frac{R_3}{R_1 + R_2}\right) \tag{3.327}$$

Resistance R_d is thus equal to:

$$R_d = \frac{V_T}{I_T} = \frac{R_3}{1 + \frac{R_2 A_{OL}}{R_1 + R_2} + \frac{R_3}{R_1 + R_2}} \tag{3.328}$$

When the op amp open-loop gain goes infinite, R_d goes to 0:

$$R_d\big|_{A_{OL} \to \infty} = 0 \tag{3.329}$$

The time constant τ_2 is also equal to 0. In other words, the transfer function expressed in (3.315) simplifies to:

$$G(s) \approx -\frac{R_1}{R_2} \tag{3.330}$$

and the output capacitor has no effect on the circuit. A Mathcad® sheet has been built and results are shown in Figure 3.130. For an open-loop gain of 10,000, the resistance R_d is 260 mΩ, and relegates

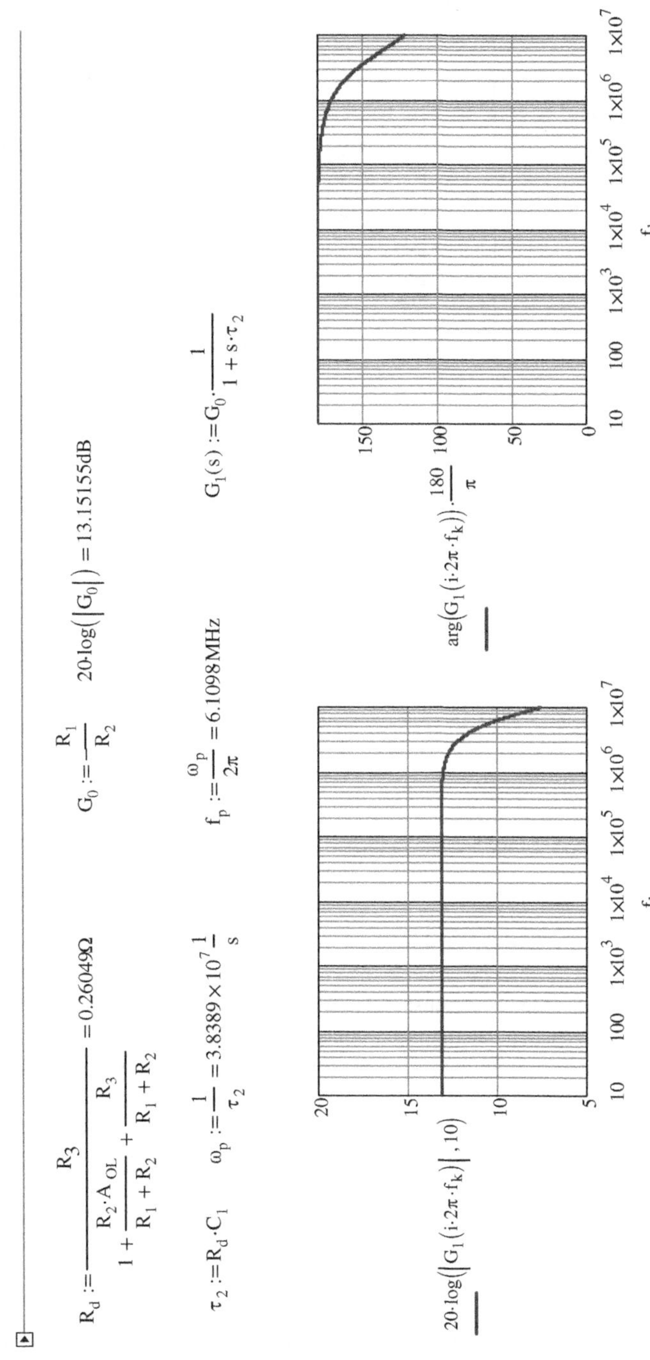

Figure 3.130 Mathcad® calculates a very small resistance driving capacitor C_1: the response is flat across the frequency range.

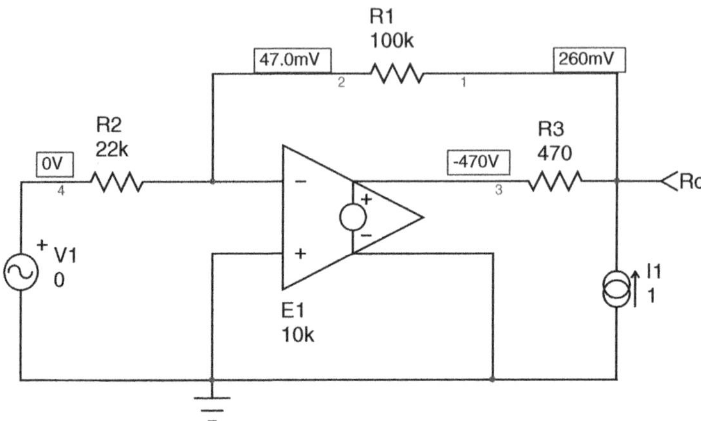

Figure 3.131 SPICE confirms a resistance R_d equal to 260 mΩ as calculated by the solver (260 mV at node 1 obtained by the 1-A source).

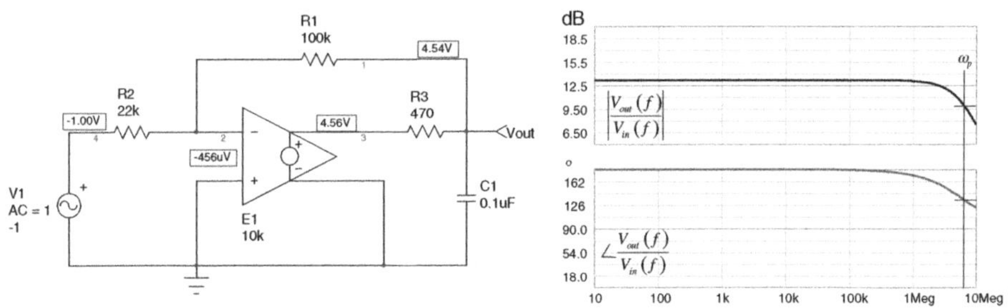

Figure 3.132 The ac response matches the Mathcad® dynamic curves and shows a pole at 6.1 MHz.

the pole to a high frequency: the ac response is flat up to the pole at 6.1 MHz. SPICE computes the same exact resistance value when calculating its bias point in Figure 3.131. Finally, the SPICE ac simulation at a given operating point confirms the pole position at 6.1 MHz (Figure 3.132).

This type of circuit is employed to stabilize an op amp when driving a heavy capacitive load. As the closed-loop output impedance of the op amp is inductive, combined with the capacitive load, it can bring instability in the circuit. An isolation resistance R_3 is added in series with the op amp so that it does not directly see the capacitor, preserving stability. More details about this technique are available from [1] and [2].

References

1. M. Pachchigar. Compensation Techniques for Driving Large-Capacitance Loads with High-Speed Amplifiers. EETime on-line publication, http://www.eetimes.com/document.asp?doc_id=1272424 (last accessed 20/12/2015).
2. D. Feucht. *Designing Dynamic Circuits*. Vol. 2 of Analog Circuit Design, Scitech Publishing, pp. 150–158. 2010.

4

Second-order Transfer Functions

In the previous chapter, we explored the heart of our subject, the Extra Element Theorem. We learned how to derive it in two different forms whether we considered the extra element as a short circuit or physically removed it from the network. If poles were obtained by turning the excitation off, zeros required a null double injection (NDI) operation which, sometimes, might be perceived as a difficult exercise to the novice. Fortunately, we were able to rephrase the EET in a different form so that NDI was no longer necessary. This approach indeed simplified the exercise but brought slightly more complex coefficients at the end. In this new chapter, we will apply the EET technique to second-order systems known as the Two Extra Element Theorem (2EET). The approach gains in complexity but the principle remains the same, applied to two elements instead of one. We will first explore the raw expression of the 2EET and further derive a simpler and more practical formula illustrated by numerous examples.

4.1 Applying the Extra Element Theorem Twice

The Extra Element Theorem considers that any 1st-order network transfer function can be broken into two terms: the leading term, or the *reference gain*, is obtained with the extra element Z put in its *reference state*. In this mode, Z is either removed from the circuit (Z is infinite) or shorted (Z is equal to 0). This first expression is then followed by a second coefficient, a correction factor k. This factor includes the pole and the zero – if any – obtained by looking at the resistance offered by the energy-storing element terminals when the network is examined in particular conditions (excitation set to 0 or NDI). This is what Figure 4.1 illustrates.

The idea behind the Two Extra Element Theorem (2EET) theory detailed in [1] is to apply the EET twice to a network containing two energy-storing elements, Z_1 and Z_2. First, one energy-storing element is identified in the network to which the classical EET is applied. The obtained expression becomes the reference gain A_{ref} for the second EET. As seen in Chapter 3, the selected 'extra' impedance in the EET is first considered as a short circuit ($Z = 0$) or removed from the circuit ($Z \to \infty$). It is its *reference state*. However, even if the EET applies to that single element, how do you treat the second element in this first step? In other words, how do you consider that element while the EET is

Linear Circuit Transfer Functions: An Introduction to Fast Analytical Techniques, First Edition. Christophe P. Basso.

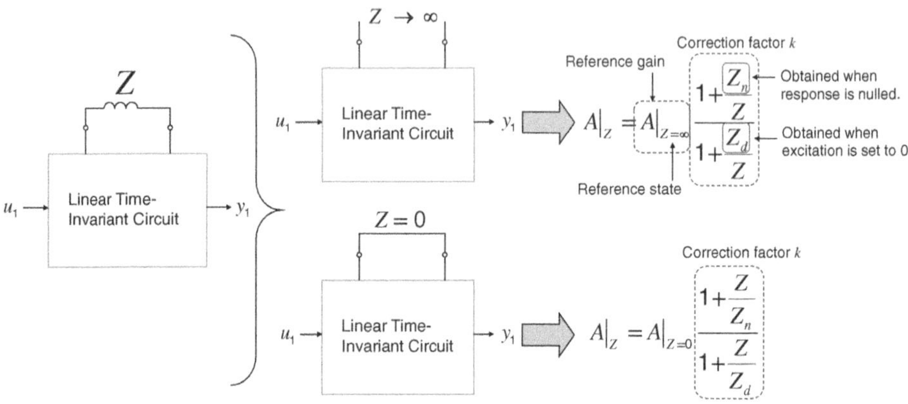

Figure 4.1 The EET places the extra element in two different states and defines an adequate correction factor, leading to the final transfer function.

carried over the first one? Well, given the presence of two energy-storing components, there are four possible ways of combining them, four possible reference states:

$$\begin{aligned}
&Z_1 = Z_2 = 0\\
&Z_1 = 0, Z_2 \to \infty\\
&Z_1 \to \infty, Z_2 = 0\\
&Z_1 \to \infty, Z_2 \to \infty
\end{aligned}$$

The number of combinations can be determined thinking of binary levels as Z_1 and Z_2 can be either open, '1', or shorted '0'. In this case, the number of possible combinations for a 2nd-order system is given by:

$$Combinations = 2^n = 2^2 = 4 \tag{4.1}$$

While you apply the first EET, you have the choice to select Z_1 and Z_2 according to the above four combinations. At the end, the 2EET will deliver four different formulations all giving identical results. This principle is illustrated by Figure 4.2 where you see the four possible combinations selected for the reference gain calculation. Then follows a correction factor k which corresponds to the EET applied a second time with the first element in place, Z_1 in the illustration.

The 2EET theorem can be formulated the following ways, covering all four different choices for Z_1 and Z_2:

$$A|_{\substack{Z_1\\Z_2}} = A|_{\substack{Z_1\to\infty\\Z_2\to\infty}} \frac{1+\dfrac{Z_{n1}|_{Z_2\to\infty}}{Z_1}}{1+\dfrac{Z_{d1}|_{Z_2\to\infty}}{Z_1}} \frac{1+\dfrac{Z_{n2}}{Z_2}}{1+\dfrac{Z_{d2}}{Z_2}} \tag{4.2}$$

$$A|_{\substack{Z_1\\Z_2}} = A|_{\substack{Z_1=0\\Z_2=0}} \frac{1+\dfrac{Z_1}{Z_{n1}|_{Z_2=0}}}{1+\dfrac{Z_1}{Z_{d1}|_{Z_2=0}}} \frac{1+\dfrac{Z_2}{Z_{n2}}}{1+\dfrac{Z_2}{Z_{d2}}} \tag{4.3}$$

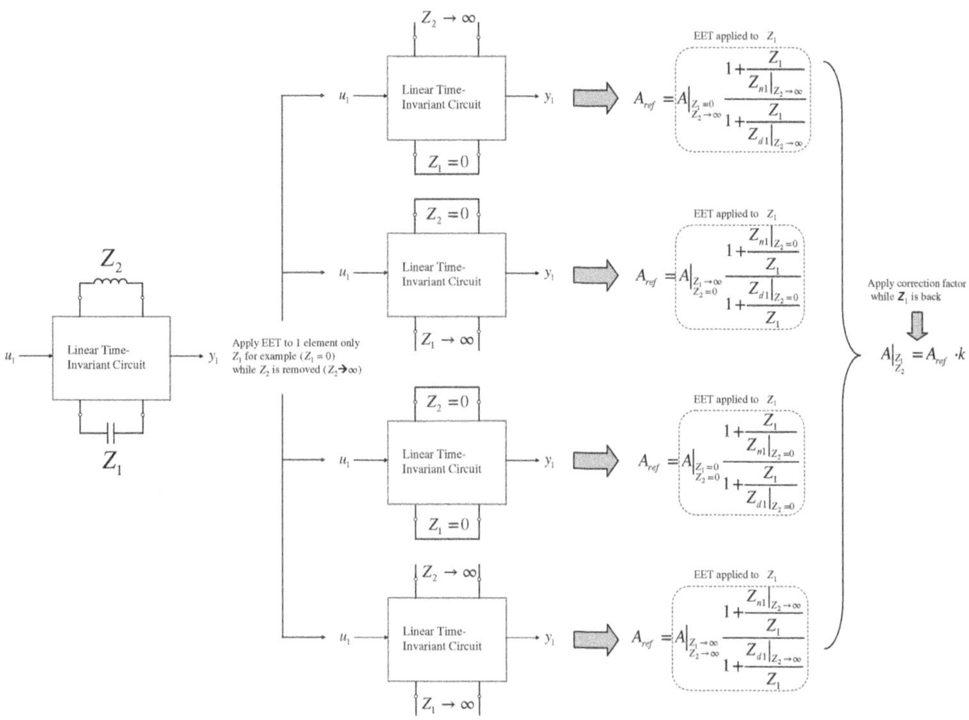

Figure 4.2 The first EET is applied over the circuit when only one element is considered (Z_1 in the drawing). The EET is then applied for a second time with one of the energy-storing components back in place. Please note that all expressions lead to identical results.

$$A|_{\substack{Z_1 \\ Z_2}} = A|_{\substack{Z_1 = 0 \\ Z_2 \to \infty}} \frac{1 + \dfrac{Z_1}{Z_{n1}|_{Z_2 \to \infty}}}{1 + \dfrac{Z_1}{Z_{d1}|_{Z_2 \to \infty}}} \frac{1 + \dfrac{Z_{n2}}{Z_2}}{1 + \dfrac{Z_{d2}}{Z_2}} \tag{4.4}$$

$$A|_{\substack{Z_1 \\ Z_2}} = A|_{\substack{Z_1 \to \infty \\ Z_2 = 0}} \frac{1 + \dfrac{Z_{n1}|_{Z_2 = 0}}{Z_1}}{1 + \dfrac{Z_{d1}|_{Z_2 = 0}}{Z_1}} \frac{1 + \dfrac{Z_2}{Z_{n2}}}{1 + \dfrac{Z_2}{Z_{d2}}} \tag{4.5}$$

In our examples, when the EET is carried over the selected element, Z_1, Z_{n1} and Z_{d1} turn into R_{n1} and R_{d1}, and are resistances respectively driving the reactance in Z_1 when the response is nulled and the excitation set to 0. In the correction factor, the selected energy-storing element is back in place. What now drives Z_2, the second element, is truly an impedance involving Z_1 as we will see in a first application example. Fortunately, network conditions to determine Z_{n2} and Z_{d2} are similar to those set for Z_{n1} and Z_{d1}.

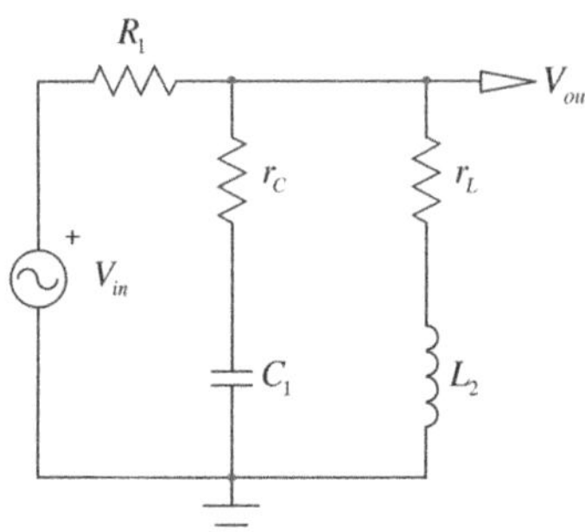

Figure 4.3 This 2nd-order circuit combines a capacitor and an inductor. There are four possible options to apply the first EET.

Consider the 2nd-order circuit in Figure 4.3. First off, without the 2EET, we can apply brute-force algebra to obtain the transfer function H of this circuit. It is given by the series-parallel combination of C_1-r_C and L_2-r_L forming an impedance divider with R_1:

$$H(s) = \frac{\left(r_C + \frac{1}{sC_1}\right) \| (r_L + sL_2)}{R_1 + \left(r_C + \frac{1}{sC_1}\right) \| (r_L + sL_2)} \tag{4.6}$$

There is absolutely no insight in that expression but its dynamic response is the one we will refer to when assessing the validity of the newly-found transfer functions using the 2EET.

If we want to apply the 2EET to this circuit, we first exercise the EET for one selected impedance, Z_1 (involving C_1) in this example. As explained, we have four possible combinations to chose from regarding the reference states for Z_1 and Z_2. Regardless of this choice, it will lead to four different equal formulations. Assume both impedances are set to infinite for the reference circuit evaluation as in (4.2). In this case, C_1 and L_2 can be physically removed from the circuit which updates to that shown in Figure 4.4.

The intermediate reference gain in this configuration is immediate:

$$H|_{\substack{Z_1 \to \infty \\ Z_2 \to \infty}} = 1 \tag{4.7}$$

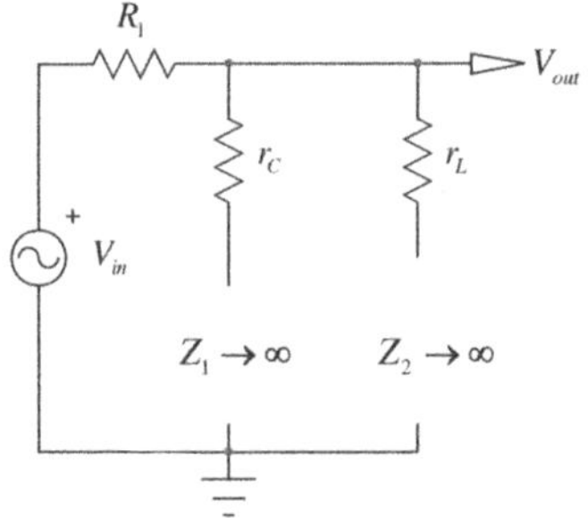

Figure 4.4 The reference circuit is chosen with Z_1 and Z_2 brought to infinite values (C_1 is zero and L_2 is infinite).

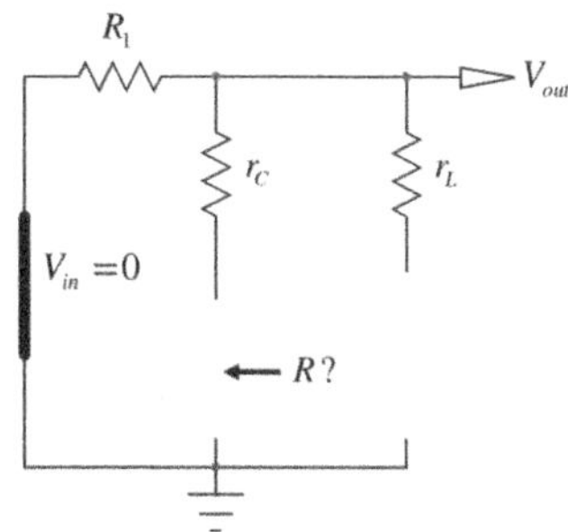

Figure 4.5 The EET is applied to Z_1 while Z_2 is removed. Here, R_d is determined.

Then, from the reference circuit, we will apply the EET considering Z_1 as the extra element while Z_2 is brought to infinity as adopted in the reference circuit. We first evaluate R_d while the excitation is set to 0 V, then we will apply a NDI to get R_n. The circuit is quite simple as shown in Figure 4.5. It is easy to see that the resistance is given by:

$$R_d|_{Z_2 \to \infty} = r_C + R_1 \tag{4.8}$$

A NDI is now performed on the circuit which turns out to be quite simple as drawn in Figure 4.6. No need for a test generator. If the upper terminal of r_C is at 0 V, the resistance seen from C_1's terminals is simply:

$$R_n|_{Z_2 \to \infty} = r_C \tag{4.9}$$

With these elements on hands, we can write the first part of the 2EET which is the reference gain H_{ref} applied to our 2nd-order network when the reference network is considered with Z_1 and Z_2 brought to infinity (physically removed from the circuit):

$$H_{ref}(s) = H|_{\substack{Z_1 \to \infty \\ Z_2 \to \infty}} \frac{1 + \frac{Z_{n1}|_{Z_2 \to \infty}}{Z_1}}{1 + \frac{Z_{d1}|_{Z_2 \to \infty}}{Z_1}} = 1 \cdot \frac{1 + \frac{r_C}{\frac{1}{sC_1}}}{1 + \frac{r_C + R_1}{\frac{1}{sC_1}}} = \frac{1 + sr_C C_1}{1 + s(r_C + R_1)C_1} \tag{4.10}$$

Up to this point, nothing too complicated compared to what we performed in Chapter 3: we applied the EET to Z_1, considering Z_2 excluded from the circuit. The correction factor that we will now

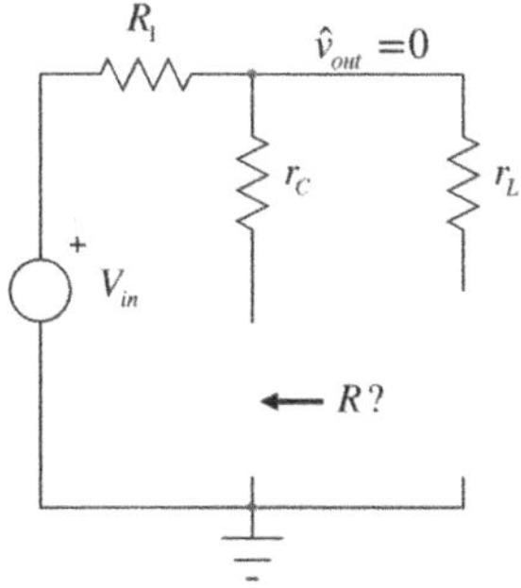

Figure 4.6 A NDI is performed while Z_2 is brought to infinity.

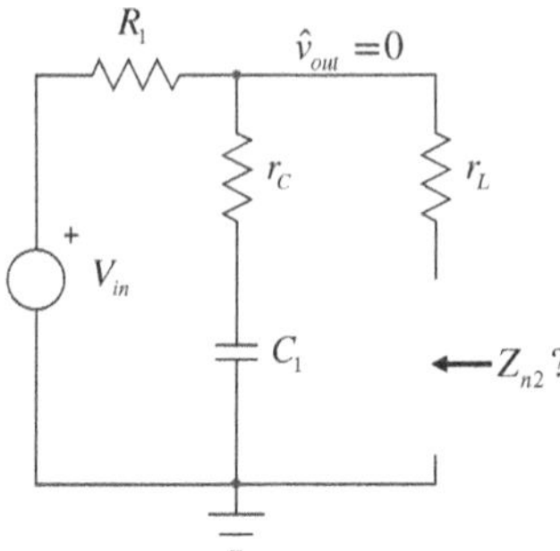

Figure 4.7 Apply now the second EET while C_1 is brought back in the circuit. Here the null in the response is considered.

determine involves Z_2 but brings Z_1 back in place. The new circuit for the correction term appears in Figure 4.7, here the NDI is considered first.

A test generator is not needed here, the impedance seen from the inductor terminals when $\hat{v}_{out} = 0$ is purely resistive in this case and equals the inductor ohmic loss:

$$Z_{n2} = r_L \tag{4.11}$$

When the excitation is turned off, as in Figure 4.8, the impedance seen from the inductor terminals now includes the capacitor impedance.

By inspection, the impedance driving L_2 in this mode is given by:

$$Z_{d2} = r_L + R_1 \| \left(\frac{1}{sC_1} + r_C\right) \tag{4.12}$$

The correction factor k is given by:

$$k = \frac{1+\dfrac{Z_{n2}}{Z_2}}{1+\dfrac{Z_{d2}}{Z_2}} = \frac{1+\dfrac{r_L}{sL_2}}{1+\dfrac{r_L + R_1 \| \left(\dfrac{1}{sC_1} + r_C\right)}{sL_2}} \tag{4.13}$$

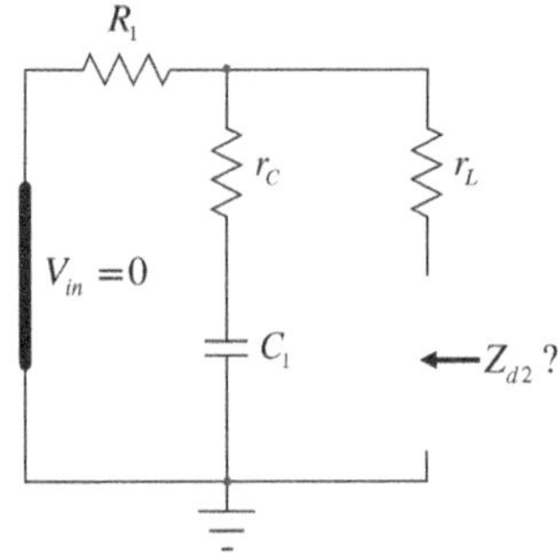

Figure 4.8 For the denominator term, turn the excitation off and look at the impedance driving L_2 while C_1 is back in the circuit.

Assembling all the pieces, we can now formulate the transfer function H linking V_{out} to V_{in} as:

$$H(s) = H_{ref}(s) \cdot k = \frac{1 + sr_C C_1}{1 + s(r_C + R_1)C_1} \frac{1 + \dfrac{r_L}{sL_2}}{1 + \dfrac{r_L + R_1 \| \left(\dfrac{1}{sC_1} + r_C\right)}{sL_2}} \tag{4.14}$$

We can check another expression in which we consider Z_1 infinite and Z_2 a short circuit. It is similar as observing the reference circuit for $s = 0$, meaning the impedances are put in the following reference states: C_1 in dc is removed from the circuit (0-valued capacitance) and L_2 is replaced by a short circuit (0-valued inductance). The 2EET formula corresponding to this situation is that of (4.5) and the path to the final transfer function is illustrated by Figure 4.9.

Assembling all the intermediate steps gives us another transfer function H:

$$H(s) = \frac{r_L}{r_L + R_1} \frac{1 + \dfrac{r_C}{\dfrac{1}{sC_1}}}{1 + \dfrac{r_C + R_1 \| r_L}{\dfrac{1}{sC_1}}} \frac{1 + \dfrac{sL_2}{r_L}}{1 + \dfrac{sL_2}{r_L + R_1 \| \left(\dfrac{1}{sC_1} + r_C\right)}} \tag{4.15}$$

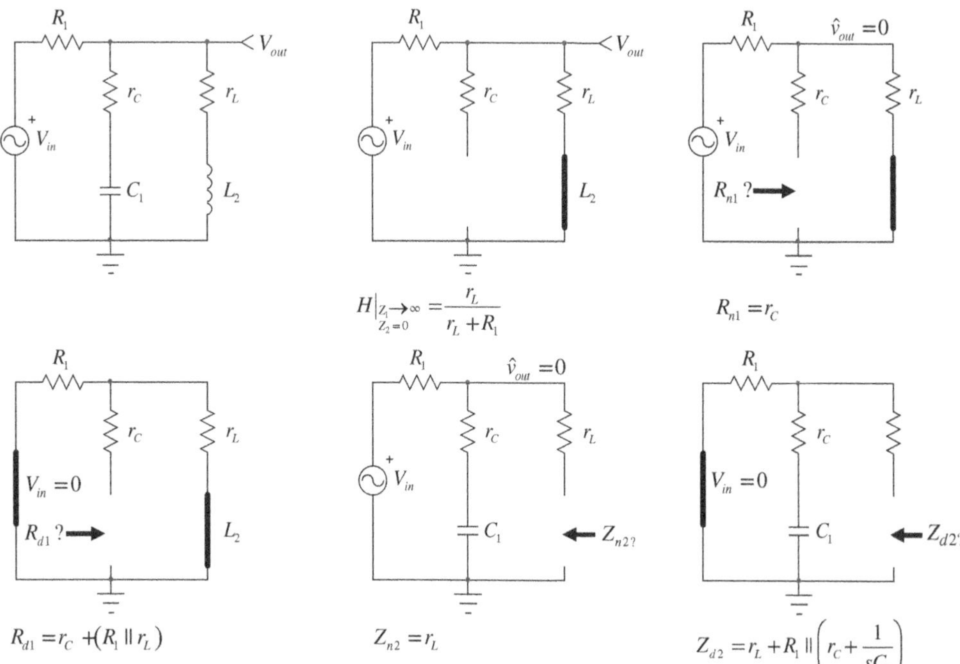

Figure 4.9 We can also consider the reference gain in dc, where C_1 is removed and L_2 is replaced by a short circuit.

For the sake of the exercise, now consider Z_2 as the extra element (instead of Z_1 in the previous cases) and apply the EET over it while Z_1 is removed. The 2EET formula in this configuration $(Z_2 = 0, Z_1 \to \infty)$ updates to:

$$A|_{\substack{Z_1 \\ Z_2}} = A|_{\substack{Z_2 = 0 \\ Z_1 \to \infty}} \frac{1 + \dfrac{Z_2}{Z_{n2}|_{Z_1 \to \infty}}}{1 + \dfrac{Z_2}{Z_{d2}|_{Z_1 \to \infty}}} \frac{1 + \dfrac{Z_{n2}}{Z_1}}{1 + \dfrac{Z_{d2}}{Z_1}} \tag{4.16}$$

If you apply the steps we have described, you should find the following transfer function expression which represents another formulation for the same transfer function characterizing Figure 4.3:

$$H(s) = \frac{r_L}{r_L + R_1} \frac{1 + \dfrac{sL_2}{r_L}}{1 + \dfrac{sL_2}{r_L + R_1}} \frac{1 + \dfrac{r_C}{\dfrac{1}{sC_1}}}{1 + \dfrac{r_C + R_1 \| (sL_2 + r_L)}{\dfrac{1}{sC_1}}} \tag{4.17}$$

To compare these formulas with the brute-force approach expression defined in (4.6), we have gathered all of them in a Mathcad® sheet and also included a SPICE simulation graph. As shown in Figure 4.10, these expressions deliver similar results, they are all identical. For these plots, we have

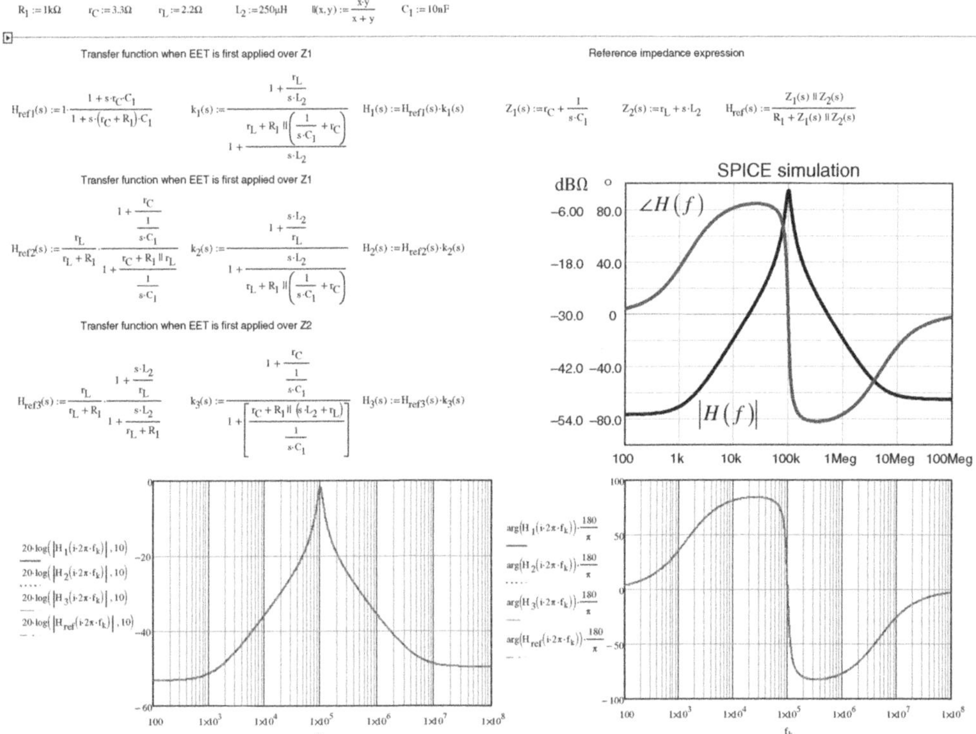

Figure 4.10 Three different expressions plotted in Mathcad® and compared to a SPICE simulation.

adopted the following component values: $R_1 = 1\,\text{k}\Omega$ $r_C = 3.3\,\Omega$ $r_L = 2.2\,\Omega$ $C_1 = 10\,\text{nF}$ $L_2 = 250\,\mu\text{H}$

4.1.1 Low-entropy 2nd-order Expressions

The formulas delivered by the 2EET are correct as confirmed by Figure 4.10. Unfortunately, raw results given by the method – regardless of the way Z_1 Z_2 are combined – do not really offer the insight we need to determine poles and zeros positions. Except for the zeros, which by inspection would have immediately been revealed: what nulls the response in the transformed version of Figure 4.3 when examined at $s = s_z$? A transformed short brought by the series arrangement of $1/sC_1$-r_C and sL_2-r_L:

$$r_C + \frac{1}{sC_1} = 0 \rightarrow s_{z_1} = -\frac{1}{r_C C_1} \tag{4.18}$$

$$sL_2 + r_L = 0 \rightarrow s_{z_2} = -\frac{r_L}{L_2} \tag{4.19}$$

Therefore

$$N(s) = (1 + sr_C C_1)\left(1 + s\frac{L_2}{r_L}\right) \tag{4.20}$$

The denominator, however, looks quite complicated. It is isolated below from (4.15):

$$D(s) = \left(1 + s\left[r_C + R_1 \| r_L\right] C_1\right)\left(1 + \frac{sL_2}{r_L + R_1 \| \left(\frac{1}{sC_1} + r_C\right)}\right) \tag{4.21}$$

Developing (4.21) gives

$$D(s) = \left(1 + sC_1\left[r_C + \left(\frac{r_L R_1}{r_L + R_1}\right)\right]\right)\left(1 + \frac{sL_2(sr_C C_1 + sR_1 C_1 + 1)}{R_1 + r_L + sC_1(r_C R_1 + r_L R_1 + r_C r_L)}\right) \tag{4.22}$$

Expanding and factoring leads to

$$D(s) = 1 + s\left[\frac{L_2}{r_L + R_1} + C_1\frac{R_1 r_C + R_1 r_L + r_C r_L}{r_L + R_1}\right] + s^2\frac{L_2}{R_1 + r_L}C_1(R_1 + r_C) \tag{4.23}$$

which is a 2nd-order polynomial expression following the form:

$$D(s) = 1 + b_1 s + b_2 s^2 \tag{4.24}$$

In a low-entropy arrangement, the denominator and the numerator are unitless but the leading term carries the unit of the transfer function, if any. In (4.24), you see a term b_1 multiplied by s while b_2 is multiplied by s^2. As demonstrated in [2], if this expression is unitless it means that the dimension of b_1 is time, [s], while the dimension of b_2 is time squared, [s^2]. We can show that b_1 is the sum of the circuit time constants while b_2 is the product of two time constants involving the two energy-storing elements. Calculations of transfer functions, regardless of the order, always start with a reference gain. Here, and for the rest of the book chapters, we will always consider the reference state of the circuit for $s = 0$ at which the reference gain H_0 is computed. Looking at Figure 4.11, the reference gain is immediate:

$$H_0 = \frac{r_L}{r_L + R_1} \tag{4.25}$$

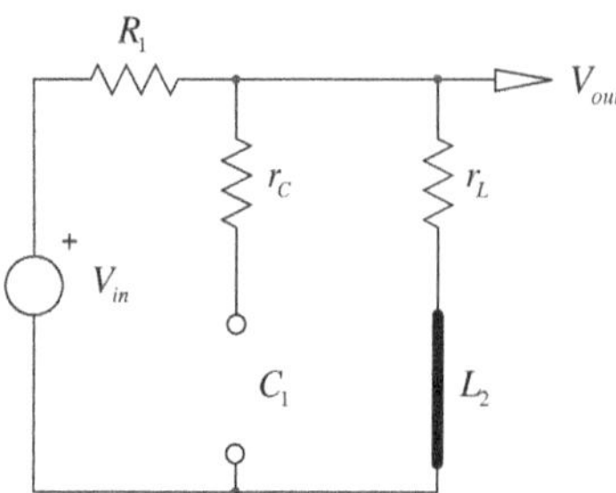

Figure 4.11 The reference gain H_0 is computed for $s=0$ where the capacitor is removed and the inductor replaced by a short circuit.

The circuit time constants τ_1 and τ_2 are then classically obtained when the excitation is turned off. To observe the circuit in dc conditions, simply remove capacitors and short inductors: these are the energy-storing components reference states, corresponding to one of the four options given by Z_1 and Z_2. Why retaining this one among the 4? Because I believe this is the most intuitive and corresponds to the way operating bias point analysis is carried, with SPICE in particular. Figure 4.12 shows the circuit in dc while the excitation is set to 0. Time constants can be extracted by looking at the resistance seen from C_1's terminals while L_2 is shorted (Figure 4.12a):

$$R = r_C + R_1 \| r_L \tag{4.26}$$

hence

$$\tau_1 = \left(r_C + R_1 \| r_L\right) C_1 \tag{4.27}$$

and by looking at the resistance seen from L_2's terminals while C_1 is removed (Figure 4.12b):

$$R = R_1 + r_L \tag{4.28}$$

leading to

$$\tau_2 = \frac{L_2}{R_1 + r_L} \tag{4.29}$$

Summing up these time constants gives b_1:

$$b_1 = \tau_1 + \tau_2 = \left(r_C + R_1 \| r_L\right) C_1 + \frac{L_2}{R_1 + r_L} \tag{4.30}$$

which is exactly what you have in (4.23) in a developed version.

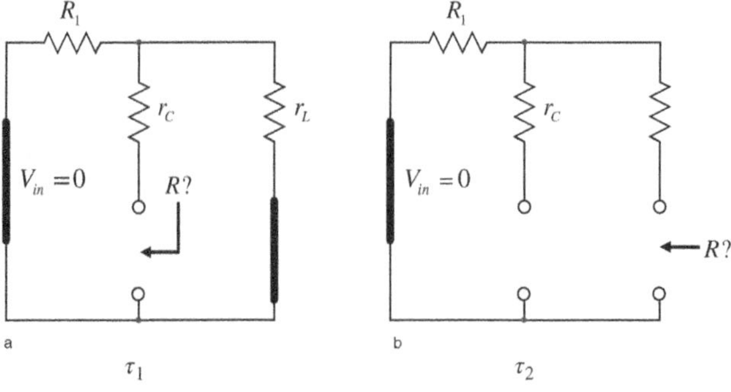

Figure 4.12 Time constants are obtained by putting the circuit in dc conditions with the excitation turned off.

The second term b_2 combines time constants in a product. One of these time constants is either τ_1 or τ_2. The second time constant is obtained while the element involved in the time constant you selected is put in its *opposite* reference state. As we will always refer to a dc reference state, that element will always be put in its high-frequency state. Assume you picked τ_1 for the first time constant, then capacitor C_1 involved in τ_1 is put in a high-frequency state or replaced by a short circuit. Should you choose τ_2 instead, then place L_2 in its high-frequency state or remove it from the circuit. Then, you look at the remaining storage element's terminals to determine the resistance that it 'sees' from its connection points. To reflect this examination mode, the definition for b_2 adopts a new formalism:

$$b_2 = \tau_1 \tau_2^1 \tag{4.31}$$

Given the fact that you could also pick τ_2 as the first time constant, a redundant expression exists for b_2 as demonstrated in [1] and [2]:

$$b_2 = \tau_2 \tau_1^2 \tag{4.32}$$

Forming and interpreting these expressions is easy as described in Figure 4.13.

The practical application to our 2nd-order circuit appears in Figure 4.14. In sketch (a), the resistance seen from C_1 terminals while L_2 is put in its high-frequency state (infinite impedance, removed from the circuit) gives $r_C + R_1$. The associated time constant is thus:

$$\tau_1^2 = C_1(r_C + R_1) \tag{4.33}$$

In sketch (b), the resistance driving inductor L_2 while capacitor C_1 is in a high-frequency state (a short circuit) is $r_L + R_1 \| r_C$ giving the second possible time constant:

$$\tau_2^1 = \frac{L_2}{r_L + R_1 \| r_C} \tag{4.34}$$

The denominator $D(s)$ can therefore be written in a first polynomial form as shown below. The s coefficient is the sum of the natural time constants determined in (4.27) and (4.29). The coefficient for

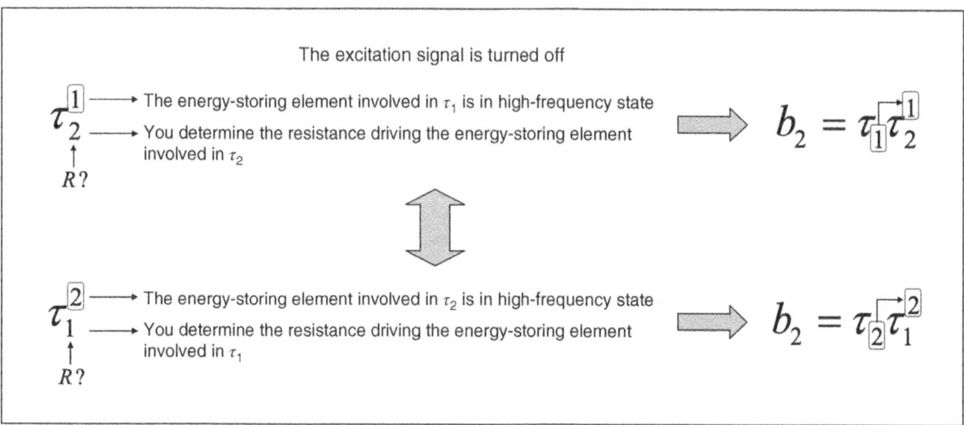

Figure 4.13 The product of time constants implies that the element involved in the first time constant is put in its high-frequency state before you determine the resistance driving the second energy-storing component.

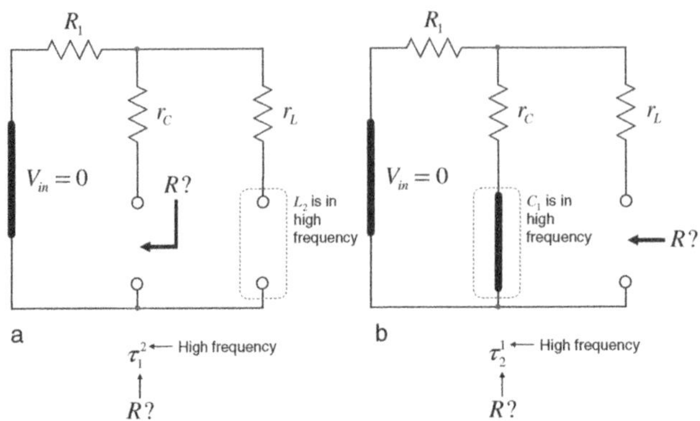

Figure 4.14 The 2nd-order denominator coefficient b_2 is found by putting one element in a state opposite to its reference state and looking at the resistance driving the other element.

s^2 is either (4.31) or (4.32):

$$D(s) = 1 + s(\tau_1 + \tau_2) + s^2 \tau_1 \tau_2^1 \tag{4.35}$$

Combining the time constants we have determined:

$$D(s) = 1 + s\left[C_1\left(r_C + R_1 \| r_L\right) + \frac{L_2}{R_1 + r_L}\right] + s^2\left[C_1\left(r_C + R_1 \| r_L\right)\frac{L_2}{r_L + R_1 \| r_C}\right] \tag{4.36}$$

also equal to

$$D(s) = 1 + s(\tau_1 + \tau_2) + s^2 \tau_2 \tau_1^2 \tag{4.37}$$

Substituting time constants values in (4.37) leads to:

$$D(s) = 1 + s\left[C_1\left(r_C + R_1 \| r_L\right) + \frac{L_2}{R_1 + r_L}\right] + s^2\left[\frac{L_2}{R_1 + r_L} C_1(r_C + R_1)\right] \tag{4.38}$$

In (4.36), b_2 is equal to

$$b_2 = C_1\left(r_C + R_1 \| r_L\right)\frac{L_2}{r_L + R_1 \| r_C} \tag{4.39}$$

while in (4.38) it is

$$b_2 = \frac{L_2}{R_1 + r_L} C_1(r_C + R_1) \tag{4.40}$$

If you develop (4.39), you obtain (4.40).

What flexibility does it bring to have two possible expressions for s^2 in $D(s)$? The first advantage is simplicity: depending on the combination, $\tau_1 \tau_2^1$ or $\tau_2 \tau_1^2$, you may find that one of the two expressions gives a complex relationship, requiring energy to simplify it. In our case, (4.40) is simpler than (4.39) at first sight. For this reason, you may sometimes engage into an option where $\tau_1 \tau_2^1$ is first selected but later realize that $\tau_2 \tau_1^2$ leads to a simpler result or a simpler circuit to analyze. The choice is yours to select either one. The second justification for this redundancy in the s^2 factor is the indeterminacy that can sometimes arise with the product of time constants. Mathematically, indeterminacies occur when you have the following combinations:

$$\frac{\infty}{\infty}$$
$$\infty - \infty$$
$$\frac{0}{0}$$
$$0 \cdot \infty$$

Please note that $0/\infty$ is not indetermined and returns 0.

Assume a time constant τ_1 equals 0 because the resistance seen by inductor L_1 for $s = 0$ is infinite (one of its branches is open for instance in the considered configuration):

$$\tau_1 = \frac{L_1}{\infty} = 0 \tag{4.41}$$

If multiplied by the second time constant τ_2^1 found when L_1 is put in its high-frequency state (removed from the circuit) a problem can arise depending on the resulting expression. If that expression is non-infinite, then all is fine and you could have:

$$b_2 = \tau_1\tau_2^1 = \frac{L_1}{\infty}R_2C_2 = 0 \tag{4.42}$$

However, if that expression is infinite, then we have an indeterminacy:

$$b_2 = \tau_1\tau_2^1 = \frac{L_1}{\infty} \cdot \infty \cdot C_2 \tag{4.43}$$

How do we remove the indeterminacy? By reshuffling the combinations and involving $\tau_2\tau_1^2$ rather than $\tau_1\tau_2^1$.

Another way exists to remove the indeterminacy. If looking through the energy-storing elements returns an infinite resistance, why not adding an extra resistance to the circuit, placed in parallel with the energy-storing device that is examined? Later on, you will be able to recalculate the transfer function while considering this element at an infinite value. Similarly, if the resistance seen from the energy-storing element is 0 and causes problems in the expressions, why not insert a small resistance in series with the connecting terminals? Once the transfer function is determined, simply turn this extra resistance to 0 Ω and simplify the expression. This extra resistance can also be placed elsewhere in the circuit as long it does the job of offering a convenient ohmic path. As you can see, even if the EET and the 2EET break a complex circuit into smaller and simpler pieces, assembling the complete puzzle still requires engineering judgment.

4.1.2 Determining the Zero Positions

The remark concerning the denominator unit also holds for the numerator which follows a 2nd-order form in case two zeros are present:

$$N(s) = 1 + a_1s + a_2s^2 \tag{4.44}$$

If N is unitless then the dimension of a_1 is time, [s], while the dimension of a_2 is time squared, [s^2]. We can show that a_1 is the sum of the circuit time constants obtained while the output is nulled and the energy-storing elements are in their reference state. Regarding a_2, it is also the product of two time constants involving the two energy-storing elements while the output is nulled and one of the elements is put in its *opposite* reference state. Figure 4.15 details this operation, very similar to what

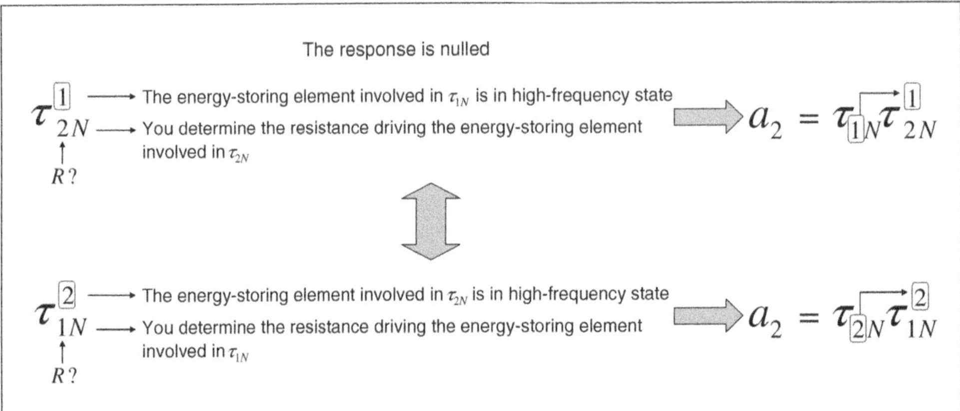

Figure 4.15 The product of time constants implies that the element involved in the first time constant is put in its high-frequency state.

we did for the denominator except that the excitation is back in place and the response is nulled. To distinguish the circuit natural time constants obtained for the numerator from those determined while the excitation is turned off, we will add the suffix N to the time constant label:

$$a_1 = \tau_{1N} + \tau_{2N} \tag{4.45}$$

and

$$a_2 = \tau_{1N}\tau_{2N}^{1} \tag{4.46}$$

also equal to

$$a_2 = \tau_{2N}\tau_{1N}^{2} \tag{4.47}$$

Figure 4.15 details how to build the 2nd-order coefficient. The process is similar to that involved in the denominator construction except that resistances are derived while the output response is nulled. Applying these concepts leads us to Figure 4.16 for the first coefficients. In sketch (a), as the output is nulled, the resistance seen from the capacitor terminals is simply r_C, giving the following time constant:

$$\tau_{1N} = r_C C_1 \tag{4.48}$$

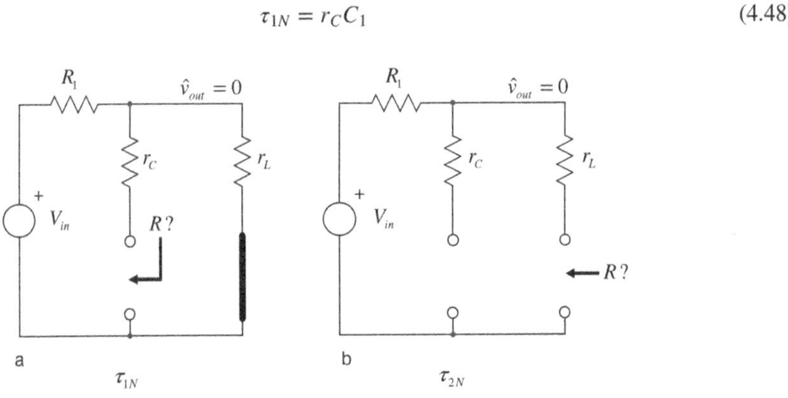

Figure 4.16 1st-order coefficients for s are obtained when the response is nulled and storage elements are in their reference states.

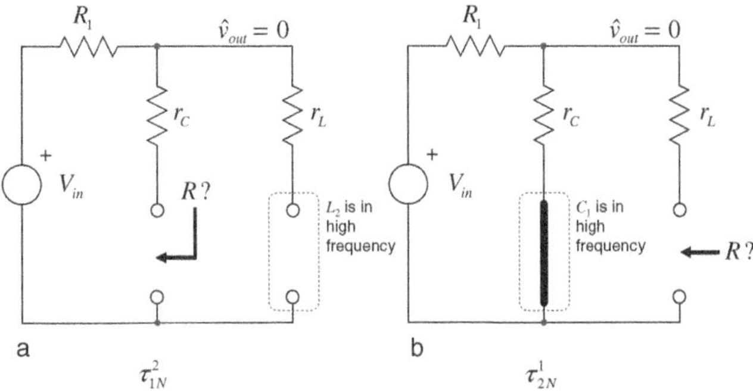

Figure 4.17 2nd-order coefficients for s^2 are also obtained when the response is nulled but one of the energy-storing element is put in its opposite reference state.

In Figure 4.16b, the resistance seen by inductor L_2 while the output is nulled is r_L. Therefore:

$$\tau_{2N} = \frac{L_2}{r_L} \tag{4.49}$$

In Figure 4.17, both sketches represent the situation for the 2nd-order coefficients determined when the output is nulled. For sketch (a), the resistance seen form the capacitor terminals while inductor L_2 is set in its high-frequency state, is, again, the series resistance r_C. The time constant is thus:

$$\tau_{1N}^2 = r_C C_1 \tag{4.50}$$

No difficulty for sketch (b) where the resistance seen from the inductor terminals while the capacitor is replaced by a short circuit (high-frequency state) is r_L:

$$\tau_{2N}^1 = \frac{L_2}{r_L} \tag{4.51}$$

Combining all these results form two possible numerator expressions combining (4.48) and (4.51) or (4.49) and (4.50) for s^2:

$$N(s) = 1 + s\left(r_C C_1 + \frac{L_2}{r_L}\right) + s^2\left(r_C C_1 \frac{L_2}{r_L}\right) \tag{4.52}$$

also equal to

$$N(s) = 1 + s\left(r_C C_1 + \frac{L_2}{r_L}\right) + s^2\left(\frac{L_2}{r_L} r_C C_1\right) \tag{4.53}$$

It can easily be factored as

$$N(s) = (1 + s r_C C_1)\left(1 + s\frac{L_2}{r_L}\right) \tag{4.54}$$

4.1.3 Rearranging and Plotting Expressions

To check our expressions, we will build a Mathcad® sheet and assemble all the time constants we found to determine the numerator and the denominator. The reference transfer function is still the one we derived in (4.6) and the curves generated by manipulating time constants will be compared to it. To write $N(s)$ and $D(s)$, I recommend capturing time constants independently in Mathcad®. Once this is done, you build D and N according to their definitions in (4.36) or (4.52), associating these time constants. That way, if a problem is identified, you can go back to the figures and identify where the

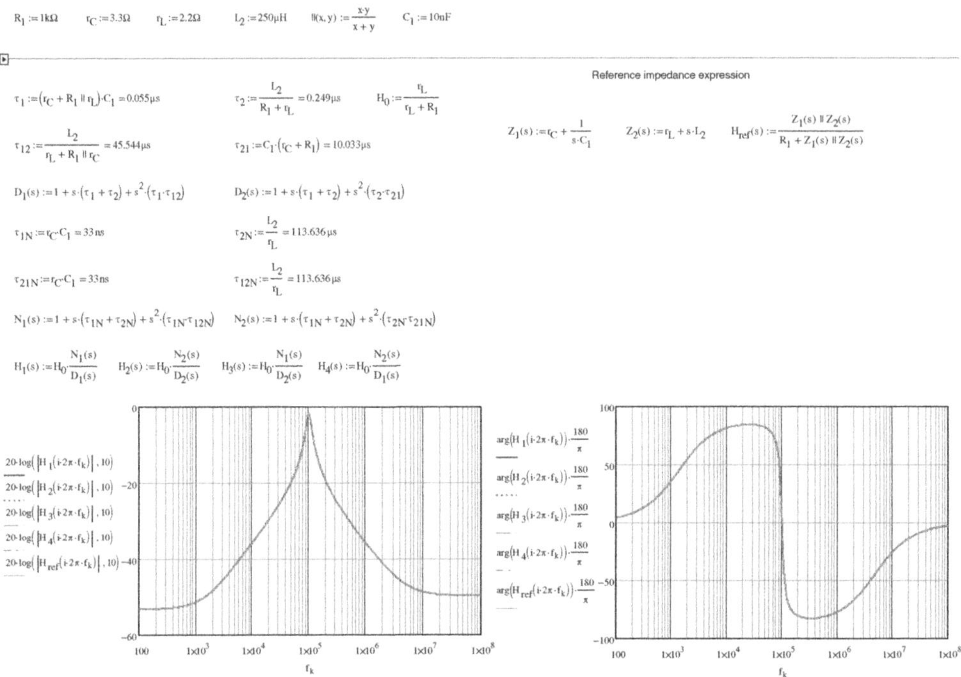

Figure 4.18 Plotting transfer functions obtained with time constants we have derived shows correct results.

mistake is: simply correct the faulty time constant and the whole thing straightens up. If you used the raw results in D and N (involving R, L or C), correcting one of the coefficients can be difficult, in particular with long expressions. By clearly expressing time constants upfront, corrections become really easy, a powerful characteristic inherent to fast analytical techniques. Figure 4.18 shows the results, in particular how D and N can take on several forms depending how you combine terms in the s^2 coefficient.

Now that we have two polynomial forms for N and D, we can rearrange them in a format suiting what we learned in Chapter 2. Starting with the numerator, we have:

$$N(s) = 1 + a_1 s + a_2 s^2 \tag{4.55}$$

Assume we did not see the simple factorization given in (4.54). We can evaluate the quality factor of this expression, given by:

$$Q_N = \frac{\sqrt{a_2}}{a_1} \tag{4.56}$$

applying component values gives a Q_N of 0.017, implying that zeros are well separated: we can apply the low-Q approximation tackled in Chapter 2. Using this tool, we can define the position of the two zeros:

$$\omega_{z_1} = \frac{a_1}{a_2} = \frac{\tau_{1N} + \tau_{2N}}{\tau_{1N}\tau_{2N}^{1}} \tag{4.57}$$

$$\omega_{z_2} = \frac{1}{a_1} = \frac{1}{\tau_{1N} + \tau_{2N}} \tag{4.58}$$

Given the component values, $\tau_{1N} = 33$ ns while $\tau_{2N} = 133.6$ µs. The two above zeros can thus be rewritten in a simpler way, as in this particular case $\tau_{2N} = \tau_{2N}^{1}$:

$$\omega_{z_1} \approx \frac{\tau_{2N}}{\tau_{1N}\tau_{2N}^{1}} = \frac{1}{\tau_{1N}} = \frac{1}{r_C C_1} \tag{4.59}$$

$$\omega_{z_2} \approx \frac{1}{\tau_{2N}} = \frac{r_L}{L_2} \tag{4.60}$$

$N(s)$ can be factored in the exact same format already obtained with (4.54):

$$N(s) = \left(1 + \frac{s}{\omega_{z_1}}\right)\left(1 + \frac{s}{\omega_{z_2}}\right) \tag{4.61}$$

Regarding $D(s)$, the same analysis applies, except that the quality factor is higher, inducing peaking in the response:

$$Q = \frac{\sqrt{b_2}}{b_1} = 5.2 \tag{4.62}$$

The resonant angular frequency ω_0 at which the double pole is positioned is defined by:

$$\omega_0 = \frac{1}{\sqrt{b_2}} = 632 \text{ krd/s} \tag{4.63}$$

or

$$f_0 = \frac{\omega_0}{2\pi} \approx 100.6 \text{ kHz} \tag{4.64}$$

With these definitions on hand, we can rewrite the denominator under the canonical form:

$$D(s) = 1 + \frac{s}{\omega_0 Q} + \left(\frac{s}{\omega_0}\right)^2 \tag{4.65}$$

and then assemble the final transfer function as:

$$H(s) = H_0 \frac{\left(1 + \frac{s}{\omega_{z_1}}\right)\left(1 + \frac{s}{\omega_{z_2}}\right)}{1 + \frac{s}{\omega_0 Q} + \left(\frac{s}{\omega_0}\right)^2} \tag{4.66}$$

Figure 4.19 shows the response given by (4.66) and confirms its agreement with the reference expression from (4.6). A summary of the method appears in Figure 4.20 for the denominator while Figure 4.21 details the steps for the numerator. In many cases, the zeros can be found by inspection, letting you write the numerator without going through the steps. You will later see that inspection leads to the simplest numerator expressions.

4.1.4 Example 1 – A Low-Pass Filter

In Figure 4.22 appears a classical 2nd-order low-pass filter. Applying the 2EET method, we will determine the transfer function linking V_{out} to V_{in} in a flash. All the steps are gathered in Figure 4.23 through individual sketches. We encourage you, at least in the beginning, to illustrate each step with a drawing so that you can later come back to it and identify a mistake in case the dynamic answer looks weird.

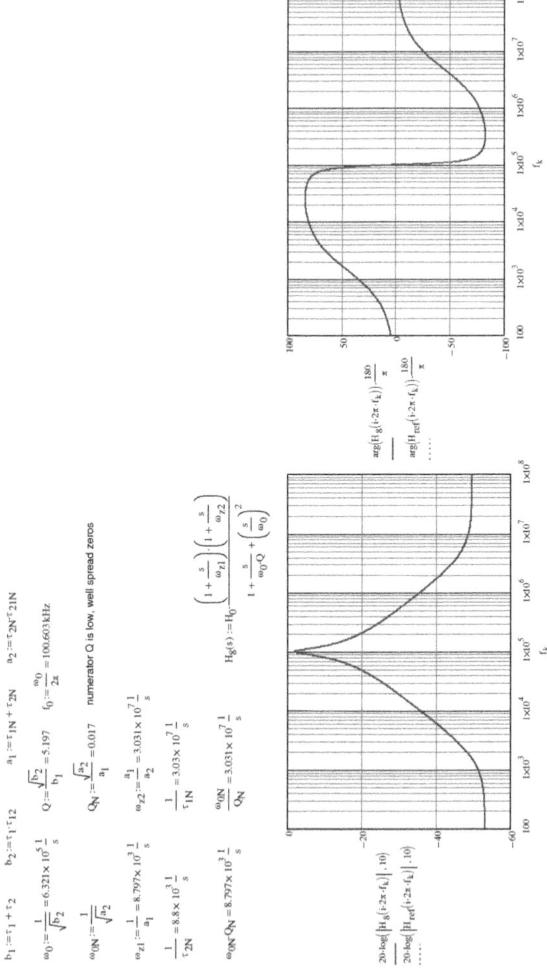

Figure 4.19 The canonical form of the transfer function matches the reference expression derived in (4.6).

Determining the reference gain

1. Draw the circuit in dc conditions, remove capacitors, short circuit inductors
2. Calculate the transfer function in this reference state, you have H_0

Determining the denominator

1. Turn the excitation off and evaluate resistances driving each energy-storing element. While you evaluate a time constant, the second energy-storing element is kept in its reference (dc) state. You obtain τ_1 and τ_2

2a. Selecting τ_1, set its energy-storing element in a high-frequency state and determine the resistance driving the second enery-storing element: τ_2^1

or

2b. Selecting τ_2, set its energy-storing element in a high-frequency state and determine the resistance driving the second enery-storing element: τ_1^2

$$D(s)=1+b_1s+b_2s^2$$
$$b_1=\tau_1+\tau_2$$
$$b_2=\tau_1\tau_2^1 \text{ or } b_2=\tau_2\tau_1^2$$
$$Q=\frac{\sqrt{b_2}}{b_1} \text{ and } \omega_0=\frac{1}{\sqrt{b_2}}$$
$$Q<<1 \quad \omega_{p_1}=\frac{1}{b_1}, \quad \omega_{p_2}=\frac{b_1}{b_2}$$

$$D(s)=1+\frac{s}{\omega_0 Q}+\left(\frac{s}{\omega_0}\right)^2$$

$Q<<1$

$$D(s)\approx\left(1+\frac{s}{\omega_{p_1}}\right)\left(1+\frac{s}{\omega_{p_2}}\right)$$

Figure 4.20 Determining the denominator requires a few steps when the circuit is considered with the excitation set to zero.

We first start with the dc state at which the reference gain is determined: C_2 is removed and L_1 is replaced by a short circuit. We are looking at Figure 4.23a. The static gain, in this mode, is immediate:

$$H_0=\frac{R_3}{r_L+R_3} \tag{4.67}$$

In Figure 4.23b, we calculate the time constant τ_2 associated with C_2. It is immediate:

$$\tau_2=C_2(r_C+r_L\|R_3) \tag{4.68}$$

In Figure 4.23c, the time constant τ_1 associated with L_1 is given by:

$$\tau_1=\frac{L_1}{r_L+R_3} \tag{4.69}$$

We have b_1 definition:

$$b_1=\tau_1+\tau_2=\frac{L_1}{r_L+R_3}+C_2(r_C+r_L\|R_3) \tag{4.70}$$

Determining the numerator

1. Bring the excitation back in place, null the response: $\hat{v}_{out} = 0$

2. The circuit is in its reference state, open capacitors, short circuit inductors.

2. Evaluate resistances driving each energy-storing element. While you evaluate a time constant, the second energy-storing element is kept in its reference (dc) state. You obtain τ_{1N} and τ_{2N}.

4a. Selecting τ_{1N}, set its energy-storing element in a high-frequency state and determine the resistance driving the second enery-storing element: τ_{2N}^{1}

or

4b. Selecting τ_{2N}, set its energy-storing element in a high-frequency state and determine the resistance driving the second enery-storing element: τ_{1N}^{2}

$$N(s) = 1 + a_1 s + a_2 s^2$$

$$a_1 = \tau_{1N} + \tau_{2N}$$

$$a_2 = \tau_{1N}\tau_{2N}^{1} \text{ or } a_2 = \tau_{2N}\tau_{1N}^{2}$$

$$Q_N = \frac{\sqrt{a_2}}{a_1} \text{ and } \omega_{0N} = \frac{1}{\sqrt{a_2}}$$

$$N(s) = 1 + \frac{s}{\omega_{0N} Q_N} + \left(\frac{s}{\omega_{0N}}\right)^2$$

$Q_N << 1$

$$Q_N << 1 \quad \omega_{z_1} = \frac{1}{a_1} \quad \omega_{z_2} = \frac{a_1}{a_2}$$

$$N(s) \approx \left(1 + \frac{s}{\omega_{z_1}}\right)\left(1 + \frac{s}{\omega_{z_2}}\right)$$

Combining the terms

Transfer function $\quad H(s) = H_0 \dfrac{N(s)}{D(s)}$

Figure 4.21 The numerator is obtained in a similar manner except that the output is nulled and the excitation back in place.

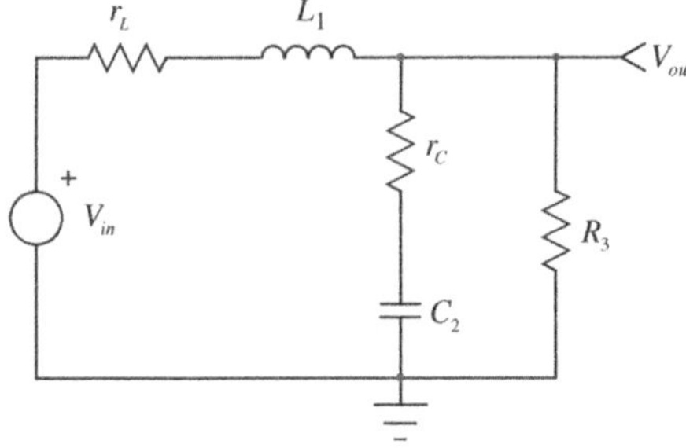

Figure 4.22 The transfer function of this low-pass filter can be found quite quickly using the method we have described.

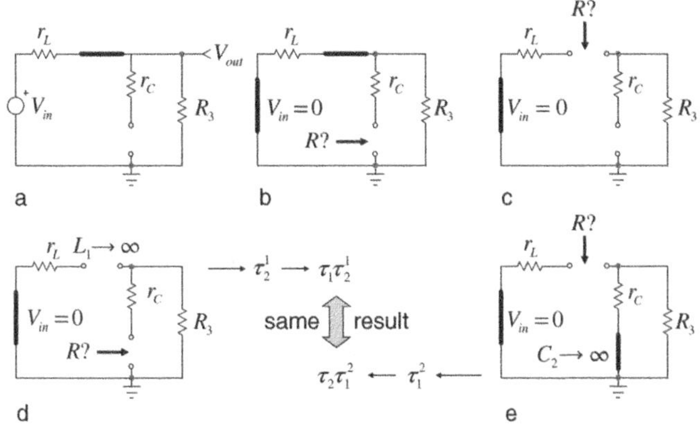

Figure 4.23 We recommend breaking the analysis steps into simple sketches you can later refer to if problems arise.

In Figure 4.23d, if we set L_1 in its high-frequency state (removed from the circuit) and determine the resistance driving C_2 in this mode, we have:

$$\tau_2^1 = C_2(r_C + R_3) \tag{4.71}$$

We have also the choice to calculate a different configuration in Figure 4.23e where:

$$\tau_1^2 = \frac{L_1}{r_L + R_3 \| r_C} \tag{4.72}$$

It looks like associating (4.71) with (4.69) will give a coefficient b_2 that is simpler than if we combine (4.72) and (4.68). Therefore, b_2 will be expressed as:

$$b_2 = \tau_1\tau_2^1 = \frac{L_1}{r_L + R_3} C_2(r_C + R_3) \tag{4.73}$$

We have our denominator $D(s)$:

$$D(s) = 1 + b_1 s + b_2 s = 1 + s\left[\frac{L_1}{r_L + R_3} + C_2\left(r_C + r_L \| R_3\right)\right] + s^2\left[\frac{L_1}{r_L + R_3} C_2(r_C + R_3)\right] \tag{4.74}$$

We can rearrange it according to the canonical form in Figure 4.20:

$$D(s) = 1 + \frac{s}{\omega_0 Q} + \left(\frac{s}{\omega_0}\right)^2 \tag{4.75}$$

in which

$$Q = \frac{\sqrt{b_2}}{b_1} = \frac{\sqrt{\frac{L_1}{r_L + R_3} C_2(r_C + R_3)}}{\frac{L_1}{r_L + R_3} + C_2\left(r_C + r_L \| R_3\right)} = \frac{\sqrt{C_2 L_1 (R_3 + r_C)(R_3 + r_L)}}{L_1 + C_2(r_C r_L + r_C R_3 + r_L R_3)} \tag{4.76}$$

and

$$\omega_0 = \frac{1}{\sqrt{b_2}} = \frac{1}{\sqrt{L_1 C_2}\sqrt{\frac{r_C + R_3}{r_L + R_3}}} \tag{4.77}$$

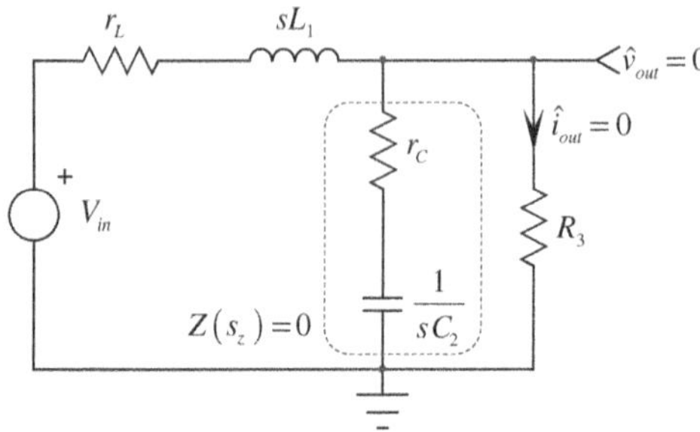

Figure 4.24 By inspecting the circuit for $s = s_z$, the response is nulled if r_C and C_2 form a transformed short circuit.

The numerator can be found by inspecting the transformed circuit as shown in Figure 4.24 at $s = s_z$. What condition brings a null in the response? In case a transformed short circuit shunts the output to ground. If the impedance created by the series connection of r_C and C_2 is a short circuit for $s = s_z$, this is the zero position:

$$Z(s) = r_C + \frac{1}{sC_2} = \frac{1 + sr_C C_2}{sC_2} = 0 \tag{4.78}$$

The root is real and equal to

$$s_z = -\frac{1}{r_C C_2} \tag{4.79}$$

Giving a LHP zero positioned at

$$\omega_z = \frac{1}{r_C C_2} \tag{4.80}$$

The numerator $N(s)$ is immediately defined as:

$$N(s) = 1 + sr_C C_2 = 1 + \frac{s}{\omega_z} \tag{4.81}$$

The final transfer function $H(s)$ combines the above results as:

$$H(s) = \frac{R_3}{R_3 + r_L} \frac{1 + sr_C C_2}{1 + s\left[\frac{L_1}{r_L + R_3} + C_2\left(r_C + r_L \| R_3\right)\right] + s^2\left[\frac{L_1}{r_L + R_3} C_2(r_C + R_3)\right]} = H_0 \frac{1 + \frac{s}{\omega_z}}{1 + \frac{s}{\omega_0 Q} + \left(\frac{s}{\omega_0}\right)^2} \tag{4.82}$$

To check the validity of this equation, we will plot its dynamic response versus that of the transfer function obtained using brute-force analysis. We have a divider formed by r_C–C_2 in parallel with R_3 and driven by r_L–L_1. The transfer function is thus:

$$H(s) = \frac{R_3 \| \left(r_C + \frac{1}{sC_2}\right)}{R_3 \| \left(r_C + \frac{1}{sC_2}\right) + r_L + sL_1} \tag{4.83}$$

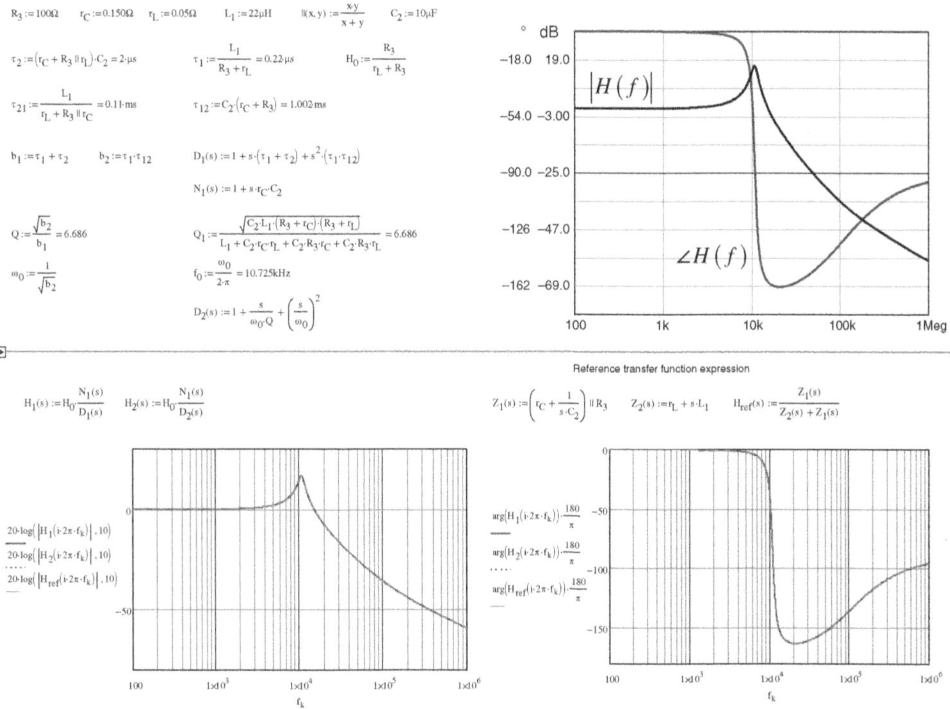

Figure 4.25 A Mathcad® sheet associates all the individual time constants to form the final expression.

Figure 4.25 shows the results gathered in the Mathcad® sheet. You can see all the individual time constants calculated and further assembled to form *D*(*s*). Should you spot a mistake somewhere, the correction is simple and immediate. As confirmed by the plots, all expressions give similar results.

4.1.5 Example 2 – A Two-capacitor Filter

Consider the capacitive filter shown in Figure 4.26. We have two capacitors whose state variables are independent, this is a 2nd-order system. Its transfer function can be determined by applying the steps given in Figure 4.20 and Figure 4.21. Starting with the denominator, Figure 4.27 gathers all the necessary stages to form *D*(*s*). Figure 4.27a sketch gives us the transfer function for $s=0$, all capacitors are removed:

$$H_0 = \frac{R_1}{R_1 + R_2} \tag{4.84}$$

The natural time constants are obtained when the input stimulus is replaced by a short circuit as in Figure 4.27b. To simplify the analysis, it is advised to redraw the sketch by folding R_3/R_2 upper terminals to the circuit ground. This is what Figure 4.27c shows. In Figure 4.27d, the time constant associated with C_2 is:

$$\tau_2 = C_2\left(R_3 + R_2 \| R_1\right) \tag{4.85}$$

In Figure 4.27e, the time constant involving C_1 is:

$$\tau_1 = C_1\left(R_2 \| R_1\right) \tag{4.86}$$

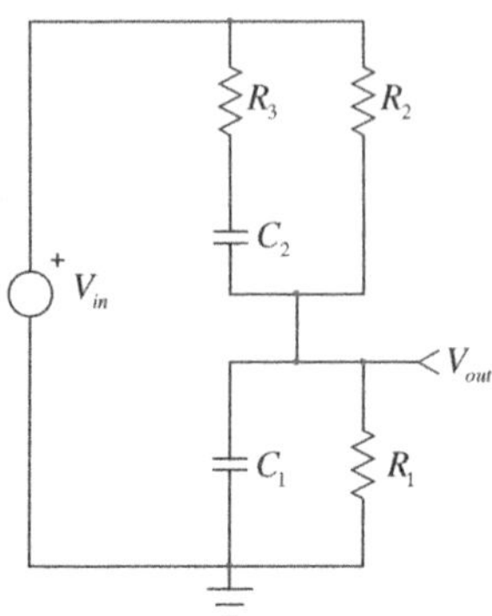

Figure 4.26 This 2-capacitor filter transfer function can be solved quickly using the 2EET.

We have coefficient b_1:

$$b_1 = \tau_1 + \tau_2 = C_1(R_2 \| R_1) + C_2(R_3 + R_2 \| R_1) \tag{4.87}$$

For coefficient b_2, Figure 4.27f shows us how to evaluate the time constant considering C_1 in its high-frequency state (replaced by a short circuit):

$$\tau_2^1 = R_3 C_2 \tag{4.88}$$

Coefficient b_2 is simply:

$$b_2 = \tau_1 \tau_2^1 = C_1(R_2 \| R_1) R_3 C_2 \tag{4.89}$$

The raw denominator is given by:

$$D(s) = 1 + b_1 s + b_2 s = 1 + s\left[C_1(R_2 \| R_1) + C_2(R_3 + R_2 \| R_1)\right] + s^2\left[C_1(R_2 \| R_1) R_3 C_2\right] \tag{4.90}$$

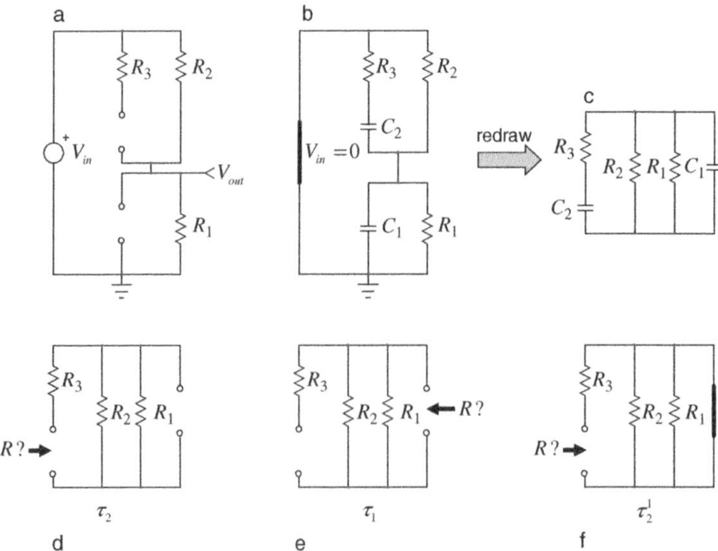

Figure 4.27 The denominator is obtained by evaluating the natural time constants when the excitation is set to 0.

We can rearrange it according to the canonical form in Figure 4.20:

$$D(s) = 1 + \frac{s}{\omega_0 Q} + \left(\frac{s}{\omega_0}\right)^2 \tag{4.91}$$

in which

$$Q = \frac{\sqrt{b_2}}{b_1} = \frac{\sqrt{C_1 C_2 (R_2 \| R_1) R_3}}{C_1 (R_2 \| R_1) + C_2 (R_3 + R_2 \| R_1)} \tag{4.92}$$

and

$$\omega_0 = \frac{1}{\sqrt{b_2}} = \frac{1}{\sqrt{C_1 C_2 (R_2 \| R_1) R_3}} \tag{4.93}$$

If Q is very small, we can consider the low-Q approximation which tells us that the 2nd-order polynomial form described in (4.91) can be replaced by two cascaded poles defined as:

$$\omega_{p_1} = \omega_0 Q = \frac{1}{b_1} \tag{4.94}$$

$$\omega_{p_2} = \frac{\omega_0}{Q} = \frac{b_1}{b_2} \tag{4.95}$$

In this case

$$D(s) \approx \left(1 + \frac{s}{\omega_{p_1}}\right)\left(1 + \frac{s}{\omega_{p_2}}\right) \tag{4.96}$$

The numerator can be derived in different ways. The fastest and simplest way is by inspection. Looking at Figure 4.28, what condition in the transformed circuit examined as $s = s_z$ would null the response? $Z_2(s)$ being a transformed short?

$$Z_2(s) = \frac{R_1}{1 + sR_1 C_1} = 0 \tag{4.97}$$

It can only happen if the denominator approaches infinity which occurs only for s approaching infinity: there is no zero associated with C_1. What about $Z_1(s)$ becoming a transformed open and blocking the stimulus propagation? We can determine its impedance in our head: imagine a current source I_T driving the parallel arrangement of R_2 with R_3 in series with C_2. The response is the voltage across I_T. For $s = 0$, I_T 'sees' R_2 as C_2 is removed at 0 Hz. What nulls the response across I_T?

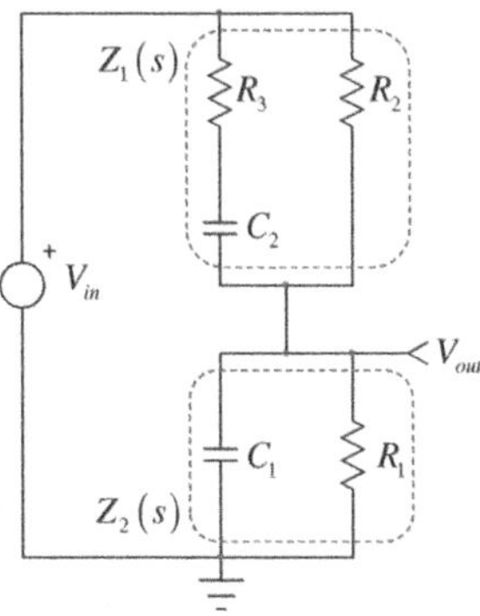

Figure 4.28 The numerator can be determined by inspection: what condition would prevent the stimulus to form a response in this circuit?

A transformed short involving R_3 and C_2. Then what resistance does C_2 'see' when the excitation I_T is turned off? R_2+R_3. Thus Z_1 is defined as:

$$Z_1(s) = R_2\frac{1+sR_3C_2}{1+sC_2(R_3+R_2)} \tag{4.98}$$

This impedance can become infinite if its denominator cancels. In other words, what is the pole affecting Z_1? Solving:

$$sC_2(R_2+R_3)+1=0 \tag{4.99}$$

we obtain

$$s_z = -\frac{1}{C_2(R_2+R_3)} \tag{4.100}$$

or

$$\omega_z = \frac{1}{C_2(R_2+R_3)} \tag{4.101}$$

The second option is to apply an NDI and look for τ_{2N}. The test circuit is presented in Figure 4.29.

As the output is nulled, the test current I_T only circulates in R_3 and R_2, giving a time constant τ_{2N} equal to

$$\tau_{2N} = C_2(R_2+R_3) \tag{4.102}$$

Leading to a zero positioned at

$$\omega_z = \frac{1}{\tau_{2N}} = \frac{1}{C_2(R_2+R_3)} \tag{4.103}$$

The final transfer function is defined as

$$H(s) = \frac{R_1}{R_1+R_2}\frac{1+sC_2(R_2+R_3)}{1+s\left[C_1\left(R_2\|R_1\right)+C_2\left(R_3+R_2\|R_1\right)\right]+s^2\left[C_1\left(R_2\|R_1\right)R_3C_2\right]} = H_0\frac{1+\frac{s}{\omega_z}}{1+\frac{s}{\omega_0 Q}+\left(\frac{s}{\omega_0}\right)^2} \tag{4.104}$$

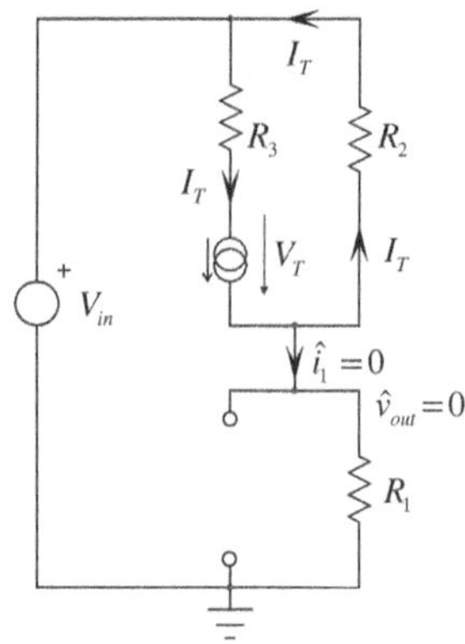

Figure 4.29 A NDI configuration lets you evaluate the zero position quickly.

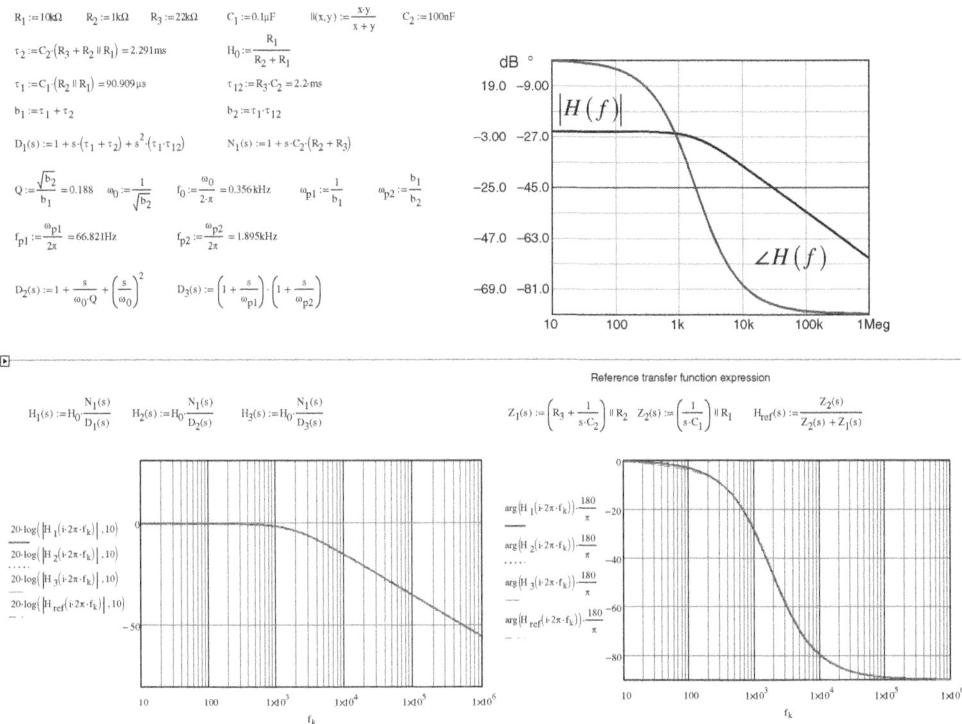

Figure 4.30 Mathcad® and SPICE agree quite well.

To check our results, we have built a Mathcad® sheet and ran a SPICE simulation. Figure 4.30 confirms our calculations are correct.

4.1.6 Example 3 – A Two-capacitor Band-stop Filter

Our third example is drawn in Figure 4.31. It is the example given in [1] and we will apply the 2EET method to determine its transfer function. The four steps necessary to derive the denominator are given in Figure 4.32. The first one is the transfer function for $s = 0$, implying that both capacitors are removed from the circuit as in Figure 4.32a.

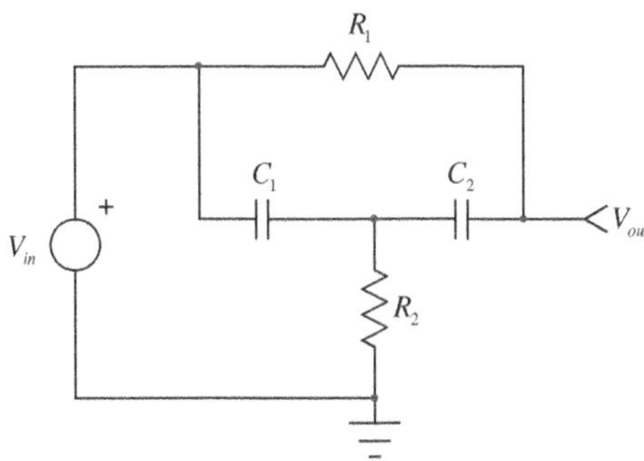

Figure 4.31 Two capacitors and two resistors form a simple band-stop filter.

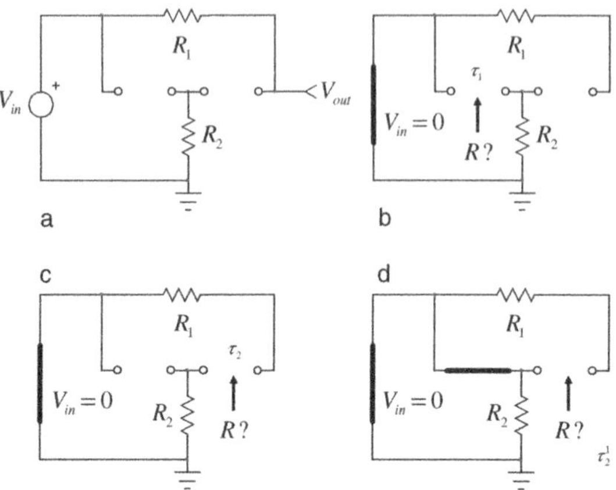

Figure 4.32 Four steps are necessary to determine the denominator coefficients.

The transfer function is immediate and equals 1:

$$H_0 = 1 \tag{4.105}$$

The first time constant τ_1 is obtained by looking into C_1's terminals while the excitation is set to 0 as in Figure 4.32b. The resistance in this configuration is R_2 as R_1 has one terminal open. The time constant is:

$$\tau_1 = R_2C_1 \tag{4.106}$$

In Figure 4.32c, we can determine the second time constant τ_2. It combines R_1 and R_2 in series to give:

$$\tau_2 = (R_1 + R_2)C_2 \tag{4.107}$$

Figure 4.32d shows how to calculate the final part of the b_2 coefficient. R_2 is shorted and R_1 remains alone. The time constant is therefore:

$$\tau_2^1 = R_1C_2 \tag{4.108}$$

The denominator gathers all these terms in the following expression:

$$D(s) = 1 + s(\tau_1 + \tau_2) + s^2\tau_1\tau_2^1 = 1 + s[R_2C_1 + (R_1 + R_2)C_2] + s^2R_1R_2C_1C_2 \tag{4.109}$$

We can rearrange it according to the canonical form in Figure 4.20:

$$D(s) = 1 + \frac{s}{\omega_0 Q} + \left(\frac{s}{\omega_0}\right)^2 \tag{4.110}$$

in which

$$Q = \frac{\sqrt{b_2}}{b_1} = \frac{\sqrt{R_1R_2C_1C_2}}{R_2C_1 + (R_1 + R_2)C_2} \tag{4.111}$$

and

$$\omega_0 = \frac{1}{\sqrt{b_2}} = \frac{1}{\sqrt{R_1R_2C_1C_2}} \tag{4.112}$$

The numerical application will tell us if Q is sufficiently low to apply the low-Q approximation, changing the 2nd-order polynomial form into two cascaded poles.

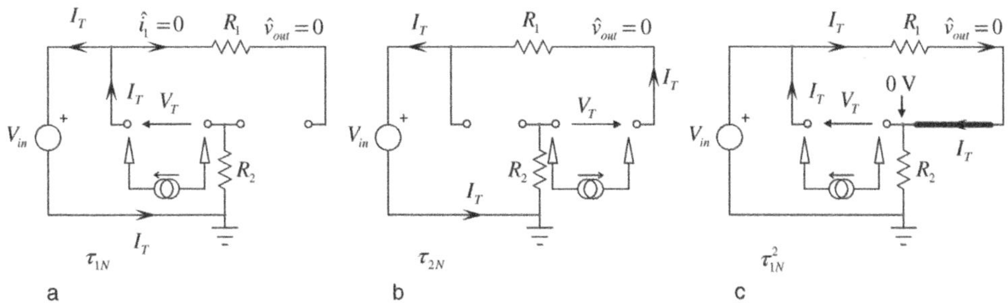

Figure 4.33 NDI is applied in three different configurations.

The zeros will be found applying NDI three times: twice to form a_1 with τ_{1N} and τ_{2N}, and a third time for a_2. The steps appear in Figure 4.33.

In Figure 4.33a, as no current circulates in R_1, the resistance offered by C_1's terminals is simply R_2:

$$\tau_{1N} = R_2 C_1 \tag{4.113}$$

In Figure 4.33b, the current I_T circulates in R_1 and R_2 however, the right-side of the current generator is grounded as $\hat{v}_{out} = 0$. So V_T is the voltage across R_2. The resistance offered by C_2's terminals is also R_2. Therefore:

$$\tau_{2N} = R_2 C_2 \tag{4.114}$$

In Figure 4.33c, C_2 is set in its high-frequency state (replaced by a short circuit) while we want to determine C_1's terminals resistance. Given the output null, the voltage across R_2 is also 0 and the only remaining resistance in the circuit is R_1. The time constant is thus:

$$\tau_{1N}^2 = R_1 C_1 \tag{4.115}$$

We can verify these expressions by capturing a SPICE circuit of Figure 4.33. This is what Figure 4.34 with the help of transconductance amplifiers. Results given at nodes R1N, R2N and R21N confirm our calculations. This is always a good practice to verify calculations with SPICE. Bias point operation is fast and results are immediate.

Before we proceed with the calculation, what is the reference transfer function we could derive using the brute-force approach? The circuit from Figure 4.31 can be redrawn splitting V_{in} in two distinct generators so that superposition can be applied. Figure 4.35a shows this idea.

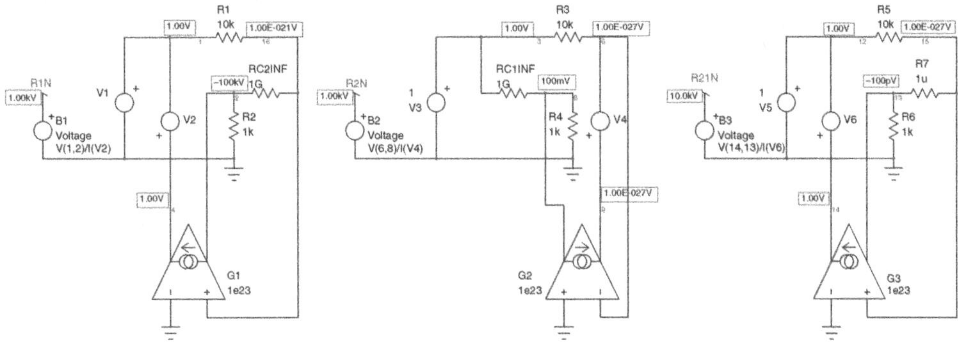

Figure 4.34 The SPICE simulation of Figure 4.33 confirms our findings.

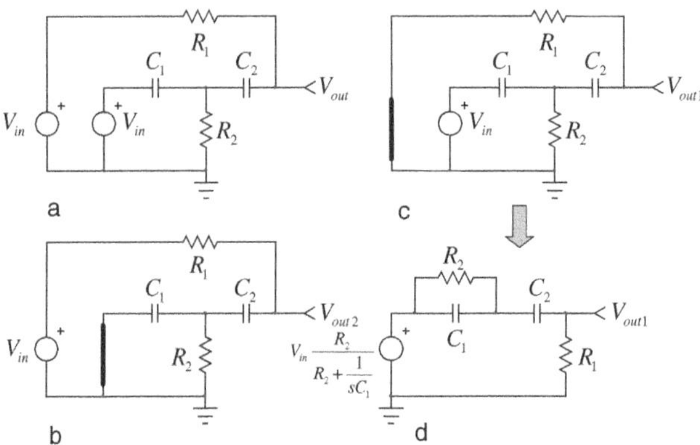

Figure 4.35 Superposition helps us find the transfer function in a quick and efficient way.

In Figure 4.35b, C_1's left terminal is grounded and the output voltage is obtained by observing a resistive divider:

$$V_{out1} = V_{in} \frac{\frac{1}{sC_2} + \frac{1}{sC_1} \| R_2}{\frac{1}{sC_2} + \frac{1}{sC_1} \| R_2 + R_1} \tag{4.116}$$

Setting the second generator to 0 V by grounding R_1's left terminal, we can rearrange Figure 4.35c in Figure 4.35d, using a Thévenin equivalent circuit featuring R_2 and C_1. The second output voltage is thus defined as:

$$V_{out2} = V_{in} \frac{R_2}{R_2 + \frac{1}{sC_1}} \cdot \frac{R_1}{\frac{1}{sC_1} \| R_2 + \frac{1}{sC_2} + R_1} \tag{4.117}$$

Assembling these two equations and factoring V_{in} leads to the transfer function we want:

$$H(s) = \frac{\frac{1}{sC_2} + \frac{1}{sC_1} \| R_2}{\frac{1}{sC_2} + \frac{1}{sC_1} \| R_2 + R_1} + \frac{R_2}{R_2 + \frac{1}{sC_1}} \cdot \frac{R_1}{\frac{1}{sC_1} \| R_2 + \frac{1}{sC_2} + R_1} \tag{4.118}$$

Needless to say, there is absolutely no insight from this expression. We can now confront all our calculations with (4.118) and see how curves compare. The Mathcad® sheet is shown in Figure 4.36 and confirms that our approach is correct. (4.118) is captured in $H_{ref}(s)$ and perfectly matches the other expressions. The denominator Q is 0.264 and poles are spread but a small discrepancy is visible when $H_3(s)$ is plotted. This is because Q is not really much smaller than 1. The SPICE simulation nicely agrees with the analytical approach.

4.1.7 Example 4 – An LC Notch Filter

A two-energy-storing element filter appears in Figure 4.37. What is the transfer function linking V_{out} to V_{in}? If we apply the brute-force method, the transfer function is given by:

$$H(s) = \frac{R_1}{\left(sL_1 \| \frac{1}{sC_2}\right) + R_1} \tag{4.119}$$

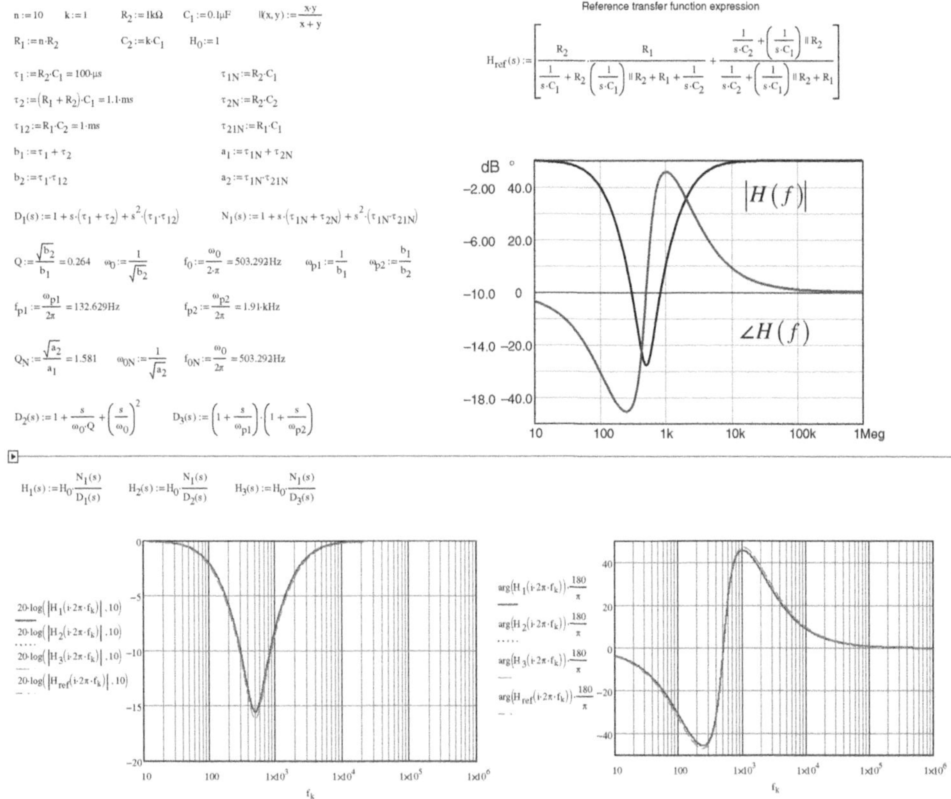

Figure 4.36 SPICE and analytical expressions deliver similar results.

It will be the reference expression we will compare our low-entropy results to. Looking at Figure 4.38a, the dc gain is quite easy here as L_1 is a short circuit for $s = 0$. We find:

$$H_0 = 1 \tag{4.120}$$

In Figure 4.38b, V_{in} is set to 0. The resistance driving capacitor C_2 while L_1 is a short circuit is 0 Ω

$$\tau_2 = 0 \tag{4.121}$$

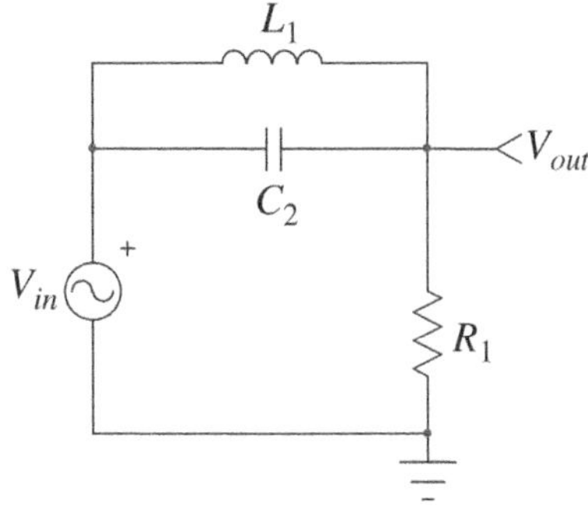

Figure 4.37 The notch filter is made of two energy-storing elements.

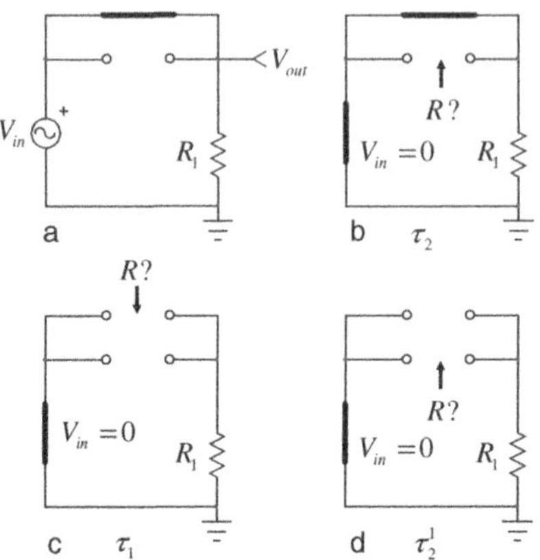

Figure 4.38 Several easy steps are taken to unveil the denominator.

In Figure 4.38c, the resistance driving L_1 while C_2 is removed is R_1:

$$\tau_1 = \frac{L_1}{R_1} \tag{4.122}$$

With these two values, coefficient b_1 equals

$$b_1 = \tau_1 + \tau_2 = \frac{L_1}{R_1} \tag{4.123}$$

The second time constant τ_2^1 is found by setting L_1 in its high-frequency state (removed from the circuit) while looking into C_2's terminals. In Figure 4.38d, if you look into C_2's connecting ports, you have R_1

$$\tau_2^1 = R_1 C_2 \tag{4.124}$$

The last coefficient b_2 is defined as

$$b_2 = \tau_1 \tau_2^1 = \frac{L_1}{R_1} R_1 C_2 = L_1 C_2 \tag{4.125}$$

The denominator $D(s)$ is equal to

$$D(s) = 1 + s\frac{L_1}{R_1} + s^2 L_1 C_2 \tag{4.126}$$

We can rearrange this expression according to the canonical form in Figure 4.20:

$$D(s) = 1 + \frac{s}{\omega_0 Q} + \left(\frac{s}{\omega_0}\right)^2 \tag{4.127}$$

in which

$$Q = \frac{\sqrt{b_2}}{b_1} = \frac{\sqrt{L_1 C_2}}{\frac{L_1}{R_1}} = R_1 \sqrt{\frac{C_2}{L_1}} \tag{4.128}$$

and

$$\omega_0 = \frac{1}{\sqrt{b_2}} = \frac{1}{\sqrt{L_1 C_2}} \tag{4.129}$$

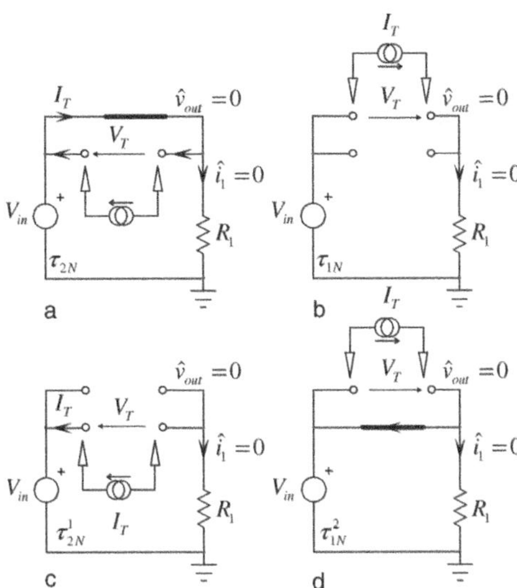

Figure 4.39 An NDI is performed three times to obtain the numerator time constants.

To obtain the numerator, we need to perform NDI three times, twice for the time constants (to determine a_1) and a third time for a_2. All these steps are gathered in Figure 4.39. The time constant involving C_2 while the output is nulled is found using Figure 4.39a. The short circuit induced by L_1 returns a zero-ohm resistance:

$$\tau_{2N} = 0 \cdot C_2 = 0 \tag{4.130}$$

In Figure 4.39b, we look at the resistance offered by L_1's terminals while the response is a null. As no current circulates, the answer is an infinite resistance:

$$\tau_{1N} = \frac{L_1}{\infty} = 0 \tag{4.131}$$

Coefficient a_1 is thus equal to

$$a_1 = \tau_{1N} + \tau_{2N} = 0 \tag{4.132}$$

Figure 4.39c should give us one answer for a part of the a_2 coefficient. Again, no current circulates so we have an infinite resistance offered by the capacitor terminals:

$$\tau_{2N}^1 = C_2 \cdot \infty \tag{4.133}$$

Let's see if the second option τ_{1N}^2 gives us a more useful answer. Figure 4.39d tells us that:

$$\tau_{1N}^2 = \frac{L_1}{0} = \infty \tag{4.134}$$

Coefficient a_2 is obtained by combining the above answers:

$$a_2 = \tau_{1N}\tau_{2N}^1 = \frac{L_1}{\infty} \cdot C_2 \cdot \infty \tag{4.135}$$

or

$$a_2 = \tau_{2N}\tau_{1N}^2 = 0 \cdot C_2 \cdot \frac{L_1}{0} \tag{4.136}$$

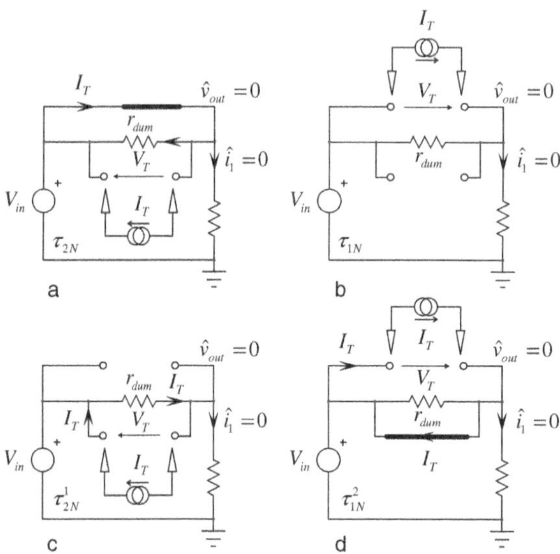

Figure 4.40 To remove the indeterminacy when the capacitor is in its dc state, an option is to add a dummy resistor r_{dum} across its terminals.

We clearly have an indeterminacy in both relationships. How could we manage to get rid of this indeterminacy in (4.135) or (4.136)? In (4.135), we can see that the problem first occurs in Figure 4.39b when no resistance is found in the current path. One solution could be to parallel C_2 with a dummy resistance r_{dum} that we will later set infinite. Another option is to add a resistance r_L in series with L_1 and have this resistance later on go to 0. We will explore both solutions for the sake of the example.

In Figure 4.40 where a resistance r_{dum} is added in parallel with C_2, the updated coefficients are as follows:

$$\tau_{2N} = 0 \tag{4.137}$$

$$\tau_{1N} = \frac{L_1}{r_{dum}} \tag{4.138}$$

$$\tau_{2N}^1 = r_{dum} C_2 \tag{4.139}$$

$$\tau_{1N}^2 = \frac{L_2}{0} = \infty \tag{4.140}$$

We can now assemble the numerator using τ_{1N} and τ_{2N}^1 as combining (4.137) and (4.140) would lead to another indeterminacy. We have:

$$N(s) = 1 + s\left(\frac{L_1}{r_{dum}} + 0\right) + s^2\left(\frac{L_1}{r_{dum}} C_2 r_{dum}\right) \tag{4.141}$$

When r_{dum} approaches infinity, N reduces to:

$$N(s) = 1 + s^2 L_1 C_2 \tag{4.142}$$

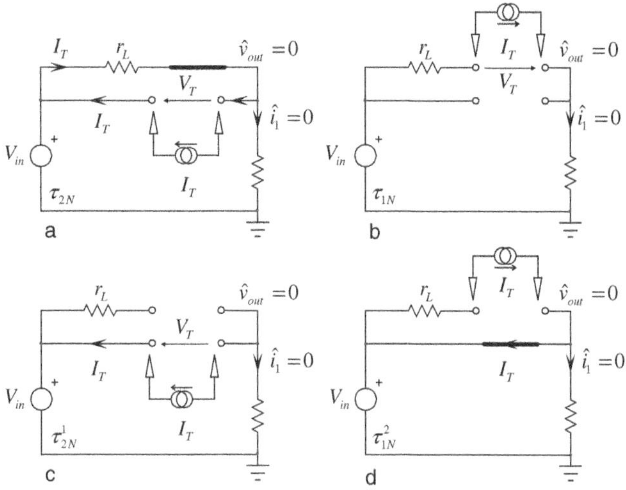

Figure 4.41 To remove the indeterminacy when the inductor is in its dc state, a solution consists of adding a dummy resistor r_L in series with its terminals.

We can run the exercise again when a resistance r_L is added in series with L_1. This is what Figure 4.41 illustrates. The time constants when this resistance is added update to:

$$\tau_{2N} = r_L C_2 \tag{4.143}$$

$$\tau_{1N} = \frac{L_1}{\infty} = 0 \tag{4.144}$$

$$\tau^1_{2N} = \infty \cdot C_2 \tag{4.145}$$

$$\tau^2_{1N} = \frac{L_1}{r_L} \tag{4.146}$$

We can now assemble the numerator using τ_{2N} and τ^2_{1N} as (4.144) and (4.145) would be indeterminate. We have:

$$N(s) = 1 + s(0 + r_L C_2) + s^2\left(C_2 r_L \frac{L_1}{r_L}\right) \tag{4.147}$$

When r_L approaches 0, N simplifies to:

$$N(s) = 1 + s^2 L_1 C_2 \tag{4.148}$$

In a canonical form, the numerator can be rearranged in the following way:

$$N(s) = 1 + \left(\frac{s}{\omega_0}\right)^2 \tag{4.149}$$

in which ω_0 is defined by (4.129). The final transfer function appears below, combining (4.120), (4.148) and (4.126):

$$H(s) = \frac{1 + s^2 L_1 C_2}{1 + s\frac{L_1}{R_1} + s^2 L_1 C_2} = \frac{1 + \left(\frac{s}{\omega_0}\right)^2}{1 + \frac{s}{\omega_0 Q} + \left(\frac{s}{\omega_0}\right)^2} \tag{4.150}$$

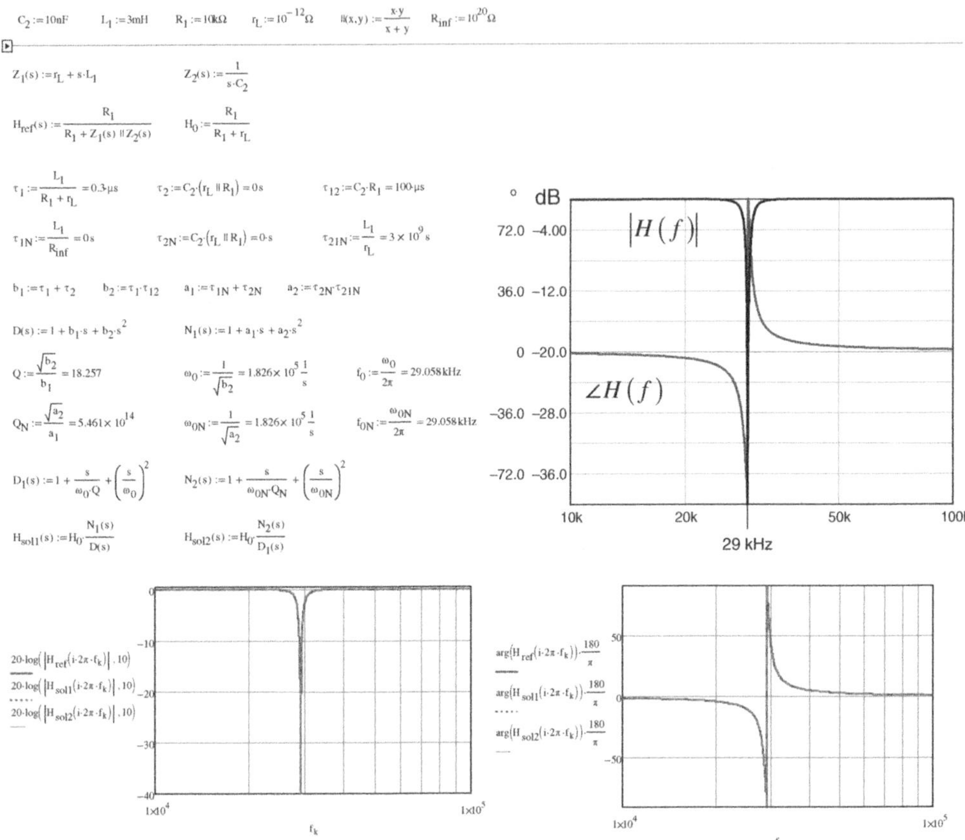

Figure 4.42 The Mathcad® sheet shows exact correlation between the brute-force approach and the 2EET.

As usual, we have gathered results obtained with (4.119) and those delivered by (4.150) in Figure 4.42. Matching is excellent between all curves and the SPICE simulation of Figure 4.37. Please note that the number of points per decade has been raised to 10 000 in order to reveal the peak at the resonant frequency.

This fourth example is interesting because you have seen how an indeterminacy can be removed by adding a series or parallel element. The method was applied to the numerator but can also very well be implemented for the denominator.

As a side note, inspection would have given the zero position in a snap shot. In the transformed Figure 4.37, what condition would block the excitation and bring an output null? If the paralleling of L_1/C_2 gives an infinite impedance at a certain s_z value. The impedance of the network is:

$$Z(s) = \frac{\frac{1}{sC_2} sL_1}{\frac{1}{sC_2} + sL_1} = \frac{sL_1}{1 + s^2 L_1 C_2} \tag{4.151}$$

The poles of this expression are the zeros of our transfer function. The impedance approaches infinity for:

$$1 + s^2 L_1 C_2 = 0 \tag{4.152}$$

which occurs for:

$$s_z = -\frac{1}{\sqrt{L_1 C_2}} \tag{4.153}$$

So $N(s) = 1 + s^2 L_1 C_2$.

4.2 A Generalized Transfer Function for 2nd-Order Systems

In Chapter 3, we have shown how the EET expressions could be rearranged to fit a format in which NDI was no longer necessary. The formula is given below and will serve to extend the method for higher order systems:

$$H(s) = \frac{H_0 + H^1 s\tau_1}{1 + s\tau_1} \tag{4.154}$$

The idea with this formula is to reuse the time constant already determined in the denominator, τ_1, and apply a corrective term H^1 to form the numerator. H^1 is the transfer function obtained by placing the energy-storing element involved in τ_1 in its high-frequency state and calculating the transfer function in this condition.

In [3], the author generalized this expression to n^{th}-order systems via simple coefficient manipulations involving two energy-storing elements for a 2nd-order network. H^1 is the transfer function obtained by setting the storage element involved in τ_1 in its high-frequency state while the element involved in τ_2 is set to its dc condition. Conversely, H^2 represents the transfer function obtained by setting the storage element involved in τ_2 in its high frequency state while that involved in τ_1 is set to its dc state. Finally, to determine H^{12} or H^{21}, put both elements in their high-frequency state and evaluate the transfer function in this condition. Figure 4.43 shows a summary of these operations. They are quite simple actually since all storage elements are either removed or replaced by a short circuit, nicely reducing the circuit complexity.

Once these elements are determined, it can be shown that the numerator of a 2nd-order transfer function can be arranged the following way:

$$N(s) = H_0 + s\left(H^1\tau_1 + H^2\tau_2\right) + s^2 H^{12}\tau_1\tau_2^1 \tag{4.155}$$

H^{1} → The energy-storing element involved in τ_1 is set to its high-frequency state.
The energy-storing element involved in τ_2 is set to its dc state.

H^{2} → The energy-storing element involved in τ_2 is set to its high-frequency state.
The energy-storing element involved in τ_1 is set to its dc state.

H^{12}, H^{21} → The energy-storing elements involved in τ_1 and τ_2 are set to their high-frequency state. H^{12} and H^{21} are the same.

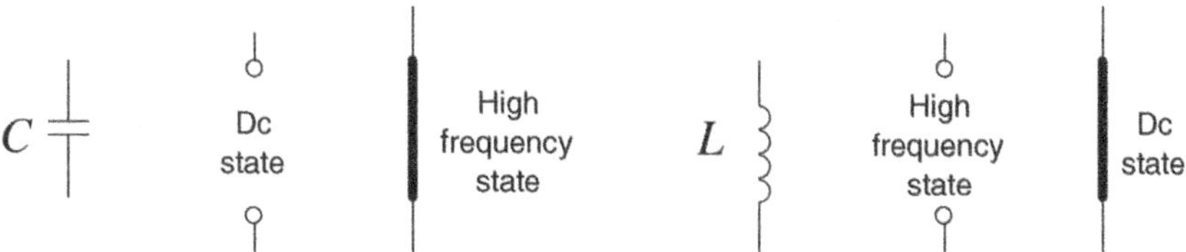

Figure 4.43 Replacing energy-storing elements by open- or short-circuits significantly simplifies the circuit.

Or if we consider $\tau_2\tau_1^2$ instead:

$$N(s) = H_0 + s\left(H^1\tau_1 + H^2\tau_2\right) + s^2H^{21}\tau_2\tau_1^2 \tag{4.156}$$

In the above expressions, τ_1, τ_2 and τ_2^1 are elements already determined for the denominator. Applying this method shields you from going through an NDI and lets you reuse what you already have. In case H_0 is a non-zero value, it can be factored so that $N(s)$ becomes:

$$N(s) = H_0\left[1 + s\left(\frac{H^1}{H_0}\tau_1 + \frac{H^2}{H_0}\tau_2\right) + s^2\frac{H^{12}}{H_0}\tau_1\tau_2^1\right] \tag{4.157}$$

Or considering $\tau_2\tau_1^2$ instead:

$$N(s) = H_0\left[1 + s\left(\frac{H^1}{H_0}\tau_1 + \frac{H^2}{H_0}\tau_2\right) + s^2\frac{H^{21}}{H_0}\tau_2\tau_1^2\right] \tag{4.158}$$

The generalized 2nd-order transfer function is defined by bringing in the denominator expression:

$$H(s) = \frac{H_0 + s\left(H^1\tau_1 + H^2\tau_2\right) + s^2H^{12}\tau_1\tau_2^1}{1 + s(\tau_1 + \tau_2) + s^2\tau_1\tau_2^1} = \frac{H_0 + s\left(H^1\tau_1 + H^2\tau_2\right) + s^2H^{21}\tau_2\tau_1^2}{1 + s(\tau_1 + \tau_2) + s^2\tau_2\tau_1^2} \tag{4.159}$$

In which H_0 can be advantageously factored in case it is a non-zero value:

$$H(s) = H_0\frac{1 + s\left(\frac{H^1}{H_0}\tau_1 + \frac{H^2}{H_0}\tau_2\right) + s^2\frac{H^{12}}{H_0}\tau_1\tau_2^1}{1 + s(\tau_1 + \tau_2) + s^2\tau_1\tau_2^1} = H_0\frac{1 + s\left(\frac{H^1}{H_0}\tau_1 + \frac{H^2}{H_0}\tau_2\right) + s^2\frac{H^{21}}{H_0}\tau_2\tau_1^2}{1 + s(\tau_1 + \tau_2) + s^2\tau_2\tau_1^2} \tag{4.160}$$

The 2EET and the generalized expression will deliver the same dynamic responses. Applying the 2EET and NDI will give the simplest form you can expect from the method to determine a transfer function. However, you will have to apply NDI several times to unveil the zeros position. On the other hand, implementing the generalized formula does not require NDI. However, the final result, despite its correctness, could be significantly more complex than what the 2EET method delivers. After all, you may take the results obtained from (4.159) or (4.160) and plot the response right away. You can also spend a little more time on the expression to seek simplifications in the numerator: the time you believed was saved by avoiding NDI is partly spent by reworking the final equation. The choice is yours to adopt either methods depending on your skills. As a third option, whenever possible, solve the numerator by inspection, this is the fastest and most efficient way you can think of. Figure 4.44 summarizes the steps to build $N(s)$ without NDI.

4.2.1 Inferring the Presence of Zeros in the Circuit

One interesting result brought by the generalized transfer function is that coefficients in the numerators exist if non-zero gains H are obtained when storage elements are alternatively or altogether set to their high-frequency states. This practical result leads to the following observation: the number of zeros in a transfer function is determined by the highest order of the numerator polynomial. Practically speaking, the number of zeros is equal to the number of energy-storing elements you can *simultaneously* put in their high-frequency state while you check that a response still exists. As a quick application, we can look at Figure 4.45 where four examples are gathered:

- Figure 4.45a: if you simultaneously open L_1 and short C_2, the gain is $r_C/(r_C + R_1)$: you have two zeros in the numerator. Imagine the same circuit in which R_1 is removed, one zero is gone.

Determining the numerator in the generalized form

1. Bring the excitation back in place

2. The circuit is in its reference state, open capacitors, short circuit inductors.

2. Calculate the reference gain H_0

3. Set the energy-storing element in τ_1 in its high-frequency state while the second element is in its dc state. Calculate H^1

4. Set the energy-storing element in τ_2 in its high-frequency state while the second element is in its dc state. Calculate H^2

5. Finally, set both energy-storing elements in their high-frequency state and calculate H^{12}

Reuse times constants from $D(s)$

$$N(s)=H_0+s\left(H^1\tau_1+H^2\tau_2\right)+s^2\tau_1\tau_2^1H^{12} \overset{H_0\neq 0}{\rightarrow} H_0\left[1+s\left(\frac{H^1}{H_0}\tau_1+\frac{H^2}{H_0}\tau_2\right)+s^2\tau_1\tau_2^1\frac{H^{12}}{H_0}\right]$$

$H_0 \neq 0$

$$a_1=\frac{H^1}{H_0}\tau_1+\frac{H^2}{H_0}\tau_2$$

$$a_2=\tau_1\tau_2^1\frac{H^{12}}{H_0} \text{ or } a_2=\tau_2\tau_1^2\frac{H^{21}}{H_0}$$

$$N(s)=1+\frac{s}{\omega_{0N}Q_N}+\left(\frac{s}{\omega_{0N}}\right)^2$$

$Q_N \ll 1$

$$\omega_{z_1}=\frac{1}{a_1} \qquad \omega_{z_2}=\frac{a_1}{a_2}$$

$$N(s)\approx\left(1+\frac{s}{\omega_{z_1}}\right)\left(1+\frac{s}{\omega_{z_2}}\right)$$

$$Q_N=\frac{\sqrt{a_2}}{a_1} \text{ and } \omega_{0N}=\frac{1}{\sqrt{a_2}}$$

Transfer function

$$H(s)=H_0\frac{N(s)}{D(s)}$$

Figure 4.44 Summary of operations to form the numerator of the 2nd-order generalized transfer function.

- Figure 4.45b: when both L_1 and C_2 are in their high-frequency state (C_2 is a short circuit and L_1 is open), there is a response and the gain is $R_3/(R_1+R_3)$: there are two zeros in the numerator.
- Figure 4.45c: if C_1 and/or C_2 are shorted individually or all together, there is no response so none of them contributes a zero in this circuit.
- Figure 4.45d: if C_1 and C_2 are simultaneously shorted, there is a response – the gain is 1 – so the numerator features two zeros.

Back to our example in Figure 4.37, opening L_1 and shorting C_2 gives a response, confirming the presence of two zeros in this circuit.

4.2.2 Generalized 2nd–order Transfer Function – Example 1

The first example appears in Figure 4.46 and shows a filter involving an inductor and a capacitor. We will apply the generalized transfer function expression to illustrate how the exercise is simplified by using this method.

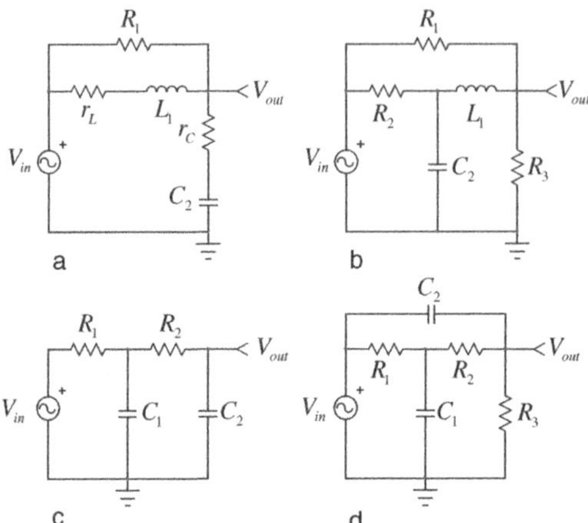

Figure 4.45 If shorting the capacitor or open-circuiting the inductor (while the other component is in its dc state) leads to a non-zero transfer function, there is a zero contributed by this component. If the two elements can be simultaneously set in their high-frequency state while a response is observed, then there is a 2nd-order term associating these components in the numerator.

We first start by determining the transfer function for $s = 0$ as shown in Figure 4.47a where all the energy-storing components are put in their dc state: L_1 is replaced by a short circuit and C_2 is open-circuited. In this case:

$$H_0 = 1 \tag{4.161}$$

The excitation is now set to 0 and 3 time constants have to be determined. We start with the resistance 'driving' inductor L_1. In Figure 4.47b, we see that this resistance equals R_3 paralleled with the sum of R_1 and R_2. Therefore:

$$\tau_1 = \frac{L_1}{R_3 \| (R_2 + R_1)} \tag{4.162}$$

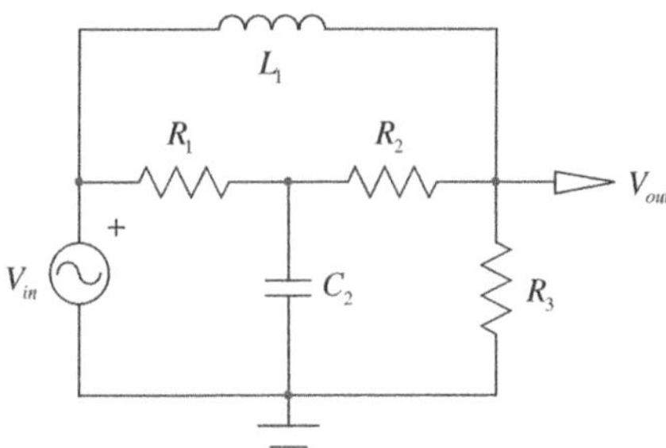

Figure 4.46 This circuit combines two energy-storage elements. What is its transfer function?

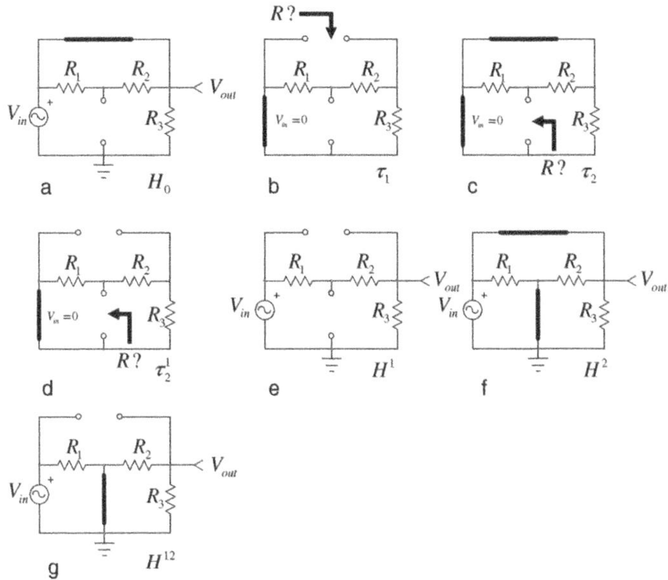

Figure 4.47 By going through simple intermediate sketches, we can quickly express the transfer function.

In Figure 4.47c, we look at the time constant involving capacitor C_2. Through its terminals, while L_1 is replaced by a short circuit, we see that R_3 is shorted and R_1 and R_2 are paralleled. Hence:

$$\tau_2 = C_2(R_1 \| R_2) \tag{4.163}$$

The last time constant is calculated by setting L_1 in its high-frequency state (open circuited) and looking at C_2's terminals to determine the driving resistance (Figure 4.47d): R_2 and R_3 are in series and come in parallel with R_1:

$$\tau_2^1 = C_2[R_1 \| (R_2 + R_3)] \tag{4.164}$$

We can now form the denominator $D(s)$:

$$\begin{aligned} D(s) &= 1 + s(\tau_1 + \tau_2) + s^2\tau_1\tau_2^1 \\ &= 1 + s\left(\frac{L_1}{R_3\|(R_1 + R_2)} + C_2(R_1\|R_2)\right) + s^2\frac{L_1}{R_3\|(R_1 + R_2)}C_2[R_1\|(R_2 + R_3)] \end{aligned} \tag{4.165}$$

We rearrange this expression according to the canonical form in Figure 4.20:

$$D(s) = 1 + \frac{s}{\omega_0 Q} + \left(\frac{s}{\omega_0}\right)^2 \tag{4.166}$$

in which

$$Q = \frac{\sqrt{b_2}}{b_1} = \sqrt{\frac{C_2}{L_1}}\frac{\sqrt{[R_1\|(R_2 + R_3)][R_3\|(R_1 + R_2)]}}{1 + \frac{C_2(R_2\|R_1)[R_3\|(R_1 + R_2)]}{L_1}} \tag{4.167}$$

and

$$\omega_0 = \frac{1}{\sqrt{b_2}} = \frac{1}{\sqrt{L_1 C_2}}\sqrt{\frac{R_3 \| (R_1 + R_2)}{R_1 \| (R_2 + R_3)}} \tag{4.168}$$

The denominator derivation requires three simple steps. The first one starts in Figure 4.47e: L_1 is set to its high-frequency state (open circuited) while C_2 is in its dc state (open circuited). From this sketch we immediately see that:

$$H^1 = \frac{R_3}{R_1 + R_2 + R_3} \tag{4.169}$$

In Figure 4.47f, L_1 is put in its dc state (short circuited) while C_2 is in its high-frequency state (a short circuit). In this configuration, the gain is:

$$H^2 = 1 \tag{4.170}$$

The last gain H^{12} is determined when both energy-storing elements are set to their high-frequency state as in Figure 4.47f. In this mode:

$$H^{12} = 0 \tag{4.171}$$

The numerator $N(s)$ is given by:

$$N(s) = H_0 + s\left(H^1\tau_1 + H^2\tau_2\right) + s^2 H^{12}\tau_1\tau_2^1 = 1 + s\left(\frac{R_3}{R_1 + R_2 + R_3}\frac{L_1}{R_3 \| (R_1 + R_2)} + C_2\left(R_1 \| R_2\right)\right) \tag{4.172}$$

There is one zero only since opening L_1 and shorting C_2 cancel the response. The final transfer function expression $H(s)$ is:

$$H(s) = \frac{1 + s\left(\frac{R_3}{R_1 + R_2 + R_3}\frac{L_1}{R_3 \| (R_1 + R_2)} + C_2\left(R_1 \| R_2\right)\right)}{1 + s\left(\frac{L_1}{R_3 \| (R_1 + R_2)} + C_2\left(R_1 \| R_2\right)\right) + s^2\frac{L_1}{R_3 \| (R_1 + R_2)}C_2\left[R_1 \| (R_2 + R_3)\right]} = \frac{1 + \frac{s}{\omega_z}}{1 + \frac{s}{\omega_0 Q} + \left(\frac{s}{\omega_0}\right)^2} \tag{4.173}$$

in which

$$\omega_z = \frac{1}{\frac{R_3}{R_1 + R_2 + R_3}\frac{L_1}{R_3 \| (R_1 + R_2)} + C_2\left(R_1 \| R_2\right)} = \frac{1}{\frac{L_1}{R_1 + R_2} + C_2\left(R_1 \| R_2\right)} \tag{4.174}$$

To check our expression, we can determine the raw transfer function by applying superposition as we did in Figure 4.35. The two intermediate drawings are shown in Figure 4.48. From these sketches, the final transfer function is given by:

$$H(s) = \frac{1}{1 + sR_1C_2}\frac{R_3 \| sL_1}{R_3 \| sL_1 + R_2 + \left(R_1 \| \frac{1}{sC_2}\right)} + \frac{R_3 \| \left[R_2 + \left(R_1 \| \frac{1}{sC_2}\right)\right]}{R_3 \| \left[R_2 + \left(R_1 \| \frac{1}{sC_2}\right)\right] + sL_1} \tag{4.175}$$

We now have everything needed to assemble a Mathcad® sheet and run a SPICE simulation. All of the results are gathered in Figure 4.49 and confirm the validity of our approach. As usual, we have separated all time constant calculations so that if you spot a difference between the curves, you can go back to the individual sketches and once the error is identified, run the correction without difficulty.

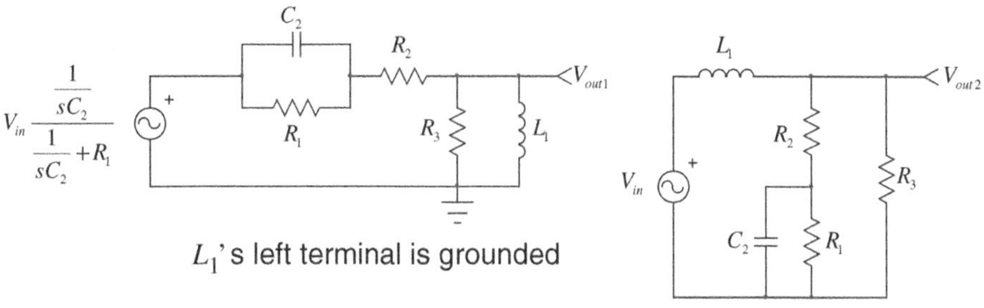

Figure 4.48 Owing to superposition, we can determine the raw transfer function and use it as a reference.

$C_2 := 10nF \quad L_1 := 3mH \quad R_1 := 10k\Omega \quad R_2 := 150\Omega \quad R_3 := 12k\Omega \quad \|(x,y) := \frac{x \cdot y}{x+y} \quad R_{inf} := 10^{20}\Omega$

$$V_1(s) := \frac{R_3 \| \left[R_2 + \left[R_1 \| \left(\frac{1}{s \cdot C_2}\right)\right]\right]}{R_3 \| \left[R_2 + \left[R_1 \| \left(\frac{1}{s \cdot C_2}\right)\right]\right] + s \cdot L_1}$$

$$V_2(s) := \frac{1}{1 + s \cdot R_1 \cdot C_2} \cdot \frac{R_3 \| (s \cdot L_1)}{R_3 \| (s \cdot L_1) + \left[R_2 + R_1 \| \left(\frac{1}{s \cdot C_2}\right)\right]}$$

$H_{ref}(s) := V_1(s) + V_2(s) \qquad H_0 := 1$

$$\tau_1 := \frac{L_1}{R_3 \| (R_1 + R_2)} = 0.546 \cdot \mu s$$

$\tau_2 := C_2 \cdot (R_2 \| R_1) = 1.478 \cdot \mu s$

$\tau_{12} := C_2 \cdot [R_1 \| (R_2 + R_3)] = 54.853 \cdot \mu s$

$H_1 := \frac{R_3}{R_1 + R_2 + R_3} \qquad H_2 := 1 \qquad H_{12} := 0$

$b_1 := \tau_1 + \tau_2 \qquad b_2 := {}_1 \cdot \tau_{12} \qquad a_1 := \tau_1 \cdot \frac{H_1}{H_0} + \tau_2 \cdot \frac{H_2}{H_0} \qquad a_2 := \tau_1 \cdot \tau_{12} \cdot \frac{H_{12}}{H_0}$

$D(s) := 1 + b_1 \cdot s + b_2 \cdot s^2 \qquad N_2(s) := 1 + a_1 \cdot s + a_2 \cdot s^2$

$Q := \frac{\sqrt{b_2}}{b_1} = 2.704 \qquad \omega_0 := \frac{1}{\sqrt{b_2}} = 1.828 \times 10^5 \frac{1}{s} \qquad f_0 := \frac{\omega_0}{2\pi} = 29.093 \cdot kHz$

$$\sqrt{\frac{C_2}{L_1}} \cdot \frac{\sqrt{[R_1 \| (R_2 + R_3)] \cdot [R_3 \| (R_1 + R_2)]}}{1 + \frac{C_2 \cdot (R_2 \| R_1) \cdot [R_3 \| (R_1 + R_2)]}{L_1}} = 2.704 \qquad \frac{1}{\sqrt{L_1 \cdot C_2}} \cdot \sqrt{\frac{R_3 \| (R_1 + R_2)}{R_1 \| (R_2 + R_3)}} = 1.828 \times 10^5 \frac{1}{s}$$

$$\omega_z := \frac{1}{\frac{R_3}{R_1 + R_2 + R_3} \cdot \frac{L_1}{R_3 \| (R_1 + R_2)} + C_2 \cdot (R_2 \| R_1)} \qquad f_z := \frac{\omega_z}{2\pi} = 89.746 \cdot kHz$$

$$\omega_{zz} := \frac{1}{\frac{L_1}{R_1 + R_2} + C_2 \cdot (R_1 \| R_2)} \qquad f_{zz} := \frac{\omega_{zz}}{2 \cdot \pi} = 89.746 \cdot kHz$$

$$D_1(s) := 1 + \frac{s}{\omega_0 \cdot Q} + \left(\frac{s}{\omega_0}\right)^2 \qquad N_1(s) := 1 + s \cdot \left(\frac{H_1}{H_0} \cdot \tau_1 + \frac{H_2}{H_0} \cdot \tau_2\right) + s^2 \cdot \frac{H_{12}}{H_0} \cdot \tau_1 \cdot \tau_{12}$$

$$H_{sol1}(s) := H_0 \cdot \frac{N_1(s)}{D(s)} \qquad H_{sol3}(s) := H_0 \cdot \frac{1 + \frac{s}{\omega_z}}{D(s)}$$

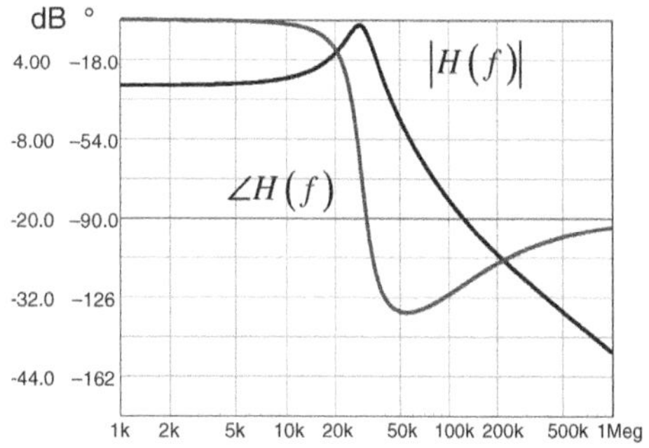

dB °
4.00 −18.0
-8.00 -54.0
-20.0 -90.0
-32.0 -126
-44.0 -162
1k 2k 5k 10k 20k 50k 100k 200k 500k 1Meg
$|H(f)|$
$\angle H(f)$

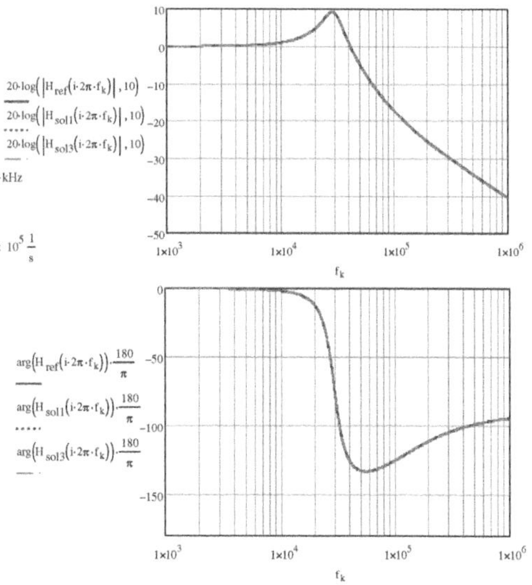

$20 \cdot \log(|H_{ref}(i \cdot 2\pi \cdot f_k)|, 10)$
$20 \cdot \log(|H_{sol1}(i \cdot 2\pi \cdot f_k)|, 10)$
$20 \cdot \log(|H_{sol3}(i \cdot 2\pi \cdot f_k)|, 10)$
10 0 -10 -20 -30 -40 -50
1×10³ 1×10⁴ 1×10⁵ 1×10⁶
f_k

$\arg(H_{ref}(i \cdot 2\pi \cdot f_k)) \cdot \frac{180}{\pi}$
$\arg(H_{sol1}(i \cdot 2\pi \cdot f_k)) \cdot \frac{180}{\pi}$
$\arg(H_{sol3}(i \cdot 2\pi \cdot f_k)) \cdot \frac{180}{\pi}$
0 -50 -100 -150
1×10³ 1×10⁴ 1×10⁵ 1×10⁶
f_k

Figure 4.49 A Mathcad® sheet helps test our equations in different configurations.

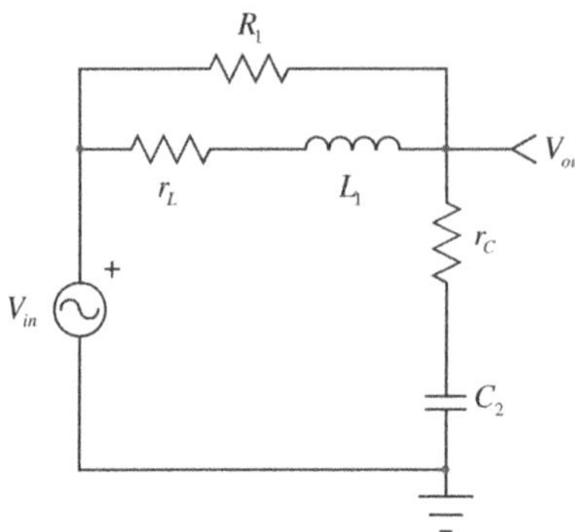

Figure 4.50 This *RLC* low-pass filter now includes a second zero.

4.2.3 Generalized 2nd–order Transfer Function – Example 2

Our next example appears in Figure 4.50. This is a *LC* filter but this time, a resistor R_1 has been added from the input to the output. Compared to the sketch shown in Figure 4.22, we immediately see in Figure 4.50 that setting L_1 in its high-frequency state while C_2 is shorted still gives a response via R_1: there are two zeros this time. As usual, we start with $s = 0$ and looking at Figure 4.51a we find that the gain is 1 in this configuration:

$$H_0 = 1 \tag{4.176}$$

The time constant involving L_1 is obtained with the help of Figure 4.51b;

$$\tau_1 = \frac{L_1}{r_L + R_1} \tag{4.177}$$

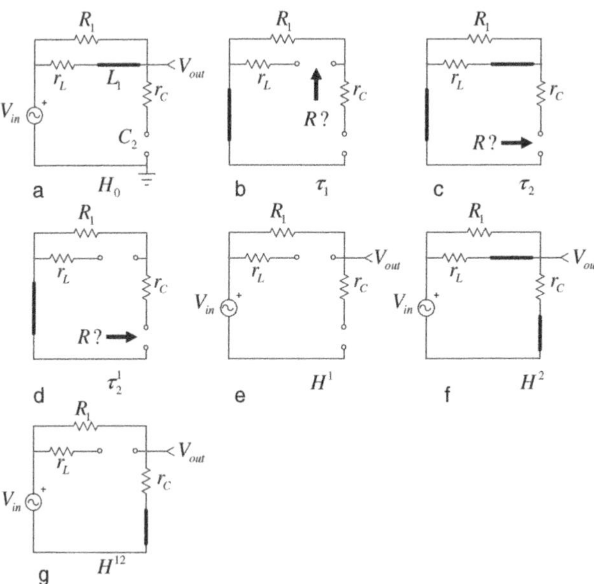

Figure 4.51 These small sketches detail the various steps we took to determine the various time constants.

The second time constant is determined by shorting L_1 and looking into C_2's terminals (Figure 4.51c):

$$\tau_2 = C_2\left(r_C + R_1 \| r_L\right) \tag{4.178}$$

Finally, we set L_1 in its high-frequency state (removed) and look into C_2's terminal (Figure 4.51d). We obtain:

$$\tau_2^1 = C_2(r_C + R_1) \tag{4.179}$$

We can now form the denominator $D(s)$:

$$D(s) = 1 + s(\tau_1 + \tau_2) + s^2\tau_1\tau_2^1 = 1 + s\left(\frac{L_1}{r_L + R_1} + C_2\left(r_C + R_1 \| r_L\right)\right) + s^2\frac{L_1}{r_L + R_1}C_2(r_C + R_1) \tag{4.180}$$

We rearrange this expression according to the canonical form in Figure 4.20:

$$D(s) = 1 + \frac{s}{\omega_0 Q} + \left(\frac{s}{\omega_0}\right)^2 \tag{4.181}$$

in which

$$Q = \frac{\sqrt{b_2}}{b_1} = \sqrt{\frac{C_2}{L_1}}\frac{\sqrt{(R_1 + r_C)(R_1 + r_L)}}{1 + \dfrac{C_2(r_C r_L + R_1 r_C + R_1 r_L)}{L_1}} \tag{4.182}$$

and

$$\omega_0 = \frac{1}{\sqrt{b_2}} = \frac{1}{\sqrt{L_1 C_2}}\sqrt{\frac{r_L + R_1}{r_C + R_1}} \tag{4.183}$$

The numerator is obtained by calculating 3 simple gains. Starting with Figure 4.51e, we find:

$$H^1 = 1 \tag{4.184}$$

Then in Figure 4.51f:

$$H^2 = \frac{r_C}{r_C + r_L \| R_1} \tag{4.185}$$

And finally, in Figure 4.51g we determine:

$$H^{12} = \frac{r_C}{r_C + R_1} \tag{4.186}$$

Considering a unity H_0, the numerator $N(s)$ is given by:

$$\begin{aligned} N(s) &= H_0 + s\left(H^1\tau_1 + H^2\tau_2\right) + s^2 H^{12}\tau_1\tau_2^1 = H_0\left[1 + s\left(\frac{H^1}{H_0}\tau_1 + \frac{H^2}{H_0}\tau_2\right) + s^2\frac{H^{12}}{H_0}\tau_1\tau_2^1\right] \\ &= 1 + s\left(\frac{L_1}{r_L + R_1} + \frac{r_C}{r_C + r_L \| R_1}C_2\left(r_C + R_1 \| r_L\right)\right) + s^2\frac{r_C}{r_C + R_1}\frac{L_1}{r_L + R_1}C_2(r_C + R_1) \end{aligned} \tag{4.187}$$

From this expression, we can extract the quality factor Q_N and a resonant angular frequency ω_{0N}:

$$Q_N = \frac{\sqrt{\frac{r_C}{r_C+R_1}\frac{L_1}{r_L+R_1}C_2(r_C+R_1)}}{\frac{L_1}{r_L+R_1}+\frac{r_C}{r_C+r_L\|R_1}C_2(r_C+R_1\|r_L)} = \frac{\sqrt{\frac{L_1C_2r_C}{R_1+r_L}}}{\frac{L_1}{R_1+r_L}+r_CC_2} \tag{4.188}$$

$$\omega_{0N} = \frac{1}{\sqrt{\frac{r_C}{r_C+R_1}\frac{L_1}{r_L+R_1}C_2(r_C+R_1)}} = \frac{1}{\sqrt{L_1C_2}}\sqrt{\frac{R_1+r_L}{r_C}} \tag{4.189}$$

With these definitions, provided that Q_N is low and much less than 1, we could express two cascaded zeros defined as:

$$\omega_{z_1} = \omega_{0N}Q_N \tag{4.190}$$

and

$$\omega_{z_2} = \frac{\omega_{0N}}{Q_N} \tag{4.191}$$

Given the coefficients complexity, the resulting expressions will be quite ugly. A better way is to check if we can determine the zeros by inspection. The transformed circuit appears in Figure 4.52.

If the response is nulled then either we have a transformed short circuit across the response or we have a transformed open blocking the excitation propagation. Can $Z_1(s)$ be a transformed open circuit? The impedance can be found immediately using fast analytical techniques: for s = 0, the impedance is R_1 paralleled with r_L. The resistance seen from L_1's terminals while the excitation is set to 0 A is r_L+R_1. Finally, we have a zero when r_L and sL_1 form a transformed short. Gathering all these elements leads to:

$$Z_1(s) = (R_1\|r_L)\cdot\frac{1+s\frac{L_1}{r_L}}{1+s\frac{L_1}{r_L+R_L}} \tag{4.192}$$

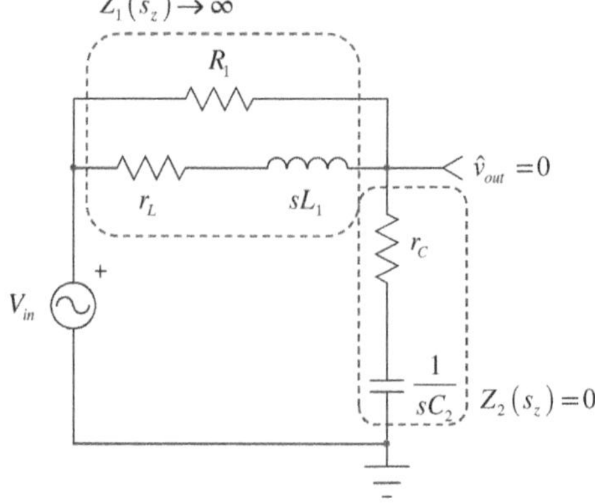

Figure 4.52 The zeros can be determined by inspection in the transformed circuit observed at $s = s_z$.

This expression approaches infinity if we cancel the denominator:

$$1 + s\frac{L_1}{r_L + R_1} = 0 \tag{4.193}$$

Leading to $s_{z_1} = -\frac{r_L + R_1}{L_1}$ or

$$\omega_{z_1} = \frac{r_L + R_1}{L_1} \tag{4.194}$$

The second zero is classically obtained when $Z_2(s_z)$ becomes a transformed short circuit:

$$Z_2(s) = \frac{1 + sr_C C_2}{sC_2} = 0 \tag{4.195}$$

It occurs for $s_{z_2} = -\frac{1}{r_C C_2}$. The second zero is thus positioned at:

$$\omega_{z_2} = \frac{1}{r_C C_2} \tag{4.196}$$

Collecting all these results leads to final transfer function expression $H(s)$:

$$H(s) = \frac{\left(1 + s\frac{r_L + R_1}{L_1}\right)(1 + sr_C C_2)}{1 + s\left(\frac{L_1}{r_L + R_1} + C_2\left(r_C + R_1 \| r_L\right)\right) + s^2\frac{L_1}{r_L + R_1}C_2(r_C + R_1)} = \frac{\left(1 + \frac{s}{\omega_{z_1}}\right)\left(1 + \frac{s}{\omega_{z_2}}\right)}{1 + \frac{s}{\omega_0 Q} + \left(\frac{s}{\omega_0}\right)^2} \tag{4.197}$$

Before we proceed with this expression, we need a reference 'high-entropy' transfer function. With these filters, superposition works well as shown in Figure 4.53.

The raw transfer function is obtained by summing V_{out1} and V_{out2}. The reference expression is thus:

$$H(s) = \frac{R_1 \| \left(r_C + \frac{1}{sC_2}\right)}{R_1 \| \left(r_C + \frac{1}{sC_2}\right) + r_L + sL_1} + \frac{\left(r_C + \frac{1}{sC_2}\right) \| (r_L + sL_1)}{\left(r_C + \frac{1}{sC_2}\right) \| (r_L + sL_1) + R_1} \tag{4.198}$$

We can now capture all these data in a Mathcad® sheet and compare all various expressions responses. This is what is shown in Figure 4.54 where all results perfectly match.

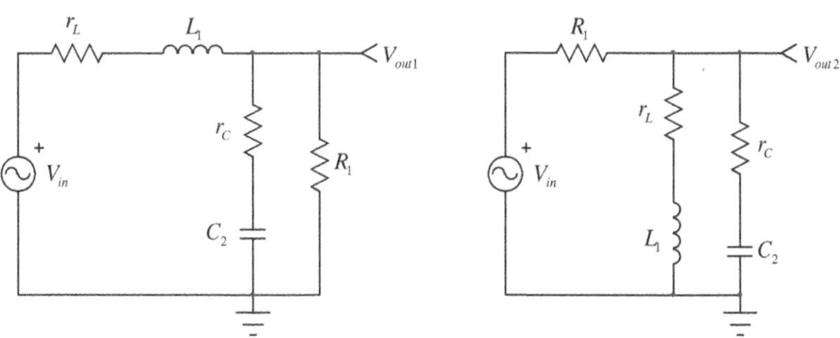

Figure 4.53 Superposition lets us derive the raw transfer function in a few lines.

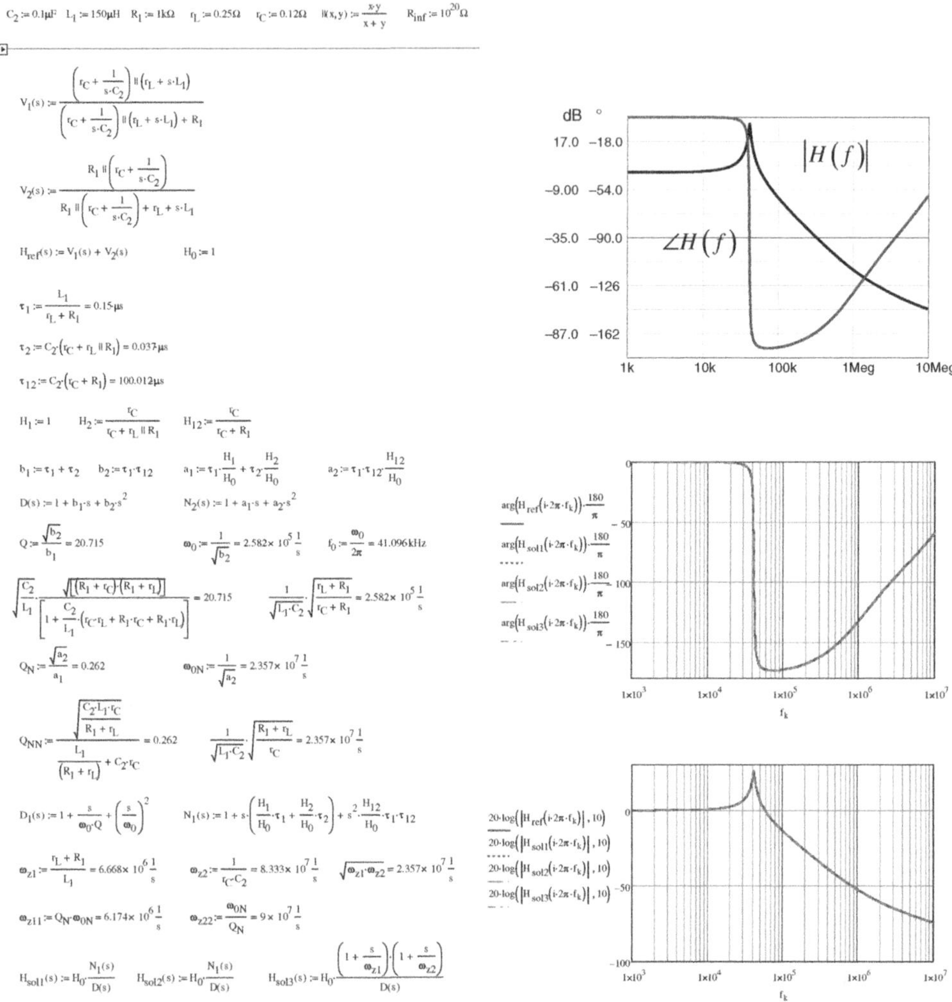

Figure 4.54 The Mathcad® sheet plots the various ac responses which all perfectly match.

4.2.4 Generalized 2nd–order Transfer Function – Example 3

For this third example, we will study a filter built around a bipolar transistor. The electrical diagram appears in Figure 4.55. What is the transfer function of this circuit?

First, we need a small-signal model obtained by replacing Q_1 by its simplified hybrid-π model as shown in Figure 4.56. We first start by determining the gain H_0 for $s=0$: C_2 opens and L_1 is replaced by a short circuit as in Figure 4.57.

The base current depends on V_{in} and the series arrangement of R_1 and r_π:

$$i_b = \frac{V_{in}}{R_1 + r_\pi} \tag{4.199}$$

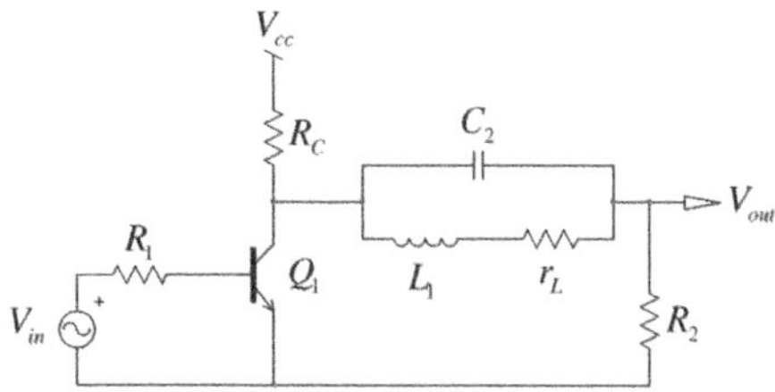

Figure 4.55 A common-emitter bipolar transistor drives a LC network.

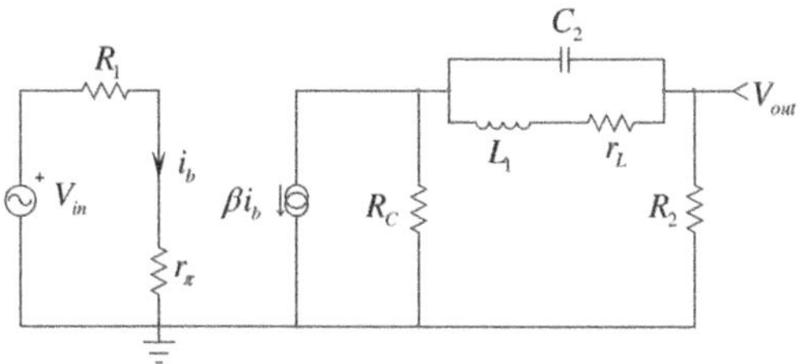

Figure 4.56 This is small-signal model on which we can carry Laplace analysis.

The output voltage across R_2 depends on the voltage across R_C, scaled down by r_L and R_2. The collector voltage $V_{(c)}$ can be expressed as:

$$V_{(c)} = -\beta i_b \left[R_C \| (r_L + R_2)\right] \tag{4.200}$$

This voltage appears in V_{out} affected by the resistive divider formed by R_2 and r_L:

$$V_{out} = -\beta i_b \left[R_C \| (r_L + R_2)\right] \frac{R_2}{r_L + R_2} \tag{4.201}$$

Substituting (4.199) in (4.201) and rearranging, we have the gain H_0 we want:

$$H_0 = -\frac{\beta}{r_\pi + R_1}\left[R_C \| (r_L + R_2)\right] \frac{R_2}{r_L + R_2} = -\frac{R_2 R_C \beta}{(R_1 + r_\pi)(R_2 + R_C + r_L)} \tag{4.202}$$

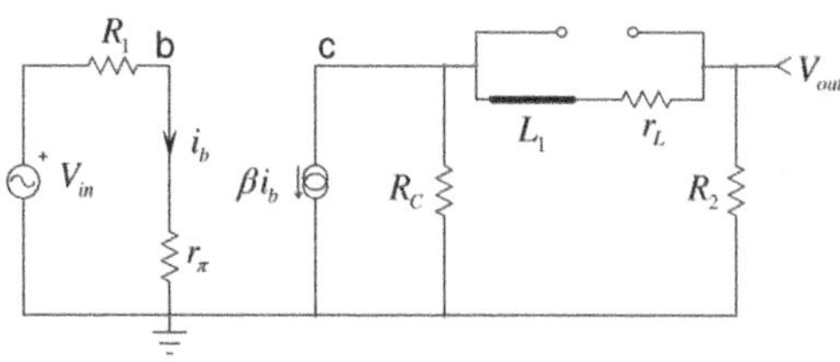

Figure 4.57 For the 0 Hz gain, C_2 is removed and L_1 replaced by a short circuit.

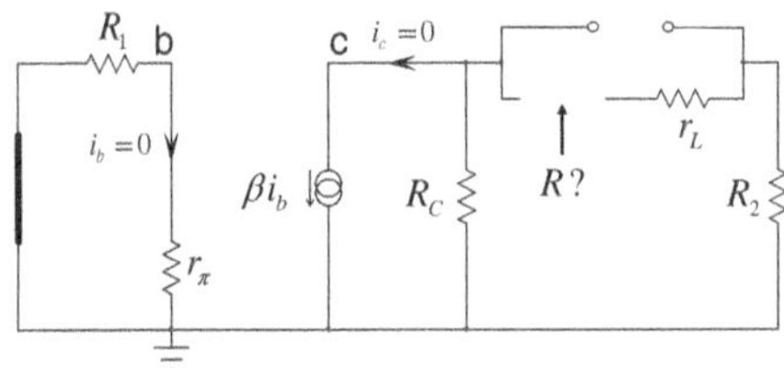

Figure 4.58 When the base current disappears because V_{in} is set to 0 V, the schematic simplifies.

For the time constant affecting L_1, we suppress the excitation ($V_{in} = 0$), so the base current disappears together with the collector current βi_b. From Figure 4.58, the resistance seen from L_1's terminals is the series arrangement of R_C, r_L and R_2:

$$\tau_1 = \frac{L_1}{r_L + R_C + R_2} \tag{4.203}$$

The second time constant is obtained by replacing L_1 by a short circuit and looking into C_2's terminals to determine τ_2. Again, it is quite simple by looking at Figure 4.59 where r_L is paralleled with R_C and R_2 in series:

$$\tau_2 = C_2\left[r_L \| (R_C + R_2)\right] \tag{4.204}$$

The final time constant we want is when L_1 is set to its high-frequency state. In this mode, we look at the resistance driving C_2. Figure 4.60 represents the circuit in this configuration. The time constant is immediate as r_L is left unconnected:

$$\tau_2^1 = C_2(R_C + R_2) \tag{4.205}$$

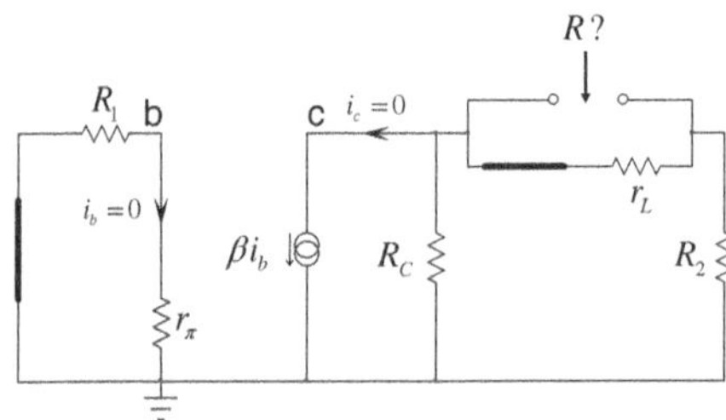

Figure 4.59 The determination of τ_2 is straightforward in this simplified sketch.

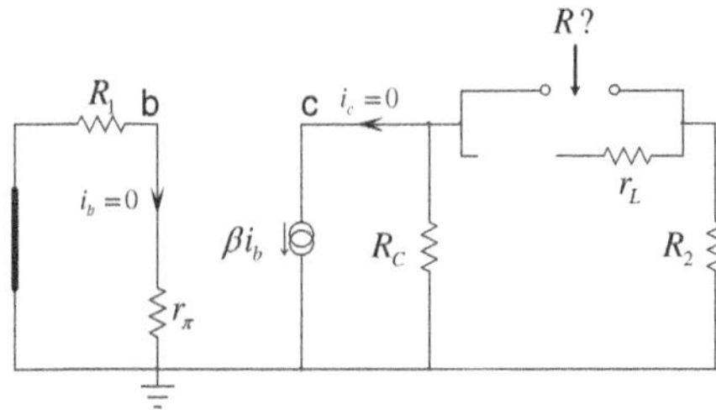

Figure 4.60 We can easily determine τ_2^1 by removing L_1 from the circuit and looking into C_2's terminals.

We now have all pieces to form our denominator $D(s)$:

$$D(s) = 1 + s(\tau_1 + \tau_2) + s^2\tau_1\tau_2^1$$
$$= 1 + s\left(\frac{L_1}{r_L + R_C + R_2} + C_2\left[r_L \| (R_C + R_2)\right]\right) + s^2\frac{L_1}{r_L + R_C + R_2}C_2(R_C + R_2) \quad (4.206)$$

This expression can be put under the canonical form:

$$D(s) = 1 + \frac{s}{\omega_0 Q} + \left(\frac{s}{\omega_0}\right)^2 \quad (4.207)$$

in which

$$Q = \frac{\sqrt{b_2}}{b_1} = \sqrt{\frac{C_2}{L_1}}\frac{\sqrt{R_2 + R_C}}{\sqrt{\frac{1}{r_L + R_C + R_2}}\left(1 + \frac{C_2 r_L(R_2 + R_C)}{L_1}\right)} \quad (4.208)$$

and

$$\omega_0 = \frac{1}{\sqrt{b_2}} = \frac{1}{\sqrt{L_1 C_2}}\sqrt{\frac{r_L + R_2 + R_C}{R_2 + R_C}} \quad (4.209)$$

The numerator now requires the calculation of three simple gains, H^1, H^2 and H^{12}. The exponent in H designates the element set in its high-frequency state while the other remains in its dc state. For H^1, L_1 is in its high-frequency state (open) while C_2 is in its dc state (open). The configuration appears in Figure 4.61 and the gain in this arrangement is simply:

$$H^1 = 0 \quad (4.210)$$

To determine H^2, capacitor C_2 is replaced by a short circuit and L_1, in its dc state, is also a short circuit. Figure 4.62 shows the resulting circuit. The transfer function in this mode corresponds to

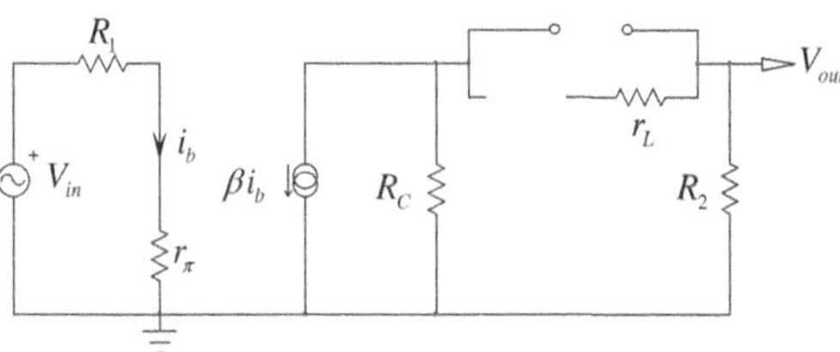

Figure 4.61 When both L_1 and C_2 are open, the circuit is really simple and the response is 0.

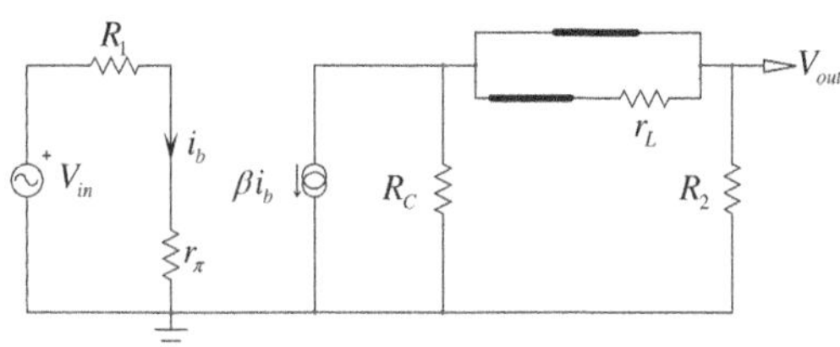

Figure 4.62 When the capacitor is open and the inductor shorted, this is the dc configuration already studied with H_0 except that r_L is shorted by C_2 in its high-frequency state.

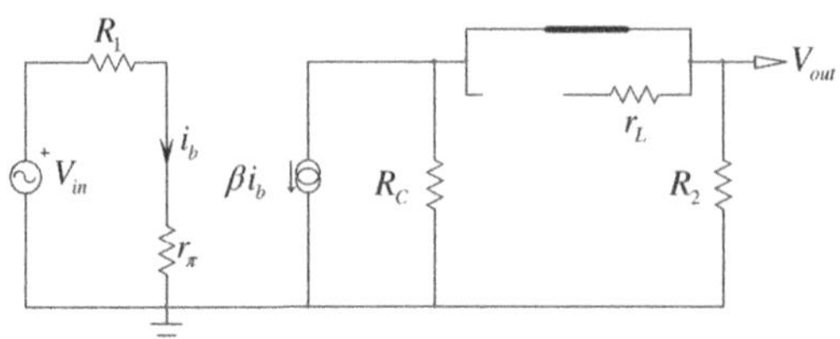

Figure 4.63 We can easily determine H^{12} by removing L_1 from the circuit and shorting C_2.

H_0 but with the term r_L set to zero because of the short circuit brought by C_2. Hence:

$$H^2 = -\frac{\beta}{r_\pi + R_1}(R_C \| R_2) \tag{4.211}$$

Finally, the gain H_{12} is determined when both energy-storing elements are put in their high-frequency state, C_2 shorting again the open-circuited inductor and its ohmic path r_L. This is what Figure 4.63 shows:

The gain is the same as that expressed by (4.211) since r_L is still short-circuited:

$$H^{12} = -\frac{\beta}{r_\pi + R_1}(R_C \| R_2) \tag{4.212}$$

The numerator can now be formed by assembling all the gains we derived:

$$\begin{aligned} N(s) &= H_0 + s\left(H^1\tau_1 + H^2\tau_2\right) + s^2 H^{12}\tau_1\tau_2^1 \\ &= -\frac{\beta}{r_\pi + R_1}\left[R_C \| (r_L + R_2)\right]\frac{R_2}{r_L + R_2} + s\left(0 - \frac{\beta}{r_\pi + R_1}(R_C \| R_2)\cdot C_2\left[r_L \| (R_C + R_2)\right]\right) \\ &+ s^2\left[-\frac{\beta}{r_\pi + R_1}(R_C \| R_2)\cdot\frac{L_1}{r_L + R_C + R_2}C_2(R_C + R_2)\right] \end{aligned} \tag{4.213}$$

If we now consider r_L to be very small compared to the other resistances, $N(s)$ greatly simplifies when factoring H_0 since in that case $H_0 \approx H^2 \approx H^{12}$:

$$N(s) \approx -\frac{\beta}{r_\pi + R_1}\left[R_C \| R_2\right]\left[1 + s^2 L_1 C_2\right] = H_0\left(1 + \frac{s^2}{\omega_{0N}{}^2}\right) \tag{4.214}$$

(4.213) is the exact numerator expression while (4.214) is an approximate version for negligible r_L values. The first expression is extremely complex while the second does not account for the loss contribution r_L. Is there a path in between? Of course, inspection can save us here. If we look at Figure 4.55, we can see that if the transformed network made of L_1, C_2 and r_L offers an infinite impedance at $s = s_z$, then the response is nulled. Therefore, the poles of this impedance will be the zeros of our transfer function. We can calculate this impedance in our head or through a quick intermediate step as in Figure 4.64.

At $s = 0$, the impedance is $R_0 = r_L$. Removing the excitation means open-circuiting the current source. In this case, in dc, C_2 'sees' r_L while L_1 'sees' an infinite value. Thus:

$$\begin{aligned} \tau_1 &= \frac{L_1}{\infty} = 0 \\ \tau_2 &= r_L C_2 \end{aligned} \tag{4.215}$$

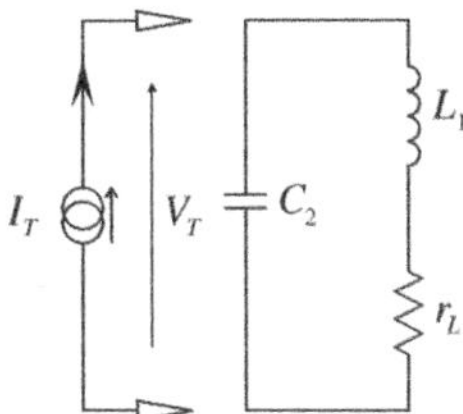

Figure 4.64 The network impedance is quickly calculated using fast analytical techniques.

Now considering τ_1^2, if C_2 is replaced by a short circuit, then L_1 'sees' r_L and we have:

$$\tau_1^2 = \frac{L_1}{r_L} \tag{4.216}$$

The denominator $D(s)$ is equal to

$$D(s) = 1 + sr_LC_2 + s^2L_1C_2 \tag{4.217}$$

In this network, the zero is found when, in the transformed circuit, a certain value $s = s_z$, nulls the response V_T in Figure 4.64. This is when r_L in series with L_1 forms a transformed short circuit:

$$sL_1 + r_L = 0 \tag{4.218}$$

implying

$$s_z = -\frac{r_L}{L_1} \tag{4.219}$$

The impedance can thus be put under the following expression:

$$Z(s) = r_L\frac{1 + s\dfrac{L_1}{r_L}}{1 + sr_LC_2 + s^2L_1C_2} = R_0\frac{1 + \dfrac{s}{\omega_z}}{1 + \dfrac{s}{\omega_0 Q} + \left(\dfrac{s}{\omega_0}\right)^2} \tag{4.220}$$

By identification:

$$\omega_0 = \frac{1}{\sqrt{L_1C_2}} \tag{4.221}$$

Q is

$$Q = \frac{1}{r_L}\sqrt{\frac{L_1}{C_2}} \tag{4.222}$$

and

$$\omega_z = \frac{r_L}{L_1} \tag{4.223}$$

The denominator $N(s)$ of our transistor-based circuit already expressed in (4.213) can advantageously be defined with (4.217):

$$N(s) = 1 + sr_LC_2 + s^2L_1C_2 = 1 + \frac{s}{\omega_{0N}Q_N} + \left(\frac{s}{\omega_{0N}Q_N}\right)^2 \tag{4.224}$$

in which ω_{0N} and Q_N are respectively defined in (4.221) and (4.222). The complete transfer function is now developed using (4.213) or (4.224) together with a common denominator defined by (4.206):

$$H(s) = -\frac{R_2 R_C \beta}{(R_1 + r_\pi)(R_2 + R_C + r_L)} \cdot \frac{1 + s r_L C_2 + s^2 L_1 C_2}{1 + s\left(\frac{L_1}{r_L + R_C + R_2} + C_2\left[r_L \| (R_C + R_2)\right]\right) + s^2 \frac{L_1}{r_L + R_C + R_2} C_2 (R_C + R_2)}$$

$$= H_0 \frac{1 + \frac{s}{\omega_{0N} Q_N} + \left(\frac{s}{\omega_{0N} Q_N}\right)^2}{1 + \frac{s}{\omega_0 Q} + \left(\frac{s}{\omega_0 Q}\right)^2} \tag{4.225}$$

One more time, you see that inspection beats all other methods by its speed and ease of determining the zeros in this case. Sometimes, inspection is not easy and you must go either through NDI or the generalized form described in this example. The generalized form gives the exact same result as NDI would, but, sometimes, more work is needed to simplify the final expression. Now that we have our transfer function obtained using different forms, we can compare responses using a Mathcad® sheet. This is what we did in Figure 4.65. All combinations lead to the exact same response!

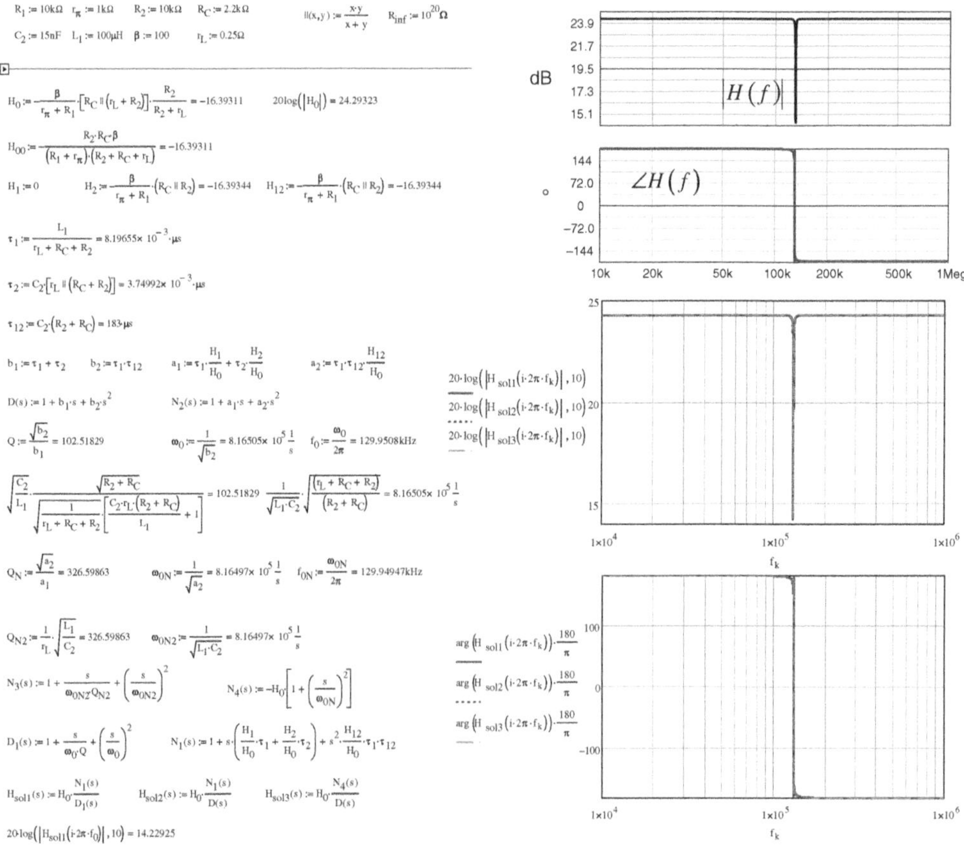

Figure 4.65 Mathcad® and SPICE deliver the exact same results.

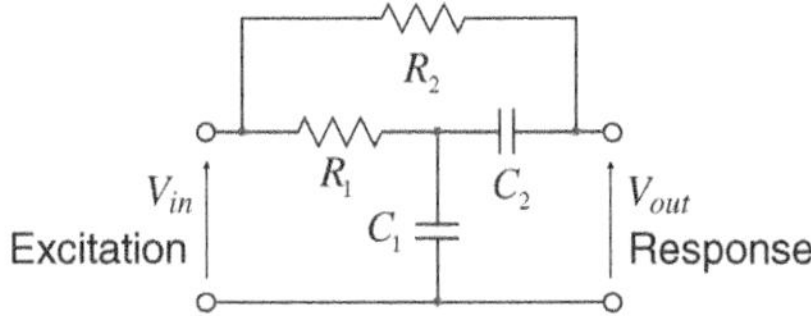

Figure 4.66 This 2-capacitor circuit peaks for a certain combination of values.

4.2.5 Generalized 2nd–order Transfer Function – Example 4

This fourth example is described in [4] and also studied in [5]. It is interesting because it is a peaking *RC* network, a *RC* network with gain. The schematic appears in Figure 4.66. What is the transfer function of this 2-capacitor circuit?

As usual, we will start with $s = 0$ and proceed with time constants τ and various gains H. The complete path is described in Figure 4.67.

From these sketches, you should find:

$$H_0 = 1 \tag{4.226}$$

$$\tau_1 = R_1 C_1 \tag{4.227}$$

$$\tau_2 = (R_1 + R_2)C_2 \tag{4.228}$$

$$\tau_2^1 = R_2 C_2 \tag{4.229}$$

$$H^1 = H^2 = 1 \tag{4.230}$$

and

$$H^{12} = 0 \tag{4.231}$$

The denominator $D(s)$ is formed by combining (4.227), (4.228) and (4.229):

$$D(s) = 1 + s(\tau_1 + \tau_2) + s^2 \tau_1 \tau_2^1 = 1 + s[R_1 C_1 + (R_1 + R_2)C_2] + s^2 R_1 R_2 C_1 C_2 \tag{4.232}$$

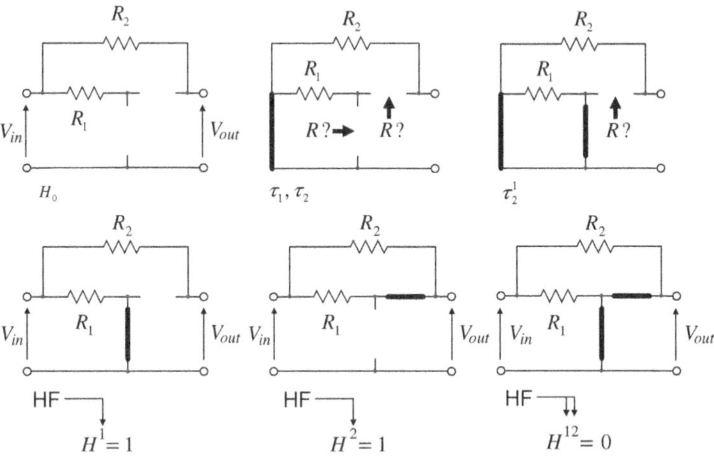

Figure 4.67 A series of six simple drawings gets you to the transfer function we want.

while $N(s)$ assembles similar equations with the gains determined in (4.230) and (4.231):

$$N(s) = H_0 + \left(\tau_1 H^1 + \tau_2 H^2\right)s + \tau_1 \tau_2^1 H^{12} s^2 = 1 + [R_1 C_1 + (R_1 + R_2)C_2]s \tag{4.233}$$

The complete transfer function H is thus defined as:

$$H(s) = \frac{1 + a_1 s}{1 + b_1 s + b_2 s^2} = \frac{1 + [R_1 C_1 + (R_1 + R_2)C_2]s}{1 + [R_1 C_1 + (R_1 + R_2)C_2]s + R_1 R_2 C_1 C_2 s^2} \tag{4.234}$$

Looking at this equation, you see that a_1 and b_1 are the same so (4.234) can be rearranged as:

$$H(s) = \frac{1 + b_1 s}{1 + b_1 s + b_2 s^2} \tag{4.235}$$

Considering a low-Q approximation, the denominator can be rearranged to fit the form $\left(1 + \frac{s}{\omega_{p1}}\right)\left(1 + \frac{s}{\omega_{p2}}\right)$. If we proceed this way, ω_{p1} will neutralize the zero (same frequency) and we build a classical low-pass filter. We could consider coincident poles at ω_p instead, with a Q equal 0.5 (see Chapter 2). In this condition, the canonical form of the denominator can be rewritten as:

$$1 + \frac{s}{\omega_0 Q} + \left(\frac{s}{\omega_0}\right)^2 = \left(1 + \frac{s}{\omega_p}\right)^2 \tag{4.236}$$

in which

$$\omega_p = \omega_0 = \frac{1}{\sqrt{b_2}} = \frac{1}{\sqrt{R_1 R_2 C_1 C_2}} \tag{4.237}$$

To write these expressions, we have arbitrarily selected a Q of 0.5. What element values among C_1, C_2, R_1 and R_2 lead to a Q of 0.5? From Chapter 2, we know that $Q = \sqrt{b_2}/b_1$. If $Q = 0.5$, then we have to satisfy:

$$0.5 = \frac{\sqrt{b_2}}{b_1} \to b_1 = 2\sqrt{b_2} \tag{4.238}$$

If we plug in (4.238) the definitions for b_1 which is $(\tau_1 + \tau_2)$ and b_2 which is $(\tau_1 \tau_2^1)$ then we must solve:

$$R_1 C_1 + (R_1 + R_2)C_2 = 2\sqrt{R_1 R_2 C_1 C_2} \tag{4.239}$$

which we can rewrite as:

$$\left[R_1 C_1 - 2\sqrt{R_1 C_1}\sqrt{R_2 C_2} + R_2 C_2\right] + R_1 C_2 = 0 \tag{4.240}$$

As indicated in [5], the left term can be factored as $(a - b)^2$which leads to:

$$\left(\sqrt{R_1 C_1} - \sqrt{R_2 C_2}\right)^2 + R_1 C_2 = 0 \tag{4.241}$$

It clearly cannot be solved for any positive values of the passive elements forming the expression. [5] offers an elegant way to make it by factoring $\sqrt{R_1 C_1}$:

$$\left[\sqrt{R_1 C_1}\left(1 - \sqrt{\frac{R_2 C_2}{R_1 C_1}}\right)\right]^2 + R_1 C_2 = 0 \tag{4.242}$$

Extracting $R_1 C_1$and factoring leads to the final equation:

$$\left(1 - \sqrt{\frac{R_2 C_2}{R_1 C_1}}\right)^2 + \frac{C_2}{C_1} = 0 \tag{4.243}$$

If $R_2C_2 = R_1C_1$ then the left term cancels and the second left term remains. Satisfying $\frac{C_2}{C_1} = \frac{R_1}{R_2} \ll 1$ is the approximate solution to solve (4.243). The final transfer function can now be approximated to:

$$H(s) \approx \frac{1 + \frac{s}{\omega_z}}{\left(1 + \frac{s}{\omega_p}\right)^2} \tag{4.244}$$

in which:

$$\omega_z = \frac{1}{b_1} = \frac{1}{R_1C_1 + R_2C_2} \tag{4.245}$$

and

$$\omega_p = \frac{1}{\sqrt{b_2}} = \frac{1}{\sqrt{R_1C_1R_2C_2}} \tag{4.246}$$

If we now consider the conditions for which (4.244) holds, $\tau_1 = \tau_2$, then you can simplify (4.245) and (4.246) as:

$$\omega_z = \frac{1}{2\tau} \tag{4.247}$$

$$\omega_p = \frac{1}{\tau} \tag{4.248}$$

with $\tau = R_1C_1 = R_2C_2$. In these conditions, the pole is twice the zero location:

$$\omega_p = 2\omega_z \tag{4.249}$$

What is the peaking frequency, or the point at which the gain is maximal? We can obtain it by differentiating (4.244) and solving for the value of ω that cancels the result:

$$\frac{d}{d\omega}|H(\omega)| = 0 \tag{4.250}$$

The magnitude of a division is the magnitude of the numerator divided by that of the denominator. Therefore, we have:

$$\left|\frac{1 + j\frac{\omega}{\omega_z}}{\left(1 + j\frac{\omega}{\omega_p}\right)^2}\right| = \frac{\sqrt{1 + \left(\frac{\omega}{\omega_z}\right)^2}}{1 + \left(\frac{\omega}{\omega_p}\right)^2} \tag{4.251}$$

proceeding with the differentiation:

$$\frac{d}{d\omega}\frac{\sqrt{1 + \left(\frac{\omega}{\omega_z}\right)^2}}{1 + \left(\frac{\omega}{\omega_p}\right)^2} = -\frac{\omega \cdot {\omega_p}^2\left(\omega^2 - {\omega_p}^2 + 2{\omega_z}^2\right)}{{\omega_z}^2\sqrt{1 + \left(\frac{\omega}{\omega_z}\right)^2}\left(\omega^2 + {\omega_p}^2\right)^2} = 0 \tag{4.252}$$

There are three roots but only one is relevant. It is:

$$\omega_{\max} = \sqrt{{\omega_p}^2 - 2{\omega_z}^2} \tag{4.253}$$

What is the value of this gain then when resonance occurs? We know the relationship between the pole and the zero with (4.249). With this result on hand, we can rewrite (4.251) with the zero only:

$$|H(\omega)| = \frac{4\omega_z^2\sqrt{\dfrac{\omega^2+\omega_z^2}{\omega_z^2}}}{\omega^2+4\omega_z^2} \tag{4.254}$$

Using (4.249) again, (4.253) becomes

$$\omega_{\max} = \sqrt{2}\omega_z \tag{4.255}$$

Substituting (4.255) for ω in (4.254) gives the maximum of the function:

$$|H(\omega_{\max})| = \frac{2\sqrt{3}}{3} = 1.155 \tag{4.256}$$

Before we proceed, we need to determine the raw transfer function of this network so that we can compare our expressions to it. Similarly to our previous examples, superposition can help as shown in Figure 4.68.

The raw transfer function is obtained by summing V_{out1} and V_{out2} respectively determined in Figure 4.69 and Figure 4.70:

$$H(s) = \frac{R_2}{R_2+\dfrac{1}{sC_2}+R_1\|\dfrac{1}{sC_1}}\frac{1}{1+sR_1C_1}+\frac{\dfrac{1}{sC_2}+R_1\|\dfrac{1}{sC_1}}{\dfrac{1}{sC_2}+R_1\|\dfrac{1}{sC_1}+R_2} \tag{4.257}$$

which can be rearranged as

$$H(s) = \frac{1}{R_2+\dfrac{1}{sC_2}+R_1\|\dfrac{1}{sC_1}}\left(\frac{R_2}{1+sR_1C_1}+\frac{1}{sC_2}+R_1\|\frac{1}{sC_1}\right) \tag{4.258}$$

Now that we have all our equations, we can capture these data in a Mathcad® sheet as we did before. This is what is offered in Figure 4.71. All equations deliver the exact same response, whether you plot the raw expressions from (4.257) and (4.258) or the approximate version of (4.244).

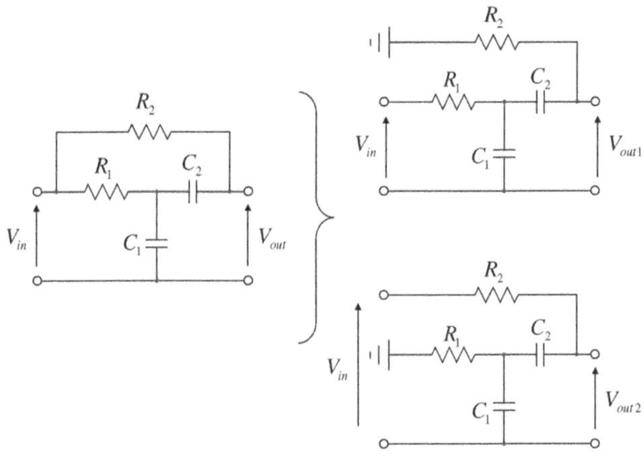

Figure 4.68 Superposition helps us to determine the raw transfer function.

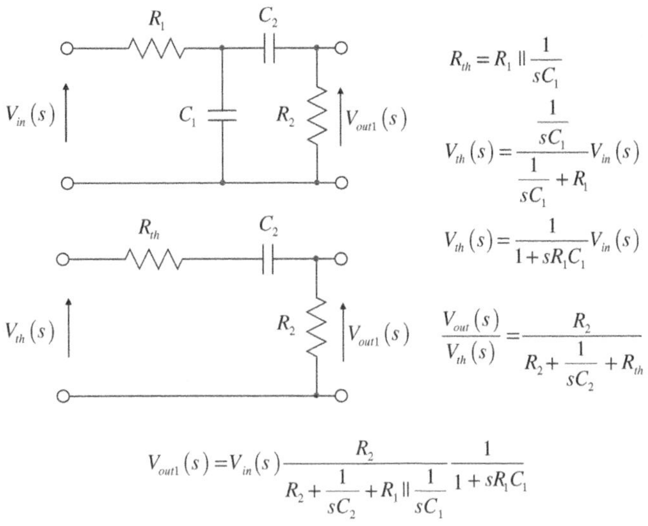

R_1
C_2
$V_{in}(s)$
C_1
R_2
$V_{out1}(s)$
R_{th}
C_2
$V_{th}(s)$
R_2
$V_{out1}(s)$
$R_{th} = R_1 \,||\, \frac{1}{sC_1}$
$V_{th}(s) = \frac{\frac{1}{sC_1}}{\frac{1}{sC_1} + R_1} V_{in}(s)$
$V_{th}(s) = \frac{1}{1 + sR_1C_1} V_{in}(s)$
$\frac{V_{out}(s)}{V_{th}(s)} = \frac{R_2}{R_2 + \frac{1}{sC_2} + R_{th}}$
$V_{out1}(s) = V_{in}(s) \frac{R_2}{R_2 + \frac{1}{sC_2} + R_1 \,||\, \frac{1}{sC_1}} \frac{1}{1 + sR_1C_1}$

Figure 4.69 The first expression uses Thévenin's theorem to solve the transfer function.

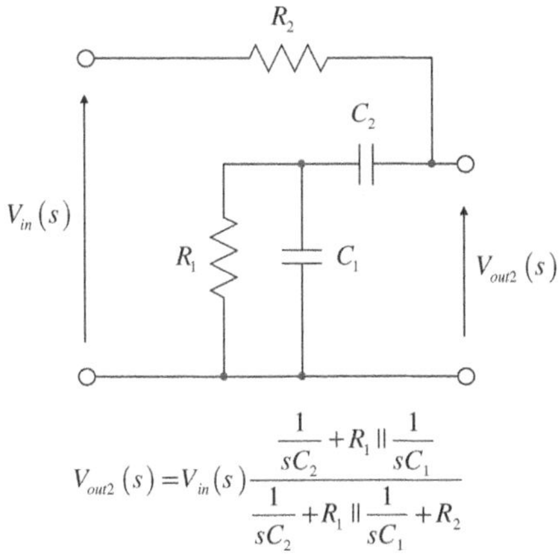

Figure 4.70 The second configuration implies an impedance divider.

This peaky RC network is interesting and I confess that I actually never thought a RC network could exhibit a magnitude higher than 1. If you want to further dig into the subject, you will find more information in [6], [7] and [8].

4.3 What Should I Retain from this Chapter ?

In this fourth chapter, we went one step farther by exploring 2nd-order circuits. Below is a summary of what we learned in this chapter:

1. The 2-EET consists of applying the EET twice to a 2nd-order system. Energy-storing elements can be combined in various states to form the reference gain. The EET is applied to a first

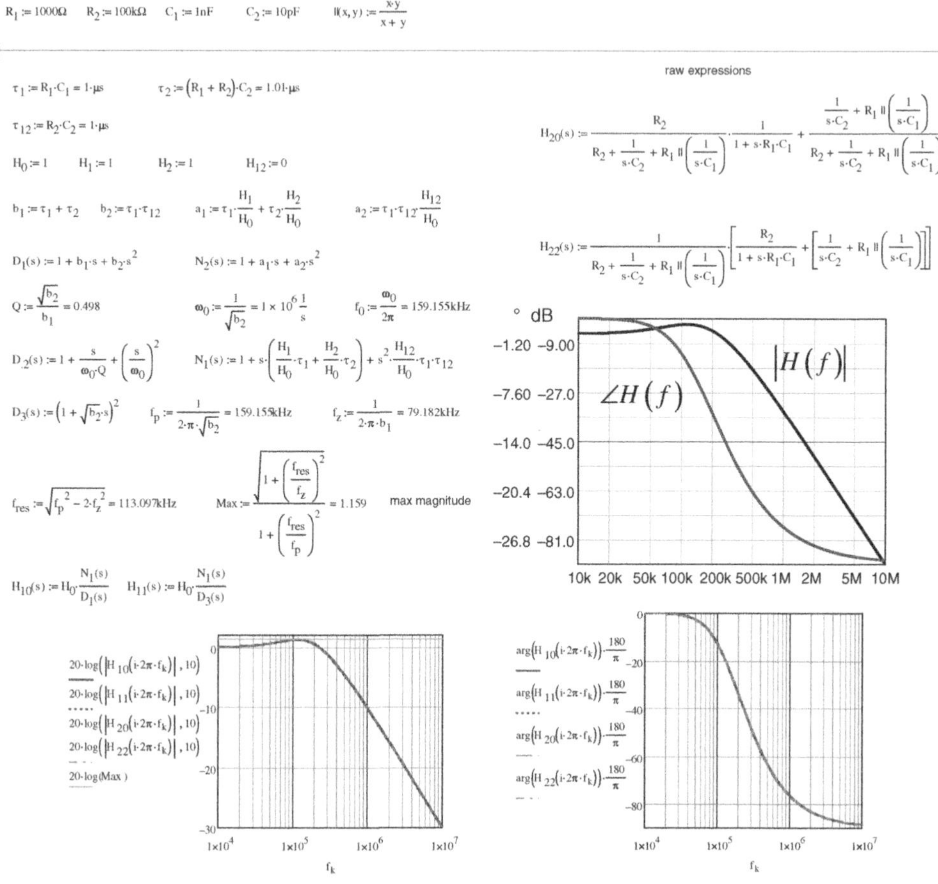

Figure 4.71 All equations deliver the exact same result which is enjoyable!

energy-storing element while the second is set in its reference state. The EET is further run a second time over the second element with the first one back in place, leading to complex driving impedances. Assembling these factors leads to the transfer function of a 2nd-order network.

2. The raw 2-EET does not offer much insight into poles and zeros locations. To develop a consistent and applicable formula, we have considered a reference (or a 0-Hz) circuit in which all energy-storing elements are put in their dc state: the circuit is first observed at $s = 0$ where capacitors are open and inductors are short circuited. This is the way SPICE calculates an operating bias point prior to launching any type of simulation.
3. As the numerator and the denominator are unitless expressions, the terms multiplied by s, a_1 or b_1, have a unit of time, [s]. a_1 or b_1 are the sum of the circuit time constants when the circuit is observed in two different conditions. a_1 coefficients are determined when the response is nulled (numerator coefficients, labeled with the N suffix, τ_{1N} and τ_{2N}) while b_1 terms are found when the excitation is set to 0 (denominator coefficients, τ_1 and τ_2).
4. The 2nd-order coefficients, a_2 or b_2, must have a unit of time squared, [s^2]. They are a product of time constants. You can pick the first time constant τ_1 and multiply it by another time constant τ_2^1obtained when the element involved in τ_1 is set in its high-frequency state while you look at

the resistance driving element 2. Owing to redundancy, you can also pick the second time constant τ_2 and multiply it by another time constantτ_1^2 obtained when the element involved in τ_2 is set in its high-frequency state while you look at the resistance driving element 1. For the numerator, you run the exercise while the response is nulled and all resulting time constants have a suffix N, $\tau_{1N}\tau_{2N}^1$ or $\tau_{2N}\tau_{1N}^2$. For the denominator, you determine $\tau_1\tau_2^1$ or $\tau_2\tau_1^2$ while the excitation is set to 0.

5. 2nd-order numerator and denominator can be rearranged in a canonical form involving a quality factor Q and a resonant frequency ω_0. When this quality factor is much smaller than 1, the numerator and/or the denominator can be respectively expressed as well-separated zeros and poles.
6. Examples show how powerful the technique is. Solving each calculation step with simple sketches lets you later easily identify and correct an error if any. Mathcad® and its individual time constant calculations represents an invaluable tool to derive and test transfer function in a swift and efficient way.
7. One more time, SPICE can also be used in a very simple manner to verify the integrity of all calculations, including the multiple zeros and poles respectfully obtained when the network is analyzed in a NDI or with the excitation set to 0.

References

1. R. D. Middlebrook. The two extra element theorem. *IEEE Proceedings Frontiers in Education, 21st Annual Conference*, Purdue Univ., Sept. 21–24 1991, pp. 702–708.
2. R. D. Middlebrook, V. Vorpérian, J. Lindal. The *N* Extra Element Theorem. *IEEE Transactions on Circuits and Systems*, **45** (9), 919–935. 1998.
3. A. Hajimiri. Generalized Time- and Transfer-Constant Circuit Analysis. *IEEE Transactions on Circuits and Systems*, **57** (6), 1105–1121. 2009.
4. J. Holbrook. *Laplace Transform for Electronic Engineers*, Pergamon Press, 1959.
5. V. Vorpérian. *Fast Analytical Techniques for Electrical and Electronic Circuits*. Cambridge University Press. 2002, pp 317–318.
6. H. Epstein. Synthesis of Passive RC Networks with Gains Greater than Unity. *Proceedings of the IRE*, July, 1951, p. 83.
7. S. Takashi, T. Sugiura. Some properties of passive RC network giving over-unity gain. http://koara.lib.keio.ac.jp/xoonips/modules/xoonips/download.php?file_id=95480 (last accessed 12/12/2015).
8. K. Castor-Perry. Ac voltage gain using just resistors and capacitors? *The Filter Wizard* issue 36, Cypress Semiconductor, http://www.cypress.com/?docID=45636 (last accessed 12/12/2015).

4.4 Appendix 4A – Problems

Below are several problems based on the information delivered in this chapter.

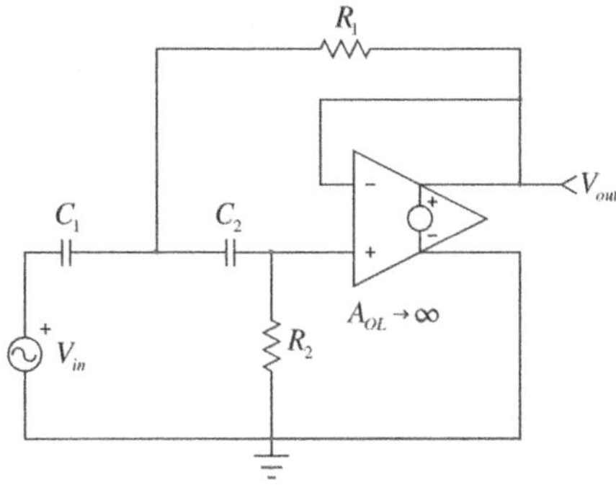

Figure 4.72 – Problem 1: What is the transfer function of this Sallen-Key filter?

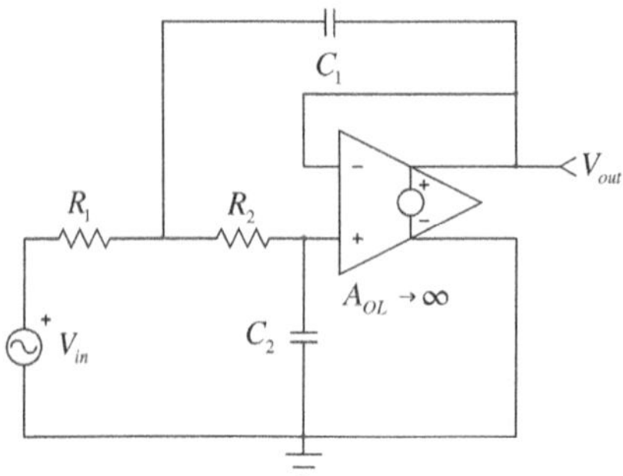

Figure 4.73 – Problem 2: What is the transfer function of this Sallen-Key filter?

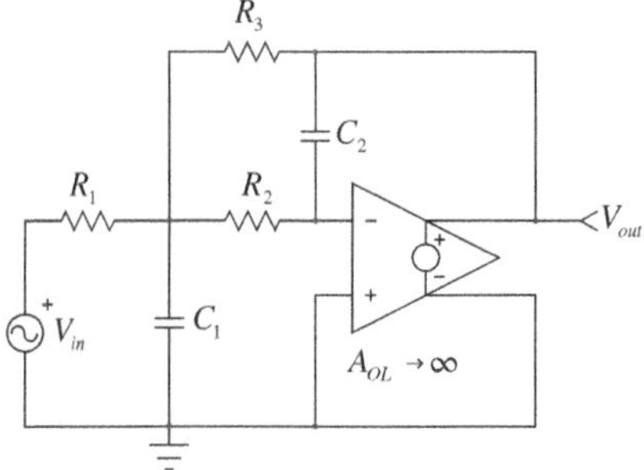

Figure 4.74 – Problem 3: What is the transfer function of this multiple feedback filter?

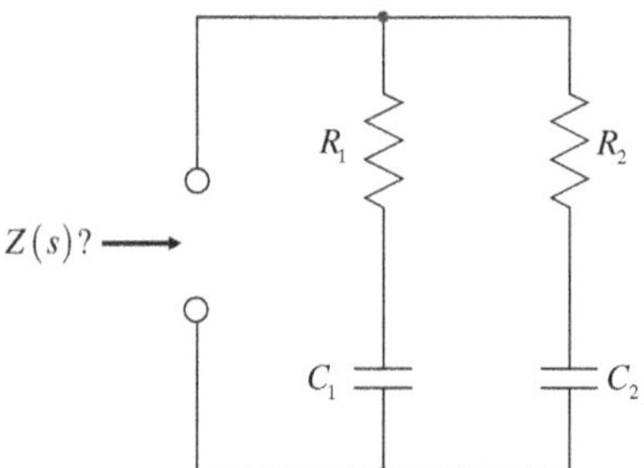

Figure 4.75 – Problem 4: What is the impedance offered by the paralleling of these elements? If you find an indeterminacy at some point, add a temporary resistor R across the connecting terminals.

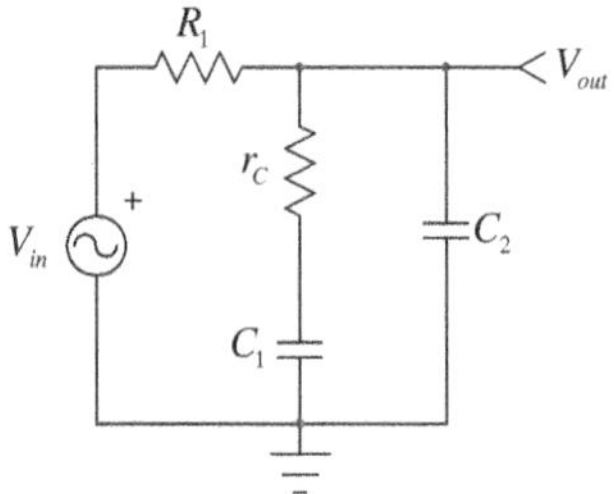

Figure 4.76 – Problem 5: Calculate the transfer function of this circuit.

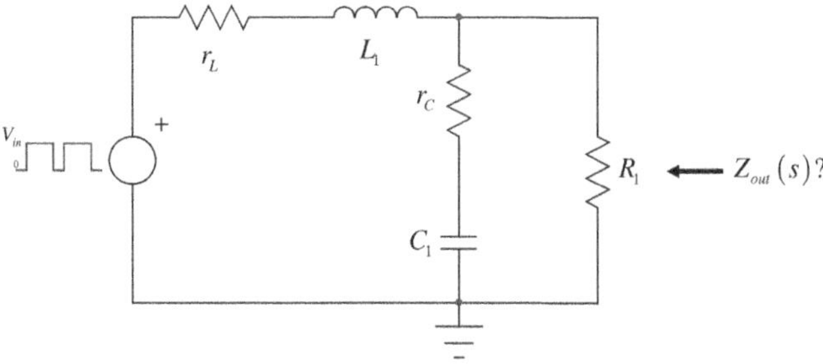

Figure 4.77 – Problem 6: Calculate the output impedance of this simplified buck converter.

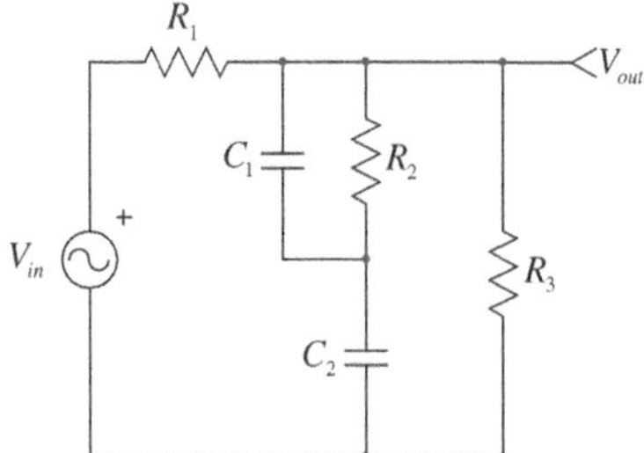

Figure 4.78 – Problem 7: What is the transfer function of this network?

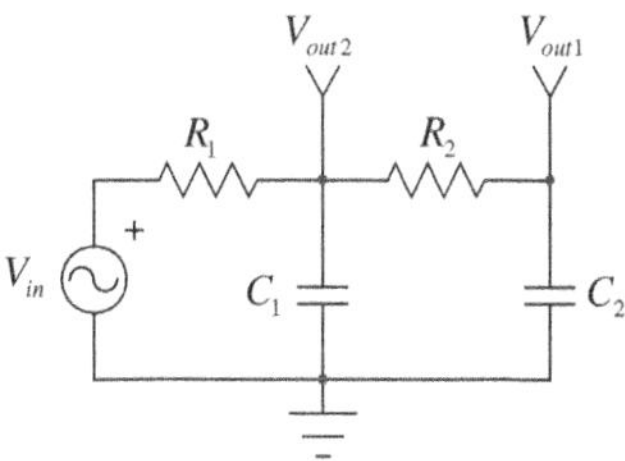

Figure 4.79 – Problem 8: What is the transfer function of these cascaded *RC* networks when V_{out1} and V_{out2} are considered?

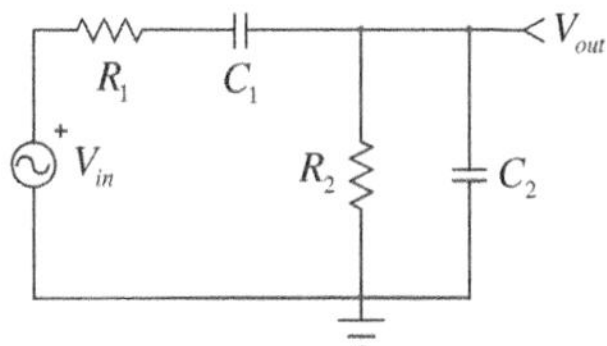

Figure 4.80 – Problem 9: What is the transfer function of this Wien bridge architecture?

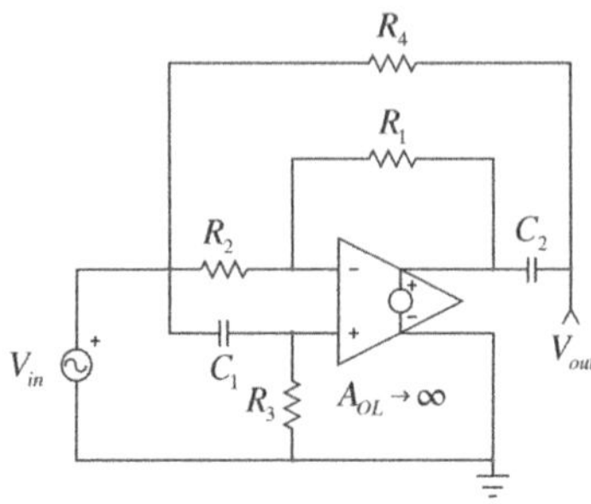

Figure 4.81 – Problem 10: What is the transfer function of this op amp-based filter?

Answers

Problem 1:

We proceed with this op amp-based filter exactly as we did before. First, check the gain for $s = 0$, then all the time constants you need to form the denominator. Then proceed with the gains for different C_1/C_2 states. Figure 4.82 gathers all these individual calculations.

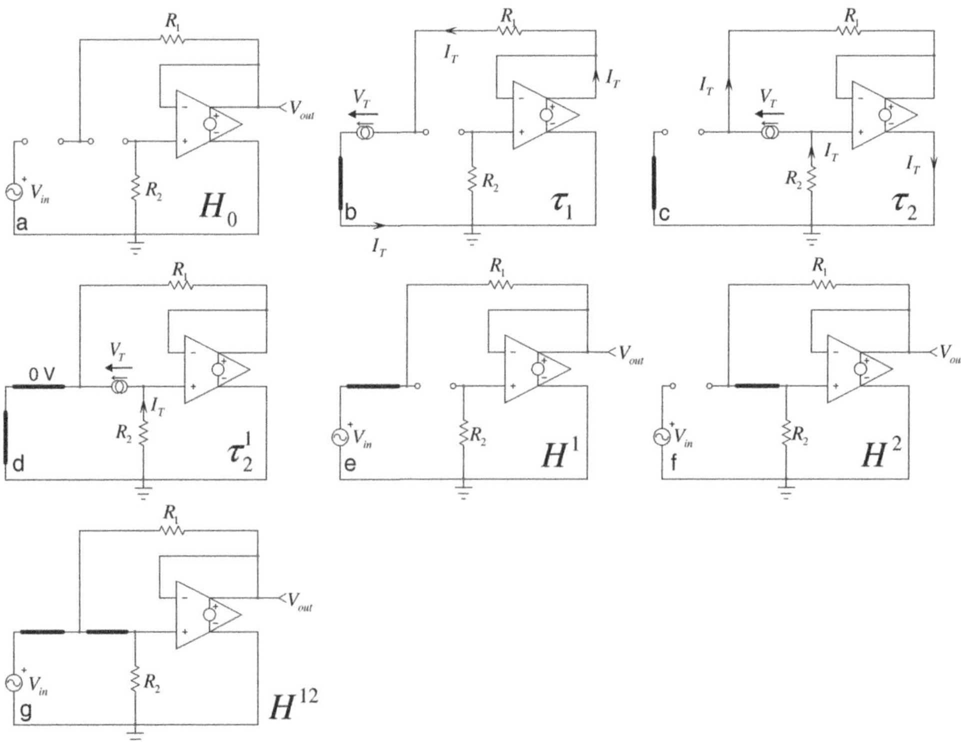

Figure 4.82 All these little drawings let us define the filter transfer function in an easy and smooth way.

The dc gain in Figure 4.82a is easy, all capacitors are open:

$$H_0 = 0 \tag{4.259}$$

In Figure 4.82b, the resistor seen by the generator is R_1, therefore:

$$\tau_1 = R_1C_1 \tag{4.260}$$

In Figure 4.82c, the right side of the current generator and the op amp output are at the same potential. This voltage, also present on the non-inverting pin is:

$$V_{right} = V_{out} = -I_TR_2 \tag{4.261}$$

The left-side of the current generator is at a potential equal to the op amp output plus the drop across R_1:

$$V_{left} = I_TR_1 + V_{out} = I_TR_1 - I_TR_2 \tag{4.262}$$

The voltage across the current generator V_T is thus $V_{left}-V_{right}$:

$$V_T = I_TR_1 - I_TR_2 + I_TR_2 = I_TR_1 \tag{4.263}$$

The resistance 'seen' by C_2 is thus R_1, leading to:

$$\tau_2 = R_1C_2 \tag{4.264}$$

In Figure 4.82d, you see that the left terminal of the current generator is grounded and forces I_T into R_2 only. Thus:

$$\tau_2^1 = R_2C_2 \tag{4.265}$$

We can form the denominator D by organizing the time constants:

$$D(s) = 1 + s(\tau_1 + \tau_2) + s^2\tau_1\tau_2^1 = 1 + sR_1(C_1 + C_2) + s^2R_1C_1R_2C_2 \tag{4.266}$$

This expression can be put under the canonical form:

$$D(s) = 1 + \frac{s}{\omega_0Q} + \left(\frac{s}{\omega_0}\right)^2 \tag{4.267}$$

in which

$$Q = \frac{\sqrt{b_2}}{b_1} = \frac{\sqrt{R_1R_2C_1C_2}}{R_1(C_1 + C_2)} \tag{4.268}$$

and

$$\omega_0 = \frac{1}{\sqrt{b_2}} = \frac{1}{\sqrt{R_1R_2C_1C_2}} \tag{4.269}$$

Regarding the numerator, Figure 4.82e and f tell that gains H^1 and H^2 are zero, therefore, a_1 is also zero. H^{12} is simple to determine, V_{out} duplicates the non-inverting pin level which is V_{in} in this configuration. Therefore:

$$H^{12} = 1 \tag{4.270}$$

The numerator is quite simple given the fact that most gains are zero:

$$N(s) = H_0 + s\left(H^1\tau_1 + H^2\tau_2\right) + s^2\tau_1\tau_2^1H^{12} = s^2R_1C_1R_2C_2 \tag{4.271}$$

The transfer function of this filter is defined by:

$$G(s) = \frac{s^2 R_1 C_1 R_2 C_2}{1 + sR_1(C_1 + C_2) + s^2 R_1 C_1 R_2 C_2} \tag{4.272}$$

For $s = 0$, the gain is zero and as s increases, the magnitude goes up with a +2-slope then the double poles kick in and flatten the curve. This type of expression can be put under a more compact form, involving the inverted poles and zeros introduced in Chapter 2. If you go back to this chapter, (4.272) can advantageously be rewritten as:

$$G(s) = \frac{1}{1 + \frac{C_1 + C_2}{sR_2 C_1 C_2} + \left(\frac{1}{s\sqrt{R_1 R_2 C_1 C_2}}\right)^2} = \frac{1}{1 + \frac{\omega_0}{sQ} + \left(\frac{\omega_0}{s}\right)^2} \tag{4.273}$$

We can now test all of our expressions in a Mathcad® sheet as shown in Figure 4.83.

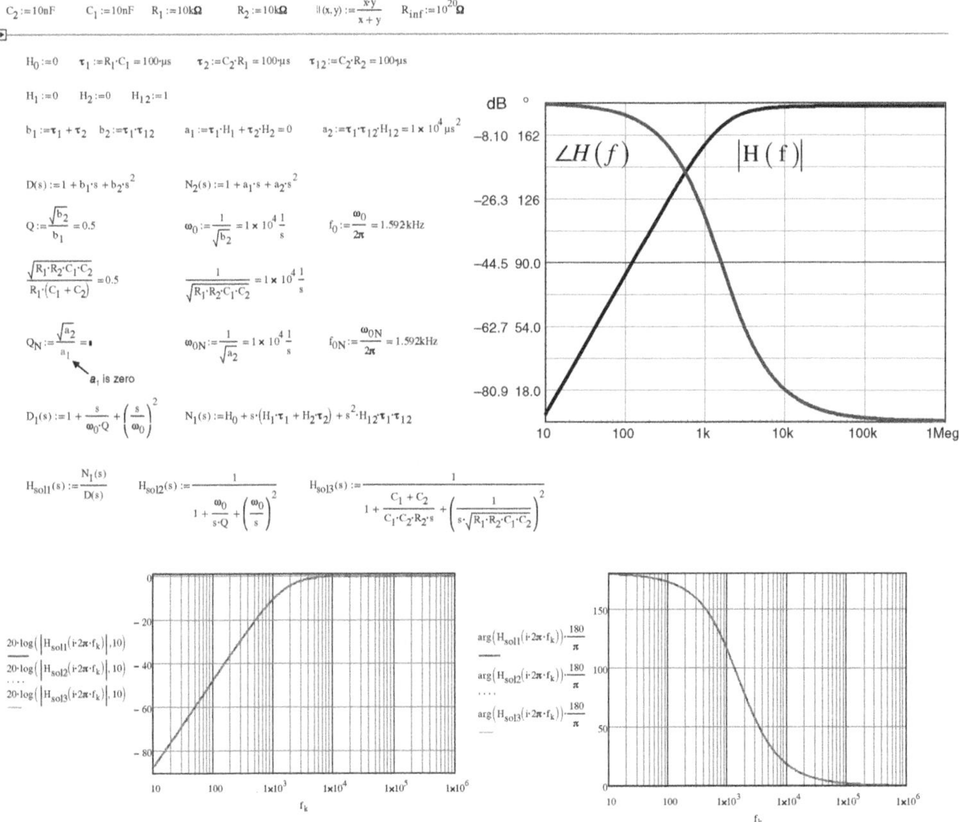

Figure 4.83 Mathcad® and SPICE confirm that our approach is correct.

Problem 2:

This op amp-based filter uses two capacitors, it is a 2nd-order filter. We will carry on exactly as we did before, starting with the gain for $s = 0$ then proceed with all the time constants. Figure 4.84 gathers the individual sketches. From sketch (a) in Figure 4.84 we immediately see that the op amp appears as a non-inverting voltage follower imposing a gain of:

$$H_0 = 1 \tag{4.274}$$

In Figure 4.84b, the test generator I_T biases R_1-R_2 junction at voltage V_1. This voltage V_1 is also present at the op amp non-inverting pin and is thus duplicated on the output. In other words, the voltage across the current generator is 0. Thus:

$$\tau_1 = 0 \cdot C_1 \tag{4.275}$$

In Figure 4.84c, the test current I_T circulates in R_1 and R_2 as C_1 is removed. Therefore:

$$\tau_2 = C_2(R_1 + R_2) \tag{4.276}$$

Regarding the s^2 term, as τ_1 is zero, we may encounter an indeterminacy if we calculate τ_2^1 so we evaluate τ_1^2 instead in Figure 4.84d. In this configuration, as the non-inverting pin is grounded by C_1 set in its high-frequency state, the bottom of the current generator is at the 0-V potential and injects

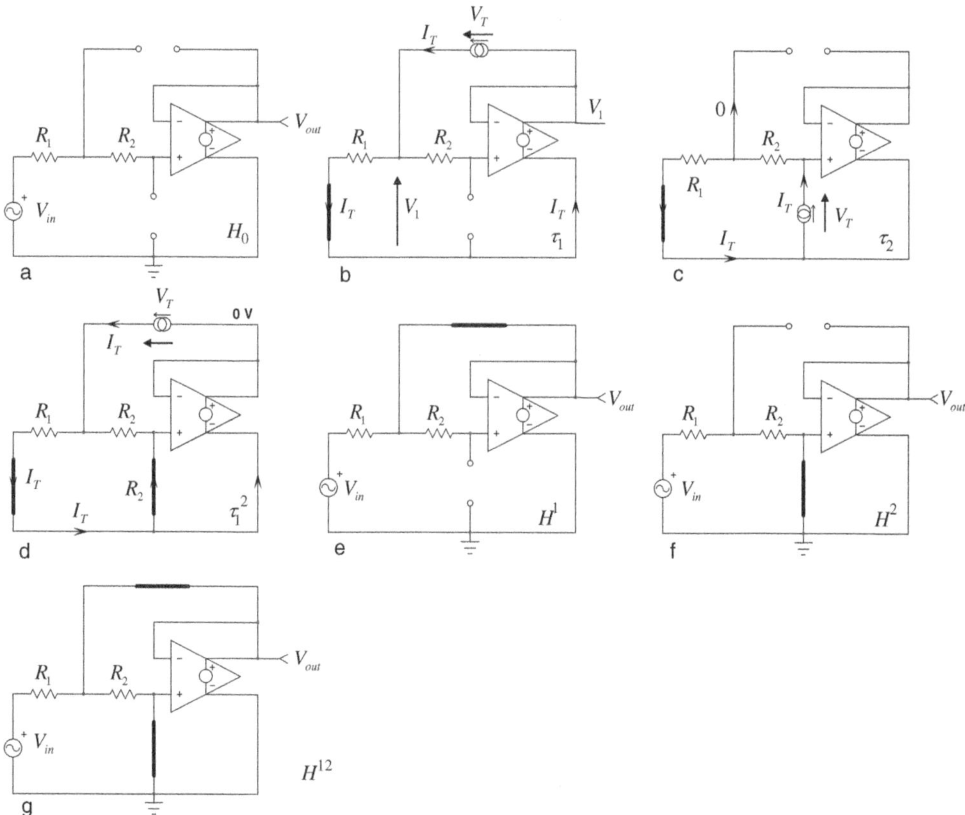

Figure 4.84 By breaking the analysis into individual drawings, the filter transfer function is quickly defined.

its current in the parallel arrangement of R_1 and R_2. This third time constant is expressed as:

$$\tau_1^2 = C_1\left(R_1 \| R_2\right) \tag{4.277}$$

We can form the denominator D by organizing the time constants. A simplification occurs in s^2 as the term $R_1 + R_2$ appears in the numerator and the denominator:

$$D(s) = 1 + s(\tau_1 + \tau_2) + s^2\tau_1\tau_2^1 = 1 + sC_2(R_1 + R_2) + s^2R_1R_2C_1C_2 \tag{4.278}$$

This expression can be put under the canonical form:

$$D(s) = 1 + \frac{s}{\omega_0 Q} + \left(\frac{s}{\omega_0}\right)^2 \tag{4.279}$$

in which

$$Q = \frac{\sqrt{b_2}}{b_1} = \frac{\sqrt{R_1R_2C_1C_2}}{C_2(R_1 + R_2)} \tag{4.280}$$

If $R_1 = R_2 = R$ and $C_1 = C_2 = C$, then Q equals 0.5.
and

$$\omega_0 = \frac{1}{\sqrt{b_2}} = \frac{1}{\sqrt{R_1R_2C_1C_2}} \tag{4.281}$$

which equals $1/RC$ if $R_1 = R_2 = R$ and $C_1 = C_2 = C$.

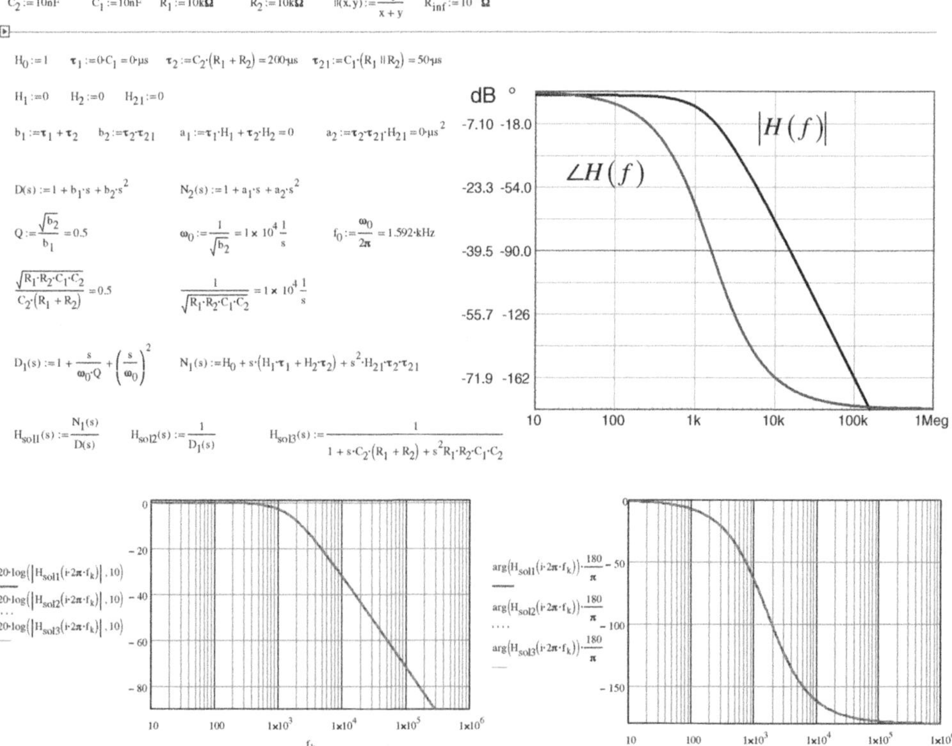

Figure 4.85 Mathcad® and SPICE show that our approach leads to the correct result.

Regarding zeros, the three sketches of Figure 4.84e to f return the same values:

$$H^1 = H^2 = H^{12} = 0 \quad (4.282)$$

Therefore:

$$N(s) = 1 \quad (4.283)$$

The transfer function of this filter is defined by:

$$G(s) = \frac{1}{1 + sC_2(R_1 + R_2) + s^2R_1C_1R_2C_2} \quad (4.284)$$

We can now test all of these expressions in a Mathcad® sheet as shown in Figure 4.85.

Problem 3:

First, check the gain for $s = 0$, then all the time constants you need to form the denominator. Then proceed with the gains for different C_1/C_2 states. However, the position of C_2 will slightly complicate the exercise as we will see. Figure 4.86 gathers all these individual calculations.

For $s = 0$ in Figure 4.86a, the gain is defined as:

$$H_0 = -\frac{R_3}{R_1} \quad (4.285)$$

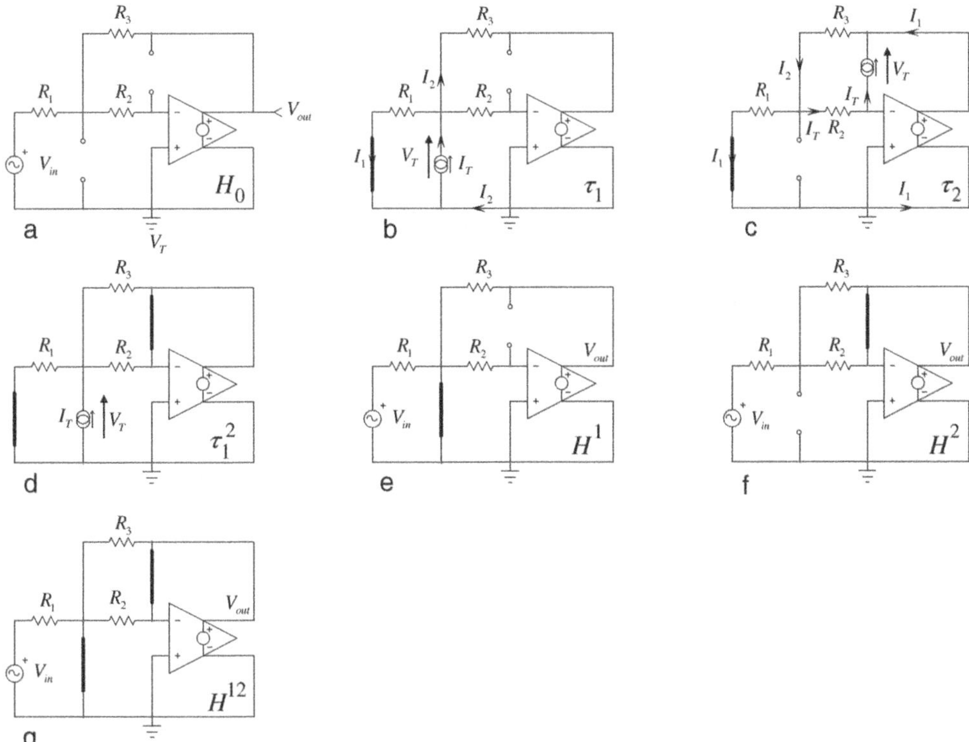

Figure 4.86 Breaking down the original circuit into simpler parts helps in determining the transfer function.

The first time constant is computed in Figure 4.86b. Assume the op amp has a finite gain, A_{OL}. The test current I_T splits into two currents, I_1 and I_2 satisfying the equation:

$$I_T = I_1 + I_2 \tag{4.286}$$

The voltage V_T is observed across resistor R_1, in which I_1 flows:

$$I_1 = \frac{V_T}{R_1} \tag{4.287}$$

The second current, I_2 depends on the voltage across R_3. The left side is biased to V_T while the right side is the op amp output voltage, V_{out}. The output voltage is also the voltage at the inverting pin, V_T, multiplied by the op amp open-loop gain A_{OL}. Thus:

$$I_2 = \frac{V_T - V_{out}}{R_3} = \frac{V_T + V_T A_{OL}}{R_3} = \frac{V_T(1 + A_{OL})}{R_3} \tag{4.288}$$

Now adding I_1 and I_2 to form I_T, we have:

$$I_T = \frac{V_T}{R_1} + V_T\left(\frac{1 + A_{OL}}{R_3}\right) \tag{4.289}$$

Factoring V_T gives:

$$I_T = V_T\left(\frac{1}{R_1} + \frac{1 + A_{OL}}{R_3}\right) \tag{4.290}$$

The resistance seen from C_1's terminals is thus:

$$R = \frac{V_T}{I_T} = \frac{1}{\frac{1}{R_1} + \frac{1 + A_{OL}}{R_3}} \tag{4.291}$$

When A_{OL} approaches infinity, this resistance becomes 0 Ω leading to:

$$\tau_1 = 0 \cdot C_1 \tag{4.292}$$

By inspection, we could also infer this result for an infinite open-loop gain given the fact that the non-inverting pin is grounded and the voltage at the inverting pin is 0 V. As no current flows in R_2, the V_T voltage is also 0 V.

The second time constant computation uses Figure 4.86c and, again starts with KCL:

$$I_T + I_1 = I_2 \tag{4.293}$$

The voltage across resistor R_1 is equal to the voltage drop across R_2 plus the voltage at the inverting pin, $V_{(-)}$:

$$R_1 I_1 = I_T R_2 + V_{(-)} \tag{4.294}$$

From this expression, we can extract the definition for I_1:

$$I_1 = \frac{V_{(-)} + I_T R_2}{R_1} \tag{4.295}$$

The voltage across R_1 also involves resistor R_3 and the op amp output:

$$R_1 I_1 = -R_3 I_2 + V_{out} = -R_3 I_2 - A_{OL} V_{(-)} \tag{4.296}$$

We can now equate (4.294) and (4.296):

$$I_T R_2 + V_{(-)} = -R_3 I_2 - A_{OL} V_{(-)} \tag{4.297}$$

and extract I_2:

$$I_2 = -\frac{V_{(-)} + A_{OL} V_{(-)} + I_T R_2}{R_3} \tag{4.298}$$

V_T, the voltage across the test generator is expressed as the op amp output voltage V_{out} minus the voltage at the inverting pin, $V_{(-)}$:

$$V_T = V_{out} - V_{(-)} = -A_{OL} V_{(-)} - V_{(-)} = -V_{(-)}(A_{OL} + 1) \tag{4.299}$$

The inverting pin voltage is obtained:

$$V_{(-)} = -\frac{V_T}{1 + A_{OL}} \tag{4.300}$$

If we substitute (4.300) in (4.298) and in (4.295) then we have:

$$I_2 = \frac{V_T - I_T R_2}{R_3} \tag{4.301}$$

and

$$I_1 = \frac{I_T R_2 - V_T + A_{OL} I_T R_2}{R_1 (A_{OL} + 1)} \tag{4.302}$$

We now substitute (4.301) and (4.302) in (4.293), we solve for $R = \frac{V_T}{I_T}$. After rearranging we have:

$$R = \frac{\frac{R_1 R_2}{A_{OL}} + \frac{R_1 R_3}{A_{OL}} + \frac{R_2 R_3}{A_{OL}} + R_1 R_2 + R_1 R_3 + R_2 R_3}{\frac{R_1}{A_{OL}} + \frac{R_3}{A_{OL}} + R_1} \tag{4.303}$$

When A_{OL} approaches infinity, the second time constant τ_2 can be expressed as:

$$\tau_2 = C_2 \left(R_2 + R_3 + \frac{R_2 R_3}{R_1} \right) \tag{4.304}$$

Regarding the s^2 term, as τ_1 is zero, we may encounter an indeterminacy if we calculate τ_2^1 so we evaluate τ_1^2instead in Figure 4.86d. In this sketch, you see that the op amp is configured in an inverting configuration whose output biases the right terminal of R_3. The gain of an op amp-based inverter is the feedback resistance divided by the resistance connected to the inverting pin, R_2 in this case. However, the feedback element is C_2 which is shorted in this mode. Therefore, the op amp gain is 0 and its output voltage is 0 V. As V_{out} and $V_{(-)}$ are connected, the inverting input is also 0 V. As a result, R_3 and R_2's right terminals are grounded. The current generator biases a node made of three paralleled resistors. Therefore:

$$\tau_1^2 = C_1 \left(R_1 \| R_2 \| R_3 \right) \tag{4.305}$$

We can form the denominator D by organizing the time constants:

$$\begin{aligned} D(s) &= 1 + s(\tau_1 + \tau_2) + s^2 \tau_1 \tau_2^1 \\ &= 1 + sC_2 \left(R_2 + R_3 + \frac{R_2 R_3}{R_1} \right) + s^2 \left[C_2 \left(R_2 + R_3 + \frac{R_2 R_3}{R_1} \right) C_1 \left(R_1 \| R_2 \| R_3 \right) \right] \end{aligned} \tag{4.306}$$

This expression can be put under the canonical form:

$$D(s) = 1 + \frac{s}{\omega_0 Q} + \left(\frac{s}{\omega_0}\right)^2 \tag{4.307}$$

in which

$$Q = \frac{\sqrt{b_2}}{b_1} = \sqrt{\frac{C_1 (R_1 \| R_2 \| R_3)}{C_2\left(R_2 + R_3 + \frac{R_2 R_3}{R_1}\right)}} \tag{4.308}$$

If $R_1 = R_2 = R_3$ then Q equals

$$Q = \frac{1}{3}\sqrt{\frac{C_1}{C_2}} \tag{4.309}$$

and if $C_1 = C_2 = C$, then Q equals 0.33:

$$\omega_0 = \frac{1}{\sqrt{b_2}} = \frac{1}{\sqrt{C_2\left(R_2 + R_3 + \frac{R_2 R_3}{R_1}\right) C_1 (R_1 \| R_2 \| R_3)}} \tag{4.310}$$

which equals $1/RC$ if $R_1 = R_2 = R_3 = R$ and $C_1 = C_2 = C$.

For the numerator, sketches in Figure 4.86f to g, once again, shows that all transfer functions H return 0:

$$H^1 = H^2 = H^{21} = 0 \tag{4.311}$$

Therefore:

$$N(s) = 1 \tag{4.312}$$

The transfer function of this filter is defined by:

$$\begin{aligned} G(s) &= -\frac{R_3}{R_1}\frac{1}{1 + sC_2\left(R_2 + R_3 + \frac{R_2 R_3}{R_1}\right) + s^2\left[C_2\left(R_2 + R_3 + \frac{R_2 R_3}{R_1}\right) C_1 (R_1 \| R_2 \| R_3)\right]} \\ &= G_0 \frac{1}{1 + \frac{s}{\omega_0 Q} + \left(\frac{s}{\omega_0}\right)^2} \end{aligned} \tag{4.313}$$

with Q and ω_0 respectively given by (4.308) and (4.310) while G_0 is defined by (4.285).

We can now test all of these expressions in a Mathcad® sheet as shown in Figure 4.87. We have purposely not tuned the filter (resistors and capacitors are not of equal values) to make it peak at a resonant frequency f_M calculated by the Mathcad® sheet. All curves nicely match the SPICE simulation.

If you are interested in filter calculations, especially these Sallen-Key filters, [1] points to a website where the author has automated calculations for these filters and also for plenty of other configurations. In [2] you will find an interesting discussion about ‘bootstrapping’, a technique widely used in active filters.

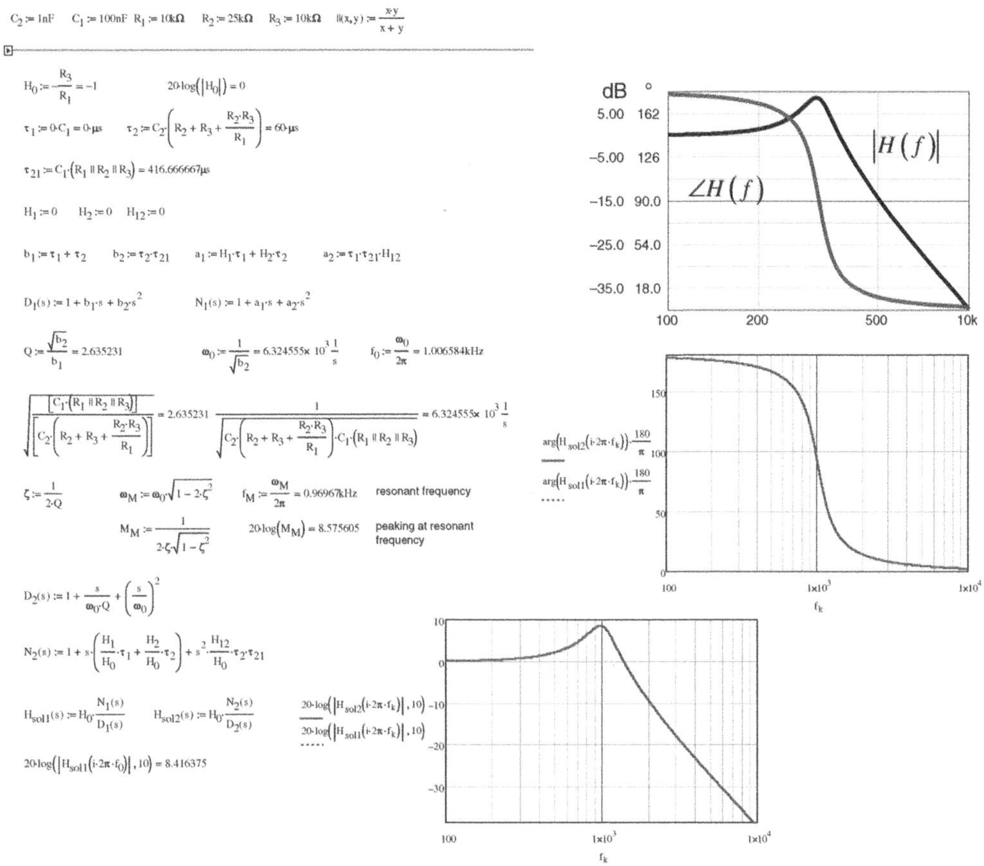

$C_2 := 1nF \quad C_1 := 100nF \quad R_1 := 10k\Omega \quad R_2 := 25k\Omega \quad R_3 := 10k\Omega \quad \|(x,y) := \frac{x \cdot y}{x+y}$

$H_0 := -\frac{R_3}{R_1} = -1 \qquad 20 \cdot \log(|H_0|) = 0$

$\tau_1 := 0 \cdot C_1 = 0 \cdot \mu s \qquad \tau_2 := C_2 \cdot \left(R_2 + R_3 + \frac{R_2 \cdot R_3}{R_1}\right) = 60 \cdot \mu s$

$\tau_{21} := C_1 \cdot (R_1 \| R_2 \| R_3) = 416.666667 \mu s$

$H_1 := 0 \quad H_2 := 0 \quad H_{12} := 0$

$b_1 := \tau_1 + \tau_2 \quad b_2 := \tau_2 \cdot \tau_{21} \quad a_1 := H_1 \cdot \tau_1 + H_2 \cdot \tau_2 \quad a_2 := \tau_1 \cdot \tau_{21} \cdot H_{12}$

$D_1(s) := 1 + b_1 \cdot s + b_2 \cdot s^2 \qquad N_1(s) := 1 + a_1 \cdot s + a_2 \cdot s^2$

$Q := \frac{\sqrt{b_2}}{b_1} = 2.635231 \qquad \omega_0 := \frac{1}{\sqrt{b_2}} = 6.324555 \times 10^3 \frac{1}{s} \qquad f_0 := \frac{\omega_0}{2\pi} = 1.006584 kHz$

$\sqrt{\frac{[C_1 \cdot (R_1 \| R_2 \| R_3)]}{\left[C_2 \cdot \left(R_2 + R_3 + \frac{R_2 \cdot R_3}{R_1}\right)\right]}} = 2.635231 \qquad \frac{1}{\sqrt{C_2 \cdot \left(R_2 + R_3 + \frac{R_2 \cdot R_3}{R_1}\right) \cdot C_1 \cdot (R_1 \| R_2 \| R_3)}} = 6.324555 \times 10^3 \frac{1}{s}$

$\zeta := \frac{1}{2 \cdot Q} \qquad \omega_M := \omega_0 \cdot \sqrt{1 - 2 \cdot \zeta^2} \qquad f_M := \frac{\omega_M}{2\pi} = 0.96967 kHz$ resonant frequency

$M_M := \frac{1}{2 \cdot \zeta \cdot \sqrt{1 - \zeta^2}} \qquad 20 \cdot \log(M_M) = 8.575605$ peaking at resonant frequency

$D_2(s) := 1 + \frac{s}{\omega_0 \cdot Q} + \left(\frac{s}{\omega_0}\right)^2$

$N_2(s) := 1 + s \cdot \left(\frac{H_1}{H_0} \cdot \tau_1 + \frac{H_2}{H_0} \cdot \tau_2\right) + s^2 \cdot \frac{H_{12}}{H_0} \cdot \tau_2 \cdot \tau_{21}$

$H_{sol1}(s) := H_0 \cdot \frac{N_1(s)}{D_1(s)} \qquad H_{sol2}(s) := H_0 \cdot \frac{N_2(s)}{D_2(s)}$

$20 \cdot \log(|H_{sol1}(i \cdot 2\pi \cdot f_0)|, 10) = 8.416375$

Figure 4.87 All expressions match the SPICE simulation. The filter is purposely not tuned to show the peaking at the resonant frequency ω_M defined in the figure.

Problem 4:

When determining an impedance expression, the excitation signal is the test current I_T driving the network while the response is the voltage across injection terminals, V_T. Figure 4.88a details this

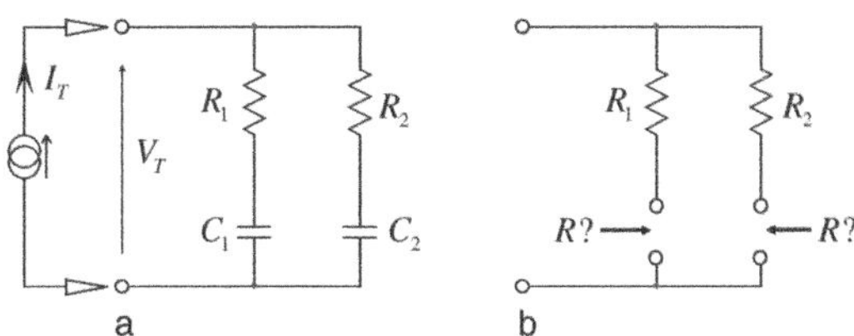

Figure 4.88 In this particular arrangement, the resistances offered by the capacitor terminals for $s = 0$ is infinite.

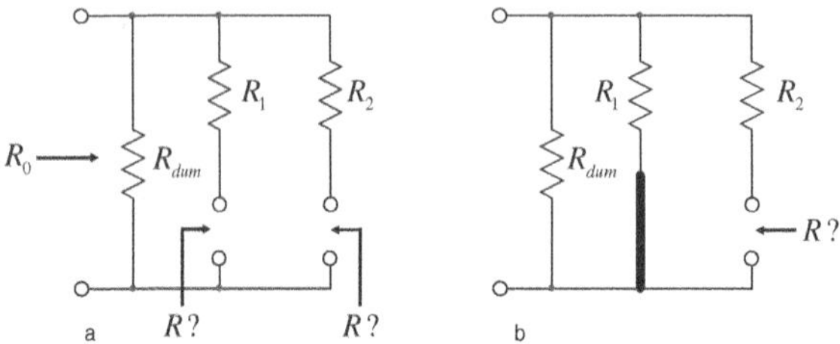

Figure 4.89 Adding an extra resistor R across the connection terminals helps fixing the resistance in dc.

concept. If we now try to proceed with the flow we have adopted so far, we can immediately see that the resistance R_0 for $s = 0$ is infinite:

$$R_0 = \infty \tag{4.314}$$

Then, if we observe the resistances offered by the capacitor's terminals while the excitation is set to 0 ($I_T = 0$ means an open-circuited generator), then both are infinite leading to infinite time constants as shown in Figure 4.88b. Infinite time constants often lead to indeterminacies depending how you later combine them. To go around this problem, we can simply install a dummy resistor R_{dum} across the excitation terminals as drawn in Figure 4.89.

This time, we can start our analysis again since the 0-Hz resistance is simply:v

$$R_0 = R_{dum} \tag{4.315}$$

The two time constants are immediate:

$$\tau_1 = C_1(R_1 + R_{dum}) \tag{4.316}$$

$$\tau_2 = C_2(R_2 + R_{dum}) \tag{4.317}$$

When C_1 is set to its high-frequency state, the network turns into that of Figure 4.89b and the time constant is simply:

$$\tau_2^1 = C_2(R_2 + R_1 \| R_{dum}) \tag{4.318}$$

The denominator D is thus written as:

$$D(s) = 1 + s[C_1(R_1 + R_{dum}) + C_2(R_2 + R_{dum})] + s^2 C_1 C_2 (R_1 + R_{dum})(R_2 + R_1 \| R_{dum}) \tag{4.319}$$

The numerator is found by inspection since the series arrangements of R_1-C_1 and R_2-C_2 in the transformed world can form a transformed short circuit nulling the response:

$$R_1 + \frac{1}{sC_1} = 0 \to s_{z_1} = -\frac{1}{R_1 C_1} \tag{4.320}$$

and

$$R_2 + \frac{1}{sC_2} = 0 \to s_{z_2} = -\frac{1}{R_2 C_2} \tag{4.321}$$

The numerator $N(s)$ is thus equal to

$$N(s) = (1 + sR_1C_1)(1 + sR_2C_2) \tag{4.322}$$

The complete transfer function is defined by gathering (4.315), (4.319) and (4.322):

$$Z(s) = R_0 \frac{N(s)}{D(s)} = R_{dum} \frac{(1 + sR_1C_1)(1 + sR_2C_2)}{1 + s[C_1(R_1 + R_{dum}) + C_2(R_2 + R_{dum})] + s^2 C_1 C_2 (R_1 + R_{dum})(R_2 + R_1 \| R_{dum})} \tag{4.323}$$

In this expression, we have the term R_{dum} which does not belong to the original schematic. We have purposely placed it to overcome the infinity terms we found. As it is placed in parallel with the network under study, we will now bring it to infinity and simplify the expression. Factoring R_{dum} in the denominator gives:

$$Z(s) = R_{dum}\frac{(1+sR_1C_1)(1+sR_2C_2)}{R_{dum}\left(\frac{1}{R_{dum}}+s\left[C_1\left(\frac{R_1}{R_{dum}}+1\right)+C_2\left(\frac{R_2}{R_{dum}}+1\right)\right]+s^2C_1C_2\left(\frac{R_1}{R_{dum}}+1\right)(R_2+R_1\|R_{dum})\right)} \tag{4.324}$$

R_{dum} is present in the numerator and the denominator so it can go away. Then, if R_{dum} approaches infinity, all terms divided by R_{dum} disappear and the term $R_1||R_{dum}$ becomes R_1. We obtain:

$$Z(s) = \frac{(1+sR_1C_1)(1+sR_2C_2)}{sC_1C_2+s^2C_1C_2(R_1+R_2)} \tag{4.325}$$

If this expression is correct, it does not fit the compact form we have seen so far. If we develop the numerator, we obtain:

$$N(s) = 1+s(R_1C_1+R_2C_2)+s^2R_1R_2C_1C_2 \tag{4.326}$$

It follows the form:

$$N(s) = 1+a_1s+a_2s^2 \tag{4.327}$$

We can determine a quality factor Q_N and a resonant angular frequency ω_{0N}:

$$Q_N = \frac{\sqrt{a_2}}{a_1} = \frac{\sqrt{C_1C_2R_1R_2}}{R_1C_1+R_2C_2} \tag{4.328}$$

$$\omega_{0N} = \frac{1}{\sqrt{C_1C_2R_1R_2}} \tag{4.329}$$

If we carry on and factor the s^2 terms in N and D, it should lead you to:

$$Z(s) = \frac{1+s(R_1C_1+R_2C_2)+s^2R_1R_2C_1C_2}{s(C_1+C_2)+s^2C_1C_2(R_1+R_2)} = \frac{s^2R_1R_2C_1C_2}{s^2C_1C_2(R_1+R_2)}\frac{1+\frac{s(R_1C_1+R_2C_2)}{s^2R_1R_2C_1C_2}+\frac{1}{s^2R_1R_2C_1C_2}}{1+\frac{s(C_1+C_2)}{s^2C_1C_2(R_1+R_2)}} \tag{4.330}$$

Running simplifications gives a function combining inversed zeros and poles:

$$Z(s) = \frac{R_1R_2}{R_1+R_2}\frac{1+\frac{R_1C_1+R_2C_2}{sR_1R_2C_1C_2}+\frac{1}{s^2R_1R_2C_1C_2}}{1+\frac{C_1+C_2}{sC_1C_2(R_1+R_2)}} \tag{4.331}$$

It follows the compact form introduced in Chapter 2:

$$Z(s) = R_\infty\frac{1+\frac{\omega_{0N}}{sQ}+\left(\frac{\omega_{0N}}{s}\right)^2}{1+\frac{\omega_p}{s}} \tag{4.332}$$

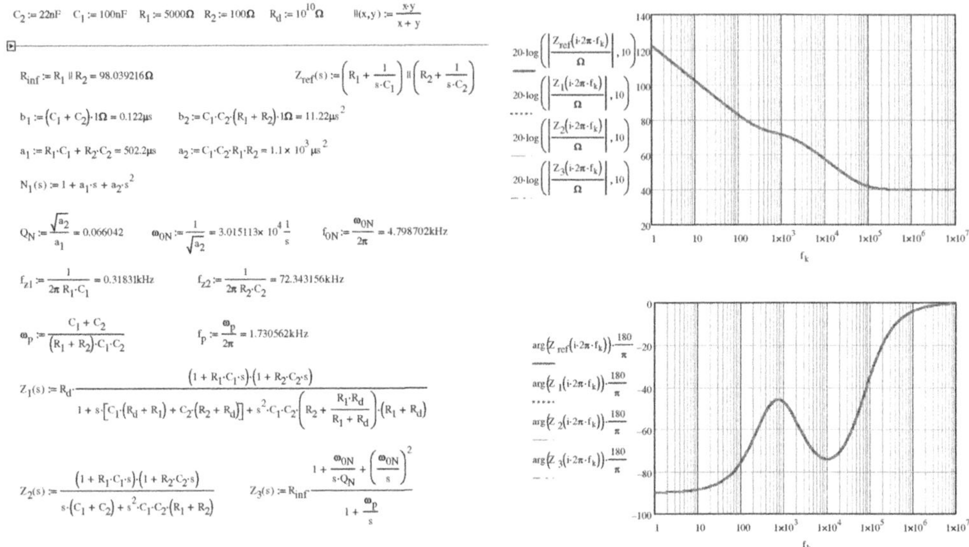

Figure 4.90 All curves perfectly superimpose with the reference expression, implying they are correct.

in which:

$$R_\infty = R_1 \| R_2 \tag{4.333}$$

$$\omega_p = \frac{C_1 + C_2}{(R_1 + R_2)C_1C_2} \tag{4.334}$$

while Q_N and ω_{ON} are respectively defined by (4.328) and (4.329). To test our expressions, we will compare their dynamic responses to the reference expression found by paralleling the two *RC* networks:

$$Z_{ref}(s) = \left(R_1 + \frac{1}{sC_1}\right) \| \left(R_2 + \frac{1}{sC_2}\right) \tag{4.335}$$

The Mathcad® sheet appears in Figure 4.90 and shows that our results are all correct.

What other option did we have in case we did not like the Figure 4.89 solution? The problem of infinity time constants occurred because the excitation was the current generator which becomes open-circuited when set to 0 A. A possibility also exists to calculate an admittance Y instead by replacing the current generator by a voltage source V_T as shown in Figure 4.91. V_T becomes the excitation while I_T becomes the response. In dc, Y_0 is now 0 and τ_1, τ_2 exist as in Figure 4.91b. Once all calculations are done, you take the inverse of the result and you have the impedance expression.

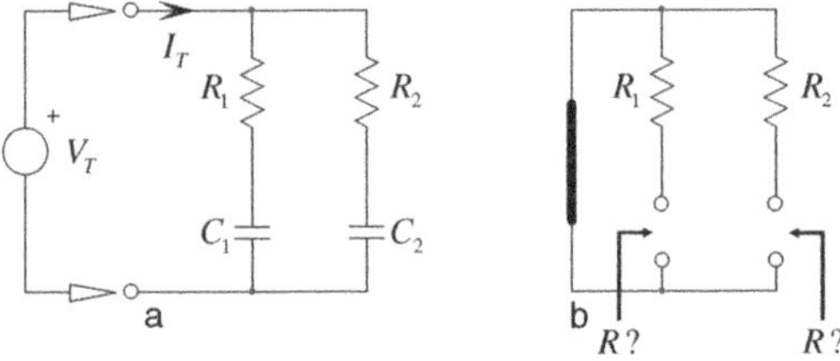

Figure 4.91 Calculating an admittance instead of an impedance helps removing the infinite resistance problem. V_T is now the driving variable while I_T is the response.

Problem 5:

This simple 2nd-order circuit transfer function can be quickly evaluated, even without going through sketches as we did before. With the experience you have gathered with the previous exercises, you should be able to derive the function by looking at the schematic only. The transfer function for $s=0$ is:

$$H_0 = 1 \tag{4.336}$$

Then, if you short the excitation and look at the resistance driving C_1, the first time constant is:

$$\tau_1 = C_1(R_1 + r_C) \tag{4.337}$$

The second time constant is even simpler as R_1 is alone in the loop:

$$\tau_2 = R_1 C_2 \tag{4.338}$$

Finally, if C_1 is shorted, what is the resistance offered by C_2's terminals? The parallel arrangement of R_1 and r_C:

$$\tau_2^1 = C_2\left(R_1 \| r_C\right) \tag{4.339}$$

The denominator $D(s)$ is thus:

$$D(s) = 1 + b_1 s + b_2 s^2 = 1 + s[C_1(R_1 + r_C) + C_2 R_1] + s^2 C_1 C_2 R_1 r_C \tag{4.340}$$

This expression can be put under the canonical form:

$$D(s) = 1 + \frac{s}{\omega_0 Q} + \left(\frac{s}{\omega_0}\right)^2 \tag{4.341}$$

in which after rearranging and simplifying expressions you will find:

$$Q = \frac{\sqrt{b_2}}{b_1} = \frac{\sqrt{C_1 C_2 R_1 r_C}}{R_1(C_1 + C_2) + C_1 r_C} \tag{4.342}$$

and

$$\omega_0 = \frac{1}{\sqrt{b_2}} = \frac{1}{\sqrt{R_1 r_C C_1 C_2}} \tag{4.343}$$

If Q is well below 1, the low-Q approximation holds and (4.341) can be rewritten as the product of two separated poles:

$$D(s) \approx \left(1 + \frac{s}{\omega_{p_1}}\right)\left(1 + \frac{s}{\omega_{p_2}}\right) \tag{4.344}$$

where

$$\omega_{p_1} = \frac{1}{b_1} = \frac{1}{R_1(C_1 + C_2) + r_C C_1} \tag{4.345}$$

$$\omega_{p_2} = \frac{b_1}{b_2} = \frac{R_1(C_1 + C_2) + C_1 r_C}{R_1 r_C C_1 C_2} \tag{4.346}$$

The numerator is determined by inspection, recognizing a transformed short at $s = s_z$ in the series combination of r_C and C_1:

$$r_C + \frac{1}{sC_1} = 0 \rightarrow s_z = -\frac{1}{r_C C_1} \tag{4.347}$$

Otherwise stated:

$$N(s) = 1 + \frac{s}{\omega_z} \tag{4.348}$$

with

$$\omega_z = \frac{1}{r_C C_1} \tag{4.349}$$

The complete transfer function is given by:

$$H(s) = \frac{1 + s r_C C_1}{1 + s[C_1(R_1 + r_C) + C_2 R_1] + s^2 C_1 C_2 R_1 r_C} \approx \frac{1 + \frac{s}{\omega_z}}{\left(1 + \frac{s}{\omega_{p_1}}\right)\left(1 + \frac{s}{\omega_{p_2}}\right)} \tag{4.350}$$

To test our expressions, the raw transfer function is obtained by considering the impedance divider brought by r_C-C_1 paralleled with C_2 and resistor R_1:

$$H(s) = \frac{\left(r_C + \frac{1}{sC_1}\right) \| \frac{1}{sC_2}}{\left(r_C + \frac{1}{sC_1}\right) \| \frac{1}{sC_2} + R_1} \tag{4.351}$$

We have captured these expressions in the Mathcad® sheet shown in Figure 4.92 and results are all in agreement.

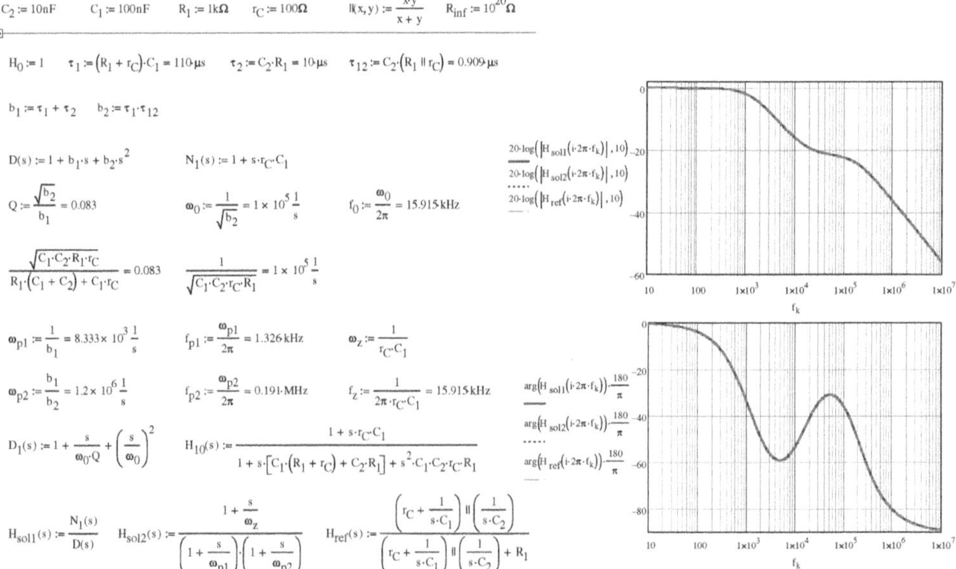

Figure 4.92 All transfer functions dynamic responses are equal.

Problem 6:

The buck converter output impedance is a classic in switching power supplies analysis. The equivalent schematic appears in Figure 4.93. The current source is the excitation signal while the response is V_T, the voltage across the connecting terminals. If we start by the numerator, you immediately identify two networks which could null the response. If Z_1 and Z_2 when observed at $s = s_z$ become a transformed short, the response disappears, $V_T = 0$. By expressing these impedances and solving for the condition at which they are a short circuit, you have the zero positions:

$$Z_1(s) = r_L + sL_1 = 0 \rightarrow s_{z_1} = -\frac{r_L}{L_1} \tag{4.352}$$

$$Z_2(s) = r_C + \frac{1}{sC_2} = 0 \rightarrow s_{z_2} = -\frac{1}{r_C C_2} \tag{4.353}$$

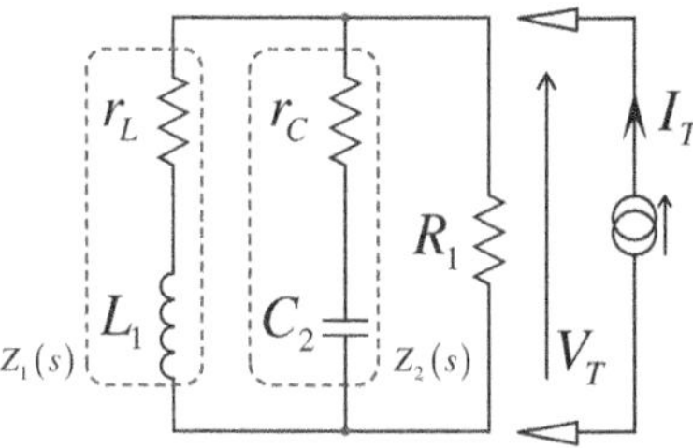

Figure 4.93 Shorting the input source lets you assess the buck output impedance.

which implies:

$$\omega_{z_1} = \frac{r_L}{L_1} \tag{4.354}$$

$$\omega_{z_2} = \frac{1}{r_C C_2} \tag{4.355}$$

The numerator $N(s)$ is thus defined as:

$$N(s) = \left(1 + s\frac{L_1}{r_L}\right)(1 + sr_C C_1) = \left(1 + \frac{s}{\omega_{z_1}}\right)\left(1 + \frac{s}{\omega_{z_2}}\right) \tag{4.356}$$

The denominator is determined following the sketches from Figure 4.94.

The resistance R_0 seen for $s = 0$ is obtained from Figure 4.94a:

$$R_0 = r_L \| R_1 \tag{4.357}$$

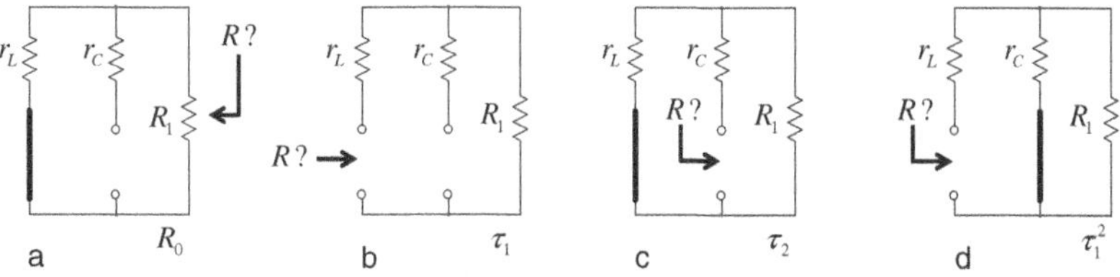

Figure 4.94 All transfer function dynamic responses are equal.

The first time constant used the Figure 4.94b sketch:

$$\tau_1 = \frac{L_1}{r_L + R_1} \tag{4.358}$$

while the second uses Figure 4.94c:

$$\tau_2 = \left(r_C + r_L \| R_1\right) C_2 \tag{4.359}$$

Finally, the high-frequency term combines τ_2 with:

$$\tau_1^2 = \frac{L_1}{r_L + r_C \| R_1} \tag{4.360}$$

We could have used τ_1 instead to find:

$$\tau_2^1 = (r_C + R_1) C_2 \tag{4.361}$$

We can now form the denominator:

$$D(s) = 1 + b_1 s + b_2 s^2 = 1 + s\left[\frac{L_1}{r_L + R_1} + C_2\left(r_C + r_L \| R_1\right)\right] + s^2 \frac{L_1}{r_L + R_1} C_2 (r_C + R_1) \tag{4.362}$$

This expression can be put under the canonical form:

$$D(s) = 1 + \frac{s}{\omega_0 Q} + \left(\frac{s}{\omega_0}\right)^2 \tag{4.363}$$

After rearranging and simplifying expressions you will find:

$$Q = \frac{\sqrt{b_2}}{b_1} = \frac{L_1 C_2 \omega_0 (r_C + R_1)}{L_1 + C_2[r_L r_C + R_1(r_L + r_C)]} \tag{4.364}$$

And:

$$\omega_0 = \frac{1}{\sqrt{b_2}} = \frac{1}{\sqrt{L_1 C_2}} \sqrt{\frac{R_1 + r_L}{R_1 + r_C}} \tag{4.365}$$

The transfer function can now be assembled and is defined by:

$$D(s) = \left(r_L \| R_1\right) \frac{\left(1 + s\frac{L_1}{r_L}\right)(1 + s r_C C_2)}{1 + s\left[\frac{L_1}{r_L + R_1} + C_2\left(r_C + r_L \| R_1\right)\right] + s^2 \frac{L_1}{r_L + R_1} C_2 (r_C + R_1)} = R_0 \frac{\left(1 + \frac{s}{\omega_{z_1}}\right)\left(1 + \frac{s}{\omega_{z_2}}\right)}{1 + \frac{s}{\omega_0 Q} + \left(\frac{s}{\omega_0}\right)^2} \tag{4.366}$$

Before checking these expressions, we need a reference transfer function obtained by observing Figure 4.93. The network impedance is the parallel arrangement of $Z_1(s)$, $Z_2(s)$ and R_1:

$$Z_{ref}(s) = (r_L + sL_1) \| \left(r_C + \frac{1}{sC_2}\right) \| R_1 \tag{4.367}$$

We now have all we need to build a Mathcad® sheet and test the validity of our results. This is what Figure 4.95 shows, confirming our calculations.

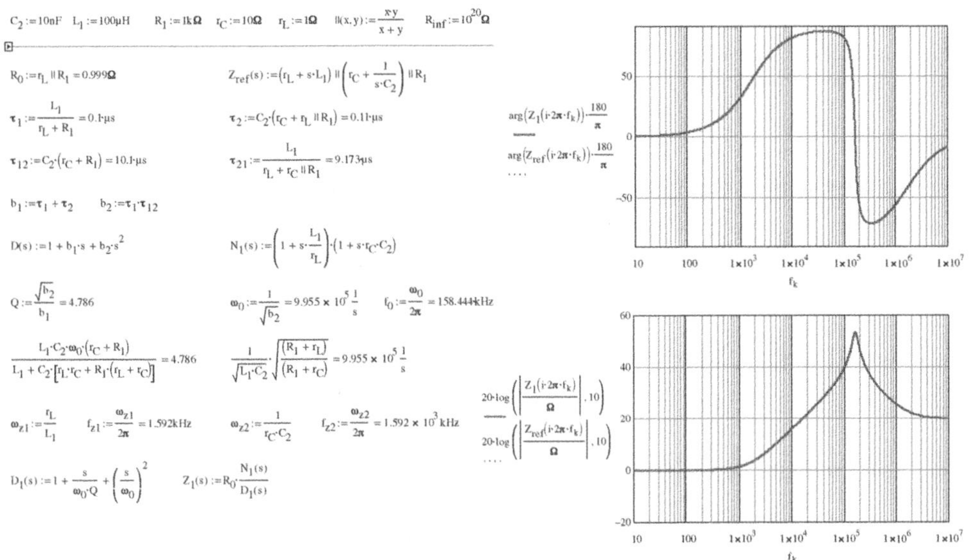

Figure 4.95 All transfer function dynamic responses are equal.

Problem 7:

For once, we will start with the raw transfer function of this network. The series-parallel arrangement of R_2-R_3 and C_1-C_2 form a divider with R_1 leading to:

$$H_{ref}(s) = \frac{\left[\left(R_2 \| \frac{1}{sC_1}\right) + \frac{1}{sC_2}\right] \| R_3}{\left[\left(R_2 \| \frac{1}{sC_1}\right) + \frac{1}{sC_2}\right] \| R_3 + R_1} \tag{4.368}$$

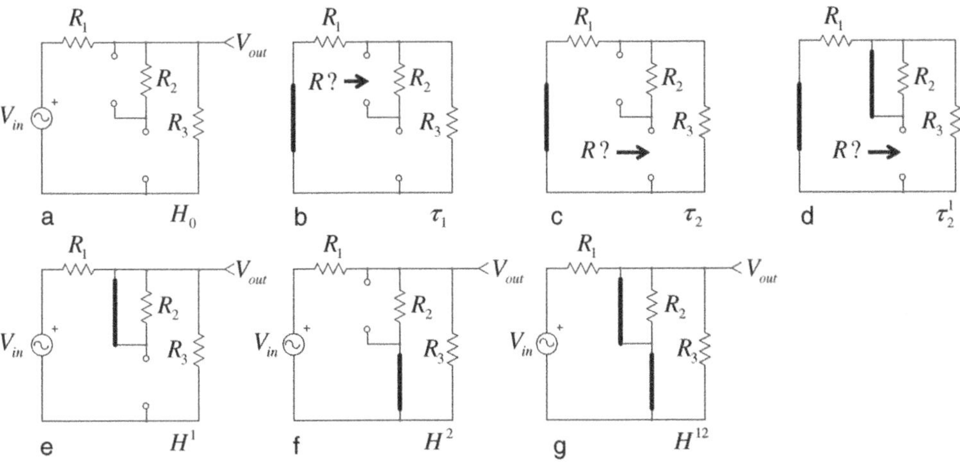

Figure 4.96 By going through a series of individual drawings, you can step-by-step identify the transfer function components and easily spot/correct an error afterwards.

Then we proceed with the series of sketches in which we start from the dc gain and calculate the various time constants we need. The drawing appears in Figure 4.96.

From Figure 4.96a, we have:

$$H_0 = \frac{R_3}{R_3 + R_1} \tag{4.369}$$

The time constant involving C_1 (Figure 4.96b) is:

$$\tau_1 = R_2 C_1 \tag{4.370}$$

For τ_2, we have:

$$\tau_2 = C_2 \left(R_2 + R_3 \| R_1\right) \tag{4.371}$$

Then when C_1 is replaced by a short circuit, if we look trough C_2's terminals, we obtain:

$$\tau_2^1 = C_2 \left(R_1 \| R_3\right) \tag{4.372}$$

We can now form the denominator:

$$D(s) = 1 + b_1 s + b_2 s^2 = 1 + s\left[R_2 C_1 + C_2 \left(R_2 + R_3 \| R_1\right)\right] + s^2 R_2 C_1 C_2 \left(R_1 \| R_3\right) \tag{4.373}$$

This expression can be put under the canonical form:

$$D(s) = 1 + \frac{s}{\omega_0 Q} + \left(\frac{s}{\omega_0}\right)^2 \tag{4.374}$$

After rearranging and simplifying expressions you will find:

$$Q = \frac{\sqrt{b_2}}{b_1} = \frac{\sqrt{C_1 C_2 R_2 \left(R_1 \| R_3\right)}}{C_1 R_2 + C_2 \left(R_2 + R_3 \| R_1\right)} \tag{4.375}$$

and

$$\omega_0 = \frac{1}{\sqrt{b_2}} = \frac{1}{\sqrt{C_1 C_2 R_2 \left(R_1 \| R_3\right)}} \tag{4.376}$$

Considering a low quality factor, the pole may be well separated. If this is the case, the denominator can be factored as two cascaded poles ω_{p1} and ω_{p2}. They are defined as:

$$\omega_{p_1} = \frac{1}{b_1} = \frac{1}{C_1 R_2 + C_2 \left(R_2 + R_3 \| R_1\right)} \tag{4.377}$$

$$\omega_{p_2} = \frac{b_1}{b_2} = \frac{C_1 R_2 + C_2 \left(R_2 + R_3 \| R_1\right)}{C_1 C_2 R_2 \left(R_1 \| R_3\right)} \tag{4.378}$$

The zeros are found by calculating three simple gains as shown in Figure 4.96e to g:

$$H^1 = \frac{R_3}{R_1 + R_3} \tag{4.379}$$

$$H^2 = \frac{R_3 \| R_2}{R_3 \| R_2 + R_1} \tag{4.380}$$

$$H^{12} = 0 \tag{4.381}$$

The numerator is quite simple given the fact that the 2nd-order term is zero:

$$N(s) = H_0\left[1 + s\left(\frac{H^1}{H_0}\tau_1 + \frac{H^2}{H_0}\tau_2\right) + s^2\frac{H^{12}}{H_0}\tau_1\tau_2^{12}\right] = H_0\left[1 + s\left(\frac{H^1}{H_0}\tau_1 + \frac{H^2}{H_0}\tau_2\right)\right] \tag{4.382}$$

If you develop and rearrange the factor of s, you should find a zero positioned at:

$$\omega_z = \frac{H_0}{H^1\tau_1 + H^2\tau_2} = \frac{1}{R_2(C_1 + C_2)} \tag{4.383}$$

We can now form the final transfer function which is equal to:

$$H(s) = \frac{R_3}{R_3 + R_1}\frac{1 + sR_2(C_1 + C_2)}{(1 + s[C_1R_2 + C_2(R_2 + R_3\|R_1)])\left(1 + s\left[\frac{C_1C_2R_2(R_1\|R_3)}{C_1R_2 + C_2(R_2 + R_3\|R_1)}\right]\right)}$$
$$= H_0\frac{1 + \frac{s}{\omega_z}}{\left(1 + \frac{s}{\omega_{p_1}}\right)\left(1 + \frac{s}{\omega_{p_2}}\right)} \tag{4.384}$$

We now have all we need to build a Mathcad® sheet and test the validity of our results. The reference function is defined in (4.368). Figure 4.97 shows all both dynamic responses, confirming our calculations.

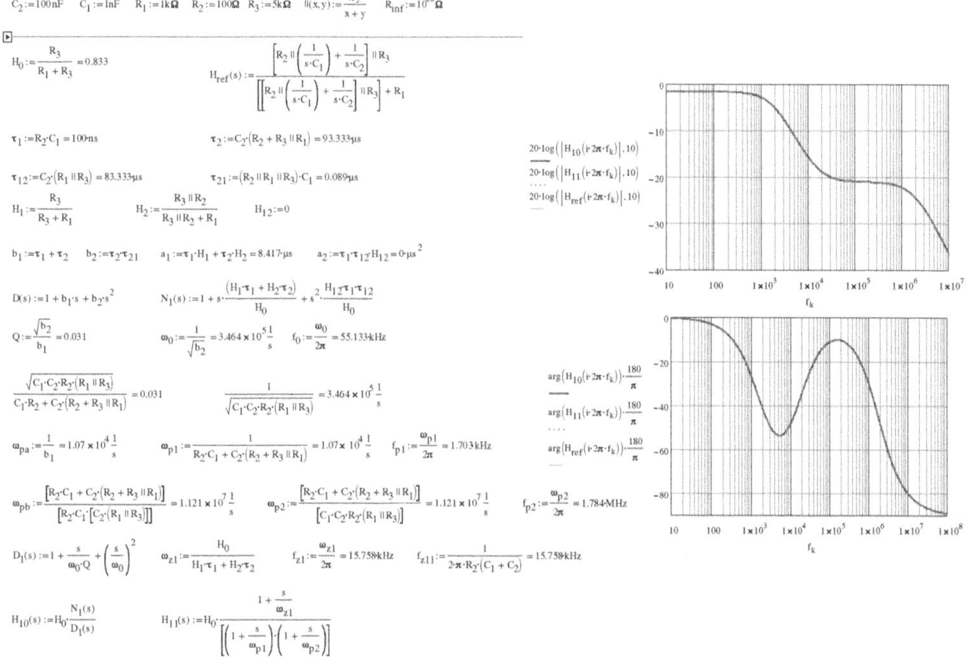

Figure 4.97 Mathcad® shows us good agreement among the different transfer functions.

Problem 8:

This is a classical circuit in which two RC networks are cascaded. What is the transfer function V_{in} to V_{out1} and V_{in} to V_{out2}? To find this using fast analytical techniques, a drawing with individual sketches is proposed in Figure 4.98. From these drawings, we can calculate the various time constants.

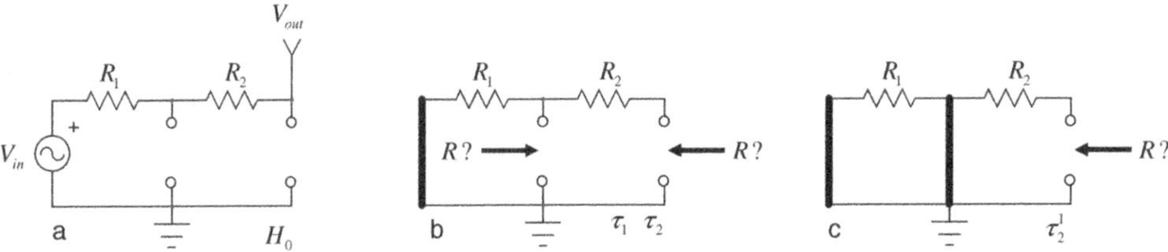

Figure 4.98 The transfer function is quickly obtained after only three small sketches.

If we start with sketch (a), the dc gain is simply:

$$H_0 = 1 \tag{4.385}$$

In Figure 4.98b, both time constants are immediately determined:

$$\tau_1 = R_1 C_1 \tag{4.386}$$

$$\tau_2 = C_2(R_1 + R_2) \tag{4.387}$$

When C_1 is set in its high-frequency state, the last time constant is obtained:

$$\tau_2^1 = R_2 C_2 \tag{4.388}$$

We can now form the denominator:

$$D(s) = 1 + b_1 s + b_2 s^2 = 1 + s[R_1 C_1 + C_2(R_1 + R_2)] + s^2 R_1 C_1 R_2 C_2 \tag{4.389}$$

This expression can be put under the canonical form:

$$D(s) = 1 + \frac{s}{\omega_0 Q} + \left(\frac{s}{\omega_0}\right)^2 \tag{4.390}$$

After rearranging and simplifying expressions you will find:

$$Q = \frac{\sqrt{b_2}}{b_1} = \frac{\sqrt{C_1 C_2 R_1 R_2}}{R_1 C_1 + C_2(R_2 + R_1)} \tag{4.391}$$

And:

$$\omega_0 = \frac{1}{\sqrt{b_2}} = \frac{1}{\sqrt{C_1 C_2 R_2 R_1}} \tag{4.392}$$

Considering a low quality factor, the pole may be well separated. If this is the case, the denominator can be factored as two cascaded poles ω_{p1} and ω_{p2}. They are defined as:

$$\omega_{p_1} = \frac{1}{b_1} = \frac{1}{R_1 C_1 + C_2(R_1 + R_2)} \tag{4.393}$$

$$\omega_{p_2} = \frac{b_1}{b_2} = \frac{R_1C_1 + C_2(R_1 + R_2)}{C_1C_2R_1R_2} \quad (4.394)$$

If C_1 or C_2 are individually or altogether shorted, there is no response thus this transfer function does not feature zeros. The complete first expression is given by:

$$H_1(s) = \frac{V_{out1}(s)}{V_{in}(s)} = \frac{1}{1 + s[R_1C_1 + C_2(R_1 + R_2)] + s^2R_1C_1R_2C_2} \approx \frac{1}{\left(1 + \frac{s}{\omega_{p_1}}\right)\left(1 + \frac{s}{\omega_{p_2}}\right)} \quad (4.395)$$

If we now consider the voltage across the first capacitor, C_1, then Figure 4.79 can be advantageously redrawn as shown in Figure 4.99.

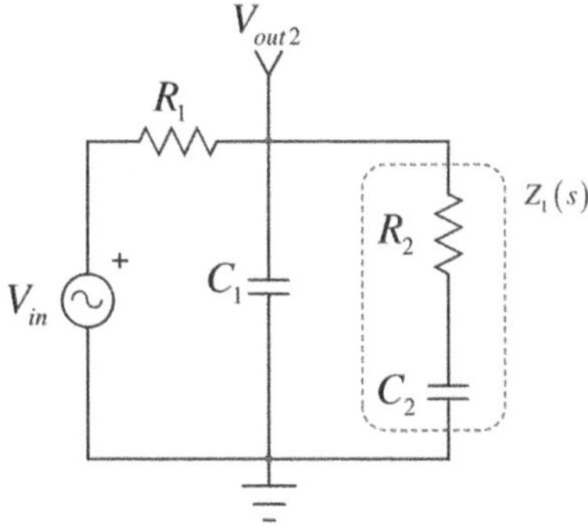

Figure 4.99 Rearranging the original figure immediately shows the presence of a zero.

In this figure, by inspection, you see that the series arrangement of R_2 and C_2 can become a transformed short at $s = s_z$. In this condition:

$$R_2 + \frac{1}{sC_2} = 0 \to s_z = -\frac{1}{R_2C_2} \quad (4.396)$$

implying a zero positioned at:

$$\omega_z = \frac{1}{R_2C_2} \quad (4.397)$$

The denominator does not change, all time constants remain the same. Natural frequencies of a network do not depend on where you select the input and output points as long as the network structure is unchanged when you turn the excitation source off. They solely depend on the circuit architecture, not on the excitation. Zeros, on the other hand, do depend on where you place the input/output points. As a result, probing V_{out} at different points in a circuit will not change the denominator but will surely affect the numerator. Here, $D(s)$ remains unchanged but given the zero defined by (4.397), the new transfer function becomes:

$$H_2(s) = \frac{V_{out2}(s)}{V_{in}(s)} = \frac{1 + sR_2C_2}{1 + s[R_1C_1 + C_2(R_1 + R_2)] + s^2R_1C_1R_2C_2} \approx \frac{1 + \frac{s}{\omega_z}}{\left(1 + \frac{s}{\omega_{p_1}}\right)\left(1 + \frac{s}{\omega_{p_2}}\right)} \quad (4.398)$$

Before we test our expressions, we need reference transfer functions. For H_1, we can obtain it by considering a Thévenin generator as in Figure 4.100. In this case, the transfer function is simply:

$$H_{ref1}(s) = \frac{1}{1 + sR_1C_1} \frac{\frac{1}{sC_2}}{\frac{1}{sC_2} + \left(R_1 \| \frac{1}{sC_1}\right) + R_2} \quad (4.399)$$

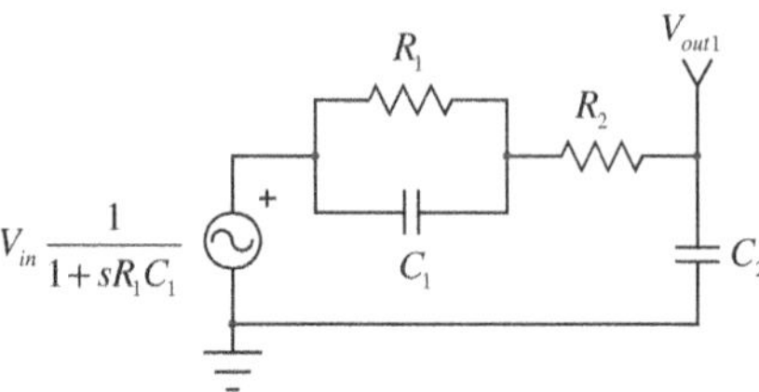

Figure 4.100 Considering a Thévenin generator helps simplify the circuit to derive the raw transfer function.

The second reference transfer function considers a simple impedance divider involving R_2-C_2 in series and paralleled with C_1 and resistor R_1:

$$H_{ref2}(s) = \frac{\left(\frac{1}{sC_1}\right) \| \left(R_2 + \frac{1}{sC_2}\right)}{\left(\frac{1}{sC_1}\right) \| \left(R_2 + \frac{1}{sC_2}\right) + R_1} \tag{4.400}$$

We have all expressions in hand now and we can capture them in a Mathcad® sheet as we have done so far. The results appear in Figure 4.101 while corresponding graphs are shown in Figure 4.102.

$$C_2 := 10\text{nF} \quad C_1 := 220\text{nF} \quad R_1 := 10\text{k}\Omega \quad R_2 := 1.5\text{k}\Omega \qquad \|(x,y) := \frac{x \cdot y}{x+y} \qquad R_{inf} := 10^{20}\Omega$$

$$H_{ref}(s) := \frac{1}{1+s \cdot R_1 \cdot C_1} \cdot \frac{\frac{1}{s \cdot C_2}}{R_2 + R_1 \| \left(\frac{1}{s \cdot C_1}\right) + \frac{1}{s \cdot C_2}} \qquad H_0 := 1 \qquad H_{ref2}(s) := \frac{\left(\frac{1}{s \cdot C_1}\right) \| \left(R_2 + \frac{1}{s \cdot C_2}\right)}{\left(\frac{1}{s \cdot C_1}\right) \| \left(R_2 + \frac{1}{s \cdot C_2}\right) + R_1}$$

$$\tau_1 := R_1 \cdot C_1 = 2.2 \times 10^3 \cdot \mu s \qquad \tau_2 := C_2 \cdot (R_1 + R_2) = 115 \cdot \mu s \qquad \tau_{12} := R_2 \cdot C_2 = 15 \cdot \mu s$$

$$b_1 := \tau_1 + \tau_2 \qquad b_2 := \tau_1 \cdot \tau_{12}$$

$$D(s) := 1 + b_1 \cdot s + b_2 \cdot s^2$$

$$Q := \frac{\sqrt{b_2}}{b_1} = 0.078 \qquad \omega_0 := \frac{1}{\sqrt{b_2}} = 5.505 \times 10^3 \frac{1}{s} \qquad f_0 := \frac{\omega_0}{2\pi} = 876.119 \cdot \text{Hz}$$

$$\omega_{p1} := Q \cdot \omega_0 \qquad f_{p1} := \frac{\omega_{p1}}{2\pi} = 68.749 \cdot \text{Hz} \qquad f_{p11} := \frac{1}{2\pi \cdot \left[R_1 \cdot C_1 + (R_1 + R_2) \cdot C_2\right]} = 68.749 \cdot \text{Hz}$$

$$\omega_{p2} := \frac{\omega_0}{Q} \qquad f_{p2} := \frac{\omega_{p2}}{2\pi} = 11.165 \cdot \text{kHz} \qquad f_{p22} := \frac{\left[R_1 \cdot C_1 + (R_1 + R_2) \cdot C_2\right]}{2 \cdot \pi \cdot R_1 \cdot C_1 \cdot R_2 \cdot C_2} = 11.165 \cdot \text{kHz}$$

$$D_1(s) := 1 + \frac{s}{\omega_0 \cdot Q} + \left(\frac{s}{\omega_0}\right)^2 \qquad D_2(s) := \left(1 + \frac{s}{\omega_{p1}}\right) \cdot \left(1 + \frac{s}{\omega_{p2}}\right) \qquad \omega_{z1} := \frac{1}{R_2 \cdot C_2}$$

$$H_{sol1}(s) := H_0 \cdot \frac{1}{D(s)} \qquad H_{sol2}(s) := H_0 \cdot \frac{1}{D_2(s)} \qquad H_2(s) := H_0 \cdot \frac{1 + \frac{s}{\omega_{z1}}}{D_2(s)}$$

Figure 4.101 The Mathcad® sheet contains all the expressions we have derived.

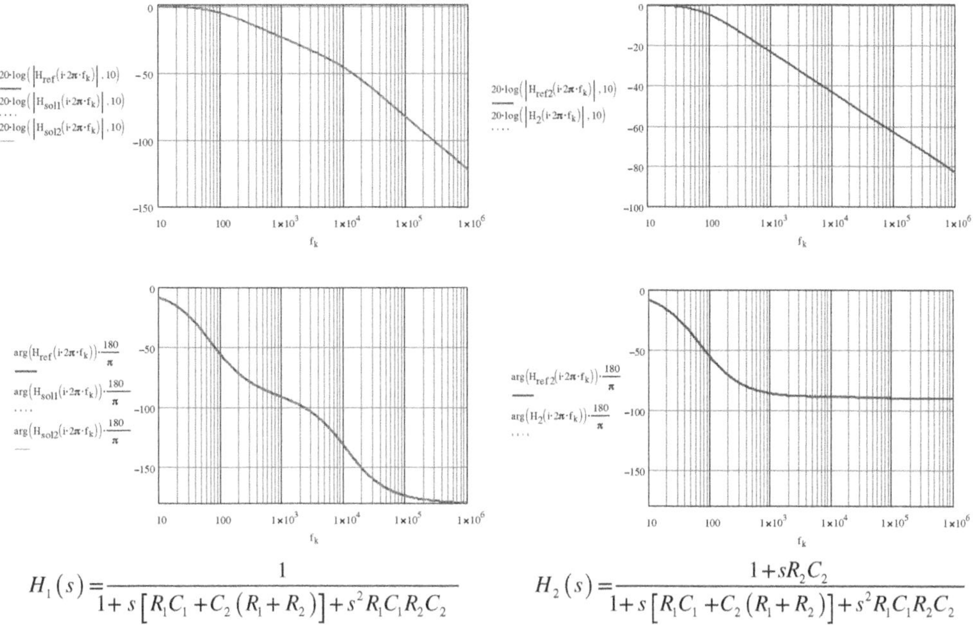

$H_1(s) = \frac{1}{1+s[R_1C_1+C_2(R_1+R_2)]+s^2R_1C_1R_2C_2}$
$H_2(s) = \frac{1+sR_2C_2}{1+s[R_1C_1+C_2(R_1+R_2)]+s^2R_1C_1R_2C_2}$

Figure 4.102 Graphs show nicely superimposed curves from the derived expressions and the reference ones.

Problem 9:

This is a circuit used in a Wien bridge oscillator. The transfer function is determined by going through the steps described in Figure 4.103.

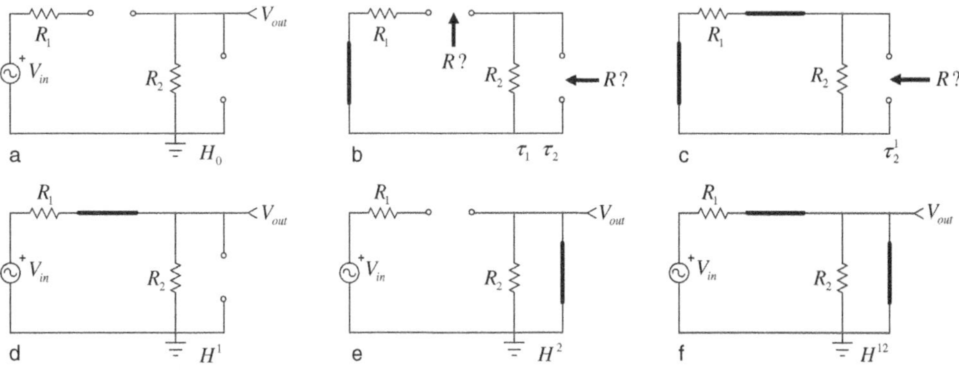

Figure 4.103 Individual sketches will get us to the transfer function quite quickly.

We start with the dc gain. Given the series capacitor C_1, the gain for $s = 0$ is:

$$H_0 = 0 \tag{4.401}$$

The time constants τ_1 and τ_2 are determined with Figure 4.103b drawing:

$$\tau_1 = C_1(R_1 + R_2) \tag{4.402}$$

$$\tau_2 = R_2C_2 \tag{4.403}$$

The higher order term is obtained easily with the help of Figure 4.103c:

$$\tau_2^1 = C_2(R_1 \| R_2) \tag{4.404}$$

We can now form the denominator and simplify the 2nd-order term:

$$D(s) = 1 + b_1 s + b_2 s^2 = 1 + s(\tau_1 + \tau_2) + s^2\tau_1\tau_2^1 = 1 + s[R_2C_2 + C_1(R_1 + R_2)] + s^2 R_1C_1R_2C_2 \tag{4.405}$$

This expression can be put under the canonical form:

$$D(s) = 1 + \frac{s}{\omega_0 Q} + \left(\frac{s}{\omega_0}\right)^2 \tag{4.406}$$

After rearranging and simplifying expressions you will find:

$$Q = \frac{\sqrt{b_2}}{b_1} = \frac{\sqrt{C_1C_2R_1R_2}}{R_2C_2 + C_1(R_1 + R_2)} \tag{4.407}$$

and

$$\omega_0 = \frac{1}{\sqrt{b_2}} = \frac{1}{\sqrt{C_1C_2R_2R_1}} \tag{4.408}$$

If we have $C_1 = C_2 = C$ and $R_1 = R_2 = R$, these definitions update to:

$$Q = \frac{RC}{3RC} = \frac{1}{3} \tag{4.409}$$

$$\omega_0 = \frac{1}{\sqrt{R^2C^2}} = \frac{1}{RC} \tag{4.410}$$

Regarding the numerator, sketches from Figure 4.103d to f lead to the following results:

$$H^1 = \frac{R_2}{R_1 + R_2} \tag{4.411}$$

$$H^2 = H^{12} = 0 \tag{4.412}$$

The numerator combines the following terms:

$$N(s) = H_0 + s(H^1\tau_1 + H^2\tau_2) + s^2 H^{12}\tau_1\tau_{12} = sH^1\tau_1 = s\frac{R_2}{R_1 + R_2}C_1(R_1 + R_2) = sR_2C_1 \tag{4.413}$$

We can now form the transfer function expression combining (4.413) and (4.405):

$$H(s) = \frac{sR_2C_1}{1 + s[R_2C_2 + C_1(R_1 + R_2)] + s^2R_1C_1R_2C_2} \tag{4.414}$$

This expression does not fit the *low-entropy* format we are used to. First, we factor sR_2C_1:

$$H(s) = \frac{1}{\frac{1}{sR_2C_1} + \frac{R_2C_2 + C_1(R_1 + R_2)}{R_2C_1} + sR_1C_2} \tag{4.415}$$

Then we factor the middle term and run simplifications:

$$H(s) = \frac{1}{\frac{R_2C_2 + C_1(R_1 + R_2)}{R_2C_1}\left[1 + \frac{1}{s[C_1(R_1 + R_2) + C_2R_2]} + s\frac{R_1C_2R_2C_1}{C_1(R_1 + R_2) + C_2R_2}\right]} \tag{4.416}$$

Revealing the leading term, we find:

$$H(s) = \frac{R_2C_1}{R_2C_2 + C_1(R_1 + R_2)} \frac{1}{1 + \frac{1}{s[C_1(R_1 + R_2) + C_2R_2]} + s\frac{R_1C_2R_2C_1}{C_1(R_1 + R_2) + C_2R_2}} \tag{4.417}$$

Implementing expressions suggested in Chapter 2, (4.417) can be rewritten in a more compact form as:

$$H(s) = H_{res}\frac{1}{1 + \left(\frac{s}{\omega_0} + \frac{\omega_0}{s}\right)Q} \tag{4.418}$$

in which:

$$H_{res} = \frac{R_2C_1}{R_2C_2 + C_1(R_1 + R_2)} \tag{4.419}$$

Again, if we have $C_1 = C_2 = C$ and $R_1 = R_2 = R$, H_{res} equals 0.333 or −9.54 dB. Before we test these expressions, we need a reference transfer function obtained from brute-force algebra. Looking at Figure 4.80, this raw transfer function is equal to:

$$H_{ref}(s) = \frac{R_2 \| \left(\frac{1}{sC_2}\right)}{R_2 \| \left(\frac{1}{sC_2}\right) + R_1 + \frac{1}{sC_1}} \tag{4.420}$$

Final curves are displayed in Figure 4.104 and show a 0° argument at the resonant frequency ω_0. If this circuit is assembled together with an op amp as shown in Figure 4.105 and the op amp exactly compensates the 9.54 dB insertion loss (a gain of 3), you have built a Wien bridge oscillator operating at around 1 kHz.

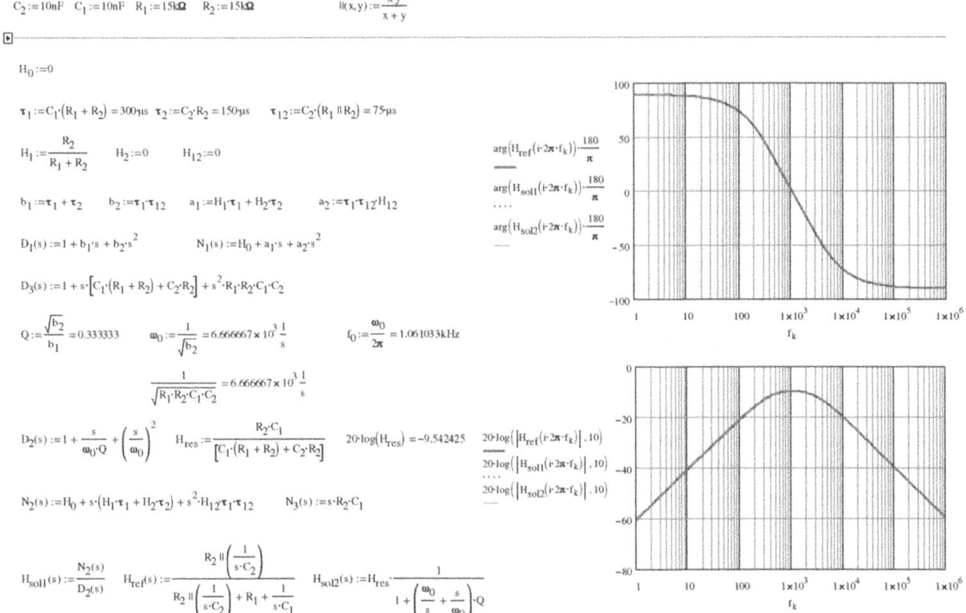

Figure 4.104 Mathcad® confirms the attenuation at the resonant frequency and shows a good agreement between the curves.

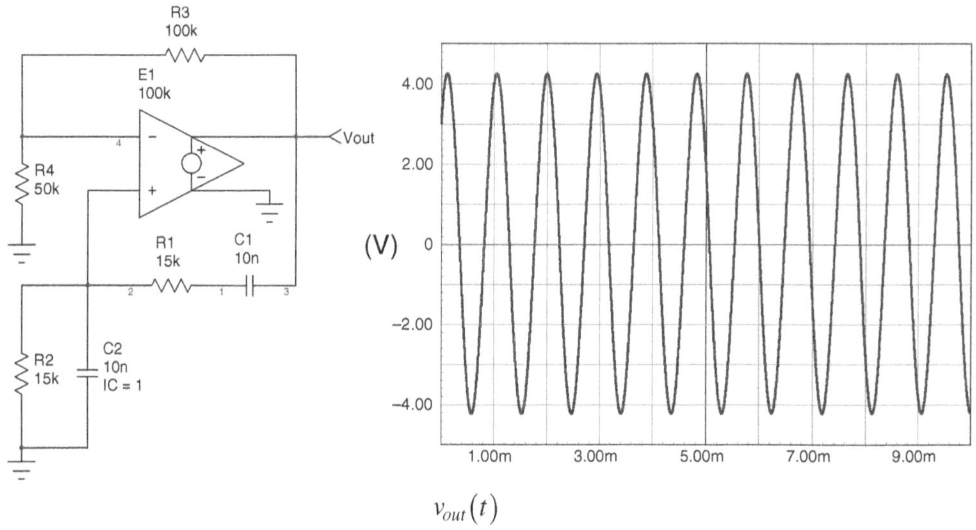

Figure 4.105 If the amplifier perfectly compensates the insertion loss, you have built an oscillator.

Problem 10:
This is our final exercise and the arrangement around the op amp looks quite complex. This is where you will see how quickly the fast analytical techniques let us derive the transfer function in a really easy way. I doubt you can beat it using KCL or KVL! Figure 4.106 gathers all the small sketches we have split the analysis into. First, the dc gain, H_0, from Figure 4.106a. As C_2 is an open circuit, V_{out} is equal to V_{in}. Thus:

$$H_0 = 1 \tag{4.421}$$

The first time constant involves capacitor C_1 and the test configuration is given in Figure 4.106b. The left side of the source is biased at 0 V while the other one connects to the low-side resistor R_3. Owing to the op amp virtual ground effect, despite the short-circuit brought by the excitation set to 0 V, current I_T splits into I_1 and I_2, to reappear in R_3. As the voltage across R_3 is V_T, the resistance 'seen' by capacitor C_1 is R_3 and the first time constant is:

$$\tau_1 = R_3C_1 \tag{4.422}$$

In Figure 4.106c, the test current circulates only through R_4 since the op amp output is 0 V. Therefore:

$$\tau_2 = R_4C_2 \tag{4.423}$$

The final time constant is selected while C_2 is a short circuit (high-frequency state) and the test circuit appears in Figure 4.106d. The test current splits in three currents. Again, all currents sum up in R_3 which implies a time constant similar to τ_1:

$$\tau_1^2 = C_1R_3 \tag{4.424}$$

We can now form the denominator by assembling the time constants we found:

$$D(s) = 1 + b_1s + b_2s^2 = 1 + s(\tau_1 + \tau_2) + s^2\tau_2\tau_1^2 = 1 + s[R_3C_1 + C_2R_4] + s^2C_2R_4C_1R_3 \tag{4.425}$$

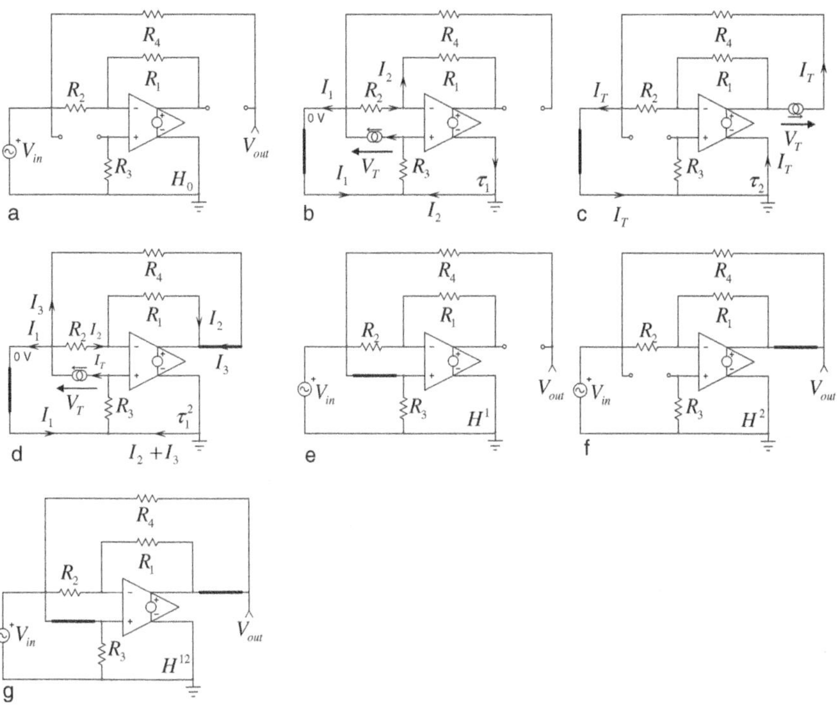

Figure 4.106 All steps are broken down into individual simple sketches.

This expression can be put under the canonical form:

$$D(s) = 1 + \frac{s}{\omega_0 Q} + \left(\frac{s}{\omega_0}\right)^2 \tag{4.426}$$

After rearranging and simplifying expressions you will find:

$$Q = \frac{\sqrt{b_2}}{b_1} = \frac{\sqrt{C_2 R_4 C_1 R_3}}{R_3 C_1 + C_2 R_4} \tag{4.427}$$

and

$$\omega_0 = \frac{1}{\sqrt{b_2}} = \frac{1}{\sqrt{C_2 R_4 C_1 R_3}} \tag{4.428}$$

If we have $C_1 = C_2 = C$ and $R_3 = R_4 = R$ then:

$$Q = \frac{RC}{2RC} = 0.5 \tag{4.429}$$

$$\omega_0 = \frac{1}{RC} \tag{4.430}$$

With a Q of 0.5, both roots are coincident and the denominator can be factored as:

$$D(s) = \left(1 + \frac{s}{\omega_0}\right)^2 \tag{4.431}$$

The numerator requires three gain calculations detailed in Figure 4.106e to g. The first one is 1 as C_2 is open:

$$H^1 = 1 \tag{4.432}$$

In Figure 4.106f, the output is that of an inverting configuration equal to

$$H^2 = -\frac{R_1}{R_2} \tag{4.433}$$

And finally, when both capacitors are set in their high-frequency states, Figure 4.106g, the gain is:

$$H^{21} = 1 \tag{4.434}$$

The denominator $N(s)$ follows the form:

$$\begin{aligned} N(s) &= H_0 + s\left(H^1\tau_1 + H^2\tau_2\right) + s^2 H^{21}\tau_2\tau_{21} \\ &= 1 + s\left(\tau_1 - \frac{R_1}{R_2}\tau_2\right) + s^2\tau_2\tau_1^2 = 1 + s\left(R_3C_1 - \frac{R_1}{R_2}R_4C_2\right) + s^2 C_2R_4C_1R_3 \end{aligned} \tag{4.435}$$

If we have $C_1 = C_2 = C$, $R_3 = R_4 = R$ and $R_1 = R_2$, then the numerator simplifies to:

$$N(s) = 1 + (sRC)^2 = 1 + \left(\frac{s}{\omega_0}\right)^2 \tag{4.436}$$

In this mode, the quality factor Q_N is infinite because the term:

$$a_1 = \tau_1 - \frac{R_1}{R_2}\tau_2 \tag{4.437}$$

is equal to 0 since τ_1 $-\tau_2$ are the same and we chose R_1 equal to R_2.

For $C_1 = C_2 = C$, $R_3 = R_4 = R$ and $R_1 = R_2$, the final transfer function can be formulated as:

$$G(s) = \frac{1 + \left(\frac{s}{\omega_0}\right)^2}{\left(1 + \frac{s}{\omega_0}\right)^2} \tag{4.438}$$

We can now capture these equations in a Mathcad® sheet and compare results with a SPICE simulation. To see the notch effect, you must increase the data point to 1000 or more. Component values adopted for the example form a 50-Hz notch filter that I found in reference [3]. As you can see in Figure 4.107, results show the perfect agreement with the SPICE graph of Figure 4.108.

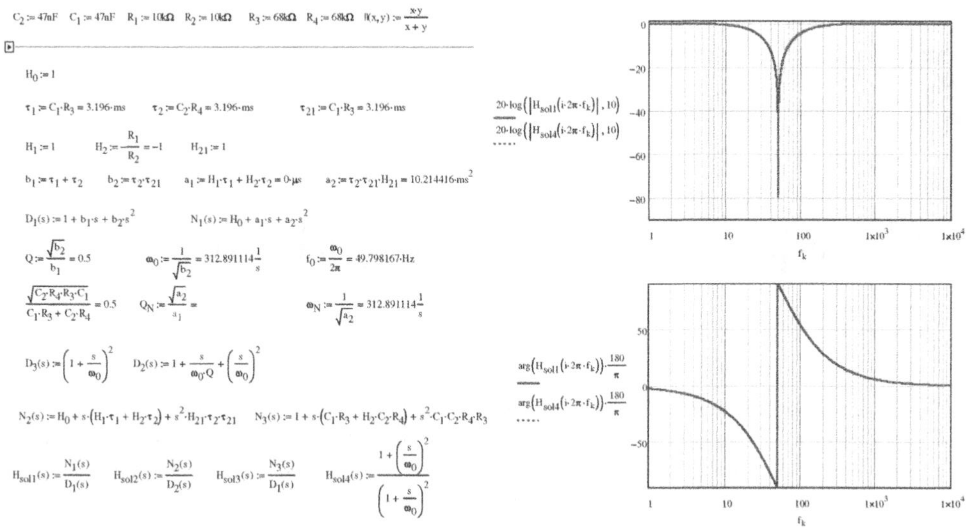

Figure 4.107 All curves perfectly agree with the SPICE simulation of this 50-Hz notch filter.

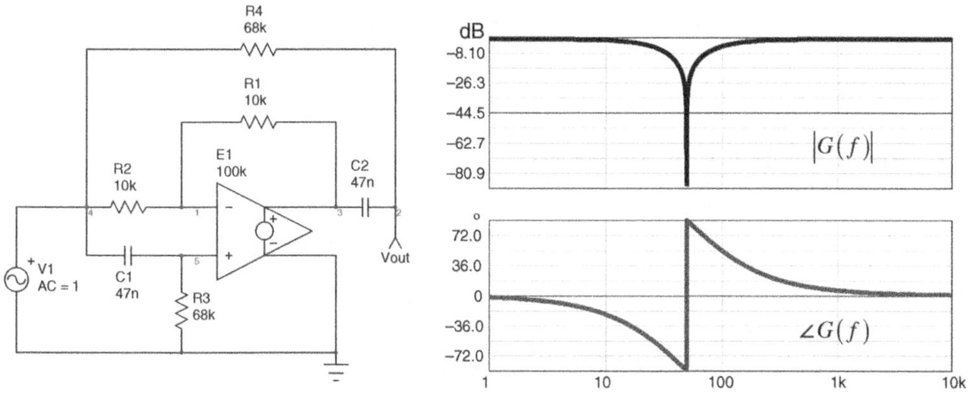

Figure 4.108 The SPICE simulation confirms the notch realization at 50 Hz.

References

1. http://sim.okawa-denshi.jp/en/Fkeisan.htm (last accessed 12/12/2015).
2. http://electronics.stackexchange.com/questions/26359/understanding-bootstrapping-in-analog-active-filters (last accessed 12/12/2015).
3. http://circuit-diagram.hqew.net/Notch-Filter$3a-The-Circuit%E2%80%99s-Diagram-and-The-Design-Formula_2729.html (last accessed 12/12/2015).

5

N^{th}-order Transfer Functions

In this chapter, we will extend the 2nd-order transfer function formulas that we introduced in Chapter 4 and apply them to higher-order networks. The idea remains the same regardless of the circuit complexity: find the various time constants when the excitation is turned off and determine the zeros by inspection, NDI or use the generalized form with simple gain calculations. As expressions combine more terms, determining high-order transfer functions requires patience and organization. Breaking the process into simple individual drawings is the way to go, helping to quickly identify a mistake when dynamic responses deviate from a reference expression or simulation. Carefully applying the aforementioned steps will lead you to complex but well-arranged expressions precisely unveiling the pole and zero positions of 3rd-order and higher order electrical circuits. We will start with a generalized n^{th}-order expression and quickly delve into complicated circuit analysis up to the 4th order.

5.1 From the 2EET to the *N*EET

The recipe for higher order circuits does not differ from what we have learned in the previous chapters. The reference gain is determined for $s = 0$ in which all capacitors are removed and inductors replaced by short circuits. The excitation is then set to zero and you 'look' into each energy-storing element to determine their driving resistance. This will lead you to the natural time constants of the circuit. An nth-order denominator obeys the following format:

$$D(s) = 1 + b_1 s + b_2 s^2 + b_3 s^3 + \dots + b_n s^n \tag{5.1}$$

In Chapter 1, we have seen that a typical transfer function combines a leading term multiplied by a ratio. In this approach, the leading term carries the unit, if any. From this fact, it follows that the ratio formed by $N(s)/D(s)$ must be unitless. If D is unitless in (5.1), then b_1 has the dimension of a time in seconds [s], b_2 has the dimension of time squared [s^2], b_3's unit is in cubic seconds [s^3], b_4's dimension is in seconds to the power 4 [s^4] and, finally, b_n's unit is in seconds to the power n.

Starting with b_1, a unit in seconds implies adding all time constants determined with the excitation signal set to 0. The generalized formula for b_1 is given as follows:

$$b_1 = \sum_{i_1=1}^{n} \tau_{i_1} \tag{5.2}$$

Linear Circuit Transfer Functions: An Introduction to Fast Analytical Techniques, First Edition. Christophe P. Basso.

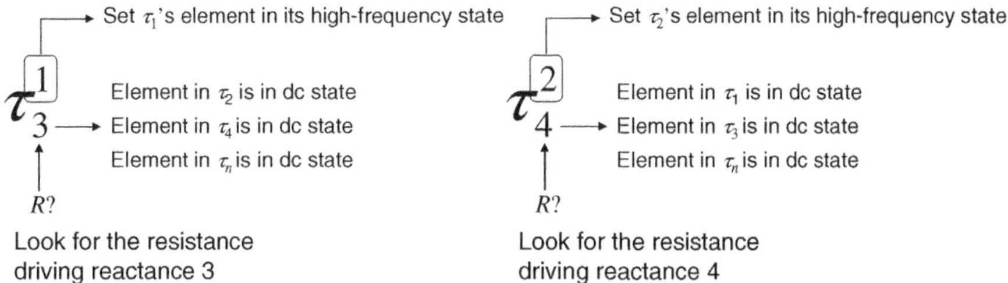

Figure 5.1 One energy-storing element is set in its high-frequency state while the others remain in their dc (reference) state.

There are n elements in b_1. Assume you work on a 4th-order network; then you combine four time constants as follows:

$$b_1 = \tau_1 + \tau_2 + \tau_3 + \tau_4 \tag{5.3}$$

For b_2, as the dimension is squared time, we sum up time constants products. We already defined what these products were in Chapter 2 such as $\tau_1\tau_2^1$ or $\tau_2\tau_1^2$: two elements among two are taken at a time to form the b_2 term in a 2nd-order network. Same principle with a high-order circuit except that all possible combinations between paired energy-storing elements must be covered; we still take two elements at a time but among n energy-storing elements. Figure 5.1 explains how to interpret this notation when several reactances are at stake: the reactance whose time constant number appears in the exponent is set in its high-frequency state and you look at the resistance driving the energy-storing element whose number appears in the subscript. During the calculation, all remaining reactances stay in their dc state.

How many combinations do we have for b_2? Because we select elements within a group of energy-storing elements, the number is determined by using a binomial coefficient. It is defined as:

$$\binom{n}{j} = \frac{n!}{j!(n-j)!} \tag{5.4}$$

in which n is the network order and j is the coefficient reference. j would be 2 for b_2, 3 for b_3 and so on until j equals n. For our b_2 term in a 4th-order network, (5.4) would give:

$$\binom{4}{2} = \frac{4!}{2!(4-2)!} = 6 \tag{5.5}$$

You start by distributing τ_1 present in (5.3):

$$b_{2a} = \tau_1\tau_2^1 + \tau_1\tau_3^1 + \tau_1\tau_4^1 \tag{5.6}$$

Then you pick τ_2. If you write $\tau_2\tau_1^2$ then you have a redundancy with $\tau_1\tau_2^1$ that you already covered in (5.6). Redundancies are easily identified if you follow Figure 5.2 illustration which also shows how to reshuffle the combinations. Reshuffling is interesting in case a) a combination leads to a complicated intermediate circuit or expression b) you have an indeterminacy.

You jump to the next terms as τ_3 and τ_4 have not been associated with τ_2 yet. Thus:

$$b_{2b} = \tau_2\tau_3^2 + \tau_2\tau_4^2 \tag{5.7}$$

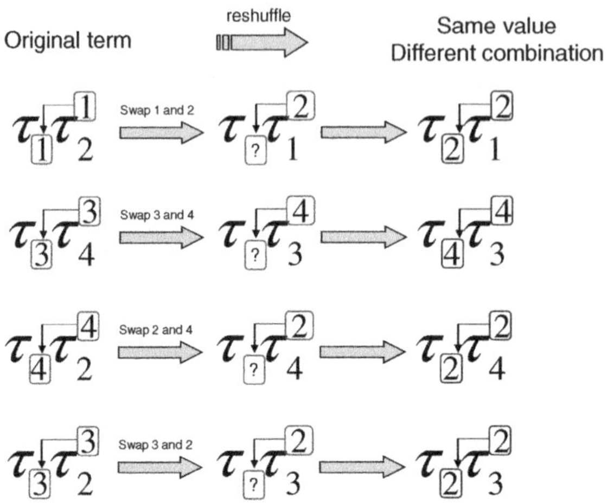

Figure 5.2 Reshuffling time constants is simple and helps remove indeterminacies or find simpler combinations.

Then go to τ_3 and distribute with the last term, τ_4:

$$b_{2c} = \tau_3\tau_4^3 \tag{5.8}$$

Should you continue distributing τ_3 with τ_2 and τ_1 for instance, respectively obtaining $\tau_3\tau_2^3$ or $\tau_3\tau_1^3$ then you would build redundancies with $\tau_2\tau_3^2$ or $\tau_1\tau_3^1$ that we already counted (see Figure 5.2). This is it, we have our elements for b_2 adding (5.6), (5.7) and (5.8):

$$b_2 = \tau_1\tau_2^1 + \tau_1\tau_3^1 + \tau_1\tau_4^1 + \tau_2\tau_3^2 + \tau_2\tau_4^2 + \tau_3\tau_4^3 \tag{5.9}$$

We have six terms as expected with (5.5). If you generalize this expression, you obtain:

$$b_2 = \sum_{i_1=1}^{n-1} \sum_{i_2=i_1+1}^{n} \tau_{i_1}\tau_{i_2}^{i_1} \tag{5.10}$$

This generalized formula does not really make sense because of the numerous reshuffling options. We give it here to show how to build the b_2 term from the beginning, leaving the possibility of altering combinations later if needed.

The third term, b_3, manages products of three time constants. Ideally, you reuse those you have already determined in b_2 but not always as explained above. From (5.5), we can calculate how many terms we have in b_3:

$$\binom{4}{3} = \frac{4!}{3!(4-3)!} = 4 \tag{5.11}$$

You combine products the following way using the time constants used in b_2 as a start. The first one is $\tau_1\tau_2^1$ meaning that the next time constant will have 1 and 2 as an 'exponent' and the following possible subscripted term is 3:

$$b_{3a} = \tau_1\tau_2^1\tau_3^{12} \tag{5.12}$$

Figure 5.3 shows how to build these terms easily. When reshuffling is necessary, you can also proceed backwards, rebuilding the previous terms from the last exponent. Figure 5.4 shows the principle in

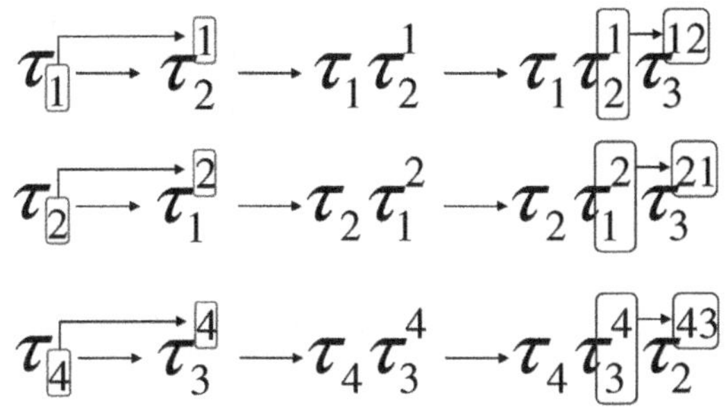

$$\tau_1 \rightarrow \tau_2^1 \rightarrow \tau_1\tau_2^1 \rightarrow \tau_1\tau_2^1\tau_3^{12}$$

$$\tau_2 \rightarrow \tau_1^2 \rightarrow \tau_2\tau_1^2 \rightarrow \tau_2\tau_1^2\tau_3^{21}$$

$$\tau_4 \rightarrow \tau_3^4 \rightarrow \tau_4\tau_3^4 \rightarrow \tau_4\tau_3^4\tau_2^{43}$$

Figure 5.3 It is easy to build the third term as it logically combines the two first ones.

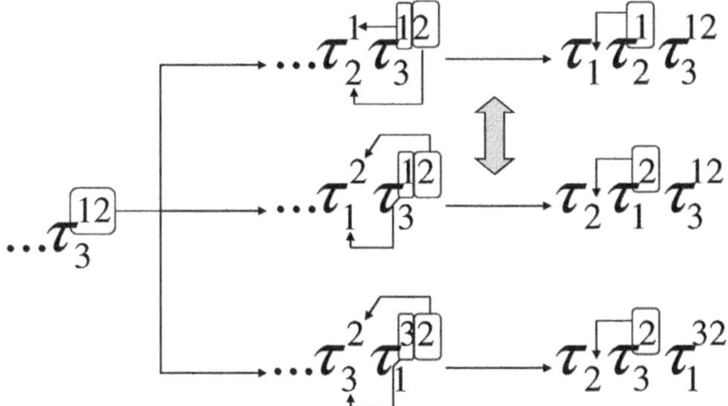

Figure 5.4 When reshuffling a 3-element term, it is sometimes easier and more convenient to start from the right side exponent and proceed backwards. Both right-side terms are identical.

which $\tau_3^{12} = \tau_3^{21}$ since both terms in the exponent similarly imply that elements in τ_1 and τ_2 are set in their high-frequency state: whether you write τ_3^{12} or τ_3^{21} designates the same configuration.

Despite two elements in the exponent place, the process is the same to calculate τ_3^{12}: set the energy-storing elements in τ_1 and τ_2 in their high-frequency state while all other elements (τ_3 and τ_4 in the example) remain in their reference state, for $s = 0$. Then look at the resistance driving element 3. Figure 5.5 graphically illustrates this notation for two examples.

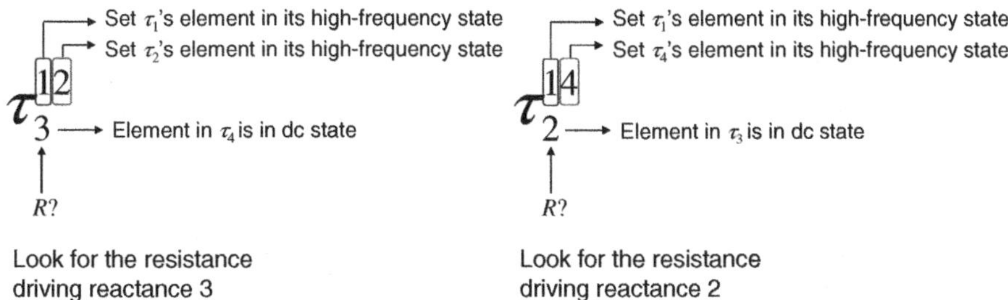

Figure 5.5 Set the two energy-storing elements referenced in the exponent in their high-frequency state and while the other elements are in their reference state ($s = 0$), look at the resistance driving the subscripted element.

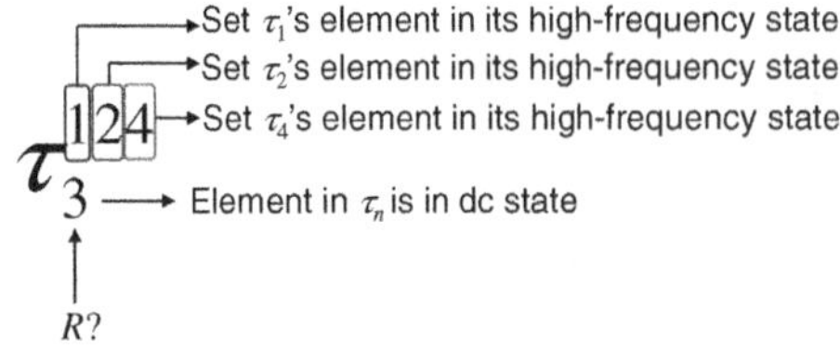

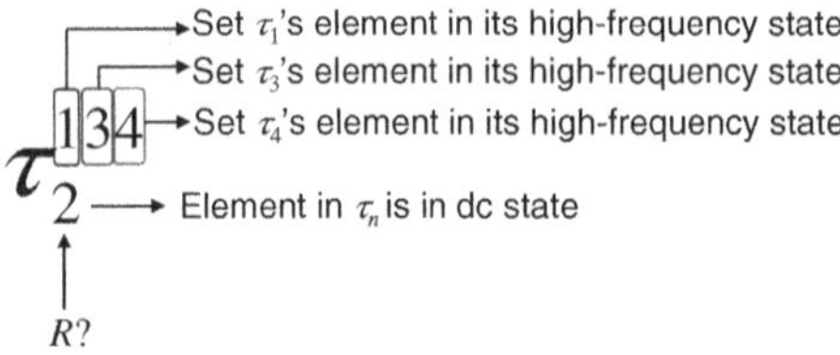

Figure 5.6 Set the three energy-storing elements referenced in the exponent in their high-frequency state and while the other elements (if any) are in their reference state ($s = 0$), look at the resistance driving the subscripted reactance.

In a 4th-order network, you can still associate $\tau_1\tau_2^1$ with τ_4 leading to:

$$b_{3b} = \tau_1\tau_2^1\tau_4^{12} \tag{5.13}$$

The next term in (5.9) is $\tau_1\tau_3^1$ which can be associated with τ_4:

$$b_{3c} = \tau_1\tau_3^1\tau_4^{13} \tag{5.14}$$

Finally, the next term is $\tau_2\tau_3^2\tau_4^{23}$:

$$b_3 = \tau_1\tau_2^1\tau_3^{12} + \tau_1\tau_2^1\tau_4^{12} + \tau_1\tau_3^1\tau_4^{13} + \tau_2\tau_3^2\tau_4^{23} \tag{5.15}$$

The generalized formula to determine b_3 is:

$$b_3 = \sum_{i_1=1}^{2}\sum_{i_2=i_1+1}^{3}\sum_{i_3=i_2+1}^{4} \tau_{i_1}\tau_{i_2}^{i_1}\tau_{i_3}^{i_1 i_2} \tag{5.16}$$

The last term, b_4, combines the product of four time constants. In this case, (5.4) returns 1 and b_4 is defined as:

$$b_4 = \tau_1\tau_2^1\tau_3^{12}\tau_4^{123} \tag{5.17}$$

Here, three elements are set in their high-frequency state while you look for the resistance driving the reactance designated by the subscripted number. Figure 5.6 gives details on this principle.

In case reshuffling is necessary, Figure 5.7 gives a few examples.

reshuffle

$$\tau_1\tau_3^1\tau_2^{13}\tau_4^{132} = \ldots\tau_4^{123} = \ldots\tau_3^{12}\,\tau_4^{123} = .\,\tau_2^{1}\,\tau_3^{12}\,\tau_4^{123} = \tau_1\,\tau_2^{1}\,\tau_3^{12}\,\tau_4^{123}$$

$$\tau_1\tau_3^1\tau_2^{13}\tau_4^{132} = \ldots\tau_3^{124} = \ldots\tau_4^{12}\,\tau_3^{124} = .\,\tau_2^{1}\,\tau_4^{12}\,\tau_3^{124} = \tau_1\,\tau_2^{1}\,\tau_4^{12}\,\tau_3^{124}$$

reshuffle

Figure 5.7 Building the 4-element term is not complicated, you can start from the left and proceed or start from the last element and go backward.

Denominators up to order 4

1st order $D(s)=1+\tau_2 s$

2nd order $D(s)=1+(\tau_1+\tau_2)s+(\tau_1\tau_2^1)s^2=1+(\tau_1+\tau_2)s+(\tau_2\tau_1^2)s^2$

3rd order $D(s)=1+(\tau_1+\tau_2+\tau_3)s+(\tau_1\tau_2^1+\tau_1\tau_3^1+\tau_2\tau_3^2)s^2+(\tau_1\tau_2^1\tau_3^{12})s^3$

4th order
$$D(s)=1+(\tau_1+\tau_2+\tau_3+\tau_4)s+(\tau_1\tau_2^1+\tau_1\tau_3^1+\tau_1\tau_4^1+\tau_2\tau_3^2+\tau_2\tau_4^2+\tau_3\tau_4^3)s^2$$
$$+(\tau_1\tau_2^1\tau_3^{12}+\tau_1\tau_2^1\tau_4^{12}+\tau_1\tau_3^1\tau_4^{13}+\tau_2\tau_3^2\tau_4^{23})s^3$$
$$+(\tau_1\tau_2^1\tau_3^{12}\tau_4^{123})s^4$$

Figure 5.8 Possible denominator expressions up to the 4th order.

The generalized formula for b_4 is:

$$b_4=\sum_{i_1=1}^{1}\sum_{i_2=i_1+1}^{2}\sum_{i_3=i_2+1}^{3}\sum_{i_4=i_3+1}^{4}\tau_{i_1}\tau_{i_2}^{i_1}\tau_{i_3}^{i_1 i_2}\tau_{i_4}^{i_1 i_2 i_3} \tag{5.18}$$

Finally, the complete generalized formula letting you express any order transfer function is given by:

$$b_j=\sum_{i_1=1}^{n+1-j}\sum_{i_2=i_1+1}^{n+2-j}\cdots\sum_{i_j=i_{j-1}+1}^{n}\tau_{i_1}\tau_{i_2}^{i_1}\tau_{i_3}^{i_1 i_2}\ldots\tau_{i_j}^{i_1 i_2\ldots i_{j-1}} \tag{5.19}$$

in which n is the circuit order and j is the polynomial coefficient subscript: b_1, b_2 . . . The generalized formula is here to guide you in writing the first set of elements constituting the polynomial terms. However, if one could think of automating (5.19) in a computer program, it could fail to deliver the correct response given the possible indeterminacies which the software could not alleviate. For this reason, it is very likely that the final expression differs from what (5.19) suggests, simply because of reshuffling that can to occur at some point in the analysis. Figure 5.8 summarizes possible denominator expressions up to the power 4.

5.1.1 3rd-order Transfer Function Example

Now that we know how to build a denominator, let's exercise our new skills on the circuit shown in Figure 5.9. There are 3 energy-storing components, this is a 3rd-order system (state variables are independent). There are no zeros since opening L_1, L_2 or shorting C_3 individually, by group or all together always cancels the response. To determine the coefficients of this network, we break the

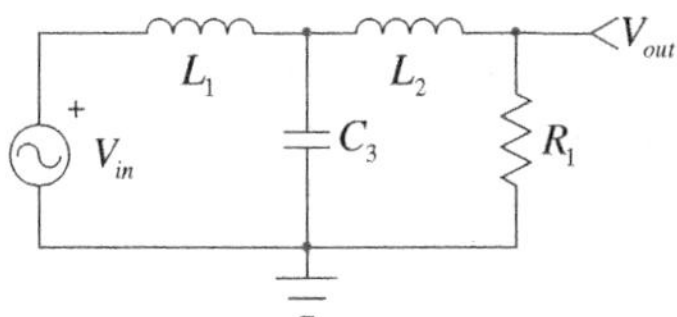

Figure 5.9 This circuit is described by a denominator of order 3.

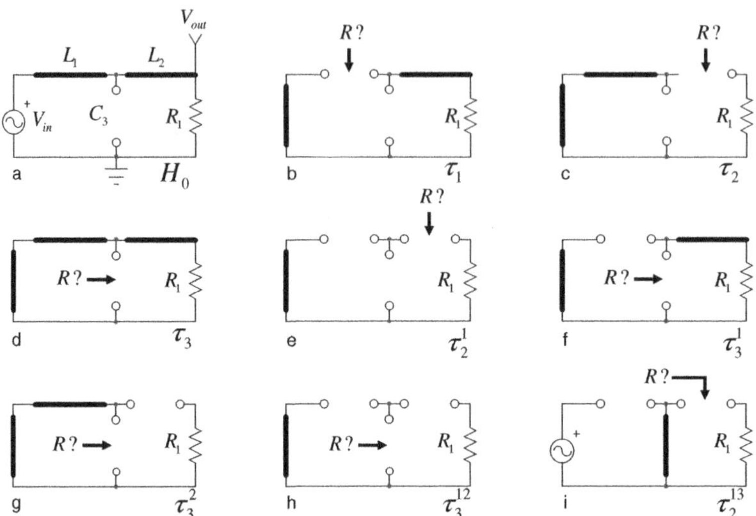

Figure 5.10 Studying a 3rd-order network is not more difficult that running the exercise for a 2nd-order circuit, it is longer and requires more care in collecting coefficients.

circuit into small pieces, each one corresponding to a particular parameter calculation. Figure 5.10 illustrates this path. For $s = 0$, we have:

$$H_0 = 1 \tag{5.20}$$

The first time constant in Figure 5.10b is simply:

$$\tau_1 = \frac{L_1}{R_1} \tag{5.21}$$

whereas in Figure 5.10c and d, you should find

$$\tau_2 = \frac{L_2}{R_1} \tag{5.22}$$

and

$$\tau_3 = 0 \cdot C_3 = 0 \tag{5.23}$$

The first term in the denominator is equal to:

$$b_1 = \tau_1 + \tau_2 + \tau_3 = \frac{L_1}{R_1} + \frac{L_2}{R_1} = \frac{L_1 + L_2}{R_1} \tag{5.24}$$

For τ_2^1, L_1 is set in its high-frequency state (removed from the circuit) and you look at the resistance driving reactance 2. In this mode, the remaining elements (C_3) are left in their reference state (C_3 is removed from the circuit). One of the branches is open thus:

$$\tau_2^1 = \frac{L_2}{\infty} = 0 \tag{5.25}$$

For τ_3^1, L_1 is still set in its high-frequency state but you look into C_3's terminals and 'see' R_1 in Figure 5.10f:

$$\tau_3^1 = R_1 C_3 \tag{5.26}$$

Now τ_3^2: L_2 is set in its high-frequency state (removed from the circuit) and you look at the resistance driving C_3. L_1 keeps its reference state and is replaced by a short circuit. From Figure 5.10g, you have:

$$\tau_3^2 = 0 \cdot C_3 = 0 \tag{5.27}$$

Combining the above definitions, we have:

$$b_2 = \tau_1 \tau_2^1 + \tau_1 \tau_3^1 + \tau_2 \tau_3^2 = \frac{L_1}{R_1} \cdot 0 + \frac{L_1}{R_1} R_1 C_3 + \frac{L_2}{R_1} \cdot 0 = L_1 C_3 \tag{5.28}$$

For the highest order term, in Figure 5.10h, we can evaluate τ_3^{12} in which both inductors are set in their high-frequency state (removed) while we look for the resistance driving reactance 3, C_3. We have:

$$\tau_3^{12} = \infty \cdot C_3 \tag{5.29}$$

which, later multiplied by τ_2^1 from (5.25) would lead to an indeterminacy. The solution is to reshuffle the last term:

$$b_3 = \tau_1 \tau_2^1 \tau_3^{12} \tag{5.30}$$

into a different combination (see Figure 5.4). We have the choice between:

$$b_3 = \tau_2 \tau_3^2 \tau_1^{23} \tag{5.31}$$

or

$$b_3 = \tau_1 \tau_3^1 \tau_2^{13} \tag{5.32}$$

We do not use $\tau_2 \tau_1^2 \tau_3^{21}$ because τ_3^{21} is already defined by (5.29). In Figure 5.10i, adopting (5.32), we have:

$$\tau_2^{13} = \frac{L_2}{R_1} \tag{5.33}$$

This leads to:

$$b_3 = \tau_1 \tau_3^1 \tau_2^{13} = \frac{L_1}{R_1} R_1 C_3 \frac{L_2}{R_1} = \frac{L_1 L_2 C_3}{R_1} \tag{5.34}$$

The denominator can now be formed by associating (5.24), (5.28) and (5.34):

$$D(s) = 1 + s\left(\frac{L_1 + L_2}{R_1}\right) + s^2 L_1 C_3 + s^3 \frac{L_1 L_2 C_3}{R_1} \tag{5.35}$$

Can we rework this expression and make it fit a different format, for instance with a 2nd-order polynomial form? We have derived different expressions for a 3rd-order formula in Chapter 2. Rearranging (5.35) depends on how the poles are spread. For instance, if (5.35) shows the domination of a single pole at low frequency then followed by a double pole at higher frequencies, we can rearrange (5.35) as:

$$D(s) \approx \left(1 + \frac{s}{\omega_p}\right)\left(1 + \frac{s}{\omega_0 Q} + \left(\frac{s}{\omega_0}\right)^2\right) \tag{5.36}$$

in which we can identify (see Chapter 2):

$$\omega_p = \frac{1}{b_1} \tag{5.37}$$

$$Q = \frac{b_1 b_3 \sqrt{\frac{b_1}{b_3}}}{b_1 b_2 - b_3} = \frac{L_2(L_1 + L_2)\sqrt{\frac{L_1 + L_2}{C_3 L_1 L_2}}}{L_1 R_1} \tag{5.38}$$

and

$$\omega_0 = \sqrt{\frac{b_1}{b_3}} = \sqrt{\frac{L_1 + L_2}{C_3 L_1 L_2}} \tag{5.39}$$

If $b_3 \ll b_1 b_2$ then (5.36) can be further simplified as:

$$D(s) \approx (1 + b_1 s)\left(1 + s\frac{b_2}{b_1} + s^2\frac{b_3}{b_1}\right) \tag{5.40}$$

It is when assigning component values to L_1, L_2, C_3 and R_1 that you will see if any of this simplified expression gives adequate results. The final transfer function linking V_{out} to V_{in} is given by:

$$H(s) = \frac{1}{1 + s\left(\frac{L_1 + L_2}{R_1}\right) + s^2 L_1 C_3 + s^3 \frac{L_1 L_2 C_3}{R_1}} \tag{5.41}$$

To test this expression, we need a reference transfer function. Applying Thévenin's and an impedance divider to Figure 5.9 should lead you to:

$$H_{ref}(s) = \frac{\frac{1}{sC_3}}{sL_1 + \frac{1}{sC_3}} \frac{R_1}{R_1 + sL_2 + \left(\frac{1}{sC_3} \| sL_1\right)} = \frac{1}{1 + s^2 L_1 C_3} \frac{R_1}{R_1 + sL_2 + \left(\frac{1}{sC_3} \| sL_1\right)} \tag{5.42}$$

We can now enter these expressions in a Mathcad® sheet and check expressions versus the dynamic response of (5.42).

With the adopted component values, curves match each other quite well in magnitude and phase. If we now test the simplified expression we gave in (5.40), the response is acceptable despite a slightly lower peaking as shown in the left side of Figure 5.12. When C_3 increases from 1 nF to 1 μF, then assumptions used to factor (5.40) are no longer valid and its response is unable to predict peaking. Should we plot (5.36) instead, curves would still match very well.

5.1.2 Transfer Functions with Zeros

All expressions currently derived to determine the poles can be equally applied for the zeros. The only difference is that all resistances driving reactances are determined while the response is nulled in a null double injection (NDI) configuration. In this mode, the excitation source is back on and a test generator I_T is installed across the terminals of the considered reactance so as to null the response. The resistance is then the voltage across the reactance's terminals V_T divided by the current source I_T. If necessary, reshuffling is possible the same way as we described for the denominator coefficients. Figure 5.13 summarizes numerator expressions from 1st order to order 4 in which the subscripted N denotes a numerator coefficient (except for the 1st-order expression in which we designated τ_1 for the numerator and τ_2 for the denominator in Chapter 2 and subsequent chapters).

$L_1 := 10\mu H \quad L_2 := 500\mu H \quad C_3 := 1nF \quad R_1 := 0.1k\Omega \quad \|(x,y) := \frac{x \cdot y}{x+y} \quad R_{inf} := 10^{23}\Omega$

$H_0 := 1 \quad \tau_1 := \frac{L_1}{R_1} = 0.1 \cdot \mu s \quad \tau_2 := \frac{L_2}{R_1} = 5 \cdot \mu s \quad \tau_3 := 0 \cdot C_3 = 0 \cdot \mu s$

$$H_{ref}(s) := \frac{\frac{1}{s \cdot C_3}}{s \cdot L_1 + \frac{1}{s \cdot C_3}} \cdot \frac{R_1}{R_1 + s \cdot L_2 + \left[\left(\frac{1}{s \cdot C_3}\right) \| (s \cdot L_1)\right]}$$

$b_1 := \tau_1 + \tau_2 + \tau_3 = 5.1 \cdot \mu s$

$\tau_{12} := \frac{L_2}{R_{inf}} = 0 \cdot \mu s \quad \tau_{13} := R_1 \cdot C_3 = 0.1 \cdot \mu s \quad \tau_{23} := 0 \cdot C_3 = 0 \cdot \mu s$

$b_2 := \tau_1 \cdot \tau_{12} + \tau_1 \cdot \tau_{13} + \tau_2 \cdot \tau_{23} = 0.01 \cdot \mu s^2$

$\tau_{132} := \frac{L_2}{R_1} = 5 \cdot \mu s \quad \tau_{123} := R_{inf} \cdot C_3 \quad b_3 := \tau_1 \cdot \tau_{13} \cdot \tau_{132} = 0.05 \mu s^3 \quad b_{33} := \tau_1 \cdot \tau_{12} \cdot \tau_{123} = 0.05 \mu s^3$

$$H_1(s) := H_0 \cdot \frac{1}{1 + b_1 \cdot s + b_2 \cdot s^2 + b_3 \cdot s^3}$$

$\omega_p := \frac{1}{b_1} \quad f_p := \frac{\omega_p}{2\pi} = 31.207 kHz$

$$Q := \frac{b_1 \cdot b_3 \cdot \sqrt{\frac{b_1}{b_3}}}{b_1 \cdot b_2 - b_3} = 2.575 \times 10^3 \qquad Q_{00} := \frac{L_2 \cdot (L_1 + L_2) \cdot \sqrt{\frac{L_1 + L_2}{C_3 \cdot L_1 \cdot L_2}}}{L_1 \cdot R_1} = 2.575 \times 10^3$$

$$\omega_0 := \sqrt{\frac{b_1}{b_3}} = 1.01 \times 10^7 \frac{1}{s} \qquad \omega_{00} := \sqrt{\frac{L_1 + L_2}{C_3 \cdot L_1 \cdot L_2}} = 1.01 \times 10^7 \frac{1}{s} \qquad f_0 := \frac{\omega_0}{2 \cdot \pi} = 1.607 MHz$$

$$H_2(s) := H_0 \cdot \frac{1}{(1 + b_1 \cdot s) \cdot \left(1 + s \cdot \frac{b_2}{b_1} + s^2 \cdot \frac{b_3}{b_1}\right)} \qquad H_3(s) := H_0 \cdot \frac{1}{(1 + b_1 \cdot s) \cdot \left[1 + \frac{s}{\omega_0 \cdot Q} + \left(\frac{s}{\omega_0}\right)^2\right]} \qquad H_5(s) := \frac{1}{1 + s \cdot \left(\frac{L_1 + L_2}{R_1}\right) + s^2 \cdot L_1 \cdot C_3 + s^3 \cdot \frac{L_1 \cdot L_2 \cdot C_3}{R_1}}$$

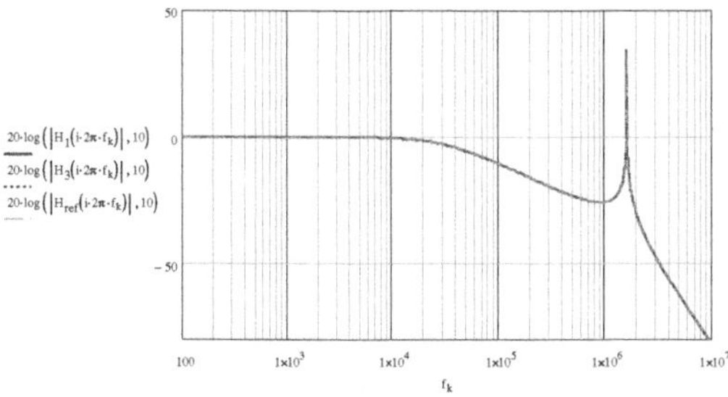

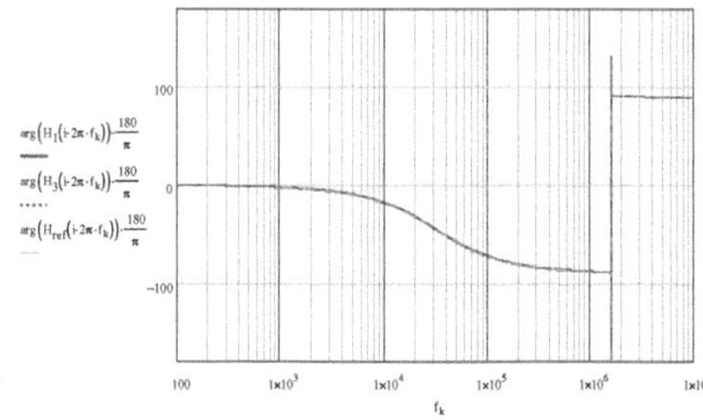

Figure 5.11 Expressions captured in Mathcad® agree well with each other given the clear separation between the double pole and the low-frequency pole: $H_1(s)$ gives the exact same response as $H_{ref}(s)$ while $H_3(s)$ is very close.

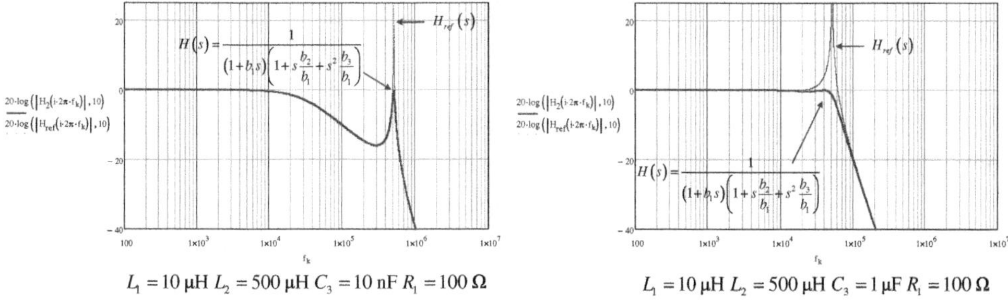

Figure 5.12 The simplified expression in (5.40) gives an approximate response compared to the reference curve in the left-side chart. It fails to predict the peak when poles are getting closer for C_3 equals 1 µF.

Numerators up to order 4

1st order $N(s) = 1 + \tau_1 s$

2nd order $N(s) = 1 + (\tau_{1N} + \tau_{2N})s + (\tau_{1N}\tau_{2N}^{1})s^2 = 1 + (\tau_{1N} + \tau_{2N})s + (\tau_{2N}\tau_{1N}^{2})s^2$

3rd order $N(s) = 1 + (\tau_{1N} + \tau_{2N} + \tau_{3N})s + (\tau_{1N}\tau_{2N}^{1} + \tau_{1N}\tau_{3N}^{1} + \tau_{2N}\tau_{3N}^{2})s^2 + (\tau_{1N}\tau_{2N}^{1}\tau_{3N}^{12})s^3$

4th order
$$\begin{aligned} N(s) = {} & 1 + (\tau_{1N} + \tau_{2N} + \tau_{3N} + \tau_{4N})s + (\tau_{1N}\tau_{2N}^{1} + \tau_{1N}\tau_{3N}^{1} + \tau_{1N}\tau_{4N}^{1} + \tau_{2N}\tau_{3N}^{2} + \tau_{2N}\tau_{4N}^{2} + \tau_{3N}\tau_{4N}^{3})s^2 \\ & + (\tau_{1N}\tau_{2N}^{1}\tau_{3N}^{12} + \tau_{1N}\tau_{2N}^{1}\tau_{4N}^{12} + \tau_{1N}\tau_{3N}^{1}\tau_{4N}^{13} + \tau_{2N}\tau_{3N}^{2}\tau_{4N}^{23})s^3 \\ & + (\tau_{1N}\tau_{2N}^{1}\tau_{3N}^{12}\tau_{4N}^{123})s^4 \end{aligned}$$

Figure 5.13 The numerator expression is determined while the response is nulled.

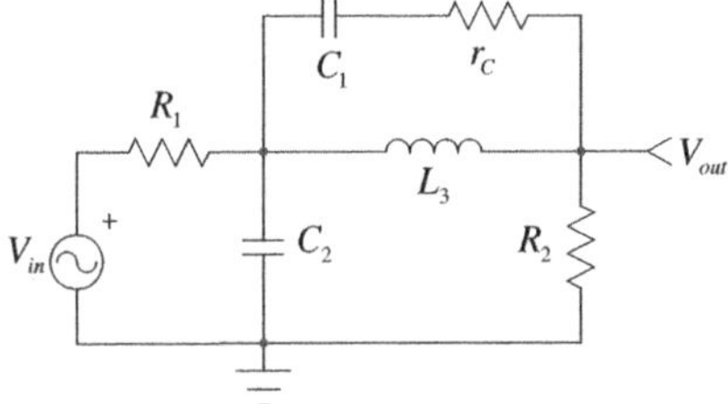

Figure 5.14 This 3rd-order circuit features several zeros that we will unveil.

The circuit shown in Figure 5.14 will be used to see how we can apply these definitions to a 3rd-order circuit. To check how many zeros exist, we can capitalize on what has been defined in Chapter 4: how many energy-storing elements can we *simultaneously* put in their high-frequency state and still observe a response?

If C_2 is shorted, regardless of L_3 and C_1's state, there is no response so no need trying associate C_2 with the other reactances. Now if L_3 is open while C_1 is a short circuit, there is response via r_C: the numerator is of 2nd order, with two zeros involving C_1 and L_3. Once we have calculated a_2, no need to calculate a_3.

To determine the zeros, we look at the resistance driving each reactance in various configurations when the output is nulled. All the steps are gathered in Figure 5.15 and are fairly simple to solve.

The first NDI time constant involves C_1 in Figure 5.15a. From that figure, given the absence of response, the test current I_T returns via resistor r_C:

$$\tau_{1N} = r_C C_1 \tag{5.43}$$

Figure 5.15b, the right terminal of r_C is at 0 V (response is nulled) but the upper terminal of the current source also via L_3 that is a short circuit at $s = 0$. Therefore $V_T = 0$ and thus:

$$\tau_{2N} = 0 \cdot C_2 \tag{5.44}$$

Regarding the inductor time constant for a nulled response, Figure 5.15c tells us that the null is satisfied if the test current I_T is equal to 0. As a result:

$$\tau_{3N} = \frac{L_3}{\infty} = 0 \tag{5.45}$$

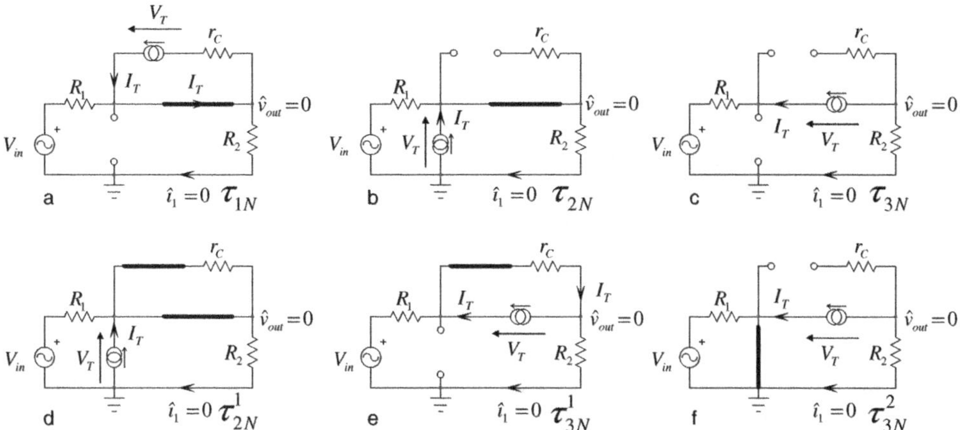

Figure 5.15 You determine the zeros by breaking the circuit into small pieces while the response is nulled.

We have the definition of a_1 equal to:

$$a_1 = \tau_{1N} + \tau_{2N} + \tau_{3N} = r_C C_1 \tag{5.46}$$

For the a_2 terms, some of the energy-storing elements will be put in their high-frequency state. In Figure 5.15d, we start with C_1, replaced by a short circuit while the test generator biases C_2's terminals. In this mode, r_C is shorted by L_3 and the only way to create a null in the output is to have V_T equals 0 V. In that case:

$$\tau_{2N}^{1} = 0 \cdot C_2 \tag{5.47}$$

Then while C_1 is still in its high-frequency state, we can explore the resistance offered by L_3's terminals with a response nulled. Figure 5.15e shows the configuration in which the test current returns to the generator via resistor r_C. We have:

$$\tau_{3N}^{1} = \frac{L_3}{r_C} \tag{5.48}$$

Finally, in Figure 5.15f, this is capacitor C_2 that is set to its high-frequency case while we determine the resistance offered by L_3's terminal in a nulled-response condition. As the I_T current cannot return via R_2 and r_C is unconnected, the only way to have a zero current in R_2 is when I_T equals 0. Now, the voltage across the current source is also 0 V and we have an indetermincacy. To get rid of this indeterminacy, simply connect a resistance R_{dum} across the current source and I_T now circulates in this resistance only as no current flows in R_2.

Figure 5.16 shows the updated schematic with its corresponding SPICE simulation. The time constant involving L_3 in NDI is thus:

$$\tau_{3N}^{2} = \frac{L_3}{R_{dum}} \tag{5.49}$$

which becomes:

$$\tau_{3N}^{2} = \frac{L_3}{\infty} \tag{5.50}$$

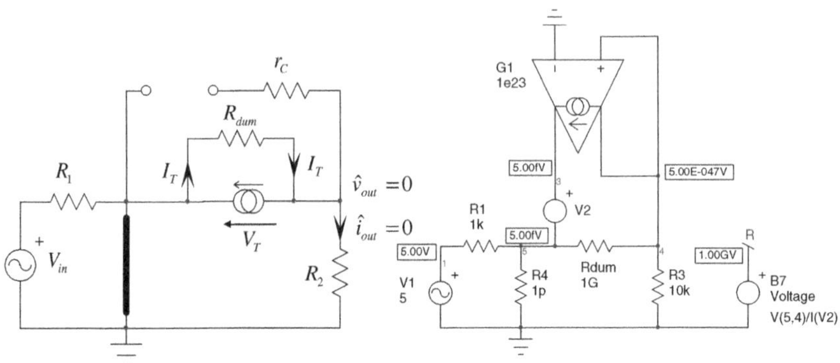

Figure 5.16 You remove the indeterminacy by adding a dummy resistance across L_3 during the NDI. The SPICE simulation confirms that the resistance seen from L_3's terminals in NDI is R_{dum}.

when R_{dum} approaches infinity. We have all the terms to form a_2 now:

$$a_2 = \tau_{1N}\tau_{2N}^1 + \tau_{1N}\tau_{3N}^1 + \tau_{2N}\tau_{3N}^2 = r_C C_1 \cdot 0 \cdot C_2 + r_C C_1 \frac{L_3}{r_C} + 0 \cdot C_2 \frac{L_3}{\infty} = L_3 C_1 \tag{5.51}$$

The numerator is defined by:

$$N(s) = 1 + a_1 s + a_2 s^2 = 1 + s r_C C_1 + s^2 L_3 C_1 \tag{5.52}$$

It is a second order polynomial that can satisfy the form:

$$N(s) = 1 + \frac{s}{\omega_{0N} Q_N} + \left(\frac{s}{\omega_{0N}}\right)^2 \tag{5.53}$$

in which

$$Q_N = \frac{\sqrt{a_2}}{a_1} = \frac{\sqrt{L_3 C_1}}{r_C C_1} = \frac{1}{r_C}\sqrt{\frac{L_3}{C_1}} \tag{5.54}$$

and

$$\omega_{0N} = \frac{1}{\sqrt{a_2}} = \frac{1}{\sqrt{L_3 C_1}} \tag{5.55}$$

We have determined the zeros in eight steps. The exercise is not overly complex but it takes time to draw all these little sketches and indeterminacies can sometimes arise. Is there any faster way to unveil the numerator? Of course: inspection! If you look at Figure 5.14 and observe the circuit for for $s = s_z$, are there conditions which would lead to a null in the response? C_2 becoming a transformed short circuit? Certainly not since it only occurs as s approaches infinity. A series impedance becoming a transformed open? Exactly, and this impedance is made of L_3 in parallel with the series combination of C_1 and r_C. This impedance becomes infinite if its denominator cancels. What we need is to find the poles of that impedance and we have the zeros of our transfer function. Figure 5.17a shows how to determine the impedance with a current generator. What we need is the denominator; that is all.

From Figure 5.17b, we determine the first time constant associated with capacitor C_1:

$$\tau_1 = r_C C_1 \tag{5.56}$$

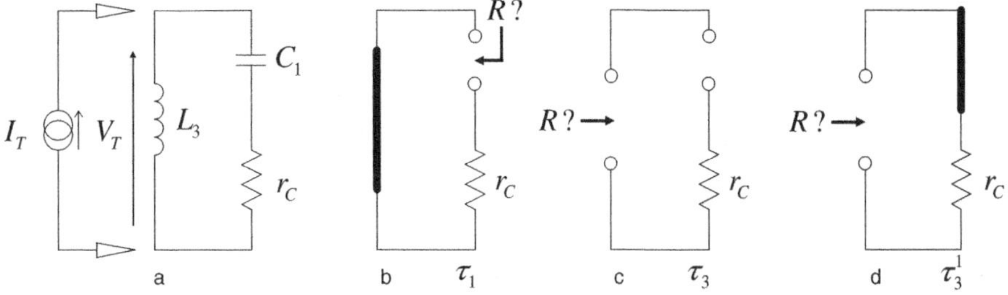

Figure 5.17 The poles of this impedance are the zeros of the transfer function we want.

In Figure 5.17c, the resistance seen from the inductor's terminals is infinite, leading to:

$$\tau_3 = \frac{L_3}{\infty} = 0 \tag{5.57}$$

Finally, Figure 5.17d shows that part of the 2nd-order term is equal to:

$$\tau_3^1 = \frac{L_3}{r_C} \tag{5.58}$$

This is it, we have all we need to construct $D(s)$:

$$D(s) = 1 + s(\tau_1 + \tau_3) + s^2\tau_1\tau_3^1 = 1 + sr_C C_1 + s^2 C_1 L_3 \tag{5.59}$$

which matches the numerator definition found in (5.52). Inspection, when possible, is always faster than NDI. Now that we have the numerator, it is time to determine the denominator. No surprise here, we follow exactly the path as in Figure 5.15 except that the excitation is now set to 0. Classically starting with the dc gain, Figure 5.15a tells us:

$$H_0 = \frac{R_2}{R_1 + R_2} \tag{5.60}$$

The three time constants are easily obtained from sketches b, c and d:

$$\tau_1 = r_C C_1 \tag{5.61}$$

$$\tau_2 = C_2(R_1 \| R_2) \tag{5.62}$$

$$\tau_3 = \frac{L_3}{R_1 + R_2} \tag{5.63}$$

The first element b_1 is defined by:

$$b_1 = \tau_1 + \tau_2 + \tau_3 = r_C C_1 + C_2(R_1 \| R_2) + \frac{L_3}{R_1 + R_2} \tag{5.64}$$

Now C_1 is set in its high-frequency state and we look at the resistance driving C_2 in Figure 5.15e. Since r_C is shorted by L_3, you see the parallel arrangement of R_1 and R_2. Thus:

$$\tau_2^1 = C_2(R_1 \| R_2) \tag{5.65}$$

Same configuration but now look at the resistance driving L_3 in Figure 5.15f:

$$\tau_3^1 = \frac{L_3}{r_C \| (R_1 + R_2)} \tag{5.66}$$

In Figure 5.15g, C_2 is set in its high-frequency state and we are interested by the resistance driving L_3, R_2 in this configuration:

$$\tau_3^2 = \frac{L_3}{R_2} \tag{5.67}$$

We can now form b_2 by adding the above terms:

$$b_2 = \tau_1\tau_2^1 + \tau_1\tau_3^1 + \tau_2\tau_3^2 = r_C C_1 C_2 (R_1 \| R_2) + r_C C_1 \frac{L_3}{r_C \| (R_1 + R_2)} + C_2 (R_1 \| R_2) \frac{L_3}{R_2} \tag{5.68}$$

The final term in $\tau_1\tau_2^1\tau_3^{12}$ is determined with the help of Figure 5.18h:

$$\tau_3^{12} = \frac{L_3}{r_C \| R_2} \tag{5.69}$$

We have everything to assemble the b_3 term as:

$$b_3 = \tau_1\tau_2^1\tau_3^{12} = r_C C_1 C_2 (R_1 \| R_2) \frac{L_3}{r_C \| R_2} \tag{5.70}$$

The denominator is equal to:

$$\begin{aligned} D(s) = {} & 1 + s\left(r_C C_1 + C_2 (R_1 \| R_2) + \frac{L_3}{R_1 + R_2}\right) \\ & + s^2\left(r_C C_1 C_2 (R_1 \| R_2) + r_C C_1 \frac{L_3}{r_C \| (R_1 + R_2)} + C_2 (R_1 \| R_2) \frac{L_3}{R_2}\right) + s^3 r_C C_1 C_2 (R_1 \| R_2) \frac{L_3}{r_C \| R_2} \end{aligned} \tag{5.71}$$

We can try to write a simplified version involving a low-frequency pole followed by a 2nd-order polynomial in which the quality factor Q is defined as:

$$Q = \frac{b_1 b_3 \sqrt{\frac{b_1}{b_3}}}{b_1 b_2 - b_3} \tag{5.72}$$

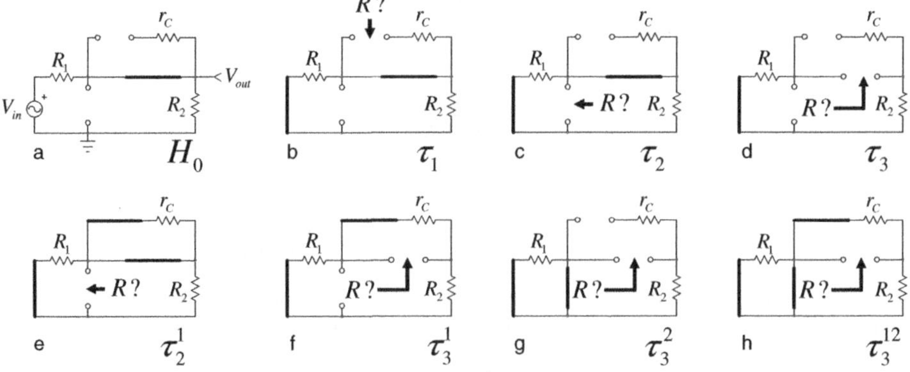

Figure 5.18 The individual time constants are determined when the excitation is set to 0 V.

a resonant frequency ω_0 equal to:

$$\omega_0 = \sqrt{\frac{b_1}{b_3}} \tag{5.73}$$

and a low-frequency pole that is

$$\omega_p = \frac{1}{b_1} \tag{5.74}$$

The simplified expression can be rearranged as

$$H(s) = \frac{N(s)}{D(s)} \approx H_0 \frac{1 + \frac{s}{\omega_{0N} Q_N} + \left(\frac{s}{\omega_{0N}}\right)^2}{\left(1 + \frac{s}{\omega_p}\right)\left[1 + \frac{s}{\omega_0 Q} + \left(\frac{s}{\omega_0}\right)^2\right]} \tag{5.75}$$

Before we test the dynamic responses, we need a raw reference transfer function. R_1 and C_2 can be part of a Thévenin voltage source. This source is affected by an output impedance made of R_1 and C_2 in parallel:

$$H_{ref}(s) = \frac{\frac{1}{sC_2}}{R_1 + \frac{1}{sC_2}} \frac{R_2}{R_2 + \left[\left(\frac{1}{sC_2}\right) \| R_1\right]\left[\left(\frac{1}{sC_1} + r_C\right) \| (sL_3)\right]} \tag{5.76}$$

We have gathered all these expressions in a Mathcad® file as shown in Figure 5.19. We have plotted the reference transfer function from (5.76) and the complete expression using (5.52) and (5.71). The agreement is excellent. (5.75) is then compared to (5.76) in Figure 5.20 and the results are good too for the selected set of components values.

5.1.3 A Generalized N^{th}-order Transfer Function

In Chapter 3, we explained how the EET could be formulated in a form where NDI was no longer necessary. More specifically, we could reuse the denominator time constants determined with the excitation source set to 0 and combine each with a gain obtained when the considered energy-storing element is set in its high-frequency state. In Chapter 4, we extended the expression to 2nd-order circuits as shown in [1]. In his paper, Ali Hajimiri generalizes the technique to n^{th}-order transfer functions, keeping a similar method: reuse denominator time constants and associate them with gains determined while some energy-storing elements are set in their high-frequency or dc states. Figure 5.21 summarizes numerators expressed in this way up to the order 4.

When the gain obtained for $s = 0$ exists, H_0, it can be factored as shown in Figure 5.22.

Building these expressions is not complicated as illustrated by Figure 5.23. The H notation implies that the energy-storing elements designated in the exponents are set in their high-frequency state while the rest are kept in their dc state. As already pointed out in Chapter 3 and 4, expressions obtained with the generalized form can sometimes lead to more complex formulas compared to those obtained with NDI or even better, by inspection. It is your choice to apply either approach, both lead to identical dynamic responses.

Now that we know how the generalized expression looks like, we can apply it to the 3rd-order example from Figure 5.24. As usual, we will draw all intermediate steps to determine the natural time constants of the circuit. The sketches are given in Figure 5.25.

$L_3 := 1\text{mH}\quad C_2 := 22\text{nF}\quad C_1 := 22\text{nF}\quad R_1 := 1\text{k}\Omega\quad R_2 := 1\text{k}\Omega\quad \|(x,y) := \frac{x\cdot y}{x+y}\quad R_{inf} := 10^{23}\Omega$

$r_C := 1.5\Omega$

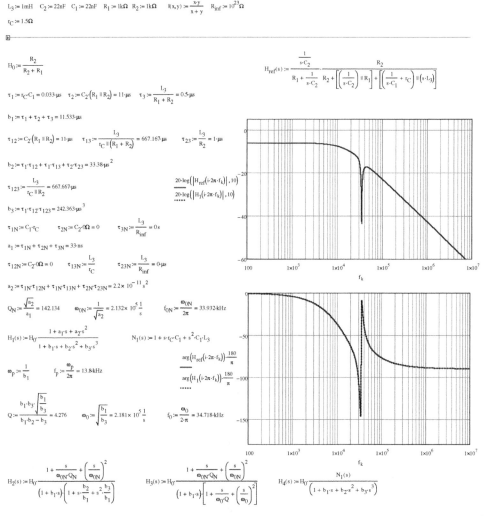

$$H_0 := \frac{R_2}{R_2+R_1}\qquad H_{ref}(s) := \frac{\frac{1}{s\cdot C_2}}{R_1+\frac{1}{s\cdot C_2}}\cdot\frac{R_2}{R_2+\left[\left(\frac{1}{s\cdot C_2}\right)\|R_1\right]+\left[\left(\frac{1}{s\cdot C_1}+r_C\right)\|(s\cdot L_3)\right]}$$

$\tau_1 := r_C\cdot C_1 = 0.033\cdot\mu s\quad \tau_2 := C_2\cdot(R_1\|R_2) = 11\cdot\mu s\quad \tau_3 := \frac{L_3}{R_1+R_2} = 0.5\cdot\mu s$

$b_1 := \tau_1+\tau_2+\tau_3 = 11.533\cdot\mu s$

$\tau_{12} := C_2\cdot(R_1\|R_2) = 11\cdot\mu s\quad \tau_{13} := \frac{L_3}{r_C\|(R_1+R_2)} = 667.167\cdot\mu s\quad \tau_{23} := \frac{L_3}{R_2} = 1\cdot\mu s$

$b_2 := \tau_1\cdot\tau_{12}+\tau_1\cdot\tau_{13}+\tau_2\cdot\tau_{23} = 33.38\cdot\mu s^2$

$\tau_{123} := \frac{L_3}{r_C\|R_2} = 667.667\cdot\mu s$

$b_3 := \tau_1\cdot\tau_{12}\cdot\tau_{123} = 242.363\cdot\mu s^3$

$\tau_{1N} := C_1\cdot r_C\quad \tau_{2N} := C_2\cdot 0\Omega = 0\quad \tau_{3N} := \frac{L_3}{R_{inf}} = 0\,s$

$a_1 := \tau_{1N}+\tau_{2N}+\tau_{3N} = 33\cdot ns$

$\tau_{12N} := C_2\cdot 0\Omega = 0\quad \tau_{13N} := \frac{L_3}{r_C}\quad \tau_{23N} := \frac{L_3}{R_{inf}} = 0\cdot\mu s$

$a_2 := \tau_{1N}\cdot\tau_{12N}+\tau_{1N}\cdot\tau_{13N}+\tau_{2N}\cdot\tau_{23N} = 2.2\times10^{-11}\,s^2$

$Q_N := \frac{\sqrt{a_2}}{a_1} = 142.134\quad \omega_{0N} := \frac{1}{\sqrt{a_2}} = 2.132\times10^5\,\frac{1}{s}\quad f_{0N} := \frac{\omega_{0N}}{2\pi} = 33.932\cdot\text{kHz}$

$H_1(s) := H_0\cdot\frac{1+a_1\cdot s+a_2\cdot s^2}{1+b_1\cdot s+b_2\cdot s^2+b_3\cdot s^3}\quad N_1(s) := 1+s\cdot r_C\cdot C_1+s^2\cdot C_1\cdot L_3$

$\omega_p := \frac{1}{b_1}\quad f_p := \frac{\omega_p}{2\pi} = 13.8\cdot\text{kHz}$

$Q := \frac{b_1\cdot b_3\cdot\sqrt{\frac{b_1}{b_3}}}{b_1\cdot b_2-b_3} = 4.276\quad \omega_0 := \sqrt{\frac{b_1}{b_3}} = 2.181\times10^5\,\frac{1}{s}\quad f_0 := \frac{\omega_0}{2\cdot\pi} = 34.718\cdot\text{kHz}$

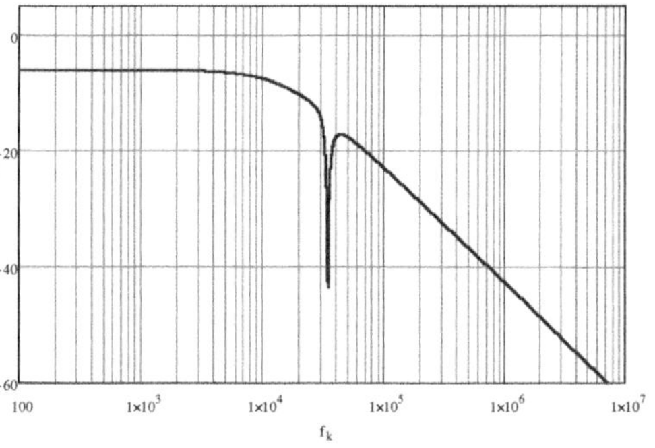

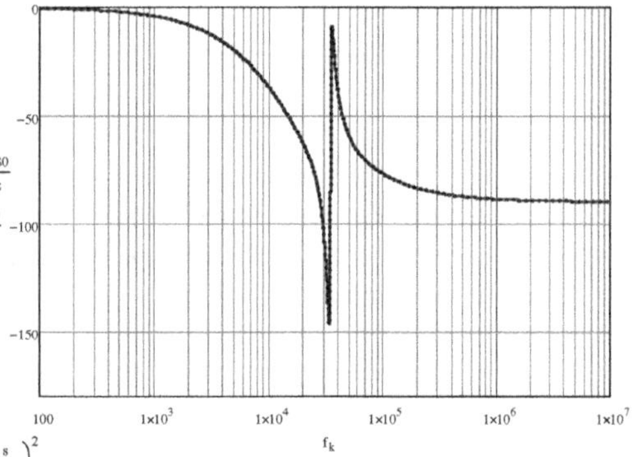

$$H_2(s) := H_0\cdot\frac{1+\frac{s}{\omega_{0N}\cdot Q_N}+\left(\frac{s}{\omega_{0N}}\right)^2}{(1+b_1\cdot s)\cdot\left(1+s\cdot\frac{b_2}{b_1}+s^2\cdot\frac{b_3}{b_1}\right)}\quad H_3(s) := H_0\cdot\frac{1+\frac{s}{\omega_{0N}\cdot Q_N}+\left(\frac{s}{\omega_{0N}}\right)^2}{(1+b_1\cdot s)\cdot\left[1+\frac{s}{\omega_0\cdot Q}+\left(\frac{s}{\omega_0}\right)^2\right]}\quad H_4(s) := H_0\cdot\frac{N_1(s)}{\left(1+b_1\cdot s+b_2\cdot s^2+b_3\cdot s^3\right)}$$

Figure 5.19 The curves agree well with each other. Here we have the reference expression $H_{ref}(s)$ versus $H_1(s)$.

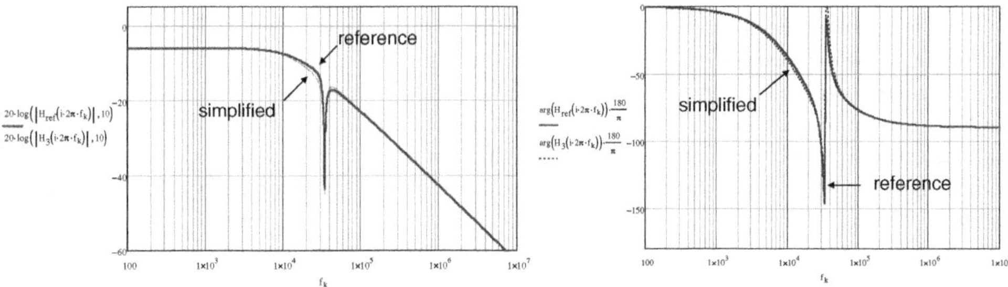

Figure 5.20 With the selected component values, the simplified expression is not that far from the true response.

1st order $N(s)=H_0+H^1\tau_1 s$ with $D(s)=1+s\tau_1$

2nd order $N(s)=H_0+\left(\tau_1H^1+\tau_2H^2\right)s+\left(\tau_1\tau_2^1H^{12}\right)s^2=H_0+\left(\tau_1H^1+\tau_2H^2\right)s+\left(\tau_2\tau_1^2H^{21}\right)s^2$

3rd order $N(s)=H_0+s\left(\tau_1H^1+\tau_2H^2+\tau_3H^3\right)+s^2\left(\tau_1\tau_2^1H^{12}+\tau_1\tau_3^1H^{13}+\tau_2\tau_3^2H^{23}\right)+s^3\left(\tau_1\tau_2^1\tau_3^{12}H^{123}\right)$

4th order
$$
\begin{aligned}
N(s)=H_0&+s\left(\tau_1H^1+\tau_2H^2+\tau_3H^3+\tau_4H^4\right)\\
&+s^2\left(\tau_1\tau_2^1H^{12}+\tau_1\tau_3^1H^{13}+\tau_1\tau_4^1H^{14}+\tau_2\tau_3^2H^{23}+\tau_2\tau_4^2H^{24}+\tau_3\tau_4^3H^{34}\right)\\
&+s^3\left(\tau_1\tau_2^1\tau_3^{12}H^{123}+\tau_1\tau_2^1\tau_4^{12}H^{124}+\tau_1\tau_3^1\tau_4^{13}H^{134}+\tau_2\tau_3^2\tau_4^{23}H^{234}\right)\\
&+s^4\left(\tau_1\tau_2^1\tau_3^{12}\tau_4^{123}H^{1234}\right)
\end{aligned}
$$

Figure 5.21 Generalized transfer functions numerators up to the order 4.

When H_0 is non-zero, it can be advantageously factored as a leading term:

$$N(s)=H_0\left(1+\frac{H^1}{H_0}\tau_1 s\right)\quad\text{with}\quad D(s)=1+s\tau_1$$

$$N(s)=H_0\left[1+\left(\tau_1\frac{H^1}{H_0}+\tau_2\frac{H^2}{H_0}\right)s+\left(\tau_1\tau_2^1\frac{H^{12}}{H_0}\right)s^2\right]$$

$$N(s)=H_0\left[1+s\left(\tau_1\frac{H^1}{H_0}+\tau_2\frac{H^2}{H_0}+\tau_3\frac{H^3}{H_0}\right)+s^2\left(\tau_1\tau_2^1\frac{H^{12}}{H_0}+\tau_1\tau_3^1\frac{H^{13}}{H_0}+\tau_2\tau_3^2\frac{H^{23}}{H_0}\right)+s^3\left(\tau_1\tau_2^1\tau_3^{12}\frac{H^{123}}{H_0}\right)\right]$$

$$
\begin{aligned}
N(s)=H_0\Bigg[1&+s\left(\tau_1\frac{H^1}{H_0}+\tau_2\frac{H^2}{H_0}+\tau_3\frac{H^3}{H_0}+\tau_4\frac{H^4}{H_0}\right)\\
&+s^2\left(\tau_1\tau_2^1\frac{H^{12}}{H_0}+\tau_1\tau_3^1\frac{H^{13}}{H_0}+\tau_1\tau_4^1\frac{H^{14}}{H_0}+\tau_2\tau_3^2\frac{H^{23}}{H_0}+\tau_2\tau_4^2\frac{H^{24}}{H_0}+\tau_3\tau_4^3\frac{H^{34}}{H_0}\right)\\
&+s^3\left(\tau_1\tau_2^1\tau_3^{12}\frac{H^{123}}{H_0}+\tau_1\tau_2^1\tau_4^{12}\frac{H^{124}}{H_0}+\tau_1\tau_3^1\tau_4^{13}\frac{H^{134}}{H_0}+\tau_2\tau_3^2\tau_4^{23}\frac{H^{234}}{H_0}\right)\\
&+s^4\left(\tau_1\tau_2^1\tau_3^{12}\tau_4^{123}\frac{H^{1234}}{H_0}\right)\Bigg]
\end{aligned}
$$

Figure 5.22 When H_0 is different than 0, it can be factored to become the leading factor, carrying the transfer function unit if any.

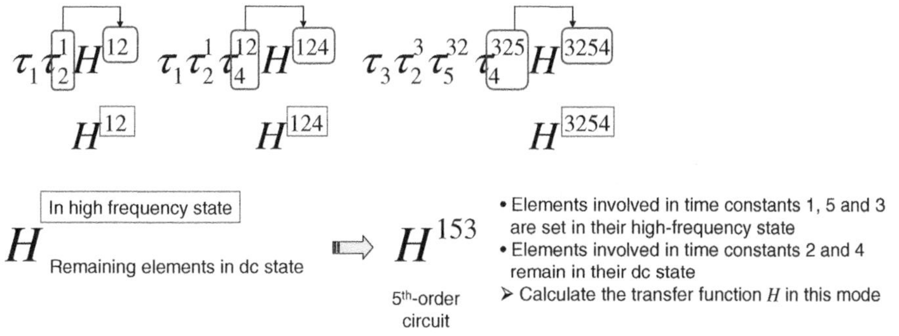

Figure 5.23 Building the generalized expression terms for $N(s)$ requires to calculate gain expressions H involving energy-storing elements set in their high-frequency or dc states.

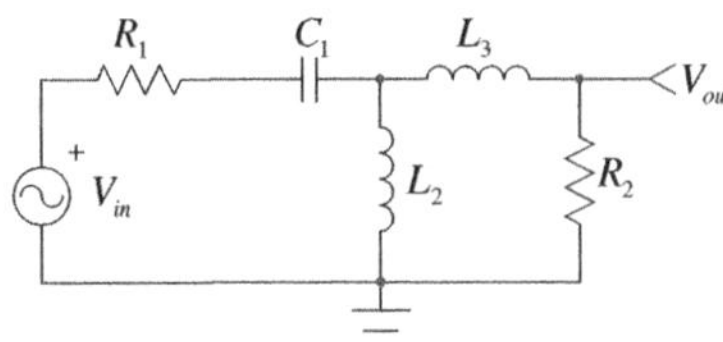

Figure 5.24 This 3rd-order circuit transfer function can be quickly obtained using the generalized expression from Figure 5.21.

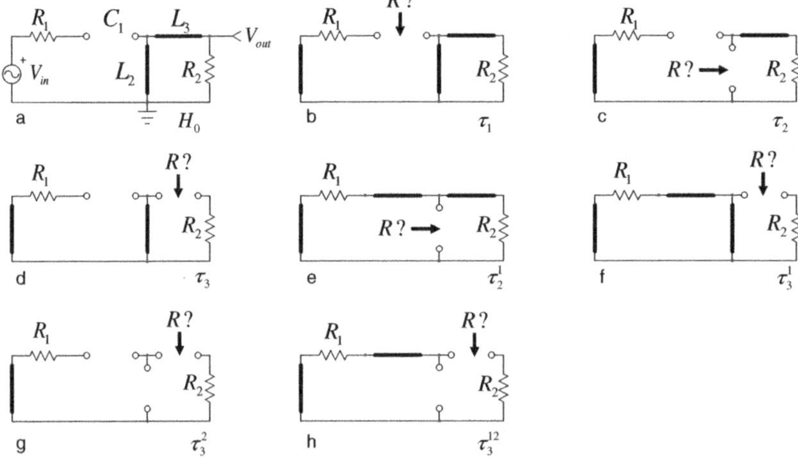

Figure 5.25 We break the circuit in intermediate steps to determine the natural time constants. Except the first case to determine H_0, the excitation is turned off for the other sketches.

If we start from the left, Figure 5.25a, we have:

$$H_0 = 0 \tag{5.77}$$

then

$$\tau_1 = R_1 C_1 \tag{5.78}$$

$$\tau_2 = \frac{L_2}{R_2} \tag{5.79}$$

$$\tau_3 = \frac{L_3}{R_2} \tag{5.80}$$

We can express b_1 as:

$$b_1 = \tau_1 + \tau_2 + \tau_3 = R_1 C_1 + \frac{L_2 + L_3}{R_2} \tag{5.81}$$

If $R_1 = R_2 = R$ this expression can be rewritten as:

$$b_1 = RC_1 + \frac{L_2 + L_3}{R} \tag{5.82}$$

If we continue with Figure 5.25e, we can determine:

$$\tau_2^1 = \frac{L_2}{R_1 \| R_2} \tag{5.83}$$

$$\tau_3^1 = \frac{L_3}{R_2} \tag{5.84}$$

$$\tau_3^2 = \frac{L_3}{\infty} = 0 \tag{5.85}$$

We can express b_2 as:

$$b_2 = \tau_1\tau_2^1 + \tau_1\tau_3^1 + \tau_2\tau_3^2 = R_1C_1\frac{L_2}{R_1 \| R_2} + R_1C_1\frac{L_3}{R_2} + \frac{L_2}{R_2}\cdot 0 = R_1C_1\frac{L_2}{R_1 \| R_2} + R_1C_1\frac{L_3}{R_2} \tag{5.86}$$

If $R_1 = R_2 = R$ this expression simplifies to:

$$b_2 = RC_1\frac{L_2}{\frac{R}{2}} + RC_1\frac{L_3}{R} = C_1(2L_2 + L_3) \tag{5.87}$$

And finally, from Figure 5.25h, we obtain:

$$\tau_3^{12} = \frac{L_3}{R_1 + R_2} \tag{5.88}$$

leading to

$$b_3 = \tau_1\tau_2^1\tau_3^{12} = R_1C_1\frac{L_2}{R_1 \| R_2}\frac{L_3}{R_1 + R_2} \tag{5.89}$$

If $R_1 = R_2 = R$ this expression simplifies to

$$b_3 = 2L_2C_1\frac{L_3}{2R} = \frac{C_1L_2L_3}{R} \tag{5.90}$$

Capitalizing on (5.82), (5.87) and (5.90), the denominator can be expressed as:

$$D(s) = 1 + s\left(RC_1 + \frac{L_2 + L_3}{R}\right) + s^2C_1(2L_2 + L_3) + s^3\frac{L_2L_3C_1}{R} \tag{5.91}$$

Now that the denominator has been found, we can apply the generalized transfer function definition by finding gains in various configurations as described by Figure 5.26.

In Figure 5.26a, capacitor C_1 is put in its high-frequency state (short circuit) while the other elements are in their dc state. In this mode, the gain is simply 0 as L_2 grounds the signal:

$$H^1 = 0 \tag{5.92}$$

For the other combinations, H^2 and H^3, the gain is also 0:

$$H^2 = H^3 = 0 \tag{5.93}$$

In Figure 5.26d, C_1 and L_2 are set in their high-frequency states (respectively short-circuited and open-circuited) while L_3 is in dc state (a short circuit). In this condition, a gain exists and is equal to:

$$H^{12} = \frac{R_2}{R_1 + R_2} \tag{5.94}$$

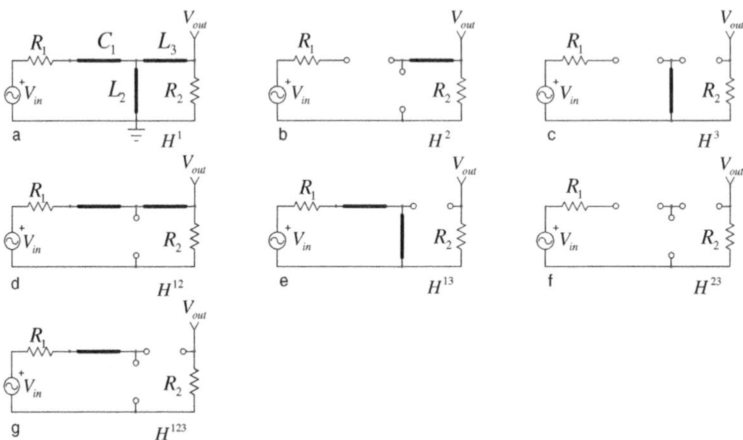

Figure 5.26 Same exercise as before except that the excitation is back and we look at the various transfer functions obtained when the energy-storing elements are alternatively set in their dc or high-frequency state.

If $R_1 = R_2 = R$ this expression simplifies to:

$$H^{12} = \frac{R}{2R} = 0.5 \tag{5.95}$$

Then, from Figure 5.26e and f, the gain is 0:

$$H^{13} = H^{23} = 0 \tag{5.96}$$

Finally, when all elements are in their high-frequency state, you see from Figure 5.26g that:

$$H^{123} = 0 \tag{5.97}$$

We can write the denominator assembling these gain expressions with the previously calculated time constants:

$$N(s) = H_0 + s\left(\tau_1 H^1 + \tau_2 H^2 + \tau_3 H^3\right) + s^2\left(\tau_1\tau_2^1 H^{12} + \tau_1\tau_3^1 H^{13} + \tau_2\tau_3^2 H^{23}\right) + s^3\left(\tau_1\tau_2^1\tau_3^{12} H^{123}\right) \tag{5.98}$$

Given numerous terms equal to 0, the numerator simplifies to:

$$N(s) = s^2\tau_1\tau_2^1 H^{12} = s^2 R_1 C_1 \frac{L_2}{R_1 \| R_2} \frac{R_2}{R_1 + R_2} = s^2 R_1 C_1 \frac{L_2(R_1 + R_2)}{R_1 R_2} \frac{R_2}{R_1 + R_2} = s^2 L_2 C_1 \tag{5.99}$$

The final transfer function combines (5.91) and (5.99) to form:

$$H(s) = \frac{s^2 L_2 C_1}{1 + s\left(RC_1 + \frac{L_2 + L_3}{R}\right) + s^2 C_1(2L_2 + L_3) + s^3 \frac{L_2 L_3 C_1}{R}} \tag{5.100}$$

We can rework this expression by factoring the numerator and simplifying:

$$H(s) = \frac{1}{\frac{1}{s^2 L_2 C_1} + \frac{s\left(RC_1 + \frac{L_2 + L_3}{R}\right)}{s^2 L_2 C_1} + \frac{s^2 C_1(2L_2 + L_3)}{s^2 L_2 C_1} + \frac{s^3 \frac{L_2 L_3 C_1}{R}}{s^2 L_2 C_1}} \tag{5.101}$$

After factoring and simplifying, we obtain:

$$H(s) = \frac{L_2}{2L_2 + L_3} \frac{1}{1 + \frac{L_2 + L_3 + C_1 R^2}{sRC_1(2L_2 + L_3)} + \frac{sL_2 L_3}{R(2L_2 + L_3)} + \frac{1}{s^2 C_1(2L_2 + L_3)}} \tag{5.102}$$

This is a first possible arrangement for our transfer function. We can also rework (5.100) by applying definitions seen in Chapter 2 in which the 3rd-order denominator can be put under the product of a low-frequency pole and a 2nd-order polynomial form:

$$D(s) \approx \left(1 + \frac{s}{\omega_p}\right)\left(1 + \frac{s}{\omega_0 Q} + \left(\frac{s}{\omega_0}\right)^2\right) \tag{5.103}$$

Applying the definitions, we have:

$$\omega_p = \frac{1}{b_1} = \frac{1}{RC_1 + \frac{L_2 + L_3}{R}} \tag{5.104}$$

$$Q = \frac{b_1 b_3 \sqrt{\frac{b_1}{b_3}}}{b_1 b_2 - b_3} = \frac{\left(L_2{}^2 L_3 + L_2 L_3{}^2 + C_1 L_2 L_3 R^2\right)\sqrt{\frac{R^2 C_1 + L_2 + L_3}{C_1 L_2 L_3}}}{2\left(L_2{}^2 R + L_2 L_3 R + C_1 L_2 R^3\right) + L_3{}^2 R + R^3 C_1 L_3} \tag{5.105}$$

$$\omega_0 = \sqrt{\frac{C_1 R^2 + L_2 + L_3}{C_1 L_2 L_3}} \tag{5.106}$$

With these expressions on hand, we obtain a new transfer function defined as follows:

$$H(s) \approx \frac{\left(\frac{s}{\omega_z}\right)^2}{\left(1 + \frac{s}{\omega_p}\right)\left(1 + \frac{s}{\omega_0 Q} + \left(\frac{s}{\omega_0}\right)^2\right)} \tag{5.107}$$

where

$$\omega_z = \frac{1}{\sqrt{L_2 C_1}} \tag{5.108}$$

In the denominator, we can factor $(s/\omega_0)^2$ and rearrange the expression. We obtain a slightly different format:

$$H(s) \approx \left(\frac{\omega_0}{\omega_z}\right)^2 \frac{1}{\left(1 + \frac{s}{\omega_p}\right)\left(1 + \frac{\omega_0}{sQ} + \left(\frac{\omega_0}{s}\right)^2\right)} \tag{5.109}$$

To compare these results with a reference equation, we need a raw expression extracted from Figure 5.24. If we apply Thévenin's followed by an impedance divider, the transfer function is:

$$H_{ref}(s) = \frac{sL_2}{sL_2 + R_1 + \frac{1}{sC_1}} \cdot \frac{R_2}{(sL_2) \| \left(R_1 + \frac{1}{sC_1}\right) + sL_3 + R_2} \tag{5.110}$$

If $R_1 = R_2 = R$ this expression does not really change and becomes:

$$H_{ref}(s) = \frac{sL_2}{sL_2 + R + \frac{1}{sC_1}} \cdot \frac{R}{(sL_2) \| \left(R + \frac{1}{sC_1}\right) + sL_3 + R} \tag{5.111}$$

We can now compare the dynamic responses given by the various formulas we have derived. All plots and equations are gathered in Figure 5.27 while the SPICE simulation results appear in Figure 5.28.

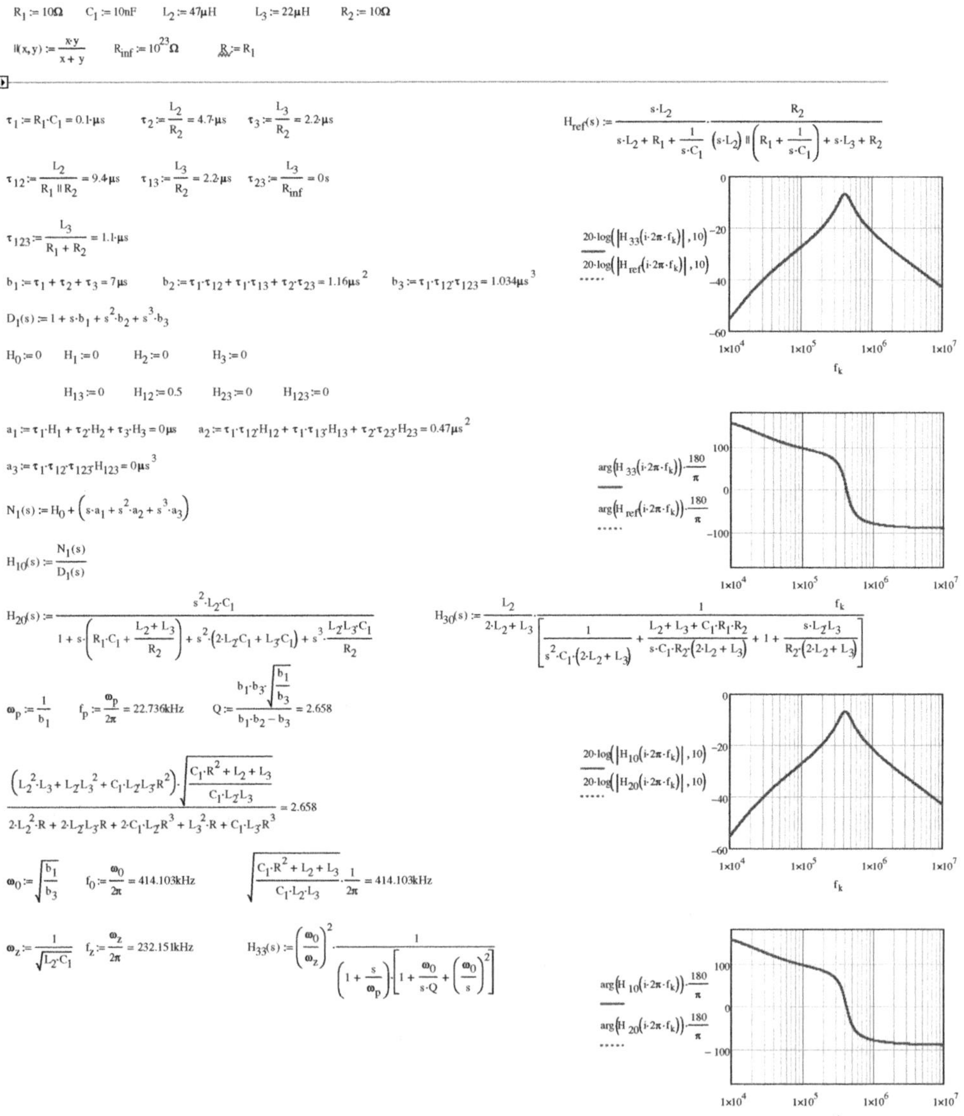

Figure 5.27 The Mathcad® sheet confirms our different derivations of the transfer function give similar responses.

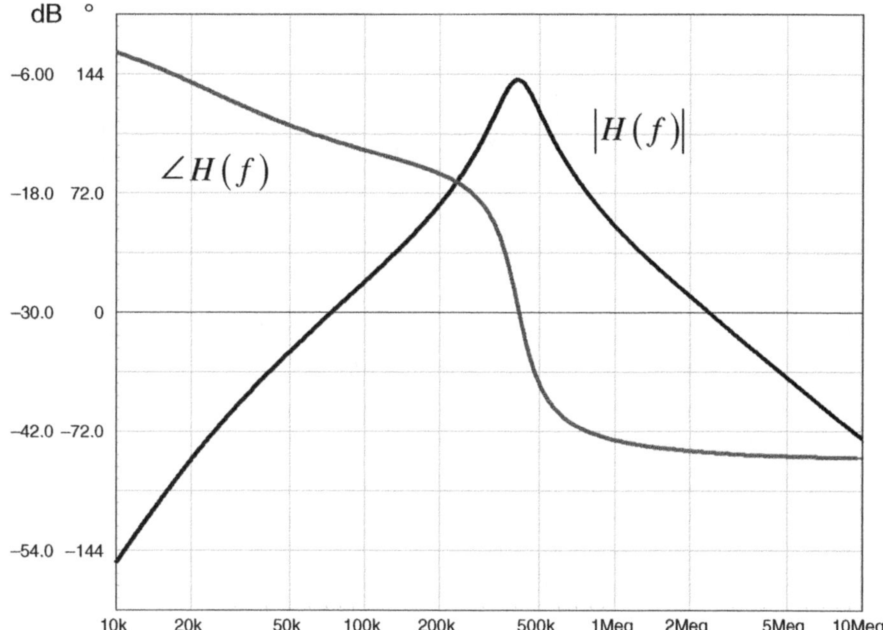

Figure 5.28 The SPICE simulation matches the Mathcad® plots very well.

All responses are identical. Please note that (5.109) does not exactly fit the format we want as the leading factor does not correspond to the peak we observe at resonance. If you see a better format for this expression, I will be happy to publish it with credits!

5.2 Five High-order Transfer Functions Examples

The first example appears in Figure 5.29. This is a circuit seen in Chapter 2 and used to explain how to find zeros by inspection. It looks like a complex circuit but we will see how easy it is to obtain its transfer function.

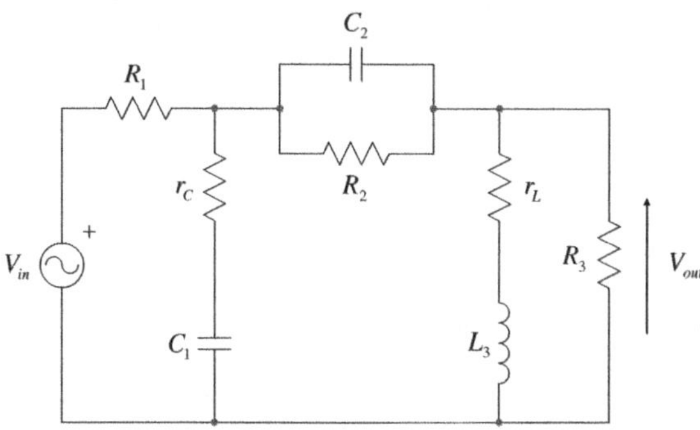

Figure 5.29 A 3rd-order circuit involving two capacitors and an inductor.

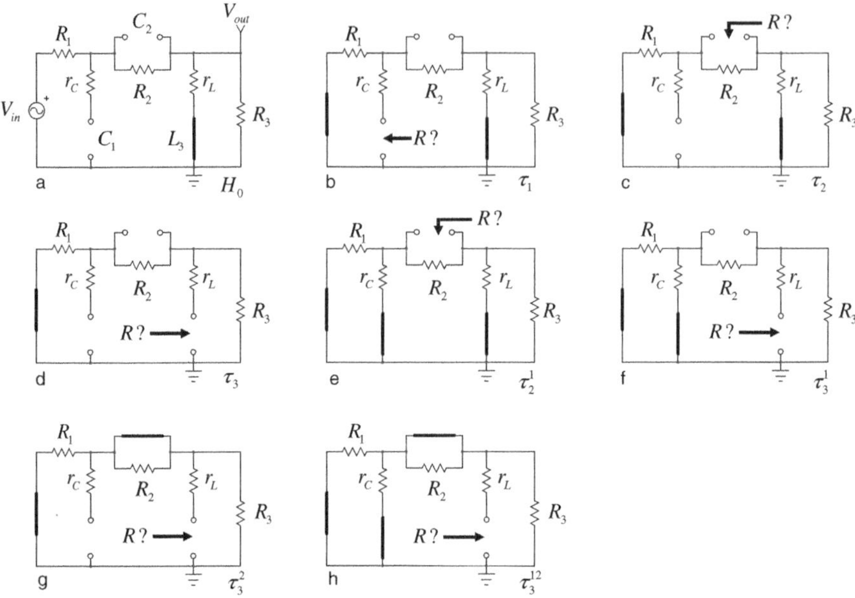

Figure 5.30 These sketches help in determining all natural time constants (when the excitation is turned off) and let you easily spot and later correct an error if any.

As usual, we have gathered all individual sketches pertinent to the determination of the natural time constants in Figure 5.30.

If we start from the left side, Figure 5.30a, the dc gain involves a simple resistive divider made of R_1 and R_2 with the paralleling of r_L and R_3:

$$H_0 = \frac{r_L \| R_3}{r_L \| R_3 + R_1 + R_2} \tag{5.112}$$

The time constants τ_1, τ_2 and τ_3 are obtained from Figure 5.30b to d:

$$\tau_1 = C_1 \left[r_C + \left(r_L \| R_3 + R_2 \right) \| R_1 \right] \tag{5.113}$$

$$\tau_2 = C_2 \left[R_2 \| \left(R_1 + r_L \| R_3 \right) \right] \tag{5.114}$$

$$\tau_3 = \frac{L_3}{r_L + R_3 \| (R_2 + R_1)} \tag{5.115}$$

We can form b_1 by adding these time constants:

$$b_1 = \tau_1 + \tau_2 + \tau_3 = C_1 \left\{ r_C + \left[\left(r_L \| R_3 \right) + R_2 \right] \| R_1 \right\} + C_2 \left[R_2 \| \left(R_1 + r_L \| R_3 \right) \right] + \frac{L_3}{r_L + R_3 \| (R_2 + R_1)} \tag{5.116}$$

In Figure 5.30e, capacitor C_1 is replaced by a short circuit and we look in C_2's terminals to determine τ_2^1:

$$\tau_2^1 = C_2[R_2 \| (R_1 \| r_C + r_L \| R_3)] \tag{5.117}$$

In Figure 5.30f, capacitor C_1 is still replaced by a short circuit but we now look in L_3's terminals to determine τ_3^1:

$$\tau_3^1 = \frac{L_3}{r_L + R_3 \| (R_2 + R_1 \| r_C)} \tag{5.118}$$

Finally, Figure 5.30g gives the last term for b_2 determined as:

$$\tau_3^2 = \frac{L_3}{r_L + R_3 \| R_1} \tag{5.119}$$

The term b_2 is obtained by adding the above time constants:

$$\begin{aligned} b_2 = \tau_1\tau_2^1 + \tau_1\tau_3^1 + \tau_2\tau_3^2 &= C_1\{r_C + [(r_L \| R_3) + R_2] \| R_1\}C_2\{R_2 \| [(R_1 \| r_C) + (r_L \| R_3)]\} \\ &+ C_1\{r_C + [(r_L \| R_3) + R_2] \| R_1\}\frac{L_3}{r_L + R_3 \| (R_2 + R_1 \| r_C)} \\ &+ C_2[R_2 \| (R_1 + r_L \| R_3)]\frac{L_3}{r_L + R_3 \| R_1} \end{aligned} \tag{5.120}$$

The last term is determined with the help of Figure 5.30h:

$$\tau_3^{12} = \frac{L_3}{r_L + R_3 \| r_C \| R_1} \tag{5.121}$$

The last term b_3 is thus equal to:

$$b_3 = \tau_1\tau_2^1\tau_3^{12} = C_1\{r_C + [(r_L \| R_3) + R_2] \| R_1\}C_2\{R_2 \| [(R_1 \| r_C) + (r_L \| R_3)]\}\frac{L_3}{r_L + R_3 \| r_C \| R_1} \tag{5.122}$$

Finally, the denominator $D(s)$ is assembled with (5.116), (5.120) and (5.122):

$$\begin{aligned} D(s) &= 1 + b_1 s + b_2 s^2 + b_3 s^3 \\ &= 1 + s\left(C_1\{r_C + [(r_L \| R_3) + R_2] \| R_1\} + C_2[R_2 \| (R_1 + r_L \| R_3)] + \frac{L_3}{r_L + R_3 \| (R_2 + R_1)}\right) \\ &+ s^2\left(\begin{aligned} &C_1\{r_C + [(r_L \| R_3) + R_2] \| R_1\}C_2\{R_2 \| [(R_1 \| r_C) + (r_L \| R_3)]\} \\ &+ C_1\{r_C + [(r_L \| R_3) + R_2] \| R_1\}\frac{L_3}{r_L + R_3 \| (R_2 + R_1 \| r_C)} \\ &+ C_2[R_2 \| (R_1 + r_L \| R_3)]\frac{L_3}{r_L + R_3 \| R_1} \end{aligned}\right) \\ &+ s^3\left(C_1\{r_C + [(r_L \| R_3) + R_2] \| R_1\}C_2\{R_2 \| [(R_1 \| r_C) + (r_L \| R_3)]\}\frac{L_3}{r_L + R_3 \| r_C \| R_1}\right) \end{aligned} \tag{5.123}$$

To determine the zeros, we have the choice between a) NDI b) the generalized approach using the natural time constants we have found or c) inspection. Needless to say that inspection, whenever possible, must be favored among all other approaches. It leads to the zeros position in the fastest way and always gives the least complex numerator. By the way, how many zeros do we have? How many energy-storing elements can we simultaneously put in their high-frequency state while observing a

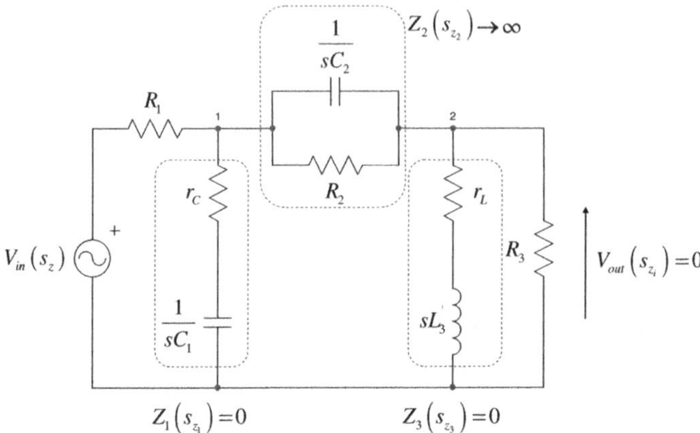

Figure 5.31 To determine the zeros by inspection, look at the transformed circuit and check combinations that could prevent the input waveform from generating a response: a transformed open with Z_2 and two transformed shorts with Z_1 and Z_3.

response in V_{out} of Figure 5.29? When C_1 and C_2 are shorted while L_3 is physically open, nothing prevents the input signal from propagating and reaching the output: there are 3 zeros in this circuit (3rd-order numerator). Figure 5.31 shows the impedance where these zeros are located. When Z_1 and Z_3 act as a transformed short, there are two zeros. The third zero is given by Z_2 which acts as a transformed open for $s = s_z$. The zeros positions are given below:

$$Z_1(s) = r_C + \frac{1}{sC_1} = \frac{1 + sr_CC_1}{sC_1} \tag{5.124}$$

What is the condition for which $Z_1(s_{z_1}) = 0$? Solve for the root when $1 + sr_CC_1 = 0$ and you find that the zero occurs for an angular frequency equal to:

$$s_{z_1} = -\frac{1}{r_CC_1} \text{ or } \omega_{z_1} = \frac{1}{r_CC_1} \tag{5.125}$$

Node 1 is connected to node 2 via the parallel arrangement of C_2 and R_2. To block the signal propagation, it would mean that Z_2 impedance evaluated at $s = s_{z2}$ becomes infinite. The impedance of the paralleled connection of C_2 and R_2 is defined as:

$$Z_2(s) = R_2 \| \frac{1}{sC_2} = \frac{R_2}{1 + sR_2C_2} \tag{5.126}$$

If we cancel the denominator, we have another zero for which Z_2 will become infinite:

$$1 + sR_2C_2 = 0 \tag{5.127}$$

implies that

$$s_{z_2} = -\frac{1}{R_2C_2} \text{ or } \omega_{z_2} = \frac{1}{R_2C_2} \tag{5.128}$$

Let's finish since we are now at node 2. In parallel with the load, we see the series connection of L_3 and its ESR labeled r_L. Can this series arrangement become a transformed short circuit for $s = s_{z3}$?

Let's check:

$$Z_3(s) = sL_3 + r_L = 0 \tag{5.129}$$

the solution is simply

$$s_{z_3} = -\frac{r_L}{L_3} \text{ or } \omega_{z_3} = \frac{r_L}{L_3} \tag{5.130}$$

Using (5.125), (5.131) and (5.130), we can immediately write the denominator polynomial as:

$$N(s) = (1 + sr_C C_1)(1 + sR_2 C_2)\left(1 + s\frac{L_3}{r_L}\right) = \left(1 + \frac{s}{\omega_{z_1}}\right)\left(1 + \frac{s}{\omega_{z_2}}\right)\left(1 + \frac{s}{\omega_{z_3}}\right) \tag{5.132}$$

The final transfer function H is obtained by dividing (5.132) by (5.123). Before we test this expression, we need a raw transfer function. By looking at Figure 5.29, we can transform the circuit using Thévenin's as shown in Figure 5.32.

From this figure, the raw transfer function H can be defined as:

$$H_{ref}(s) = \frac{r_C + \dfrac{1}{sC_1}}{r_C + \dfrac{1}{sC_1} + R_1} \cdot \frac{Z_3(s)}{R_{th}(s) + Z_2(s) + Z_3(s)} = \frac{1 + sr_C C_1}{1 + sC_1(r_C + R_1)} \cdot \frac{Z_3(s)}{R_{th}(s) + Z_2(s) + Z_3(s)} \tag{5.133}$$

with

$$R_{th}(s) = R_1 \parallel \left(r_C + \frac{1}{sC_1}\right) \tag{5.134}$$

$$Z_2(s) = \frac{1}{sC_2} \parallel R_2 \tag{5.135}$$

$$Z_3(s) = (r_L + sL_3) \parallel R_3 \tag{5.136}$$

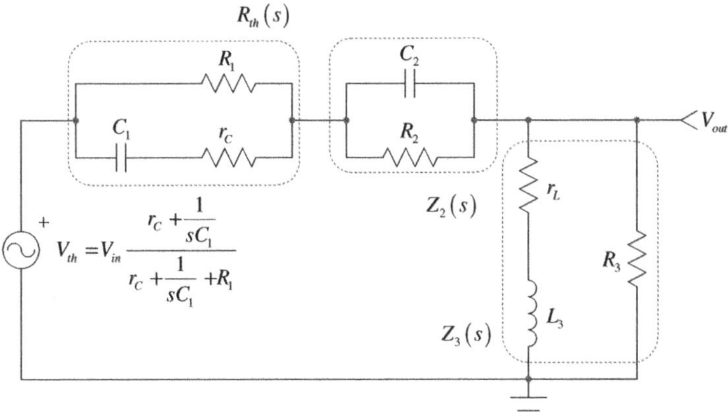

Figure 5.32 Using Thévenin's theorem leads us to the exact but extremely complex transfer function expression.

We have everything to build a Mathcad® sheet and compare the various dynamic responses. All results appear in Figure 5.33 and prove the correctness of the approach. The matching with SPICE is excellent. Please note that I had made a mistake at first and a clear divergence existed between H_1 and the raw response from (5.133). I spotted the mistakes in H_0 and τ_1 and corrected them in a few seconds to unveil the exact response. The organization of separated time constants as shown in Figure 5.33 lends itself very well to these late corrections and this presentation format must be encouraged to

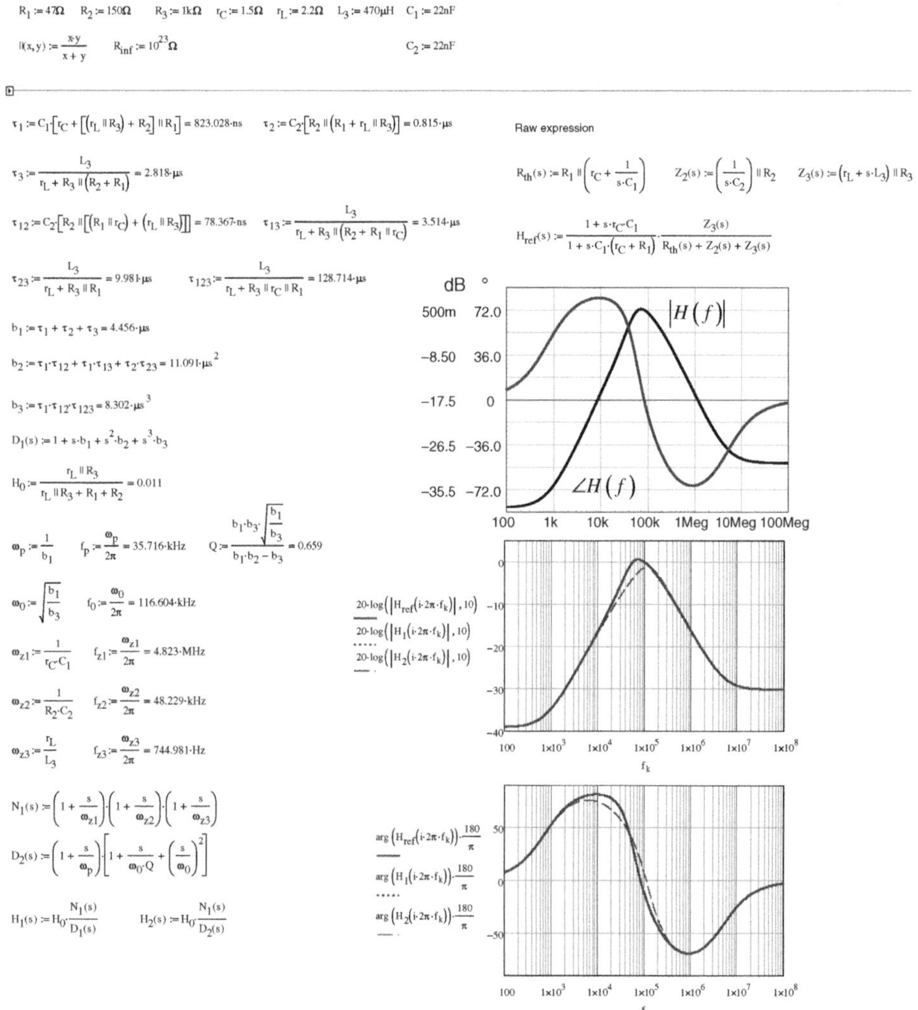

Figure 5.33 Excellent matching between the raw and low-entropy expressions. The dynamic response is similar to that of SPICE. Please note that the approximated response H_2 slightly diverges from the exact response near the peak.

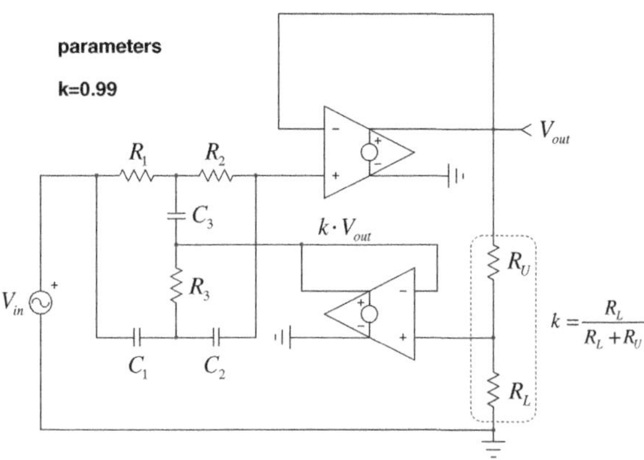

Figure 5.34 This active twin-T filter can be used to reject any unwanted frequency with a theoretical infinite Q.

follow. I have included an approximate transfer function, H_2, transforming the denominator into the product of a pole with a 2nd-order polynomial form. The overall shape is in good agreement but slightly deviates near the peak.

5.2.1 *Example 2: A 3rd-order Active Notch Circuit*

The circuit appears in Figure 5.34. It differs from the classical passive notch circuit in which the junction of C_3-R_3 is grounded. Here, a portion of V_{out} is actually biasing this junction and helps building an extremely sharp rejecter. By selecting k between 0 (grounded, similar to the classical passive notch) and less than 1, you can tweak the notch width as you need to. For instance, for k values approaching 1, we will see that the dynamic response is extremely sharp, really rejecting the center frequency without affecting frequencies before or after resonance. As k reduces, the funnel widens and affects frequencies around resonance. We will start by determining the natural circuit time constants as detailed in Figure 5.35 and Figure 5.36.

In Figure 5.35a, the quasi-static gain is immediate and is equal to 1 since only R_1 and R_2 are involved in the input path:

$$H_0 = 1 \tag{5.137}$$

In Figure 5.35b, the test generator is grounded and its right terminal returns to ground via R_3 and the source kV_{out}. In this case, V_{out} is zero therefore the resistance seen from C_1's terminals is R_3:

$$\tau_1 = C_1 R_3 \tag{5.138}$$

In Figure 5.35c, we need to invoke KCL and KVL. The test generator voltage V_T is equal to:

$$V_T = I_T R_3 + kV_{out} - V_{out} = I_T R_3 + V_{out}(k - 1) \tag{5.139}$$

By observing the drawing, V_{out} is also present across R_1 and R_2 but given the circulation of I_T, the sign is negative:

$$V_{out} = -I_T(R_1 + R_2) \tag{5.140}$$

If we substitute (5.140) into (5.139), we have:

$$V_T = I_T R_3 - I_T(R_1 + R_2)(k - 1) = I_T[R_3 + (1 - k)(R_1 + R_2)] \tag{5.141}$$

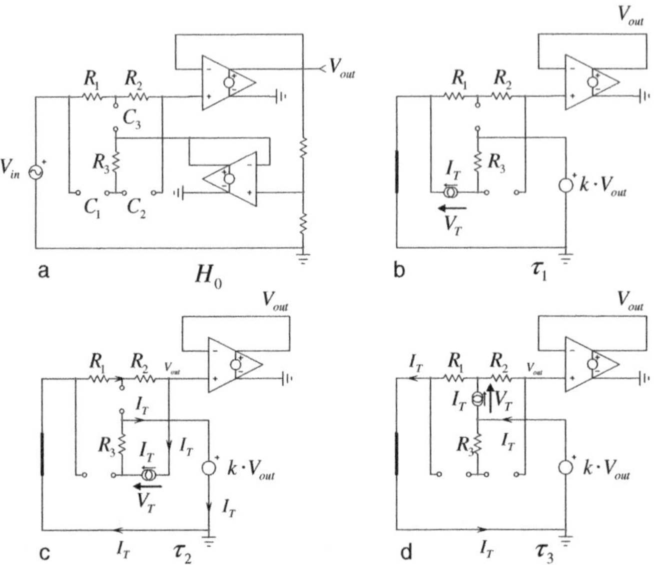

Figure 5.35 Excitation is set to 0 to calculate the first time constants.

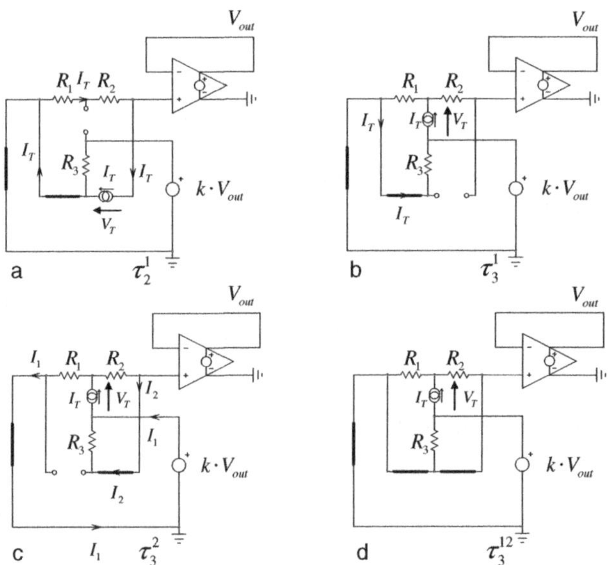

Figure 5.36 Higher-order time constants require some simple manipulations to determine the resistance driving the selected reactance.

The second time constant is:

$$\tau_2 = C_2[R_3 + (1-k)(R_1 + R_2)] \quad (5.142)$$

With Figure 5.35d, we should be able to determine τ_3. The voltage across the generator is equal to:

$$V_T = I_T R_1 - kV_{out} \quad (5.143)$$

The voltage drop across R_2 is zero since C_2 is open and no current flows in this branch. As R_2's right terminal is equal to V_{out}, we have:

$$V_{out} = I_T R_1 \quad (5.144)$$

If we update (5.143) with (5.144), we have:

$$V_T = I_T R_1 - kI_T R_1 = I_T[R_1(1-k)] \quad (5.145)$$

The third time constant is thus:

$$\tau_3 = C_3 R_1(1-k) \quad (5.146)$$

We can form b_1 equal to:

$$b_1 = \tau_1 + \tau_2 + \tau_3 = R_3C_1 + C_2[R_3 + (1-k)(R_1 + R_2)] + C_3R_1(1-k) \quad (5.147)$$

In practice, the notch design considers equal resistance values for R_1 and R_2, while R_3 is half this value:

$$\begin{aligned} R_1 &= R_2 = R \\ R_3 &= \frac{R}{2} \end{aligned} \quad (5.148)$$

then

$$\begin{aligned} C_1 &= C_2 = C \\ C_3 &= 2C \end{aligned} \quad (5.149)$$

If we consider these values and substitute them in (5.147), we obtain:

$$b_1 = RC(5-4k) \quad (5.150)$$

In Figure 5.36a, C_1 is replaced by a short circuit and we look through C_2's terminals. Current I_T flows in R_1 and R_2 implying that:

$$V_{out} = -I_T(R_1 + R_2) \quad (5.151)$$

As the current generator's left terminal is grounded, the voltage across it is $-V_{out}$, therefore:

$$V_T = I_T(R_1 + R_2) \quad (5.152)$$

Otherwise stated:

$$\tau_2^1 = C_2(R_1 + R_2) \quad (5.153)$$

In Figure 5.36b, C_1 is still a short circuit but we now look through C_3's terminals. From the figure as we see no current flows in R_2 (C_2 is open):

$$V_{out} = I_T R_1 \quad (5.154)$$

The low-side terminal of the current generator is thus biased at:

$$kV_{out} = kI_T R_1 \tag{5.155}$$

The current generator voltage is thus:

$$V_T = V_{out} - kV_{out} = I_T R_1 - kI_T R_1 = I_T R_1(1-k) \tag{5.156}$$

The second time constant τ_3^1 is thus equal to:

$$\tau_3^1 = C_3 R_1(1-k) \tag{5.157}$$

The last time constant τ_3^2 is rather more complex to determine if we look at Figure 5.36c. The current I_T splits into I_1 and I_2. We have:

$$I_T = I_1 + I_2 \tag{5.158}$$

I_2 flows in R_3 therefore:

$$I_2 = \frac{V_{out} - kV_{out}}{R_3} = V_{out}\frac{1-k}{R_3} \tag{5.159}$$

The voltage at the current generator upper terminal is equal to:

$$I_1 R_1 = V_{out} + R_2 I_2 \tag{5.160}$$

If you substitute (5.159) in (5.160) and solve for I_1, you find:

$$I_1 = \frac{V_{out}\left(1 + R_2\frac{1-k}{R_3}\right)}{R_1} \tag{5.161}$$

From (5.158):

$$I_T = V_{out}\frac{1-k}{R_3} + \frac{V_{out}\left(1 + R_2\frac{1-k}{R_3}\right)}{R_1} = \frac{V_{out}[R_1 + R_2 + R_3 - k(R_1 + R_2)]}{R_1 R_3} \tag{5.162}$$

therefore

$$V_{out} = \frac{R_1 R_3}{R_1 + R_2 + R_3 - k(R_1 + R_2)} I_T \tag{5.163}$$

The generator voltage V_T is defined as:

$$V_T = I_1 R_1 - kV_{out} \tag{5.164}$$

Substituting (5.161) in (5.164), then replacing V_{out} by (5.163) gives us:

$$V_T = R_1\frac{V_{out}\left(1 + R_2\frac{1-k}{R_3}\right)}{R_1} - kV_{out} \tag{5.165}$$

$$V_T = I_T\frac{R_1(1-k)(R_2 + R_3)}{R_1 + R_2 + R_3 - k(R_1 + R_2)} \tag{5.166}$$

The time constant τ_3^2 is thus equal to:

$$\tau_3^2 = C_3\left[\frac{R_1(1-k)(R_2+R_3)}{R_1+R_2+R_3-k(R_1+R_2)}\right] \tag{5.167}$$

We can now express coefficient b_2 by adding (5.153), (5.157) and (5.167):

$$\begin{aligned} b_2 &= \tau_1\tau_2^1 + \tau_1\tau_3^1 + \tau_2\tau_3^2 \\ &= C_1R_3C_2(R_1+R_2) + C_1R_3C_3R_1(1-k) \\ &+ C_2[R_3+(1-k)(R_1+R_2)]C_3\left[\frac{R_1(1-k)(R_2+R_3)}{R_1+R_2+R_3-k(R_1+R_2)}\right] \end{aligned} \tag{5.168}$$

If we consider (5.148) and (5.149), this expression simplifies to:

$$b_2 = (RC)^2(5-4k) \tag{5.169}$$

The last coefficient, b_3, is obtained with the help of Figure 5.36d. Here, the output is equal to 0 V, implying that the low-side terminal of the current generator is also grounded. Current I_T splits between R_1 and R_2 implying that:

$$\tau_3^{12} = C_3\left(R_1 \| R_2\right) \tag{5.170}$$

Reusing (5.138) and (5.153), we have:

$$b_3 = \tau_1\tau_2^1\tau_3^{12} = C_1R_3C_2(R_1+R_2)C_3\left(R_1 \| R_2\right) \tag{5.171}$$

If we consider (5.148) and (5.149), this expression simplifies to:

$$b_3 = (RC)^3 \tag{5.172}$$

The denominator considering (5.148) and (5.149) component values is thus expressed by:

$$D(s) = 1 + sRC(5-4k) + s^2R^2C^2(5-4k) + s^3R^3C^3 \tag{5.173}$$

We can determine the numerator using NDI or the generalized transfer function expression for a 3rd-order system:

$$N(s) = H_0 + s\left(\tau_1H^1 + \tau_2H^2 + \tau_3H^3\right) + s^2\left(\tau_1\tau_2^1H^{12} + \tau_1\tau_3^1H^{13} + \tau_2\tau_3^2H^{23}\right) + s^3\left(\tau_1\tau_2^1\tau_3^{12}H^{123}\right) \tag{5.174}$$

H_0 has already been determined and is equal to 1. All remaining transfer functions are described in the series of sketches from Figure 5.37. In sketch (a), capacitor C_1 is replaced by a short circuit while the other elements are kept in their dc state (all capacitors open). kV_{out} plays no role and the gain is 1:

$$H^1 = 1 \tag{5.175}$$

In Figure 5.37b, the configuration requires a quick intermediate state as shown in Figure 5.38 which represents a simplified version. We can apply superposition theorem to express the ratio V_{out} to V_{in}. If V_{in} is equal to 0, we have:

$$V_{out1}\big|_{V_{in}=0} = kV_{out}\frac{R_1+R_2}{R_1+R_2+R_3} \tag{5.176}$$

Now, if kV_{out} is made 0, we have:

$$V_{out2}\big|_{kV_{out}=0} = V_{in}\frac{R_3}{R_1+R_2+R_3} \tag{5.177}$$

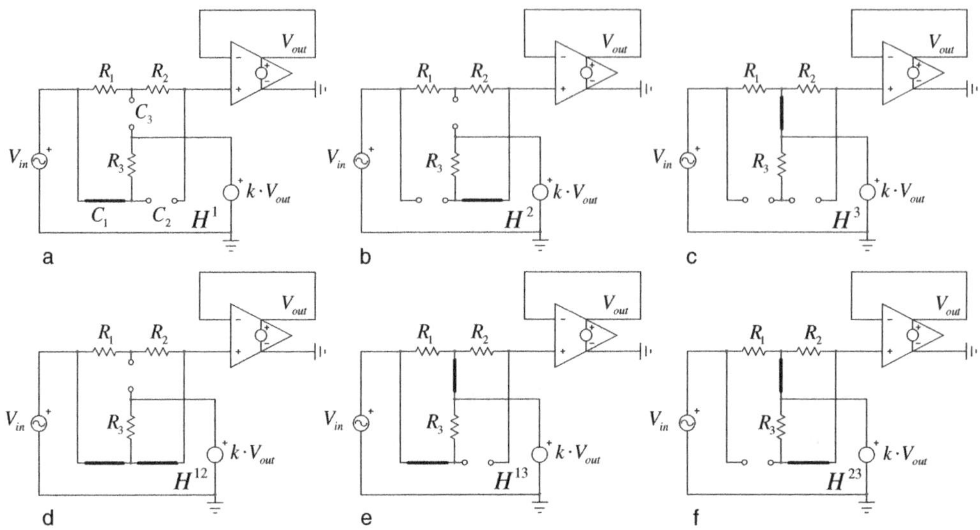

Figure 5.37 Gains are easily calculated by setting energy-storing elements in their high-frequency state or in their dc state.

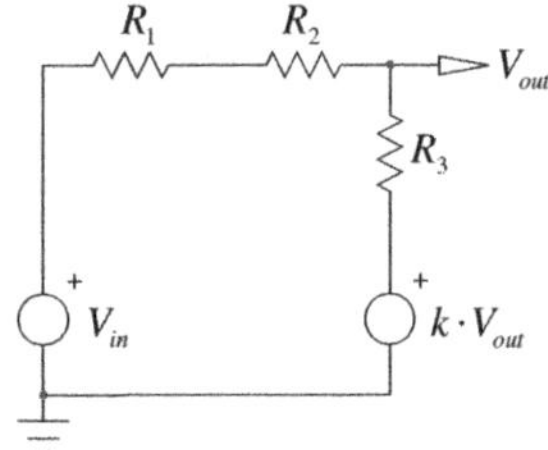

Figure 5.38 With this intermediate sketch, the gain when C_2 is replaced by a short circuit is quickly obtained.

The output voltage is the sum of these two expressions:

$$V_{out} = kV_{out}\frac{R_1 + R_2}{R_1 + R_2 + R_3} + V_{in}\frac{R_3}{R_1 + R_2 + R_3} \tag{5.178}$$

Rearranging and factoring, we find:

$$H^2 = \frac{R_3}{R_1 + R_2 + R_3}\frac{1}{1 - k\frac{R_1 + R_2}{R_1 + R_2 + R_3}} = \frac{R_3}{R_1 + R_2 + R_3 - k(R_1 + R_2)} \tag{5.179}$$

H_3 is found with the help of Figure 5.37c. If you observe the figure, no current flows in R_2 but its right terminal is at V_{out} while its left terminal is biased at kV_{out}. Without current, this condition can only be satisfied if V_{out} equals 0. Therefore:

$$H^3 = 0 \tag{5.180}$$

We can now form coefficient a_1 by combining the natural time constants found for the denominator and the above gain definitions. As H^3 is zero, we have only two terms:

$$\begin{aligned} a_1 &= \tau_1 H^1 + \tau_2 H^2 + \tau_3 H^3 \\ &= C_1 R_3 + C_2[R_3 + (1-k)(R_1 + R_2)]\frac{R_3}{R_1 + R_2 + R_3 - k(R_1 + R_2)} + C_3 R_1 (1-k)\cdot 0 \\ &= R_3(C_1 + C_2) \end{aligned} \tag{5.181}$$

If we replace R_1, R_2, R_3, C_1 and C_2 by the below values:

$$\begin{aligned} R_1 &= R_2 = R \\ R_3 &= \frac{R}{2} \\ C_1 &= C_2 = C \\ C_3 &= 2C \end{aligned} \tag{5.182}$$

Then (5.181) nicely simplifies to:

$$a_1 = RC \tag{5.183}$$

For H^{12} in Figure 5.37d, C_1 and C_2 short R_1 and R_2 leading to:

$$H^{12} = 1 \tag{5.184}$$

For H^{13} in Figure 5.37e, same situation as for H^3 where R_2 has a bias while no current crosses it, thus:

$$H^{13} = 0 \tag{5.185}$$

H^{23} depicted in Figure 5.37f is also zero since no current cross R_2 and R_3 despite a bias at kV_{out} for R_2's left terminal and V_{out} at its right terminal:

$$H^{23} = 0 \tag{5.186}$$

The second term a_2 can be assembled by associating the 2nd-order time constants values found for D. Just one term because H^{13} and H^{23} are 0:

$$a_2 = \tau_1\tau_2^1 H^{12} + \tau_1\tau_3^1 H^{13} + \tau_2\tau_3^2 H^{23} = C_1 R_3 C_2 (R_1 + R_2) \tag{5.187}$$

Considering the correspondence between elements described by (5.182), this terms simplifies to:

$$a_2 = R^2 C^2 \tag{5.188}$$

Finally, Figure 5.39 depicts the configuration for the last gain in which all capacitors are replaced by short circuits.

As R_1 and R_2 are shorted, the gain of this network is 1:

$$H^{123} = 1 \tag{5.189}$$

The last term is a_3 and is defined as:

$$a_3 = \tau_1\tau_2^1\tau_3^{12} H^{123} = C_1 R_3 C_2 (R_1 + R_3) C_3 \left(R_1 \| R_2\right) \tag{5.190}$$

Considering the correspondence between elements described by (5.182), this terms simplifies to:

$$a_3 = R^3 C^3 \tag{5.191}$$

The numerator $N(s)$ can thus expressed as:

$$N(s) = 1 + sRC + s^2 R^2 C^2 + s^3 R^3 C^3 \tag{5.192}$$

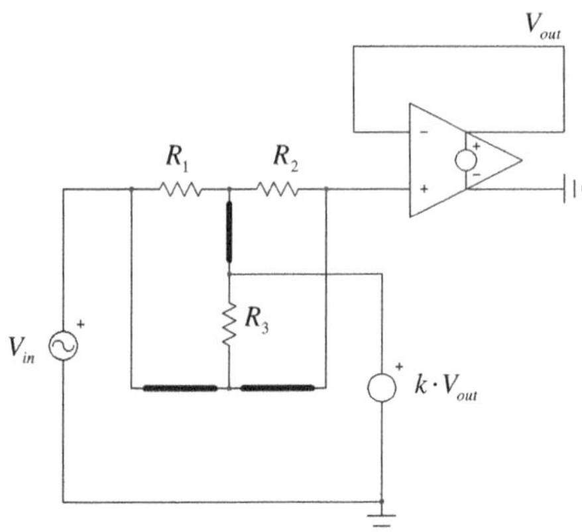

Figure 5.39 The last gain H^{123} is determined when all capacitors are set to their high-frequency state: a short circuit.

With $D(s)$ defined in (5.173), we have our transfer function:

$$H(s) = \frac{1 + sRC + s^2R^2C^2 + s^3R^3C^3}{1 + sRC(5 - 4k) + s^2R^2C^2(5 - 4k) + s^3R^3C^3} \tag{5.193}$$

Can we easily rearrange it in a simpler, more compact format? We can first try to factor the numerator. Ideally, we would like to put under the following form:

$$N(s) = (1 + x_1 s)\left(1 + sy_1 + s^2 y_2\right) \tag{5.194}$$

If we develop this expression, we obtain:

$$(1 + x_1 s)\left(1 + sy_1 + s^2 y_2\right) = 1 + s(x_1 + y_1) + s^2(x_1 y_1 + y_2) + s^3 x_1 y_2 \tag{5.195}$$

If we identify all terms with $N(s)$ from (5.192), we have the following system of equations:

$$\begin{aligned} x_1 + y_1 &= RC \\ x_1 y_1 + y_2 &= R^2C^2 \\ x_1 y_2 &= R^3C^3 \end{aligned} \tag{5.196}$$

From this result, $N(s)$ can be factored into:

$$N(s) = (1 + sRC)\left(1 + s^2R^2C^2\right) \tag{5.197}$$

If we write:

$$\omega_0 = \frac{1}{RC} \tag{5.198}$$

then

$$N(s) = \left(1 + \frac{s}{\omega_0}\right)\left(1 + \left(\frac{s}{\omega_0}\right)^2\right) \tag{5.199}$$

In this expression as the term a_1 is 0 in the second polynomial, it simply means that Q_N approaches infinity.

For the denominator, we follow the exact same path in an attempt to factor it as in (5.194). If we identify all terms in (5.195) with $D(s)$ from (5.173), we have the following system of equations:

$$\begin{aligned} x_1 + y_1 &= RC(5-4k) \\ x_1 y_1 + y_2 &= R^2C^2(5-4k) \\ x_1 y_2 &= R^3C^3 \end{aligned} \tag{5.200}$$

Solving this three-unknown system, the denominator can be nicely rearranged as:

$$D(s) = (1 + sRC)\left(1 + 4sRC(1-k) + s^2R^2C^2\right) \tag{5.201}$$

If we adopt (5.198), (5.201) can be rewritten as:

$$D(s) = \left(1 + \frac{s}{\omega_0}\right)\left(1 + 4\frac{s}{\omega_0}(1-k) + \left(\frac{s}{\omega_0}\right)^2\right) \tag{5.202}$$

The quality factor Q is simply defined by solving:

$$4\frac{s}{\omega_0}(1-k) = \frac{s}{\omega_0 Q} \tag{5.203}$$

giving

$$Q = \frac{1}{4(1-k)} \tag{5.204}$$

Finally, the transfer function greatly simplifies since $1 + \frac{s}{\omega_0}$ disappears in N and D:

$$H(s) = \frac{\left(1 + \frac{s}{\omega_0}\right)\left(1 + \left(\frac{s}{\omega_0}\right)^2\right)}{\left(1 + \frac{s}{\omega_0}\right)\left(1 + 4\frac{s}{\omega_0}(1-k) + \left(\frac{s}{\omega_0}\right)^2\right)} = \frac{1 + \left(\frac{s}{\omega_0}\right)^2}{1 + 4\frac{s}{\omega_0}(1-k) + \left(\frac{s}{\omega_0}\right)^2} \tag{5.205}$$

You see that if k reaches 1, Q approaches infinity and (5.205) simplifies to 1: the notch disappears. We now have everything to run a series of tests and check how our transfer functions behave. Figure 5.40 gathers all Mathcad® equations. The left side corresponds to equations involving individually-selected elements while the right side considers values following (5.182). As you can see, results are identical. The left-side equations are interesting in case you need to later assign component tolerances or have some of them individually vary. Time constant values correspond to a 60-Hz notch filter. Figure 5.41 shows the various responses you can get as k varies between 0 and 0.99. For k equals 0, C_3-R_3 junction is grounded and the notch response is that of a passive circuit. Finally, rather than deriving a raw transfer function, we have run a SPICE simulation for k equals 0.5 and superimposed Mathcad® curves with that of SPICE: they nicely agree as shown in Figure 5.42.

5.2.2 *Example 3: A 4th-order* LC *Passive Filter*

We are now going to derive the transfer function of Figure 5.43 circuit. You can identify two cascaded LC networks loaded by resistor R_1. There are four storage elements, this is a 4th-order system. There are no zeros as none of the four storage elements could be set in its high-frequency state without nulling the response: L_1 and L_3 open the series path while set in their high-frequency state and C_2 or C_4 would shunt the path to ground also nulling the response.

$$f_0 := 60\,\text{Hz} \qquad m := 0.99$$

$$R := 10\,\text{M}\Omega \qquad C := \frac{1}{2\pi \cdot f_0 \cdot R} = 265.258238\ \text{pF}$$

$$C_2 := C \quad C_1 := C \quad C_3 := 2\cdot C \quad R_1 := R \quad R_2 := R \quad R_3 := \frac{R}{2} \qquad \|(x,y) := \frac{x\cdot y}{x+y}$$

$$\tau_1 := C_1\cdot R_3 = 1.326291\cdot\text{ms}$$

$$\tau_2 := C_2\cdot\left[R_3 + \left(R_1 + R_2\right)\cdot(1-m)\right] = 1.379343\cdot\text{ms}$$

$$\tau_3 := C_3\cdot R_1\cdot(1-m) = 53.051648\cdot\mu\text{s}$$

$$\tau_{12} := C_2\cdot\left(R_1 + R_2\right) = 5.305165\cdot\text{ms}$$

$$\tau_{13} := C_3\cdot R_1\cdot(1-m) = 53.051648\cdot\mu\text{s}$$

$$\tau_{23} := C_3\cdot\left[\frac{R_1\cdot(1-m)\cdot\left(R_2+R_3\right)}{R_1+R_2+R_3-m\cdot\left(R_1+R_2\right)}\right] = 0.153034\cdot\text{ms}$$

$$\tau_{123} := C_3\cdot\left(R_1 \,\|\, R_2\right) = 2.652582\cdot\text{ms}$$

$$b_1 := \tau_1 + \tau_2 + \tau_3 = 2.758686\cdot\text{ms}$$

$$b_2 := \tau_1\cdot\tau_{12} + \tau_1\cdot\tau_{13} + \tau_2\cdot\tau_{23} = 7.317641\times 10^{-6}\,\text{s}^2$$

$$b_3 := \tau_1\cdot\tau_{12}\cdot\tau_{123} = 1.866408\times 10^{-8}\,\text{s}^3$$

$$H_0 := 1 \qquad H_1 := 1 \qquad H_2 := \frac{R_3}{R_1+R_2+R_3-m\cdot\left(R_1+R_2\right)} = 0.961538 \qquad H_3 := 0$$

$$H_{12} := 1 \qquad H_{13} := 0 \qquad H_{23} := 0 \qquad H_{123} := 1$$

$$a_1 := \tau_1\cdot H_1 + \tau_2\cdot H_2 + \tau_3\cdot H_3 = 2.652582\cdot\text{ms}$$

$$a_2 := \tau_1\cdot\tau_{12}\cdot H_{12} + \tau_1\cdot\tau_{13}\cdot H_{13} + \tau_2\cdot\tau_{23}\cdot H_{23} = 7.036193\times 10^{-6}\,\text{s}^2$$

$$a_3 := \tau_1\cdot\tau_{12}\cdot\tau_{123}\cdot H_{123} = 1.866408\times 10^{-8}\,\text{s}^3$$

$$D_1(s) := 1 + s\cdot b_1 + s^2\cdot b_2 + s^3\cdot b_3$$

$$N_1(s) := H_0 + a_1\cdot s + a_2\cdot s^2 + a_3\cdot s^3 \qquad H_{10}(s) := \frac{N_1(s)}{D_1(s)}$$

$$\tau_{1a} := \frac{C\cdot R}{2} = 1.326291\cdot\text{ms}$$

$$\tau_{2a} := C\cdot\left[\frac{R}{2} + 2R\cdot(1-m)\right] = 1.379343\cdot\text{ms}$$

$$\tau_{3a} := 2\cdot C\cdot R\cdot(1-m) = 53.051648\cdot\mu\text{s}$$

$$\tau_{12a} := 2C\cdot R = 5.305165\cdot\text{ms}$$

$$\tau_{13a} := 2R\cdot C\cdot(1-m) = 53.051648\cdot\mu\text{s}$$

$$\tau_{23a} := 2\cdot C\cdot\left[\frac{R\cdot(1-m)\cdot\left(R+\frac{R}{2}\right)}{2R+\frac{R}{2}-m\cdot 2R}\right] = 0.153034\cdot\text{ms} \qquad \tau_{23aa} := \frac{6\cdot C\cdot R\cdot(m-1)}{4\cdot m-5} = 0.153034\cdot\text{ms}$$

$$\tau_{123a} := C\cdot R = 2.652582\cdot\text{ms}$$

$$b_{1a} := \frac{C\cdot R}{2} + C\cdot\left[\frac{R}{2} + 2R\cdot(1-m)\right] + 2\cdot C\cdot R\cdot(1-m) = 2.758686\cdot\text{ms} \qquad b_{1aa} := R\cdot C\cdot(5-4m) = 2.758686\cdot\text{ms}$$

$$b_{2aa} := (R\cdot C)^2\cdot(5-4\cdot m) = 7.317641\times 10^{-6}\,\text{s}^2$$

$$b_{3a} := \frac{C\cdot R}{2}\cdot(2C\cdot R)\cdot(C\cdot R) = 1.866408\times 10^{-8}\,\text{s}^3 \qquad b_{3aa} := (R\cdot C)^3 = 1.866408\times 10^{-8}\,\text{s}^3$$

$$H_{2a} := \frac{1}{5-4\cdot m} = 0.961538$$

H3 equals 0

$$a_{1a} := \frac{C\cdot R}{2} + C\cdot\left[\frac{R}{2} + 2R\cdot(1-m)\right]\cdot\frac{1}{5-4\cdot m} = 2.652582\cdot\text{ms} \qquad a_{1aa} := R\cdot C = 2.652582\cdot\text{ms}$$

$$a_{2a} := \frac{C\cdot R}{2}\cdot(2C\cdot R) = 7.036193\times 10^{-6}\,\text{s}^2 \qquad a_{2aa} := (R\cdot C)^2 = 7.036193\times 10^{-6}\,\text{s}^2$$

$$a_{3a} := \frac{C\cdot R}{2}\cdot(2C\cdot R)\cdot(C\cdot R) = 1.866408\times 10^{-8}\,\text{s}^3 \qquad a_{3aa} := (R\cdot C)^3 = 1.866408\times 10^{-8}\,\text{s}^3$$

$$\omega_0 := \frac{1}{R\cdot C} \qquad Q := \frac{1}{4(1-m)} = 25$$

$$H_{20}(s) := \frac{1 + s\cdot R\cdot C + s^2\cdot(R\cdot C)^2 + s^3\cdot(R\cdot C)^3}{1 + s\cdot R\cdot C\cdot(5-4m) + s^2\cdot(R\cdot C)^2\cdot(5-4m) + s^3\cdot(R\cdot C)^3} \qquad H_{30}(s) := \frac{1+\left(\frac{s}{\omega_0}\right)^2}{1+\frac{s}{\omega_0\cdot Q}+\left(\frac{s}{\omega_0}\right)^2}$$

Figure 5.40 The Mathcad® sheet shows literal expressions involving all independently-selected resistors and capacitors (left side) while the right side confirms results when (5.182) is adopted.

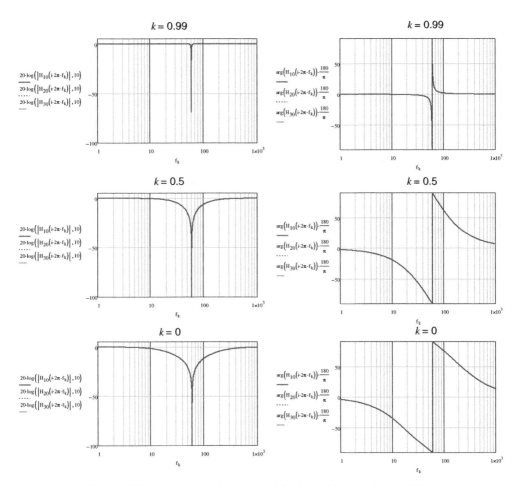

Figure 5.41 The funnel width varies as k is changed between 0 up to 0.99.

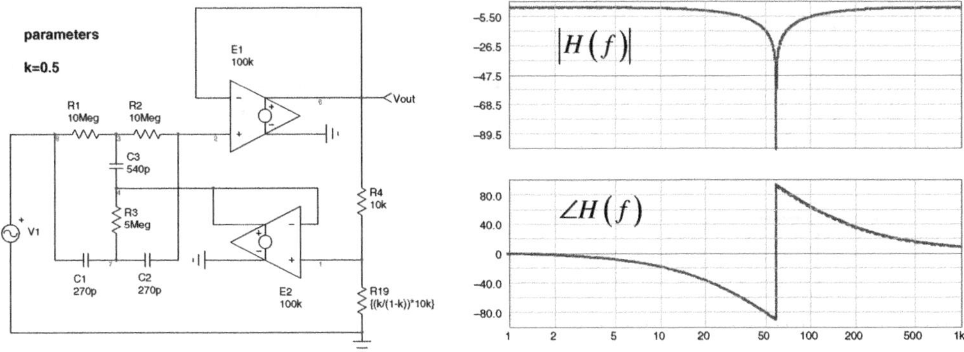

Figure 5.42 Results between simulation and SPICE match very well.

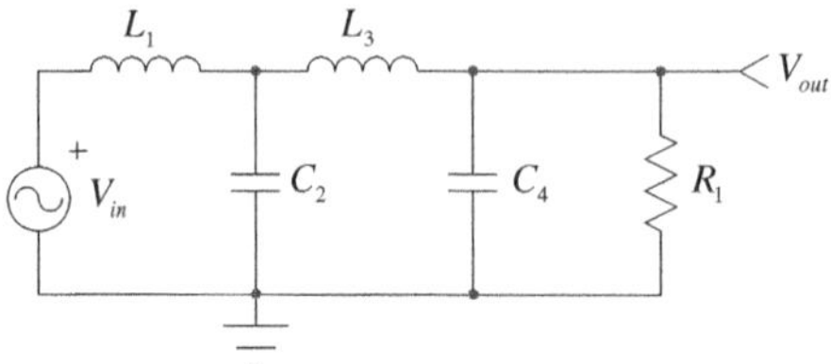

Figure 5.43 What is the transfer function of these cascaded *LC* networks?

The process remains the same despite the number of energy-storing elements. All the steps are gathered in Figure 5.44. The dc gain is immediate and equal to 1:

$$H_0 = 1 \tag{5.206}$$

The excitation is now turned off (replaced by a short circuit) and Figure 5.44b tells us that:

$$\tau_1 = \frac{L_1}{R_1} \tag{5.207}$$

The other time constants are easy to find. Some of them are equal to 0 so watch for some potential indeterminacy conflicts later:

$$\tau_2 = C_2 \cdot 0 = 0 \tag{5.208}$$

$$\tau_3 = \frac{L_3}{R_1} \tag{5.209}$$

and

$$\tau_4 = C_4 \cdot 0 = 0 \tag{5.210}$$

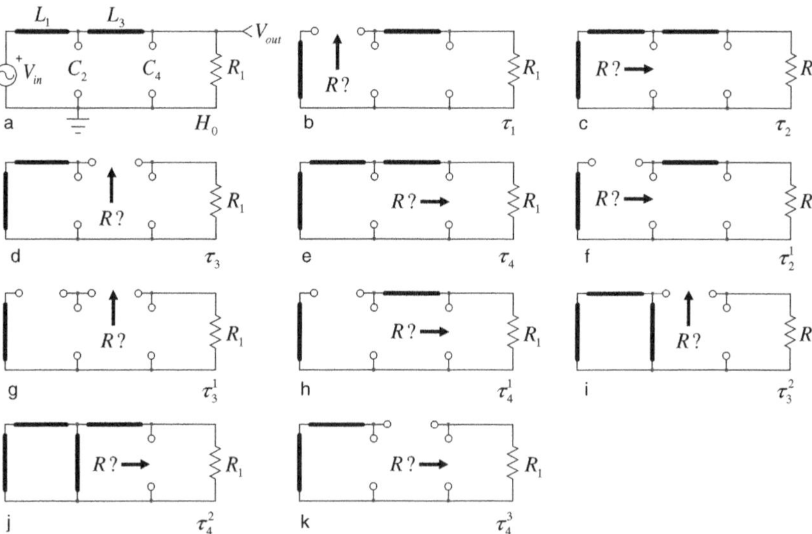

Figure 5.44 The same process as before: look at the resistances seen from the energy-storing elements terminals to determine the circuit natural time constants.

The first denominator coefficient b_1 equals:

$$b_1 = \tau_1 + \tau_2 + \tau_3 + \tau_4 = \frac{L_1 + L_3}{R_1} \tag{5.211}$$

How many coefficients do we have in b_2? Simply reuse (5.5) which tells us that we have six terms:

$$b_2 = \tau_1\tau_2^1 + \tau_1\tau_3^1 + \tau_1\tau_4^1 + \tau_2\tau_3^2 + \tau_2\tau_4^2 + \tau_3\tau_4^3 \tag{5.212}$$

Starting with the first one in Figure 5.44f, we have:

$$\tau_2^1 = C_2R_1 \tag{5.213}$$

Then, following with Figure 5.44g:

$$\tau_3^1 = \frac{L_3}{\infty} \tag{5.214}$$

From Figure 5.44h:

$$\tau_4^1 = C_4R_1 \tag{5.215}$$

The fourth term in b_2 is equal to:

$$\tau_3^2 = \frac{L_3}{R_1} \tag{5.216}$$

Following with Figure 5.44j:

$$\tau_4^2 = C_4{\cdot}0 = 0 \tag{5.217}$$

and finally

$$\tau_4^3 = C_4R_1 \tag{5.218}$$

If we assemble these terms as indicated by (5.212), we have:

$$b_2 = \frac{L_1}{R_1}C_2R_1 + \frac{L_1}{R_1}\frac{L_3}{\infty} + \frac{L_1}{R_1}C_4R_1 + 0{\cdot}\frac{L_3}{R_1} + 0{\cdot}0 + \frac{L_3}{R_1}C_4R_1 \tag{5.219}$$

We finally do not have indeterminacies and (5.219) simplifies to:

$$b_2 = L_1C_2 + L_1C_4 + L_3C_4 = L_1(C_2 + C_4) + L_3C_4 \tag{5.220}$$

The third coefficient b_3 is defined by:

$$b_3 = \tau_1\tau_2^1\tau_3^{12} + \tau_1\tau_2^1\tau_4^{12} + \tau_1\tau_3^1\tau_4^{13} + \tau_2\tau_3^2\tau_4^{23} \tag{5.221}$$

We start in Figure 5.45a from which we need to find the resistance seen from L_3 while L_1 and C_2 are in their high-frequency state. We see that:

$$\tau_3^{12} = \frac{L_3}{R_1} \tag{5.222}$$

The second term is a 0-Ω resistance given the short circuits brought by C_1 and L_3 (Figure 5.45b):

$$\tau_4^{12} = C_4{\cdot}0 = 0 \tag{5.223}$$

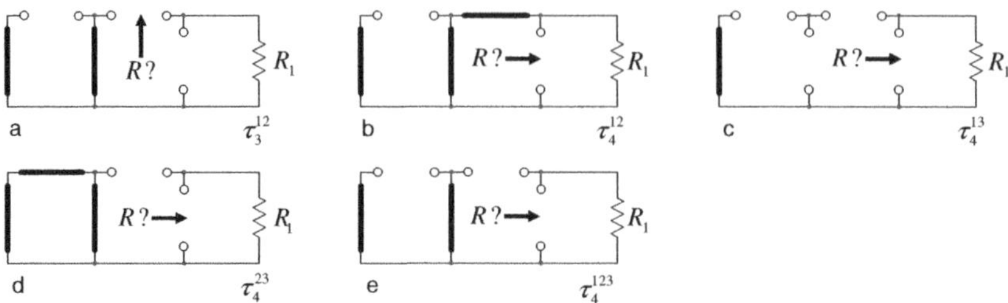

Figure 5.45 No particular problems for this 4th-order network.

Following with Figure 5.45c

$$\tau_4^{13} = C_4 R_1 \tag{5.224}$$

and Figure 5.45d:

$$\tau_4^{23} = C_4 R_1 \tag{5.225}$$

Assembling these terms to form b_3 leads to:

$$b_3 = \frac{L_1}{R_1} C_2 R_1 \frac{L_3}{R_1} + \frac{L_1}{R_1} C_2 R_1 \cdot 0 + \frac{L_1}{R_1} \frac{L_3}{\infty} C_4 R_1 + 0 \cdot \frac{L_3}{R_1} C_4 R_1 \tag{5.226}$$

After simplifying, we obtain:

$$b_3 = \frac{L_1}{R_1} C_2 R_1 \frac{L_3}{R_1} = \frac{L_1 L_3 C_2}{R_1} \tag{5.227}$$

The last coefficient b_4 features 1 single term as shown below:

$$b_4 = \tau_1 \tau_2^1 \tau_3^{12} \tau_4^{123} \tag{5.228}$$

L_1, L_3 and C_2 are put in their high-frequency state while you look at the resistance offered by C_4's terminals as shown in Figure 5.45e:

$$\tau_4^{123} = C_4 R_1 \tag{5.229}$$

Assembling the terms, we have:

$$b_4 = \frac{L_1}{R_1} C_2 R_1 \frac{L_3}{R_1} C_4 R_1 = L_1 L_3 C_2 C_4 \tag{5.230}$$

There we go, we can now express the denominator of our 4th-order network:

$$\begin{aligned} D(s) &= 1 + a_1 s + a_2 s^2 + a_3 s^3 + a_4 s^4 \\ &= 1 + \left(\frac{L_1 + L_3}{R_1}\right) s + [L_1(C_2 + C_4) + L_3 C_4] s^2 + \frac{L_1 L_3 C_2}{R_1} s^3 + L_1 L_3 C_2 C_4 s^4 \end{aligned} \tag{5.231}$$

The expression describing Figure 5.43's transfer function is:

$$H(s) = \frac{1}{1 + \left(\frac{L_1 + L_3}{R_1}\right) s + [L_1(C_2 + C_4) + L_3 C_4] s^2 + \frac{L_1 L_3 C_2}{R_1} s^3 + L_1 L_3 C_2 C_4 s^4} \tag{5.232}$$

What we can do now is rearrange this expression to make it fit a 4th-order Butterworth polynomial form. In the literature, such a transfer function is described by:

$$H(s)=\frac{1}{\left(1+0.7654\frac{s}{\omega_0}+\left(\frac{s}{\omega_0}\right)^2\right)\left(1+1.8478\frac{s}{\omega_0}+\left(\frac{s}{\omega_0}\right)^2\right)} \tag{5.233}$$

If we concentrate on the denominator and develop it, we obtain after factorization:

$$D(s)=1+s\left(\frac{0.7654}{\omega_0}+\frac{1.8478}{\omega_0}\right)+s^2\left(\frac{1.41430612}{\omega_0{}^2}+\frac{2}{\omega_0{}^2}\right)+s^3\left(\frac{0.7654}{\omega_0{}^3}+\frac{1.8478}{\omega_0{}^3}\right)+\left(\frac{s}{\omega_0}\right)^4 \tag{5.234}$$

Identifying each term with those of (5.232), we come up with the following system of equations:

$$\begin{aligned}\frac{L_1+L_3}{R_1}&=\frac{0.7654}{\omega_0}+\frac{1.8478}{\omega_0}\\ L_1(C_2+C_4)+L_3C_4&=\frac{1.41430612}{\omega_0{}^2}+\frac{2}{\omega_0{}^2}\\ \frac{L_1L_3C_2}{R_1}&=\frac{0.7654}{\omega_0{}^3}+\frac{1.8478}{\omega_0{}^3}\\ L_1L_3C_2C_4&=\frac{1}{\omega_0{}^4}\end{aligned} \tag{5.235}$$

With the help of Mathcad® we obtain for a 100-kHz cutoff frequency the below theoretical values – pardon the decimals for the picofarad values ☺ – to match the Butterworth response. R_1 is set to 4.7 kΩ:

$$L_1=\frac{R_1}{\omega_0}\times 1.53081=11.45\,\text{mH} \tag{5.236}$$

$$L_3=\frac{R_1}{\omega_0}\times 1.08238=8.10\,\text{mH} \tag{5.237}$$

$$C_2=\frac{1.57713}{\omega_0 R_1}=534.06\,\text{pF} \tag{5.238}$$

$$C_4=\frac{0.38267}{R_1\omega_0}=129.58\,\text{pF} \tag{5.239}$$

Before we test these results, we need a raw transfer function to check our transfer functions integrity. Looking at Figure 5.43, we can build a Thévenin's generator around L_1 and C_2, followed by an impedance divider. Applying this technique, we determine the raw transfer function as:

$$H_{res}(s)=\frac{1}{1+s^2L_1C_2}\frac{R_1\|\left(\frac{1}{sC_4}\right)}{R_1\|\left(\frac{1}{sC_4}\right)+sL_3+(sL_1)\|\left(\frac{1}{sC_2}\right)} \tag{5.240}$$

We can now compare results from (5.240) and those of (5.232) and (5.233). As confirmed in Figure 5.46, all curves are similar. The response is maximally flat and shows no peaking at all. The drop in magnitude is 80 dB over a decade, confirming the 4th-order low-pass filter response.

5.2.3 *Example 4: A 4th-order Band-pass Active Filter*

While browsing documents on active filtering, I found an interesting structure in [2] shown in Figure 5.47. There are four energy-storing elements with independent state variables: this is a

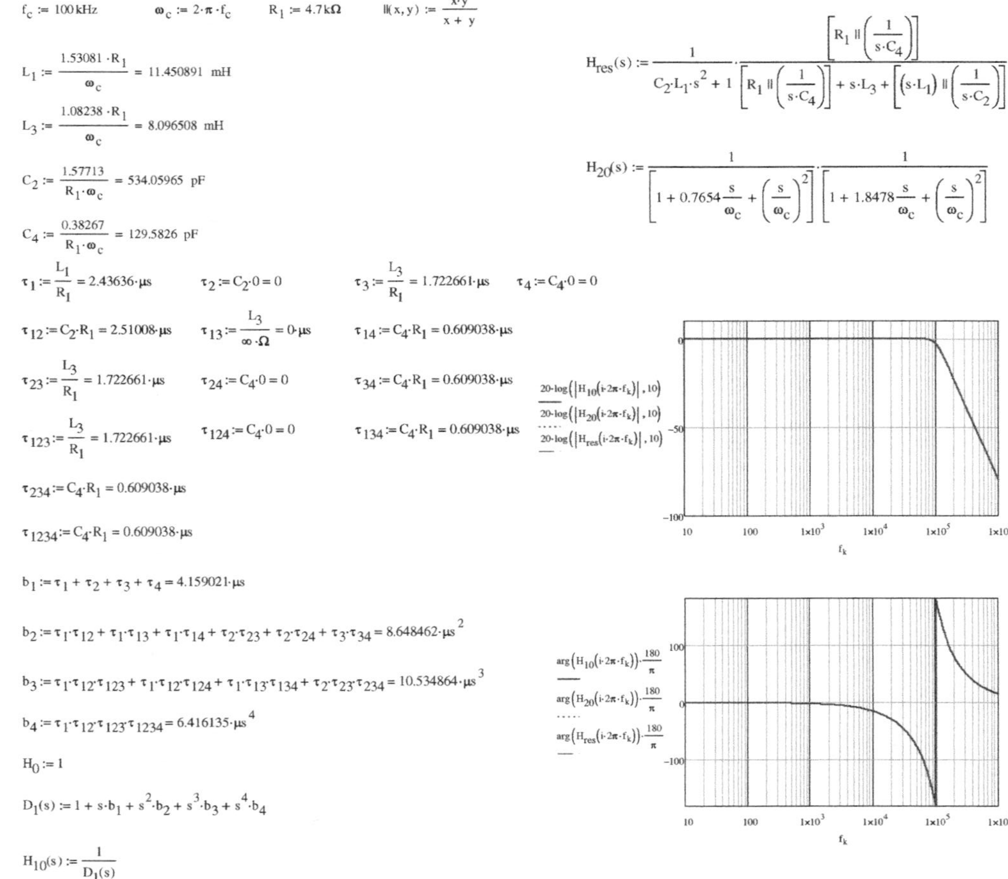

Figure 5.46 Mathcad® confirms the validity of our approach and shows identical dynamic responses.

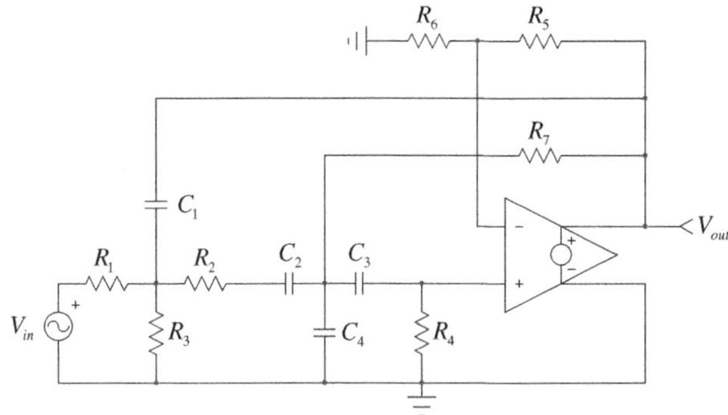

Figure 5.47 This is a 4th-order active filter. What is its transfer function?

4th-order network. We can see that the op amp is wired in a non-inverting configuration with a gain k equal to $(R_5/R_6 + 1)$. For the sake of a simpler analysis, we can redraw Figure 5.47 as suggested in Figure 5.48. The node 'p' for 'plus' is multiplied by k, the gain imposed by R_5 and R_6 in the non-inverting configuration. We will first start by determining the natural time constants to build the denominator $D(s)$. Then, the numerator $N(s)$ will be obtained using the generalized 4th-order transfer function described in Figure 5.21.

In Figure 5.49a, we have gathered the first simplified schematic corresponding to the various cases we need to explore. In some occasions, determining the resistance seen from the energy-storing element is easy and inspection works well. In other situations, you need to install a test generator I_T and use KCL/KVL to determine the resistance. Starting with the quasi-static gain, a series capacitor blocks the response for $s = 0$, therefore:

$$H_0 = 0 \tag{5.241}$$

In Figure 5.49b, the node 'p' is at 0 V given the lack of connection to a bias point (C_3 is removed in dc). Therefore, $k \cdot V_{(p)}$ is also 0 V and R_7's upper terminal is grounded. Here, the only resistors you can see are R_1 parallel with R_3. Thus:

$$\tau_1 = C_1(R_1 \| R_3) \tag{5.242}$$

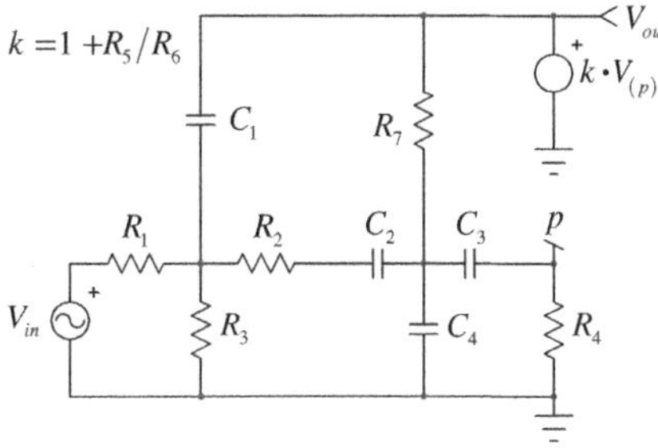

Figure 5.48 Without the op amp, the circuit looks simpler to analyze.

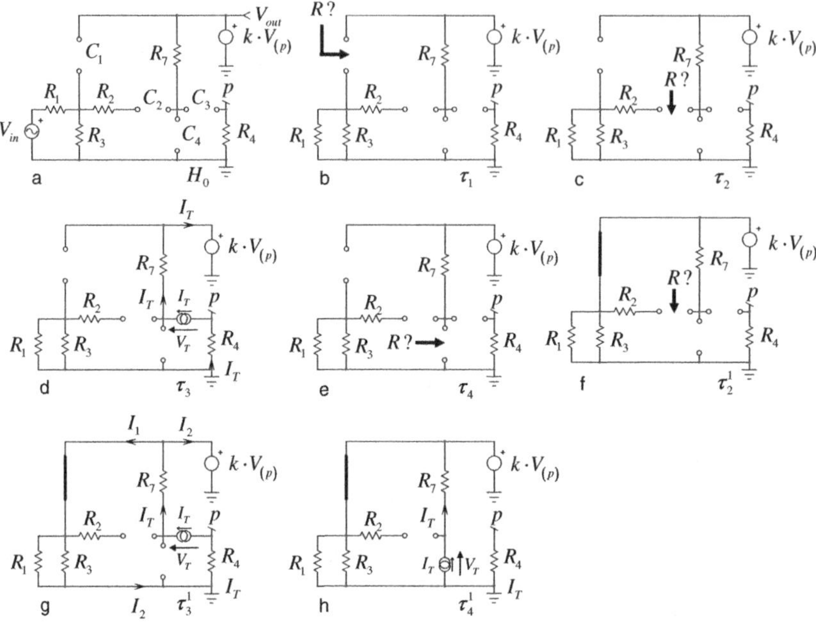

Figure 5.49 Setting the excitation to 0 leads to determining the circuit natural time constants.

The second time constant involves C_2 in Figure 5.49c. The resistance seen from the capacitor's terminals is R_7 in series with R_2 and the parallel combination of R_1/R_3. We have:

$$\tau_2 = C_2(R_7 + R_2 + R_1 \| R_3) \tag{5.243}$$

The third time constant requires the addition of a test generator as drawn in Figure 5.49d. The voltage at node p is the current I_T times R_4:

$$V_{(p)} = -I_T R_4 \tag{5.244}$$

This voltage is scaled by k and appears on R_7's upper terminal. We can write the following expression for V_T:

$$V_T = I_T R_7 + k \cdot V_{(p)} + R_4 I_T \tag{5.245}$$

Substituting (5.244) into (5.245) and factoring I_T, we have:

$$V_T = I_T[R_7 + R_4(1 - k)] \tag{5.246}$$

The time constant involving capacitor C_3 is defined as:

$$\tau_3 = C_3[R_7 + R_4(1 - k)] \tag{5.247}$$

The fourth time constant is easy to find by looking at Figure 5.49e: R_7 is the only resistor seen by C_4:

$$\tau_4 = C_4 R_7 \tag{5.248}$$

The first term b_1 is formed by adding all these time constants:

$$b_1 = \tau_1 + \tau_2 + \tau_3 + \tau_4 = C_1(R_1 \| R_3) + C_2(R_7 + R_2 + R_1 \| R_3) + C_3[R_7 + R_4(1-k)] + C_4 R_7 \quad (5.249)$$

We can now start combining various energy-storing elements states. We begin with τ_2^1 described by Figure 5.49f. Again, R_7 upper terminal is grounded since $V_{(p)} = 0$. The resistance is simply the series combination of R_7 and R_2:

$$\tau_2^1 = C_2(R_2 + R_7) \quad (5.250)$$

The next time constant looks rather more complex in Figure 5.49g. Actually, since $k \cdot V_{(p)}$ biases R_7's upper terminal, the resistive network made of R_1 and R_3 has no effect on the resistance seen from C_3's terminals: the configuration is similar to that of Figure 5.49d for τ_3:

$$\tau_3^1 = C_3[R_7 + R_4(1-k)] \quad (5.251)$$

In Figure 5.49h, even if C_1 is replaced by a short circuit, the time constant is similar to that already determined in Figure 5.49e:

$$\tau_4^1 = C_4 R_7 \quad (5.252)$$

The next term is calculated using sketches from Figure 5.50. In sketch (a), we need to determine τ_3^2 capitalizing on the split of current I_T between I_1 and I_2. To ease our exercise, I have redrawn the sketch in a more familiar format as shown in Figure 5.51.

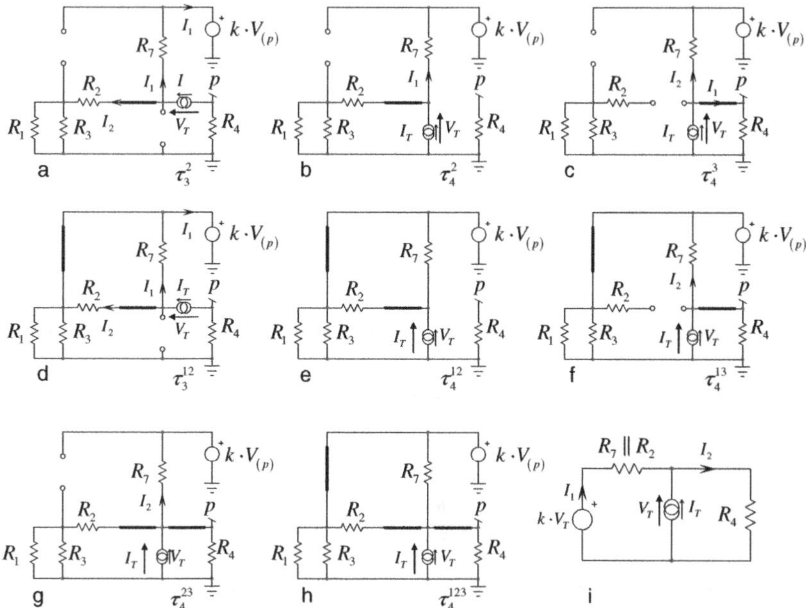

Figure 5.50 The excitation is still set to 0 to determine the remaining time constants.

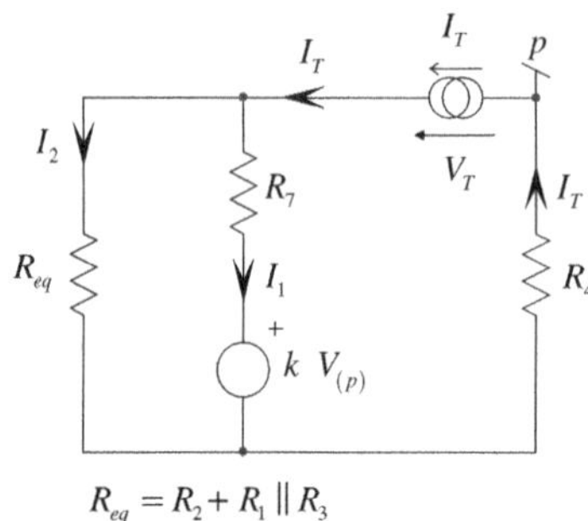

Figure 5.51 A rearranged sketch helps determining the resistance in an easier way.

The current circulating in R_7 is equal to the voltage across that resistor divided by the resistance value. Applying KVL, we have:

$$I_1 = \frac{R_{eq}I_2 - k \cdot V_{(p)}}{R_7} \tag{5.253}$$

The voltage at node 'p' equals $-I_T$ times R_4. Substituting this definition in (5.253) gives:

$$I_1 = \frac{R_{eq}I_2 + k \cdot I_T R_4}{R_7} \tag{5.254}$$

Current I_2 is equal to:

$$I_2 = \frac{V_T - I_T R_4}{R_{eq}} \tag{5.255}$$

Substituting (5.255) in (5.254) leads to:

$$I_1 = \frac{V_T - I_T R_4 + I_T R_4 \cdot k}{R_7} \tag{5.256}$$

The test current I_T is the sum of I_1 and I_2. Therefore:

$$I_T = \frac{V_T - I_T R_4 + I_T R_4 \cdot k}{R_7} + \frac{V_T - I_T R_4}{R_{eq}} \tag{5.257}$$

Rearranging, factoring I_T and isolating V_T gives us the resistance seen from C_3's terminals:

$$R = \frac{R_4 R_7 + (R_2 + R_3 \| R_1)(R_4 + R_7 - k \cdot R_4)}{R_7 + R_2 + R_3 \| R_1} = \frac{R_7 - k \cdot R_4}{R_7 + R_2 + R_3 \| R_1}(R_2 + R_3 \| R_1) + R_4 \tag{5.258}$$

The time constant we want is thus:

$$\tau_3^2 = C_3 \left[\frac{R_7 - k \cdot R_4}{R_7 + R_2 + R_3 \| R_1}(R_2 + R_3 \| R_1) + R_4 \right] \tag{5.259}$$

The next time constant τ_4^2 is determined using Figure 5.50b. Here, despite the presence of the test generator, we can determine the resistance by inspection. R_4 being open, node 'p' is grounded and so is R_7's upper terminal. R_7 thus appears across C_4's connections, in parallel with R_2 in series with R_1

parallel with R_3. The time constant is then:

$$\tau_4^2 = C_4\left[R_7 \,\|\, \left(R_2 + R_3 \,\|\, R_1\right)\right] \tag{5.260}$$

In Figure 5.50c, the test current I_T splits into I_1 and I_2. They are defined as follows:

$$I_1 = \frac{V_T}{R_4} \tag{5.261}$$

$$I_2 = \frac{V_T - I_1 R_4 \cdot k}{R_7} \tag{5.262}$$

Substituting (5.261) in (5.262), we obtain:

$$I_2 = \frac{V_T(1-k)}{R_7} \tag{5.263}$$

Finally,

$$I_T = I_1 + I_2 = \frac{V_T}{R_4} + \frac{V_T(1-k)}{R_7} = V_T\left(\frac{1}{R_4} + \frac{1-k}{R_7}\right) \tag{5.264}$$

Factoring and rearranging V_T/I_T, we obtain the following time constant definition:

$$\tau_4^3 = C_4\left(\frac{R_4 R_7}{R_4(1-k) + R_7}\right) \tag{5.265}$$

We can form term b_2 defined as:

$$\begin{aligned} b_2 &= \tau_1\tau_2^1 + \tau_1\tau_3^1 + \tau_1\tau_4^1 + \tau_2\tau_3^2 + \tau_2\tau_4^2 + \tau_3\tau_4^3 \\ &= C_1\left(R_1 \,\|\, R_3\right)C_2(R_2 + R_7) + C_1\left(R_1 \,\|\, R_3\right)C_3[R_7 + R_4(1-k)] + C_1\left(R_1 \,\|\, R_3\right)C_4 R_7 \\ &\quad + C_2\left(R_7 + R_2 + R_1 \,\|\, R_3\right)C_3\left[\frac{R_7 - k \cdot R_4}{R_7 + R_2 + R_3 \,\|\, R_1}\left(R_2 + R_3 \,\|\, R_1\right) + R_4\right] \\ &\quad + C_2\left(R_7 + R_2 + R_1 \,\|\, R_3\right)C_4\left[R_7 \,\|\, \left(R_2 + R_3 \,\|\, R_1\right)\right] \\ &\quad + C_3[R_7 + R_4(1-k)]C_4\left(\frac{R_4 R_7}{R_4(1-k) + R_7}\right) \end{aligned} \tag{5.266}$$

In Figure 5.50d, we see that R_2 is paralleled with R_7 as C_1 is a short circuit. R_1 and R_3 play no role as their upper terminal is at a fixed bias, $k \cdot V_{(p)}$. A simplified version of this sketch is shown in Figure 5.52.

From there, we can write:

$$-k \cdot R_4 I_T + I_T\left(R_2 \,\|\, R_7\right) + R_4 I_T = V_T \tag{5.267}$$

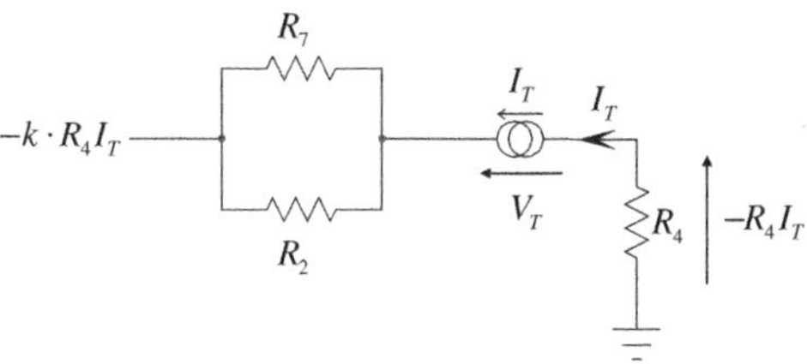

Figure 5.52 The sketch simplifies and lets you determine the resistance in an easy way.

Factoring I_T and isolating V_T, we obtain:

$$\tau_3^{12} = C_3\left[R_2 \| R_7 + R_4(1-k)\right] \tag{5.268}$$

The sketch in Figure 5.50e lets you determine τ_4^{12} quite easily. You can see that the voltage at node 'p' is 0 V, implying that R_7's upper terminal is grounded: R_1 and R_3 are shorted and leave the picture. The resistance seen from C_4's terminals is simply R_7 parallel with R_2:

$$\tau_4^{12} = C_4\left(R_2 \| R_7\right) \tag{5.269}$$

In Figure 5.50f, despite C_1 being shorted, C_2 is in dc state and disconnect R_1, R_2 and R_3. The remaining network is similar to that of Figure 5.50c and the time constant is that of (5.265):

$$\tau_4^{13} = C_4\left(\frac{R_4R_7}{R_4(1-k)+R_7}\right) \tag{5.270}$$

When C_2 and C_3 are shorted, you obtain Figure 5.50g circuit. You can observe that is the same as in Figure 5.50c with R_2 plus R_1 paralleled R_3 connected across C_4's terminals. Therefore:

$$\tau_4^{23} = C_4\left[\left(R_2 + R_1 \| R_3\right) \| \left(\frac{R_4R_7}{R_4(1-k)+R_7}\right)\right] \tag{5.271}$$

The third term b_3 is equal to:

$$\begin{aligned} b_3 &= \tau_1\tau_2^1\tau_3^{12} + \tau_1\tau_2^1\tau_4^{12} + \tau_1\tau_3^1\tau_4^{13} + \tau_2\tau_3^2\tau_4^{23} \\ &= C_1\left(R_1 \| R_3\right)C_2(R_2+R_7)C_3\left[R_2 \| R_7 + R_4(1-k)\right] + C_1\left(R_1 \| R_3\right)C_2(R_2+R_7)C_4\left(R_2 \| R_7\right) \\ &\quad + C_1\left(R_1 \| R_3\right)C_3[R_7 + R_4(1-k)]C_4\left(\frac{R_4R_7}{R_4(1-k)+R_7}\right) \\ &\quad + C_2\left(R_7 + R_2 + R_1 \| R_3\right)C_3\left[\frac{R_7 - k\cdot R_4}{R_7 + R_2 + R_3 \| R_1}\left(R_2 + R_3 \| R_1\right) + R_4\right]C_4\left[\left(R_2 + R_1 \| R_3\right) \| \left(\frac{R_4R_7}{R_4(1-k)+R_7}\right)\right] \end{aligned} \tag{5.272}$$

Final lap with τ_4^{123} in Figure 5.50h! To ease the determination of this last time constant, Figure 5.50i offers a simpler view. R_1 and R_3 again disappear from the picture because $k\cdot V_{(p)}$ is applied across them. From that sketch:

$$I_2 = \frac{V_T}{R_4} \tag{5.273}$$

$$I_1 = \frac{k\cdot V_T - V_T}{R_7 \| R_2} = \frac{V_T(k-1)}{R_7 \| R_2} \tag{5.274}$$

I_T is equal to $I_2 - I_1$

$$I_T = \frac{V_T}{R_4} - \frac{V_T(k-1)}{R_7 \| R_2} = V_T\left(\frac{1}{R_4} + \frac{1-k}{R_7 \| R_2}\right) \tag{5.275}$$

The time constant is thus:

$$\tau_4^{123} = C_4\left(\frac{1}{\frac{1}{R_4} + \frac{1-k}{R_7 \| R_2}}\right) = C_4\left[\frac{R_4\left(R_2 \| R_7\right)}{R_4(1-k) + R_2 \| R_7}\right] \tag{5.276}$$

and b_4 equal to:

$$b_4 = \tau_1\tau_2^1\tau_3^{12}\tau_4^{123} = C_1\left(R_1 \| R_3\right)C_2(R_2+R_7)C_3\left[R_2 \| R_7 + R_4(1-k)\right]C_4\left[\frac{R_4\left(R_2 \| R_7\right)}{R_4(1-k)+R_2 \| R_7}\right] \tag{5.277}$$

The denominator is then defined by assembling the various b terms:

$$D(s) = 1 + b_1 s + b_2 s^2 + b_3 s^3 + b_4 s^4 \tag{5.278}$$

In the above sequence, we have determined driving resistances using inspection of KCL/KVL techniques. To make sure no mistake was made, it is always safer to confront calculation results with a simple SPICE dc point analysis. You capture the original schematic and replace the energy-storing elements by short circuits or opens according to the time constant you want. The voltage across the 1-A current generator is the resistance you look for. Figure 5.53 gives you an example for the first time constants in our example.

Now that the denominator is derived, we can concentrate on the numerator. The excitation signal is back and we look at the output signal while energy-storing elements are set to either high-frequency or dc states. The first batch of configurations is shown in Figure 5.54. H_0 is null as expressed by (5.241) and confirmed by Figure 5.54a sketch.

In Figure 5.54b, the non-inverting pin is pulled to ground via R_4, so the op amp output is 0 V. In sketch (c), still no bias on the non-inverting pin, imposing a 0-V output. In d, the op amp delivers a 0-V output as in sketch (e) where R_4 grounds the '+' pin. This situation reproduces up to sketch (h) where the op amp output is still 0 V. Capitalizing on these findings, we have:

$$H_0 = H^1 = H^2 = H^3 = H^4 = H^{12} = H^{13} = H^{14} = 0 \tag{5.279}$$

In Figure 5.54i, the op amp output is not zero and a simpler schematic as in the left side of Figure 5.55 will help us solve the dc transfer function. A Thévenin's transformation reduces the circuit to three resistors.

Applying KCL at node 'p', we have:

$$I_2 = I_1 + I_3 \tag{5.280}$$

Each current is defined as:

$$I_1 = \frac{V_{in}\dfrac{R_3}{R_3+R_1} - V_{(p)}}{R_2 + R_3 \| R_1} \tag{5.281}$$

$$I_2 = \frac{V_{(p)}}{R_4} \tag{5.282}$$

$$I_3 = \frac{k \cdot V_{(p)} - V_{(p)}}{R_7} \tag{5.283}$$

Adding these results and extracting $V_{(p)}$ then $k \cdot V_{(p)}$ for V_{out} leads to:

$$\frac{V_{out}}{V_{in}} = H^{23} = \frac{k \cdot R_3}{\left(R_1+R_3\right)\left(R_2+R_1 \| R_3\right)\left(\dfrac{1}{R_4}+\dfrac{1}{R_2+R_1 \| R_3}-\dfrac{k-1}{R_7}\right)} \tag{5.284}$$

The second part of the dc transfer function sketches appears in Figure 5.56.

In cases where the non-inverting pin is pulled alone to ground via R_4, the output is 0 V (Figure 5.56a, b and d). In Figure 5.56c, the circuit reduces to a simpler circuit shown in

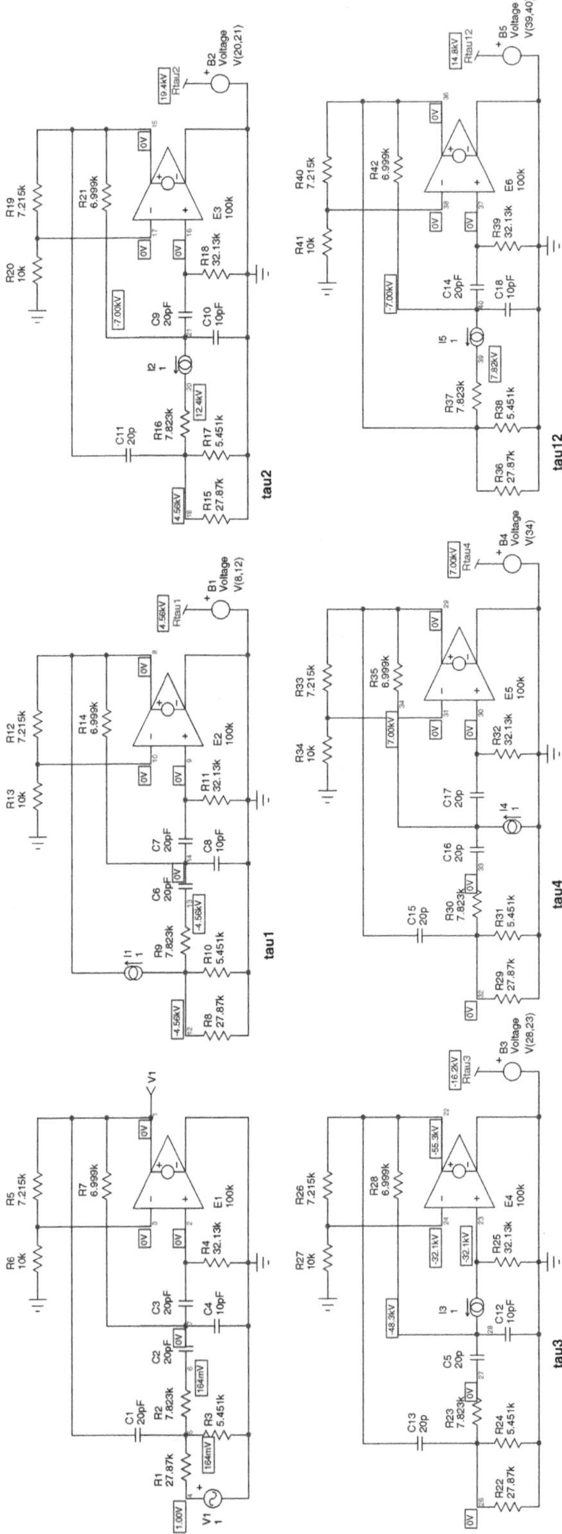

Figure 5.53 A simple SPICE simulation helps confirm the resistances found by inspection or with KVL/KCL analysis.

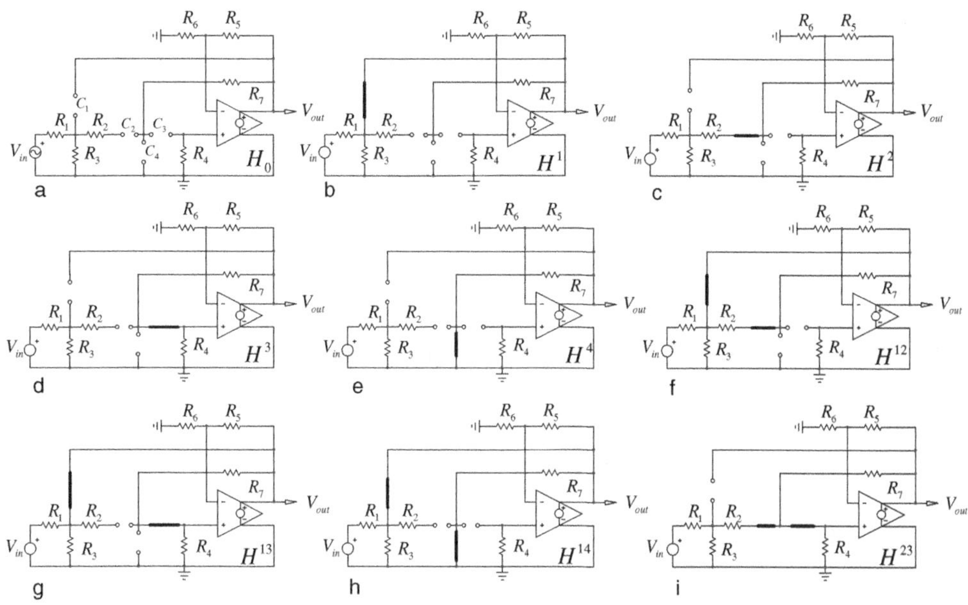

Figure 5.54 Dc gains are determined with energy-storing elements alternatively set to different states: open- or short-circuited.

Figure 5.56h. The potential at node 'p' equals the potential at node 1 divided by k. If the division ratio brought by R_4 and R_2 is different than $1/k$, then the only condition where this is true is when $V_{out} = 0$ V as indicated by the operating point. The rest of the configurations also return 0. Therefore, we have:

$$H^{24} = H^{34} = H^{123} = H^{124} = H^{134} = H^{234} = H^{1234} = 0 \tag{5.285}$$

which is confirmed by the bias point in Figure 5.57.

The numerator coefficients are almost all equal to 0 except one term, a_2:

$$a_1 = \tau_1 H^1 + \tau_2 H^2 + \tau_3 H^3 + \tau_4 H^4 = 0 \tag{5.286}$$

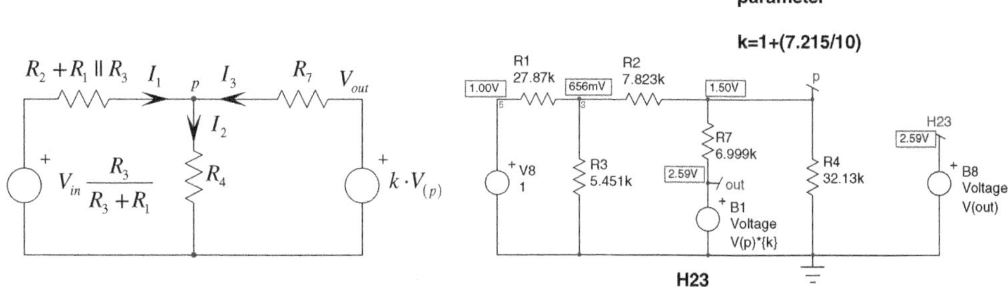

Figure 5.55 A simpler representation of Figure 5.54i helps to work out the gain in this configuration.

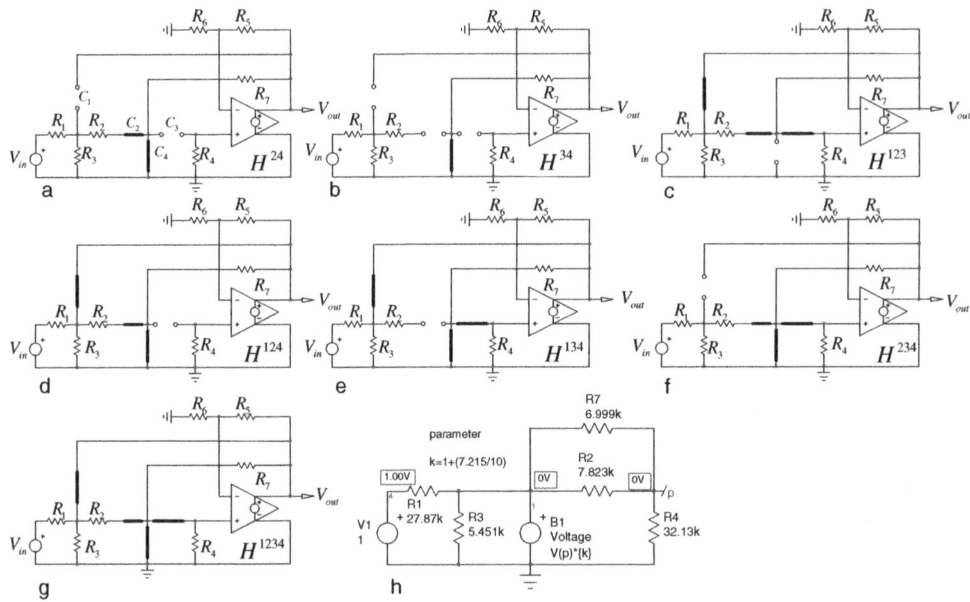

Figure 5.56 This last part of dc transfer functions study shows that all outputs are zeroed.

$$\begin{aligned} a_2 &= \tau_1\tau_2^1H^{12} + \tau_1\tau_3^1H^{13} + \tau_1\tau_4^1H^{14} + \tau_2\tau_3^2H^{23} + \tau_2\tau_4^2H^{24} + \tau_3\tau_4^3H^{34} = \tau_2\tau_3^2H^{23} \\ &= C_2\left(R_7 + R_2 + R_1 \| R_3\right)C_3\left[\frac{R_7 - k\cdot R_4}{R_7 + R_2 + R_3 \| R_1}\left(R_2 + R_3 \| R_1\right) + R_4\right] \\ &\times \frac{k\cdot R_3}{\left(R_1 + R_3\right)\left(R_2 + R_1 \| R_3\right)\left(\frac{1}{R_4} + \frac{1}{R_2 + R_1 \| R_3} - \frac{k-1}{R_7}\right)} \end{aligned} \tag{5.287}$$

$$a_3 = \tau_1\tau_2^1\tau_3^{12}H^{123} + \tau_1\tau_2^1\tau_4^{12}H^{124} + \tau_1\tau_3^1\tau_4^{13}H^{134} + \tau_2\tau_3^2\tau_4^{23}H^{234} = 0 \tag{5.288}$$

$$a_4 = \tau_1\tau_2^1\tau_3^{12}\tau_4^{123}H^{1234} = 0 \tag{5.289}$$

Simplifying (5.287) with Mathcad® leads to the final numerator expression:

$$N(s) = a_2s^2 = k\frac{C_2C_3R_3R_4R_7}{R_1 + R_3}s^2 \tag{5.290}$$

In which k is $\frac{R_5}{R_6} + 1$. The transfer function H is defined by:

$$H(s) = \frac{a_2s^2}{1 + b_1s + b_2s^2 + b_3s^3 + b_4s^4} \tag{5.291}$$

By factoring a_2s^2 and then a_2/b_2 in the denominator, it is possible to rearrange H the following way:

$$H(s) = \frac{a_2}{b_2}\frac{1}{1 + \frac{b_1}{b_2s} + s\frac{b_3}{b_2} + s^2\frac{b_4}{b_2} + \frac{1}{b_2s^2}} \tag{5.292}$$

However, in this expression, the unitless leading term does not correspond to the filter gain at the resonant frequency. Reference [3] explores filter transfer functions and there is a 4th-order band-pass

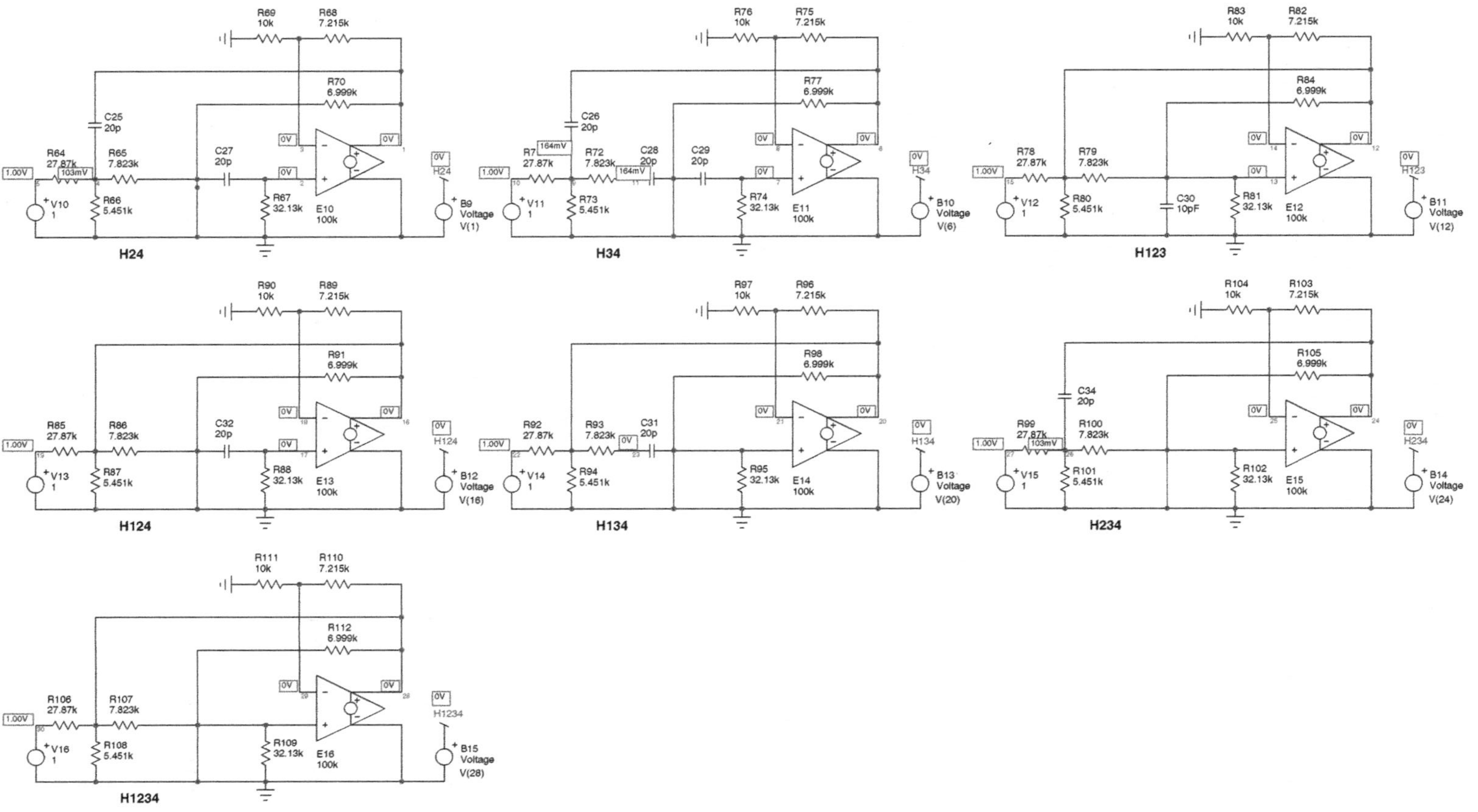

Figure 5.57 A simple dc operating point confirms that all output for all Figure 5.56 configurations deliver 0 V.

filter example on page 16-32. Equation (16-12) in reference [3] presents a possible factorized solution. If a reader friend succeeds in rearranging (5.292) so that it matches eq. (16-12) in reference [3], I will be glad to publish the expression with credits to the valorous reader!

Figure 5.58 gathers all Mathcad® equations with the corresponding dynamic response. Component values are those of [2] and impose a 1-MHz resonant frequency with a flat 0-dB gain. The comparison between the analytical response and that of simulation appears in Figure 5.59: perfect matching!

5.2.4 Example 5: A 3rd-order Low-pass Active GIC Filter

Figure 5.60 shows a generalized impedance converter (GIC) circuit using two op amps. The circuit represents a filter whose transfer function is to be determined considering a non-infinite open-loop gain A_{OL}. Needless to say the circuit is intimidating given the unusual configuration of the two op amps. As usual, we will break the circuit in small pieces, step-by-step gathering all time constants. Here, I systematically adopted the usage of SPICE to confirm that each of our steps leads to the correct result. We start with τ_1 in Figure 5.61.

This first time constant has been determined using a simplified circuit as shown in the right side of the figure. It is important to confirm that operating points from the original circuit and that of the equivalent schematic diagram perfectly match several digits after the decimal part. A small deviation would indicate a mistake. Here, I have purposely set the op amp open-loop gain A_{OL} to a low value – 100 or 40 dB – so that simulation shows me these low-gain effects. Later on, changing A_{OL} and increasing it to higher levels (100k for instance) should not produce differences between simulation data and Mathcad® analytical results. We can write a few equations involving currents and voltages. What we want is the ratio V_T/I_T which determine the resistance driving capacitor C_1. The first equation is:

$$V_T = I_T R_4 + I_1(R_1 + R_2) - \left(V_{(A)} - V_T\right)A_{OL} \tag{5.293}$$

The current I_1 is defined by:

$$I_1 = -\frac{V_T A_{OL}}{R_1 + R_2} \tag{5.294}$$

While the voltage at node A is given by:

$$V_{(A)} = I_1 R_1 - V_{(A)} A_{OL} + V_T A_{OL} \tag{5.295}$$

Extracting $V_{(A)}$ from this equation:

$$V_{(A)} = \frac{R_1 I_1 + V_T A_{OL}}{1 + A_{OL}} \tag{5.296}$$

Combining (5.294) and (5.296) in (5.293) then solving and rearranging for V_T and I_T, we obtain:

$$\frac{V_T}{I_T} = \frac{R_4(R_1 + R_2) + A_{OL} R_4 (R_1 + R_2)}{R_1 + R_2 + A_{OL}(R_1 + R_2) + {A_{OL}}^2 R_2} \tag{5.297}$$

When A_{OL} approaches infinity, this equation returns 0. The first time constant involving A_{OL} is thus defined by:

$$\tau_1 = C_1 \frac{R_4(R_1 + R_2) + A_{OL} R_4 (R_1 + R_2)}{R_1 + R_2 + A_{OL}(R_1 + R_2) + {A_{OL}}^2 R_2} \tag{5.298}$$

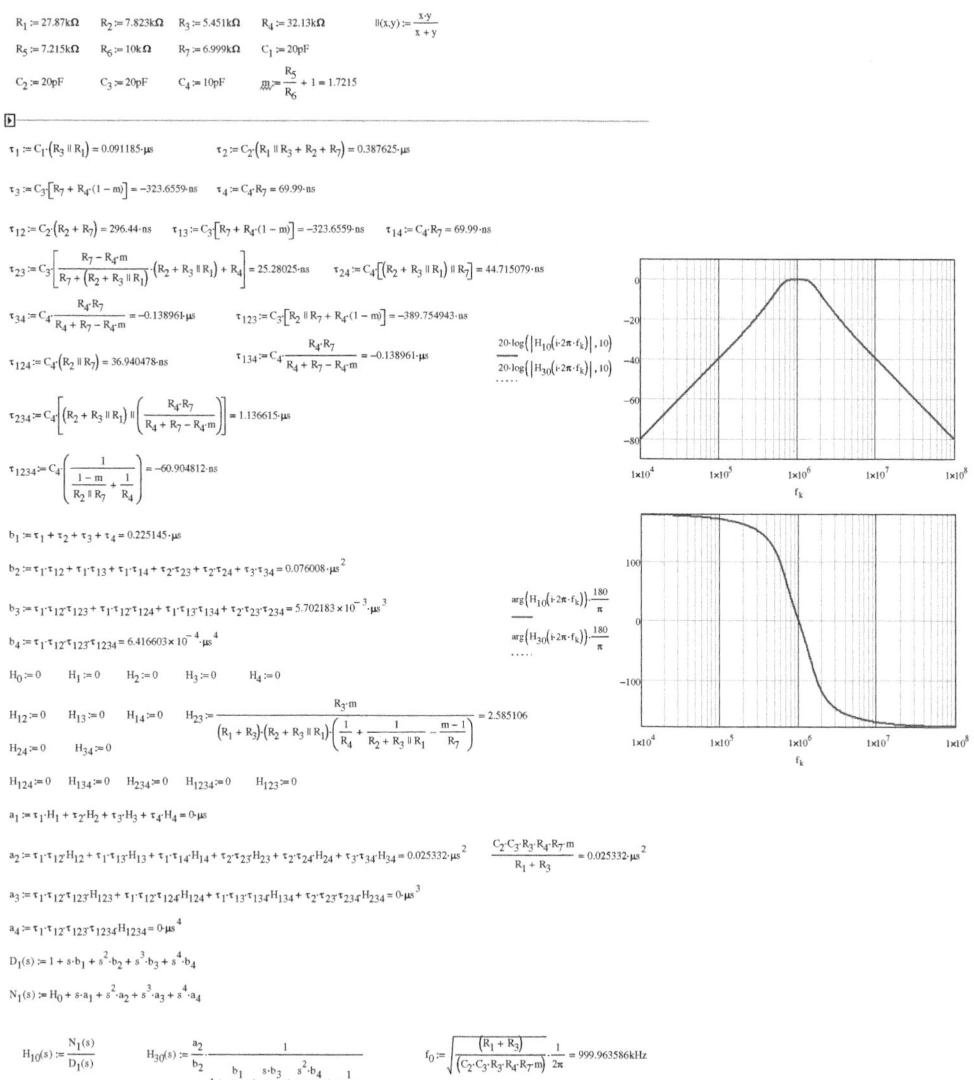

Figure 5.58 The Mathcad® shows the various time constants we have derived.

Figure 5.62 describes the circuit to obtain τ_2, the second time constant. From the simplified version in the right-side of the figure, we can write:

$$V_T = I_T R_5 + V_{(A)} A_{OL} - V_{(op)} A_{OL} \tag{5.299}$$

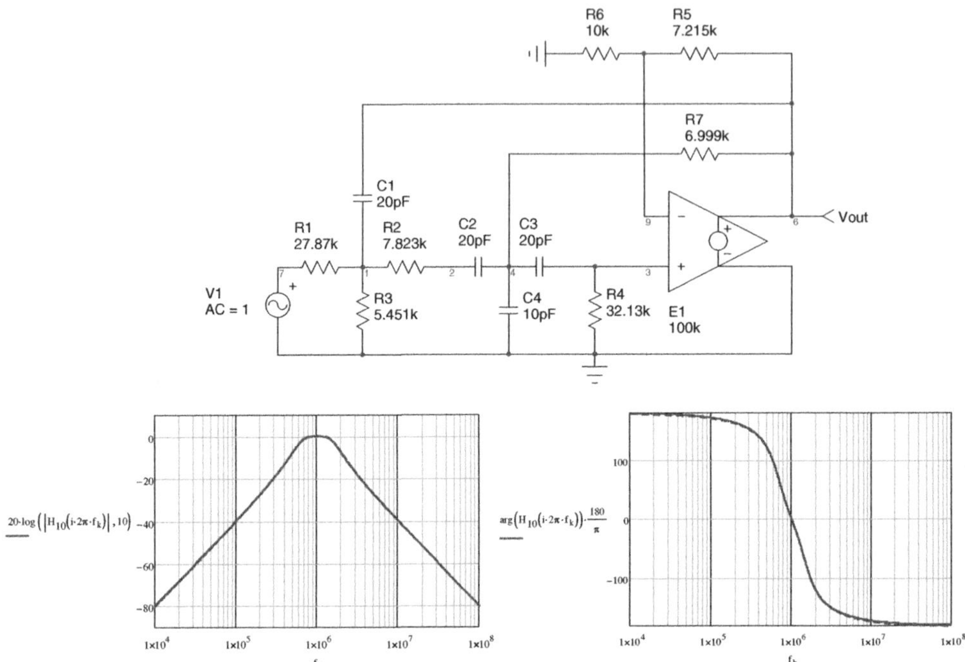

Figure 5.59 A SPICE circuit confirms the perfect matching between the dynamic response analytically obtained and the simulation results. Component values lead to a 1-MHz band-pass frequency.

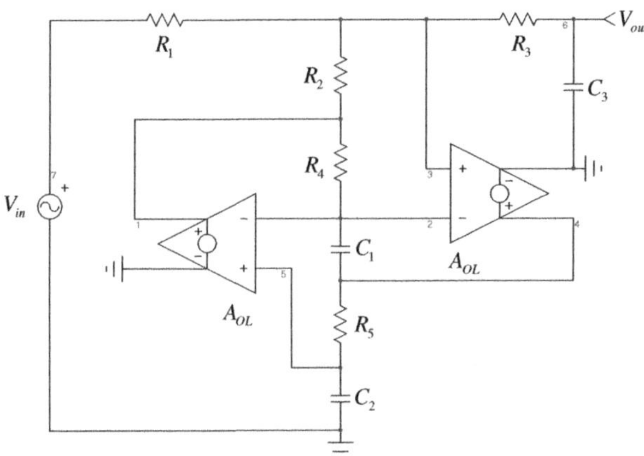

Figure 5.60 A generalized impedance converter filter uses two op amps and three capacitors to form a 3rd-order network.

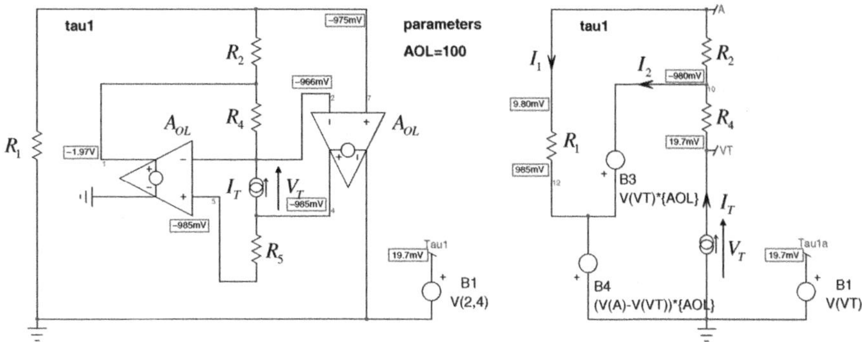

Figure 5.61 Determining time constants for a complex circuit like the GIC filter is eased by using simplified diagrams. Here we look at τ_1.

The voltage at node A can be expressed as:

$$V_{(A)} = \left[V_T A_{OL} - V_{(op)} A_{OL}\right] \frac{R_1}{R_1 + R_2} \tag{5.300}$$

Then the voltage at node op is determined:

$$V_{(op)} = V_T A_{OL} - V_{(op)} A_{OL} \tag{5.301}$$

From which you can extract:

$$V_{(op)} = \frac{V_T A_{OL}}{1 + A_{OL}} \tag{5.302}$$

Now substituting (5.302) in (5.300) and (5.299) then rearranging, we have:

$$\frac{V_T}{I_T} = \frac{R_5(R_1 + R_2) + A_{OL} R_5(R_1 + R_2)}{R_1 + R_2 + A_{OL}(R_1 + R_2) + A_{OL}{}^2 R_2} \tag{5.303}$$

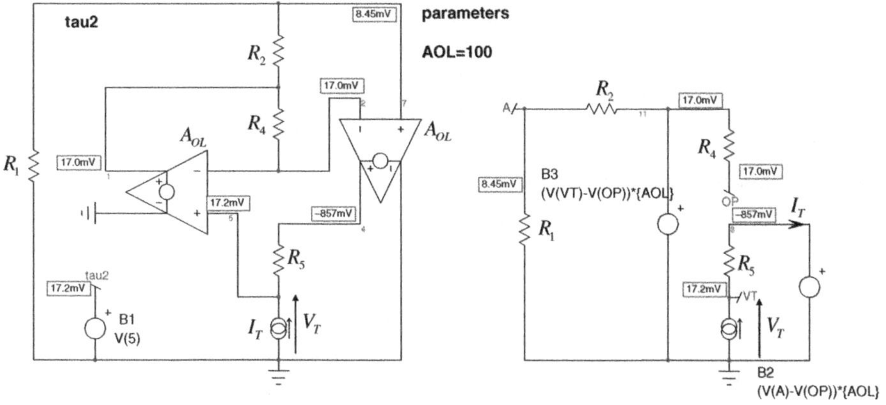

Figure 5.62 τ_2 follows a similar principle.

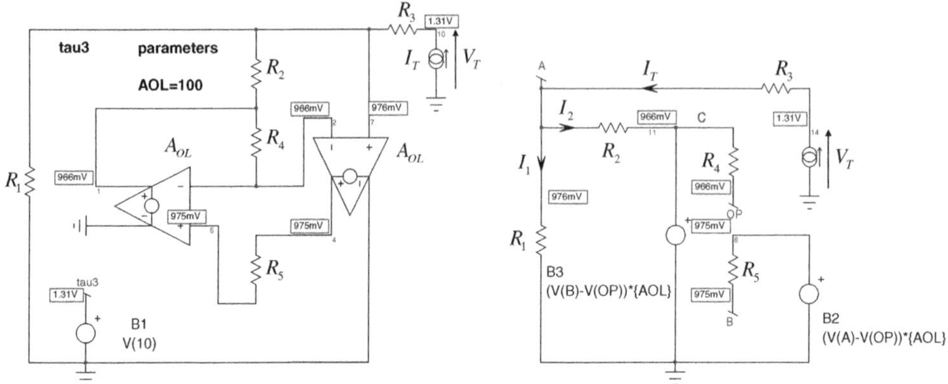

Figure 5.63 The third time constant is also determined using a simplified circuit.

When A_{OL} is large, this equation returns 0. The second time constant is defined by:

$$\tau_2 = C_2 \frac{R_5(R_1 + R_2) + A_{OL} R_5(R_1 + R_2)}{R_1 + R_2 + A_{OL}(R_1 + R_2) + A_{OL}^2 R_2} \tag{5.304}$$

The third time constant is determined using Figure 5.63. We can write a few equations from reading the schematic diagram:

$$V_{(B)} = \left(V_{(A)} - V_{(op)}\right) A_{OL} \tag{5.305}$$

$$V_{(C)} = \left(V_{(B)} - V_{(op)}\right) A_{OL} \tag{5.306}$$

Because no current flows in R_4:

$$V_{(C)} = V_{(op)} \tag{5.307}$$

Combining the above equations and solving for $V_{(C)}$ leads you to:

$$V_{(C)} = \frac{A_{OL}^2 V_{(A)}}{A_{OL}^2 + A_{OL} + 1} \tag{5.308}$$

Current I_2 is obtained knowing the voltage across R_2:

$$I_2 = \frac{V_{(A)} - V_{(C)}}{R_2} = \frac{V_{(A)} - \dfrac{A_{OL}^2 V_{(A)}}{1 + A_{OL} + A_{OL}^2}}{R_2} = \frac{(1 + A_{OL})(V_T - I_T R_3)}{R_2\left(1 + A_{OL} + A_{OL}^2\right)} \tag{5.309}$$

Current I_1 is flowing in R_1 then the voltage at node A is:

$$V_{(A)} = R_1 I_1 \tag{5.310}$$

then

$$V_T = V_{(A)} + I_T R_3 \tag{5.311}$$

from which we have

$$V_{(A)} = V_T - I_T R_3 \tag{5.312}$$

Equating (5.310) and (5.312) gives:

$$I_1 = \frac{V_T - I_T R_3}{R_1} \tag{5.313}$$

Current I_T is the sum of I_1 and I_2. Therefore:

$$I_T = I_1 + I_2 = \frac{V_T - I_T R_3}{R_1} + \frac{(1 + A_{OL})(V_T - I_T R_3)}{R_2\left(1 + A_{OL} + A_{OL}^2\right)} \tag{5.314}$$

Solving for V_T and rearranging gives:

$$\frac{V_T}{I_T} = \frac{1 + \frac{R_3}{R_1} + \frac{R_3(1 + A_{OL})}{R_2\left(1 + A_{OL} + A_{OL}^2\right)}}{\frac{1}{R_1} + \frac{1 + A_{OL}}{R_2\left(1 + A_{OL} + A_{OL}^2\right)}} \tag{5.315}$$

which gives us the third time constant:

$$\tau_3 = C_3 \frac{1 + \frac{R_3}{R_1} + \frac{R_3(1 + A_{OL})}{R_2\left(1 + A_{OL} + A_{OL}^2\right)}}{\frac{1}{R_1} + \frac{1 + A_{OL}}{R_2\left(1 + A_{OL} + A_{OL}^2\right)}} \tag{5.316}$$

When A_{OL} approaches infinity, this expression nicely simplifies to:

$$\tau_3 = C_3(R_1 + R_3) \tag{5.317}$$

We can build our first coefficient b_1 equal to:

$$\begin{aligned} b_1 = \tau_1 + \tau_2 + \tau_3 = C_1 &\frac{R_4(R_1 + R_2) + A_{OL} R_4(R_1 + R_2)}{R_1 + R_2 + A_{OL}(R_1 + R_2) + A_{OL}^2 R_2} \\ + C_2 \frac{R_5(R_1 + R_2) + A_{OL} R_5(R_1 + R_2)}{R_1 + R_2 + A_{OL}(R_1 + R_2) + A_{OL}^2 R_2} &+ C_3 \frac{1 + \frac{R_3}{R_1} + \frac{R_3(1 + A_{OL})}{R_2\left(1 + A_{OL} + A_{OL}^2\right)}}{\frac{1}{R_1} + \frac{1 + A_{OL}}{R_2\left(1 + A_{OL} + A_{OL}^2\right)}} \end{aligned} \tag{5.318}$$

When A_{OL} is large, this expression simplifies to:

$$b_1 = C_3(R_1 + R_3) \tag{5.319}$$

In Figure 5.64, capacitors C_1 is now shorted and we have to determine the resistance seen from C_2's terminals to obtain τ_2^1 (C_3 in its dc state is removed from the figure).

The first mesh equation is:

$$V_T = R_5 I_T + \left(V_{(\text{A})} - V_{(\text{op})}\right) A_{OL} \tag{5.320}$$

The voltage at node A is determined using a simple voltage divider implying R_1 and R_2:

$$V_{(\text{A})} = \left[V_T A_{OL} - V_{(\text{op})} A_{OL}\right] \frac{R_1}{R_1 + R_2} \tag{5.321}$$

For node op, its level can be obtained by:

$$V_{(\text{op})} = V_T - R_5 I_T \tag{5.322}$$

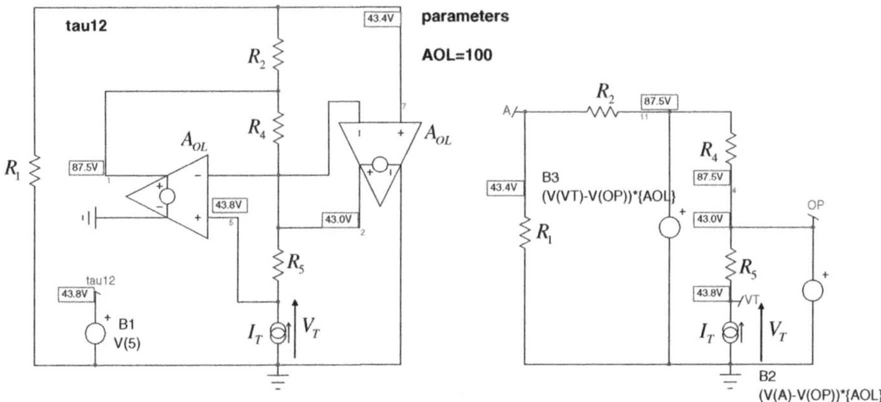

Figure 5.64 Capacitors 1 is shorted, C_3 is removed while C_2's terminals are biased by the current generator I_T.

but it is also equal to

$$V_{(\text{op})} = V_{(\text{A})}A_{OL} - V_{(\text{op})}A_{OL} \tag{5.323}$$

from which we extract

$$V_{(\text{op})} = \frac{V_{(\text{A})}A_{OL}}{1 + A_{OL}} \tag{5.324}$$

Equating equations (5.322) and (5.324) then extracting $V_{(\text{A})}$ leads to:

$$V_{(\text{A})} = \frac{(V_T - I_T R_5)(1 + A_{OL})}{A_{OL}} \tag{5.325}$$

If we now equate (5.325) and (5.321), we have a definition for $V_{(\text{op})}$:

$$V_{(\text{op})} = -\frac{\left(\frac{(A_{OL}+1)(V_T - I_T R_5)}{A_{OL}} - \frac{A_{OL} R_1 V_T}{R_1 + R_2}\right)(R_1 + R_2)}{A_{OL} R_1} \tag{5.326}$$

recognizing that (5.322) is equal to (5.326) and solving for V_T leads after rearranging to:

$$\frac{V_T}{I_T} = \frac{R_5\left(R_1 + R_2 + A_{OL}(R_1 + R_2) + A_{OL}{}^2 R_1\right)}{(1 + A_{OL})(R_1 + R_2)} = \frac{R_5\left(1 + A_{OL} + A_{OL}{}^2 \frac{R_1}{R_1 + R_2}\right)}{1 + A_{OL}} \tag{5.327}$$

leading to the following time constant:

$$\tau_2^1 = C_2 R_5\left(1 + \frac{A_{OL}{}^2}{1 + A_{OL}} \frac{R_1}{R_1 + R_2}\right) \tag{5.328}$$

This time constant is approaching infinity as A_{OL} increases. We may have an indeterminacy when multiplied with τ_1. Now looking at Figure 5.65, we can determine τ_3^1 by shorting C_1 and looking into C_3's terminals. The configuration is quite simple since $V_{(\text{OP})}$ and $V_{(\text{B})}$ being at similar potentials then B_3 returns 0 V and grounds R_2's right terminal. Therefore:

$$\tau_3^1 = C_3\left(R_3 + R_1 \| R_2\right) \tag{5.329}$$

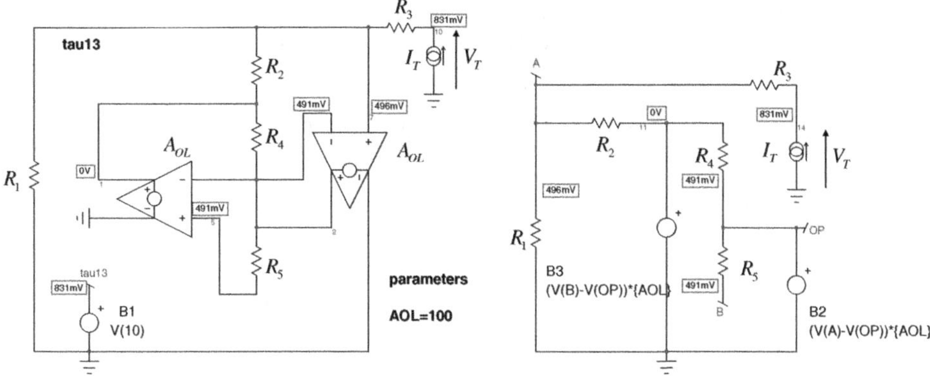

Figure 5.65 The resistance seen from C_3's terminals can be determined by inspection.

The configuration for τ_3^2 in Figure 5.66 is not different from that of Figure 5.65 and time constants for both diagrams are similar:

$$\tau_3^2 = C_3\left(R_3 + R_1 \| R_2\right) \tag{5.330}$$

The second coefficient b_2 is now assembled as follows:

$$\begin{aligned} b_2 = \tau_1\tau_2^1 + \tau_1\tau_3^1 + \tau_2\tau_3^2 = {} & C_1\frac{R_4(R_1+R_2) + A_{OL}R_4(R_1+R_2)}{R_1+R_2+A_{OL}(R_1+R_2)+{A_{OL}}^2R_2}\cdot C_2R_5\left(1+\frac{{A_{OL}}^2}{1+A_{OL}}\frac{R_1}{R_1+R_2}\right) \\ & +C_1\frac{R_4(R_1+R_2) + A_{OL}R_4(R_1+R_2)}{R_1+R_2+A_{OL}(R_1+R_2)+{A_{OL}}^2R_2}\cdot C_3\left(R_3+R_1\|R_2\right) \\ & +C_2\frac{R_5(R_1+R_2) + A_{OL}R_5(R_1+R_2)}{R_1+R_2+A_{OL}(R_1+R_2)+{A_{OL}}^2R_2}\cdot C_3\left(R_3+R_1\|R_2\right) \end{aligned} \tag{5.331}$$

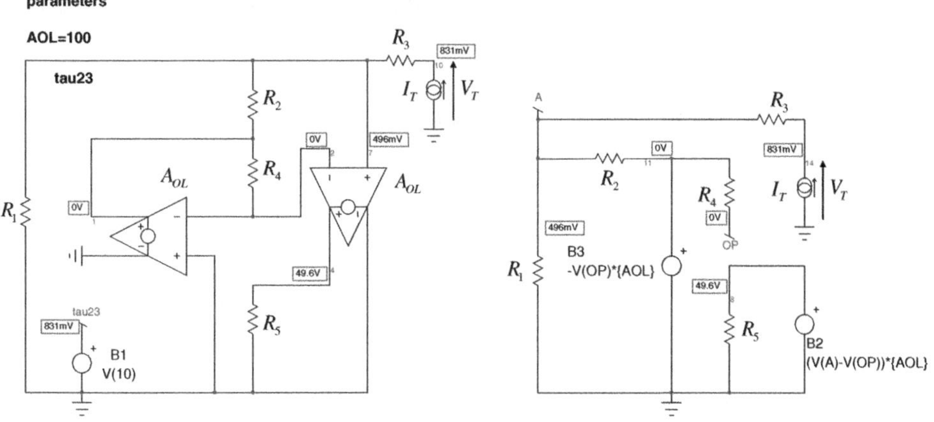

Figure 5.66 This configuration is not different from the preceding one and returns a similar resistance from C_3's terminals.

How does this expression behaves when A_{OL} approaches infinity? There is no ambiguity for the terms $\tau_1\tau_3^1$ and $\tau_2\tau_3^2$ as both go to zero in this case. However $\tau_1\tau_2^1$ has to be reworked. Let's expand it. We have:

$$\tau_1\tau_2^1 = \frac{R_4R_5\left(R_1 + R_2 + A_{OL}(R_1 + R_2) + A_{OL}{}^2R_1\right)}{R_1 + R_2 + A_{OL}(R_1 + R_2) + A_{OL}{}^2R_2}C_1C_2 \tag{5.332}$$

If we factor $A_{OL}{}^2$, then:

$$\tau_1\tau_2^1 = \frac{R_4R_5\left(\frac{R_1 + R_2}{A_{OL}{}^2} + \frac{R_1 + R_2}{A_{OL}} + R_1\right)A_{OL}{}^2}{\left(\frac{R_1 + R_2}{A_{OL}{}^2} + \frac{R_1 + R_2}{A_{OL}} + R_2\right)A_{OL}{}^2}C_1C_2 \tag{5.333}$$

Considering A_{OL} a large value, this expression simplifies to:

$$\tau_1\tau_2^1 \approx \frac{R_4R_5R_1}{R_2}C_1C_2 \tag{5.334}$$

We can now rewrite b_2 definition for A_{OL} approaching infinity:

$$b_2 \approx \frac{R_4R_5R_1}{R_2}C_1C_2 \tag{5.335}$$

The term b_3 requires the determination of τ_3^{12} to associate it with τ_1 and τ_2^1. However, τ_1 being 0 when A_{OL} approaches infinity, we may end up in an indeterminacy. Why not attempt to reshuffle the time constants? If rather than solving for τ_3^{12} we look at τ_2^{13} as shown in Figure 5.67, we can see a resemblance with Figure 5.64 where R_3 now comes in parallel with R_1. It saves us time since we simply update (5.328) as

$$\tau_2^{13} = C_2R_5\left(1 + \frac{A_{OL}{}^2}{1 + A_{OL}} \frac{R_1 \| R_3}{R_1 \| R_3 + R_2}\right) \tag{5.336}$$

To check how b_3 behaves as A_{OL} approaches infinity, let's develop $\tau_1\tau_3^1\tau_2^{13}$ and see what simplification can be brought:

$$\tau_1\tau_3^1\tau_2^{13} = \frac{C_1C_2C_3R_4R_5\left(R_1R_2 + R_1R_3 + R_2R_3 + A_{OL}{}^2R_1R_3 + A_{OL}R_1R_2 + A_{OL}R_1R_3 + A_{OL}R_2R_3\right)}{R_1 + R_2 + A_{OL}(R_1 + R_2) + A_{OL}{}^2R_2} \tag{5.337}$$

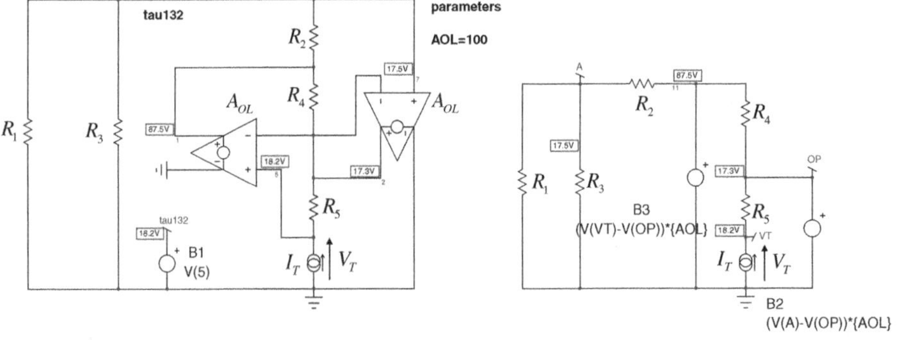

Figure 5.67 This last configuration builds on Figure 5.64, simply adding R_3 in parallel with R_1:

Factoring A_{OL}^2 we have:

$$\tau_1\tau_3^1\tau_2^{13}=\frac{C_1C_2C_3R_4R_5\left(\frac{R_1R_2+R_1R_3+R_2R_3}{A_{OL}^2}+R_1R_3+\frac{A_{OL}R_1R_2}{A_{OL}^2}+\frac{A_{OL}R_1R_3}{A_{OL}^2}+\frac{A_{OL}R_2R_3}{A_{OL}^2}\right)A_{OL}^2}{\left(\frac{R_1+R_2}{A_{OL}^2}+\frac{A_{OL}(R_1+R_2)}{A_{OL}^2}+R_2\right)A_{OL}^2} \tag{5.338}$$

When A_{OL} approaches infinity, this expression simplifies to:

$$\tau_1\tau_3^1\tau_2^{13}\approx\frac{C_1C_2C_3R_4R_5R_1R_3}{R_2} \tag{5.339}$$

We can now assemble all the pieces to form the final coefficient b_3 including the op amp open-loop gain effects:

$$b_3=\tau_1\tau_3^1\tau_2^{13}=C_1\frac{R_4(R_1+R_2)+A_{OL}R_4(R_1+R_2)}{R_1+R_2+A_{OL}(R_1+R_2)+A_{OL}^2R_2}C_3\left(R_3+R_1\|R_2\right)C_2R_5\left(1+\frac{A_{OL}^2}{1+A_{OL}}\frac{R_1\|R_3}{R_1\|R_3+R_2}\right) \tag{5.340}$$

or if we consider A_{OL} approaching infinity:

$$b_3\approx\frac{C_1C_2C_3R_4R_5R_1R_3}{R_2} \tag{5.341}$$

The complete denominator D is then obtained by combining all b coefficients:

$$\begin{aligned}D(s)&=1+s(\tau_1+\tau_2+\tau_3)+s^2\left(\tau_1\tau_2^1+\tau_1\tau_3^1+\tau_2\tau_3^2\right)+s^3\tau_1\tau_3^1\tau_2^{13}\\&=1+s\left(\begin{array}{l}C_1\frac{R_4(R_1+R_2)+A_{OL}R_4(R_1+R_2)}{R_1+R_2+A_{OL}(R_1+R_2)+A_{OL}^2R_2}\\+C_2\frac{R_5(R_1+R_2)+A_{OL}R_5(R_1+R_2)}{R_1+R_2+A_{OL}(R_1+R_2)+A_{OL}^2R_2}+C_3\frac{1+\frac{R_3}{R_1}+\frac{R_3(1+A_{OL})}{R_2\left(1+A_{OL}+A_{OL}^2\right)}}{\frac{1}{R_1}+\frac{1+A_{OL}}{R_2\left(1+A_{OL}+A_{OL}^2\right)}}\end{array}\right)\\&+s^2\left(\begin{array}{l}C_1\frac{R_4(R_1+R_2)+A_{OL}R_4(R_1+R_2)}{R_1+R_2+A_{OL}(R_1+R_2)+A_{OL}^2R_2}\cdot C_2R_5\left(1+\frac{A_{OL}^2}{1+A_{OL}}\frac{R_1}{R_1+R_2}\right)\\+C_1\frac{R_4(R_1+R_2)+A_{OL}R_4(R_1+R_2)}{R_1+R_2+A_{OL}(R_1+R_2)+A_{OL}^2R_2}\cdot C_3\left(R_3+R_1\|R_2\right)\\+C_2\frac{R_5(R_1+R_2)+A_{OL}R_5(R_1+R_2)}{R_1+R_2+A_{OL}(R_1+R_2)+A_{OL}^2R_2}\cdot C_3\left(R_3+R_1\|R_2\right)\end{array}\right)\\&+s^3\left(C_1\frac{R_4(R_1+R_2)+A_{OL}R_4(R_1+R_2)}{R_1+R_2+A_{OL}(R_1+R_2)+A_{OL}^2R_2}C_3\left(R_3+R_1\|R_2\right)C_2R_5\left(1+\frac{A_{OL}^2}{1+A_{OL}}\frac{R_1\|R_3}{R_1\|R_3+R_2}\right)\right)\end{aligned} \tag{5.342}$$

When considering a large open-loop gain A_{OL}, this expression simplifies to:

$$D(s)\approx1+C_3(R_1+R_3)s+\frac{R_4R_5R_1}{R_2}C_1C_2s^2+\frac{C_1C_3^2R_4R_5R_1R_3}{R_2}s^3 \tag{5.343}$$

It is time to look at the numerator expression now and, in particular, the quasi-static gain H_0. The sketch for this gain is given in Figure 5.68.

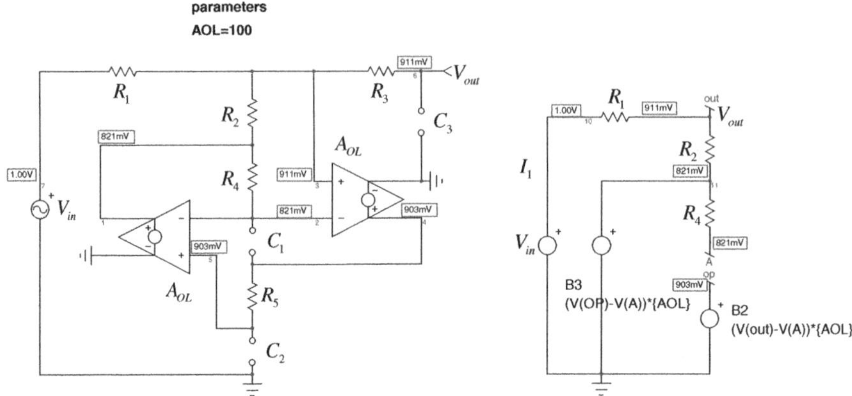

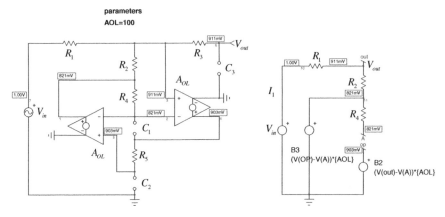

Figure 5.68 The gain H_0 is determined while all capacitors are open-circuited.

From this drawing, current I_1 is equal to:

$$I_1 = \frac{V_{in} - V_{out}}{R_1} \tag{5.344}$$

which is also equal to

$$I_1 = \frac{V_{out} - A_{OL}\left(V_{(\mathrm{op})} - V_{(\mathrm{A})}\right)}{R_2} \tag{5.345}$$

as no current flows in R_4.

Voltage at node op is defined by:

$$V_{(\mathrm{op})} = V_{out}A_{OL} - V_{(\mathrm{A})}A_{OL} \tag{5.346}$$

The voltage at node A is V_{out} minus the drop across R_2. Using (5.344) for I_1, we have:

$$V_{(\mathrm{A})} = V_{out} - R_2 I_1 = V_{out} - R_2\frac{V_{in} - V_{out}}{R_1} \tag{5.347}$$

Substituting (5.347) in (5.346) leads to the op node definition:

$$V_{(\mathrm{op})} = \frac{A_{OL}R_2(V_{in} - V_{out})}{R_1} \tag{5.348}$$

Then, if we equate (5.344) and (5.345) while replacing voltage at node op by (5.348):

$$\frac{V_{in} - V_{out}}{R_1} = \frac{R_1 V_{out} - A_{OL}^2 R_2 V_{in} + A_{OL}^2 R_2 V_{out} + A_{OL}R_1 V_{(\mathrm{A})}}{R_1 R_2} \tag{5.349}$$

Now replacing $V_{(\mathrm{A})}$ by (5.347) and factoring, we obtain:

$$H_0 = \frac{R_2\left(1 + A_{OL} + A_{OL}^2\right)}{R_1 + R_2 + A_{OL}(R_1 + R_2) + A_{OL}^2 R_2} \tag{5.350}$$

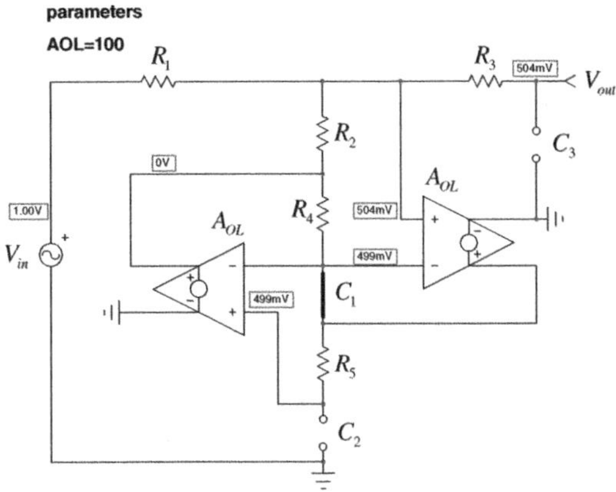

Figure 5.69 Capacitor C_1 is shorted to determine H^1, bringing both inputs of the left-side op amp at the same potential.

If we factor $A_{OL}{}^2$, we have:

$$H_0 = \frac{A_{OL}{}^2 R_2 \left(\frac{1}{A_{OL}{}^2} + \frac{1}{A_{OL}} + 1 \right)}{A_{OL}{}^2 \left(\frac{R_1 + R_2}{A_{OL}{}^2} + \frac{R_1 + R_2}{A_{OL}} + R_2 \right)} \tag{5.351}$$

Considering a large A_{OL}, this equation becomes:

$$H_0 \approx 1 \tag{5.352}$$

The first gain H^1 is obtained by shorting C_1 as shown in Figure 5.69. As both inputs of the left-side op amp are at a similar potential, its output is 0 V and grounds R_2's low-side terminal. R_2 and R_1 form a simple resistive divider leading to:

$$H^1 = \frac{R_2}{R_1 + R_2} \tag{5.353}$$

H^2 is not different since C_2 being shorted, the left-side op amp non-inverting input is grounded, bringing R_2's lower-side terminal at 0 V. The gain is thus:

$$H^2 = \frac{R_2}{R_1 + R_2} \tag{5.354}$$

For H^3 and all configurations involving a short-circuited C_3, the output is always zero. Therefore:

$$H^3 = H^{13} = H^{23} = H^{123} = 0 \tag{5.355}$$

We just have to determine H^{12} with the sketch shown in Figure 5.71. The circuit equation relates to current I_1 which circulates in R_1 and R_2:

$$I_1 = \frac{V_{in} - V_{out}}{R_1} = \frac{V_{out} - V_{(A)}}{R_2} \tag{5.356}$$

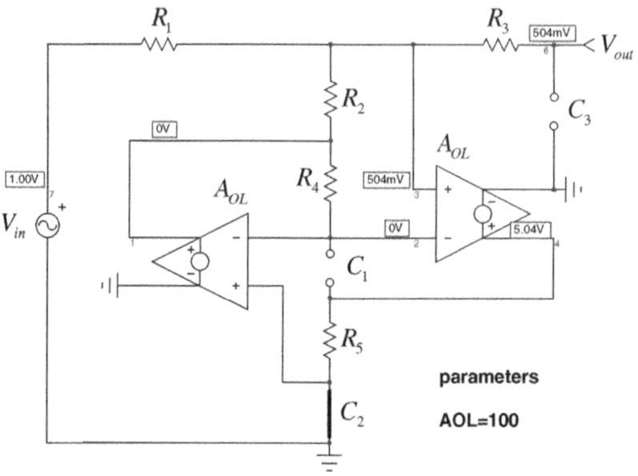

Figure 5.70 Capacitor C_2 is shorted to determine H^2.

From this equation, the voltage at node A comes as

$$V_{(A)} = \frac{R_1 V_{out} - R_2 V_{in} + R_2 V_{out}}{R_1} \tag{5.357}$$

The voltage at node A can also be obtained by

$$V_{(A)} = -V_{(op)} A_{OL} \tag{5.358}$$

Node op is determined with the following equation:

$$V_{(op)} = A_{OL}\left(V_{out} - V_{(op)}\right) \tag{5.359}$$

from which we obtain

$$V_{(op)} = \frac{A_{OL} V_{out}}{1 + A_{OL}} \tag{5.360}$$

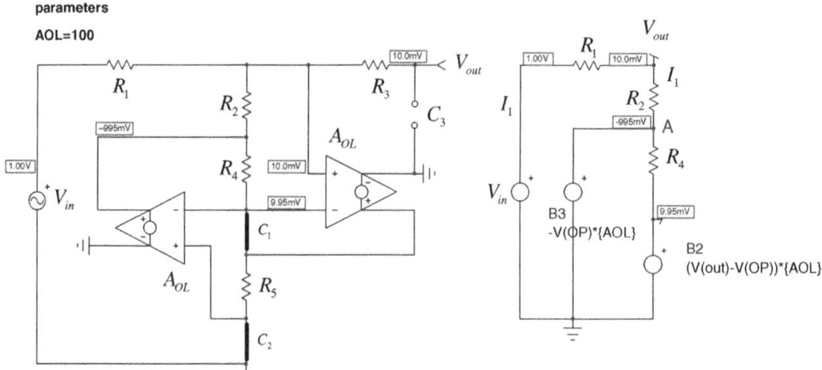

Figure 5.71 H^{12} requires a dedicated simplified sketch to determine its definition.

If we substitute (5.360) in (5.358), we have

$$V_{(A)} = -\frac{{A_{OL}}^2 V_{out}}{1 + A_{OL}} \tag{5.361}$$

Now equating (5.361) and (5.357) then rearranging for H^{12}, we obtain

$$H^{12} = \frac{R_2}{R_1\left(\frac{R_1 + R_2}{R_1} + \frac{{A_{OL}}^2}{1 + A_{OL}}\right)} \tag{5.362}$$

When the op amp open-loop gain increases, this term goes to 0. For H^{123}, as capacitor C_3 is replaced by a short circuit, then

$$H^{123} = 0 \tag{5.363}$$

As a summary, we can now assemble our expressions to form the numerator N. It appears in the below formula:

$$N(s) = H_0 + s\left(\tau_1 H^1 + \tau_2 H^2 + \tau_3 H^3\right) + s^2\left(\tau_1 \tau_2^1 H^{12} + \tau_1 \tau_3^1 H^{13} + \tau_2 \tau_3^2 H^{23}\right) + s^3\left(\tau_1 \tau_3^1 \tau_2^{13} H^{123}\right) \tag{5.364}$$

You can include in the above expression all the raw definitions we have determined for the various time constants. In this case, you will see the impact of the op amp gain, A_{OL} on the dynamic response. Considering all the zeroed-gain expressions, (5.364) finally reduces to:

$$N(s) = H_0 + s\left(\tau_1 H^1 + \tau_2 H^2\right) + s^2 \tau_1 \tau_2^1 H^{12} \tag{5.365}$$

If you consider a large gain A_{OL}, then $N(s)$ is simply equal to 1. We now have two transfer functions, one comprehensive including the effects of the op amp open-loop gain, and one simplified in which A_{OL} is considered infinite:

$$H(s) = H_0 \frac{1 + s\left(\frac{\tau_1 H^1 + \tau_2 H^2}{H_0}\right) + s^2 \frac{\tau_1 \tau_2^1 H^{12}}{H_0}}{1 + s(\tau_1 + \tau_2 + \tau_3) + s^2\left(\tau_1 \tau_2^1 + \tau_1 \tau_3^1 + \tau_2 \tau_3^2\right) + s^3 \tau_1 \tau_3^1 \tau_2^{13}} \tag{5.366}$$

When A_{OL} approaches infinity, this expression simplifies to:

$$H(s) \approx \frac{1}{1 + C_3(R_1 + R_3)s + \frac{R_4 R_5 R_1}{R_2} C_1 C_2 s^2 + \frac{C_1 C_2 C_3 R_4 R_5 R_1 R_3}{R_2} s^3} \tag{5.367}$$

The 3rd-order denominator can be rearranged under the product of a dominant low-frequency pole and two coincident poles as shown in Chapter 2. Final plots will show how this expression deviates from the original one, (5.366). Following Chapter 2 definitions we have:

$$\omega_p = \frac{1}{b_1} = \frac{1}{C_3(R_1 + R_3)} \tag{5.368}$$

$$\omega_0 = \sqrt{\frac{b_1}{b_3}} = \sqrt{\frac{R_2(R_1 + R_3)}{C_1 C_3 R_1 R_3 R_4 R_5}} \tag{5.369}$$

$$Q = \frac{b_1 b_3 \sqrt{\frac{b_1}{b_3}}}{b_1 b_2 - b_3} = \frac{C_2 C_3 R_1 R_3 R_4 R_5 (R_1 + R_3)\sqrt{\frac{R_2(R_1 + R_3)}{C_1 C_3 R_1 R_3 R_4 R_5}}}{C_2 {R_1}^2 R_4 R_5 + R_1 R_3 R_4 R_5 (C_2 - C_3)} \tag{5.370}$$

$R_1 := 0.9852\ \Omega \quad R_2 := 1\Omega \quad R_3 := 0.335\ \Omega \quad R_4 := 1\Omega \quad R_5 := 0.8746\ \Omega \quad A_{OL} := 10^5$

$C_1 := 3.98\ \mu F \quad C_2 := 3.98\ \mu F \quad C_3 := 3.98\ \mu F \quad \|(x,y) := \frac{x \cdot y}{x + y}$

$$\tau_1 := C_1 \cdot \frac{R_1 \cdot R_4 + R_2 \cdot R_4 + A_{OL} \cdot R_4 \cdot (R_1 + R_2)}{R_1 + R_2 + A_{OL} \cdot (R_1 + R_2) + A_{OL}^2 \cdot R_2} = 7.90102 \times 10^{-5} \cdot \mu s$$

$$\tau_2 := \frac{R_5 \cdot (R_1 + R_2) + A_{OL} \cdot R_5 \cdot (R_1 + R_2)}{R_1 + R_2 + A_{OL} \cdot (R_1 + R_2) + A_{OL}^2 \cdot R_2} \cdot C_2 = 6.91023 \times 10^{-5} \cdot \mu s$$

$$\tau_3 := C_3 \cdot \left[\frac{1 + \frac{R_3}{R_1} + \frac{R_3 \cdot (A_{OL} + 1)}{R_2 \cdot (A_{OL}^2 + A_{OL} + 1)}}{\frac{1}{R_1} + \frac{A_{OL} + 1}{R_2 \cdot (A_{OL}^2 + A_{OL} + 1)}} \right] = 5.25436 \cdot \mu s$$

$$\tau_{12} := C_2 \cdot \left[R_5 \cdot \left(1 + \frac{A_{OL}^2}{1 + A_{OL}} \cdot \frac{R_1}{R_1 + R_2} \right) \right] = 1.7275 \times 10^5 \cdot \mu s$$

$$\tau_{13} := [R_3 + (R_1 \| R_2)] \cdot C_3 = 3.30846 \cdot \mu s$$

$$\tau_{23} := [R_3 + (R_1 \| R_2)] \cdot C_3 = 3.30846 \cdot \mu s$$

$$\tau_{132} := \left[R_5 \cdot \left(1 + \frac{A_{OL}^2}{1 + A_{OL}} \cdot \frac{R_1 \| R_3}{R_1 \| R_3 + R_2} \right) \right] \cdot C_2 = 6.96196 \times 10^4 \cdot \mu s$$

$b_1 := \tau_1 + \tau_2 + \tau_3 = 5.25451 \cdot \mu s \qquad b_{11} := C_3 \cdot (R_1 + R_3) = 5.2544 \cdot \mu s$

$b_2 := \tau_1 \cdot \tau_{12} + \tau_1 \cdot \tau_{13} + \tau_2 \cdot \tau_{23} = 13.64947 \cdot \mu s^2 \qquad b_{22} := \frac{R_4 \cdot R_5 \cdot R_1}{R_2} \cdot C_1 \cdot C_2 = 13.64897 \cdot \mu s^2$

$b_3 := \tau_1 \cdot \tau_{13} \cdot \tau_{132} = 18.19873 \cdot \mu s^3 \qquad b_{33} := \frac{C_1 \cdot C_2 \cdot C_3 \cdot R_4 \cdot R_5 \cdot R_1 \cdot R_3}{R_2} = 18.19818 \cdot \mu s^3$

$D_1(s) := 1 + b_1 \cdot s + b_2 \cdot s^2 + b_3 \cdot s^3 \qquad D_2(s) := 1 + b_{11} \cdot s + b_{22} \cdot s^2 + b_{33} \cdot s^3$

$$H_0 := \frac{R_2 \cdot (A_{OL}^2 + A_{OL} + 1)}{R_1 + R_2 + A_{OL} \cdot R_1 + A_{OL} \cdot R_2 + A_{OL}^2 \cdot R_2} = 0.99999$$

$H_1 := \frac{R_2}{R_1 + R_2} = 0.50373 \qquad H_2 := \frac{R_2}{R_1 + R_2} = 0.50373 \qquad H_3 := 0$

$H_{12} := \frac{R_2}{R_1 \cdot \left(\frac{R_1 + R_2}{R_1} + \frac{A_{OL}^2}{A_{OL} + 1} \right)} = 1.01501 \times 10^{-5} \qquad H_{13} := 0 \qquad H_{23} := 0 \qquad H_{123} := 0$

$$a_1 := \tau_1 \cdot H_1 + \tau_2 \cdot H_2 + \tau_3 \cdot H_3 = 0.07461 \cdot ns$$

$$a_2 := \tau_1 \cdot \tau_{12} \cdot H_{12} + \tau_1 \cdot \tau_{13} \cdot H_{13} + \tau_2 \cdot \tau_{23} \cdot H_{23} = 1.38539 \times 10^{-4} \cdot \mu s^2$$

$$a_3 := \tau_1 \cdot \tau_{13} \cdot \tau_{132} \cdot H_{123} = 0$$

$N_1(s) := H_0 + a_1 \cdot s + a_2 \cdot s^2 + a_3 \cdot s^3 \qquad N_2(s) := H_0 + s \cdot (\tau_1 \cdot H_1 + \tau_2 \cdot H_2) + s^2 \cdot \tau_1 \cdot \tau_{12} \cdot H_{12} + s^3 \cdot \tau_1 \cdot \tau_{13} \cdot \tau_{132} \cdot H_{123}$

$H_{10}(s) := \frac{N_1(s)}{D_1(s)} \qquad H_{20}(s) := \frac{N_2(s)}{D_2(s)} \qquad H_{30}(s) := \frac{1}{D_2(s)}$

$\omega_p := \frac{1}{b_{11}} \qquad f_p := \frac{\omega_p}{2\pi} = 30.28986 \cdot kHz \qquad Q := \frac{b_{11} \cdot b_{33} \cdot \sqrt{\frac{b_{11}}{b_{33}}}}{b_{11} \cdot b_{22} - b_{33}} = 0.96004 \qquad \omega_0 := \sqrt{\frac{b_{11}}{b_{33}}} \qquad f_0 := \frac{\omega_0}{2\pi} = 85.51998 \cdot kHz$

$$\frac{C_2 \cdot C_3 \cdot R_1 \cdot R_3 \cdot R_4 \cdot R_5 \cdot (R_1 + R_3) \cdot \sqrt{\frac{R_2 \cdot (R_1 + R_3)}{C_1 \cdot C_3 \cdot R_1 \cdot R_3 \cdot R_4 \cdot R_5}}}{C_2 \cdot R_1^2 \cdot R_4 \cdot R_5 + (R_1 \cdot R_3 \cdot R_4 \cdot R_5) \cdot (C_2 - C_3)} = 0.96004 \qquad \sqrt{\frac{R_2 \cdot (R_1 + R_3)}{C_1 \cdot C_3 \cdot R_1 \cdot R_3 \cdot R_4 \cdot R_5}} \cdot \frac{1}{2\pi} = 85.51998 \cdot kHz$$

$D_3(s) := \left(1 + \frac{s}{\omega_p} \right) \cdot \left[1 + \frac{s}{\omega_0 \cdot Q} + \left(\frac{s}{\omega_0} \right)^2 \right] \qquad H_{40}(s) := \frac{1}{D_3(s)}$

Figure 5.72 Mathcad® calculates all time constants and lets us check simplified expressions.

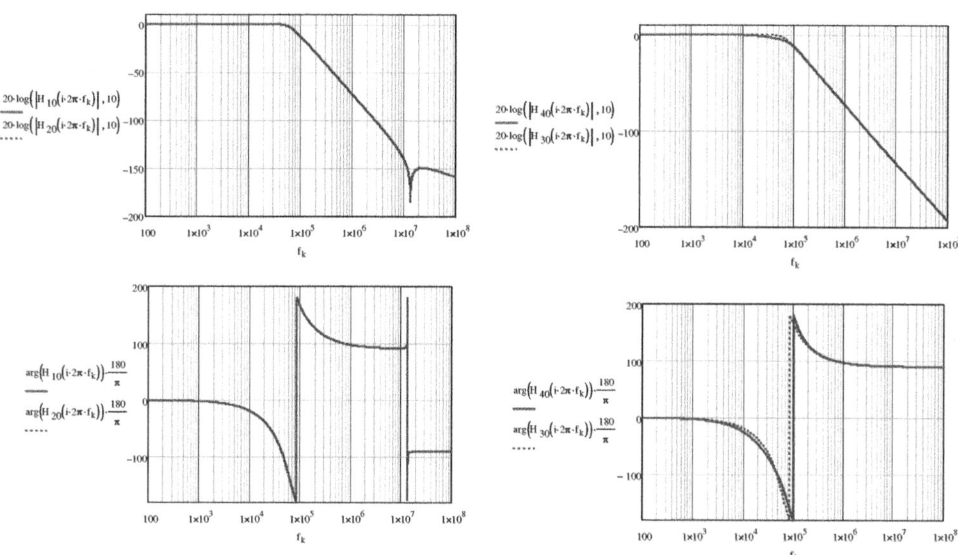

Figure 5.73 The complete expression reveals the flat response of a low-pass filter affected by a high-frequency notch.

The simplified transfer function is thus:

$$H(s) \approx \frac{1}{\left(1+\frac{s}{\omega_p}\right)\left(1+\frac{s}{\omega_0 Q}+\left(\frac{s}{\omega_0}\right)^2\right)} \tag{5.371}$$

The Mathcad® sheet is given in Figure 5.72 while curves appear in Figure 5.73 for an open-loop gain of 100k or 100 dB. The complete expression from H_{10} and its simplified version H_{20} predict a low-pass filter without peaking and reveal the presence of a high-frequency notch. When $N(s)$ is considered 1, H_{30} gives excellent results without the notch. Finally, if we plot (5.371) under H_{40}, there is a small gain deviation with a mismatch in the cutoff frequency.

The final check is with SPICE and we captured the GIC filter configuration in Figure 5.74. The figure includes simulation and analytical results. They match very well, validating our calculations.

5.3 What Should I Retain from this Chapter ?

In this fifth and last chapter, we went one step farther, exploring n^{th}-order circuits. Below is a summary of what we learned in this section:

1. Dealing with a high-order network is not overly complex compared to what we learned in previous chapters. There are more terms in the numerator and the denominator but the exercise remains the same. The key to success lies in slicing the network into numerous individual sketches, each representing a particular time constant configuration for D and N.
2. When determining these higher-order transfer functions, we considered a reference circuit in which all energy-storing elements are put in their dc state: the circuit is first observed at $s = 0$ where capacitors are open and inductors are short circuited. This is the way SPICE calculates an operating bias point prior to launching any type of simulation. Please note that we could do the

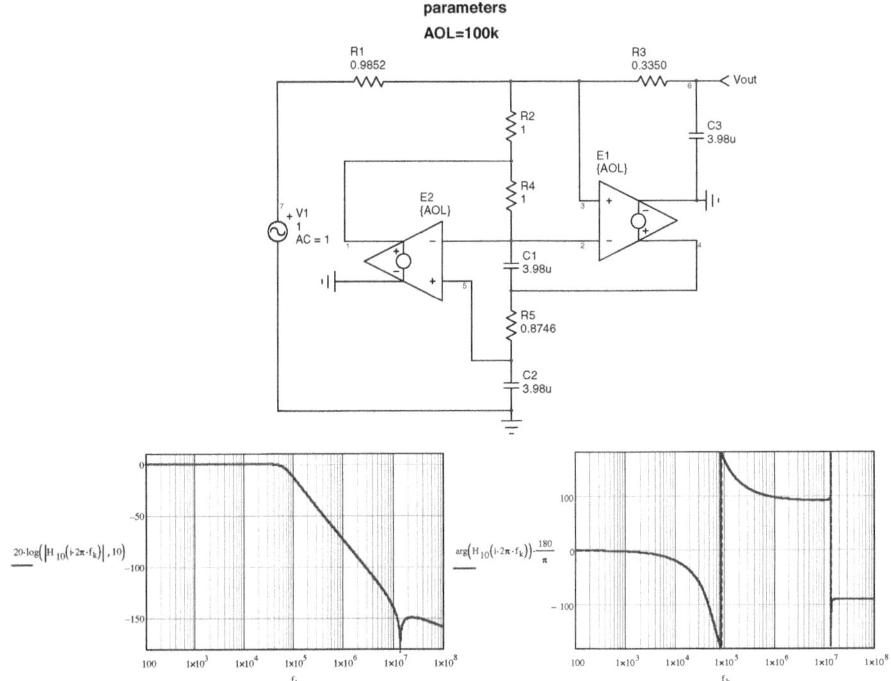

Figure 5.74 SPICE simulations and analytical expressions deliver the exact same responses.

other way around, considering the reference state as s approaches infinity, later alternatively setting energy-storing elements in their dc state. This is a possible option but I felt that $s = 0$ was more intuitive for engineers.

3. The number of terms in $b_1, b_2 \ldots b_n$ depend on the order n of the circuit. A binomial expression helps define how many terms form these coefficients. We gave generic formulas to determine the combined time constants present in the coefficients. However, redundancies naturally exist and you have the possibility to reshuffle time constant combinations to a) find an easier step to solve intermediate circuit b) get rid of an indeterminacy.
4. n^{th}-order numerator and denominator can sometimes be rearranged in a canonical form involving a quality factor Q and a resonant frequency ω_0. Approximate expressions are possible when neglecting certain elements, but often to the detriment of the final dynamic response precision.
5. Complex examples show how the technique can lead you to the result in a fast way. To succeed in your endeavor of determining a particular network transfer function, I recommend you take advantage of existing tools such as Mathcad® and SPICE. The first one lets you organize individual time constants and later associate them to form the numerator and denominator coefficients. Should you identify a mistake at some point, fixing the individual guilty time constant is easy and efficient. Finally, in complex architectures where controlled sources are involved, simple dc operating points in SPICE let you immediately check if the expression you derived is correct or not.

References

1. A. Hajimiri. Generalized time- and transfer-constant circuit analysis. *IEEE Transactions on Circuits and Systems*, **57** (6), 1105–1121. 2009.
2. http://www.filter-solutions.com/active.html (last accessed 12/12/2015).
3. http://www.ti.com/lit/ml/sloa088/sloa088.pdf (last accessed 12/12/2015).

5.4 Appendix 5A – Problems

Below are several problems based on the information delivered in this chapter.

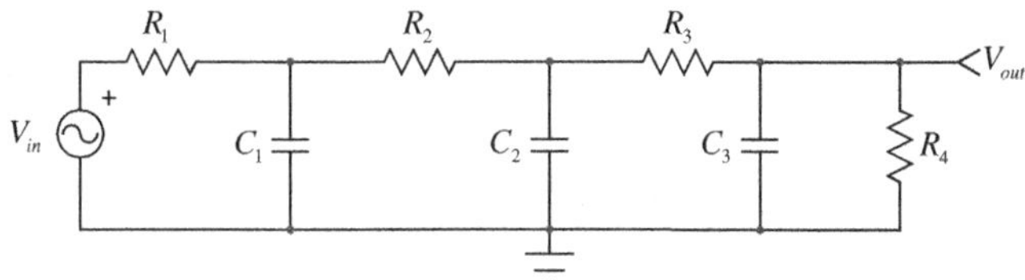

Figure 5.75 – Problem 1: What is the transfer function of this network?

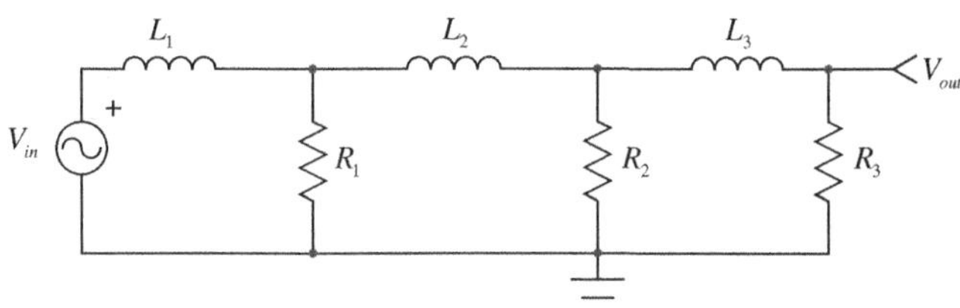

Figure 5.76 – Problem 2: What is the transfer function of this network?

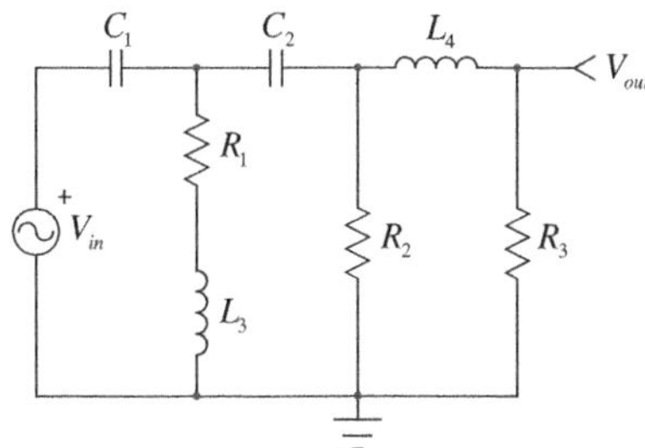

Figure 5.77 – Problem 3: What is the transfer function of this loudspeaker filter?

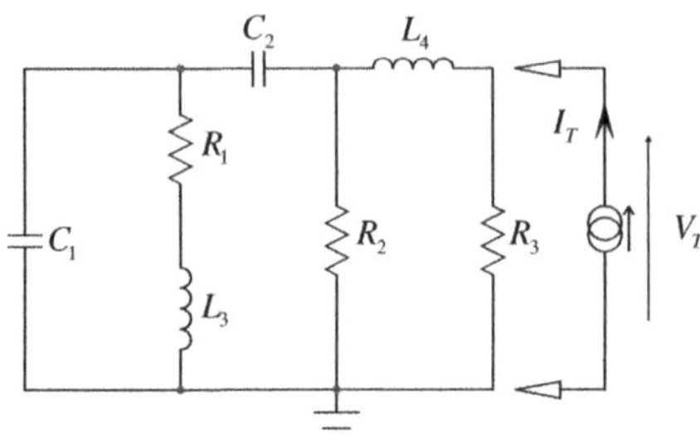

Figure 5.78 – Problem 4: What is the output impedance of this filter?

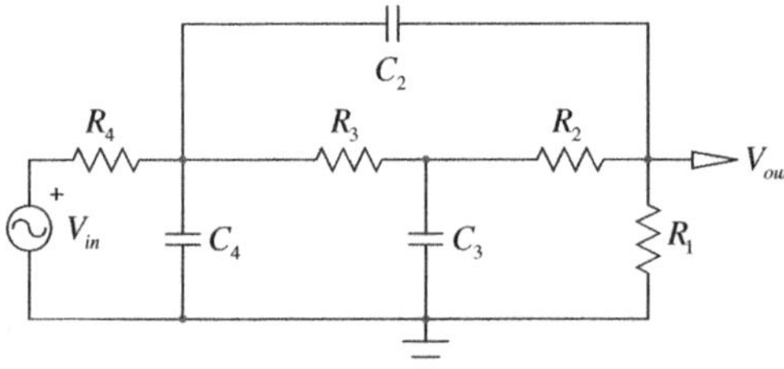

Figure 5.79 – Problem 5: What is the transfer function of this network?

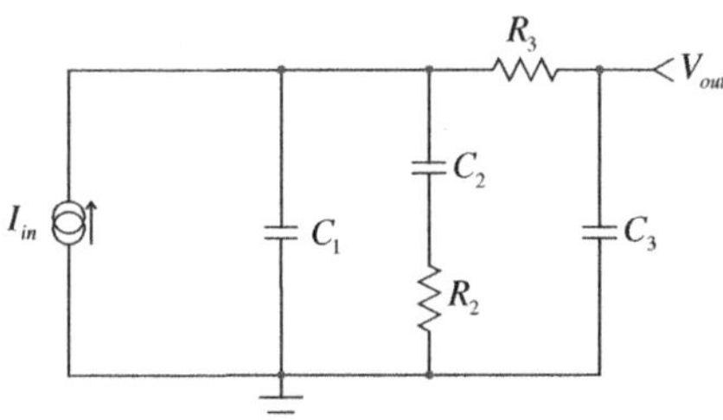

Figure 5.80 – Problem 6: What relationship links the output voltage to the input current?

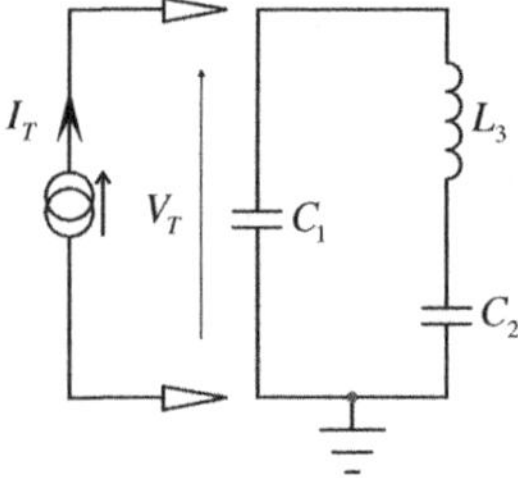

Figure 5.81 – Problem 7: What is the impedance offered by the network? Despite its simplicity, use fast analytical techniques – yes, I insist ☺.

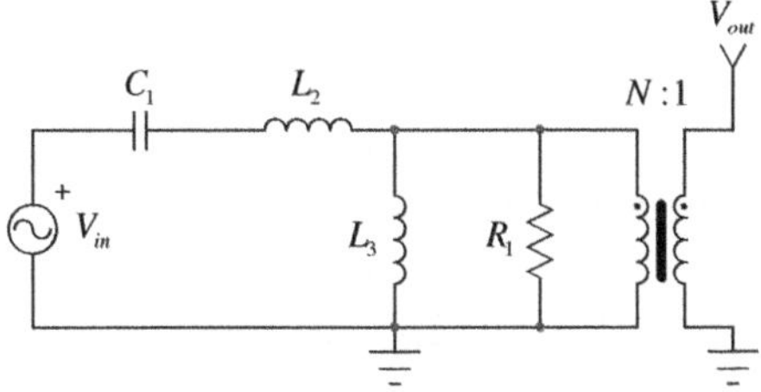

Figure 5.82 – Problem 8: What is the transfer function of this network?

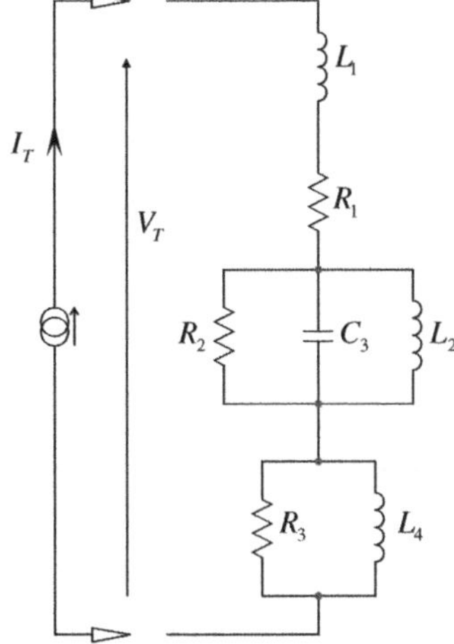

Figure 5.83 – Problem 9: What is the impedance of this loudspeaker model?

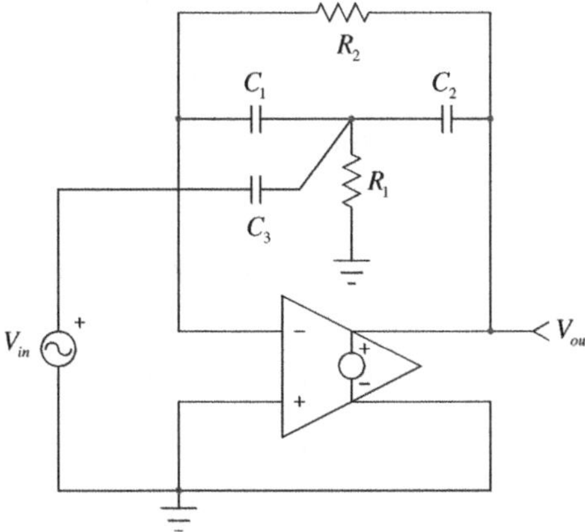

Figure 5.84 – Problem 10: What is the transfer function of this circuit?

Answers

Problem 1:

Figure 5.85 shows the stages to determine the network denominator of this 3rd-order filter. The quasi-static gain H_0 involves a resistive divider made of R_4 and the other series resistors:

$$H_0 = \frac{R_4}{R_1 + R_2 + R_3 + R_4} \tag{5.372}$$

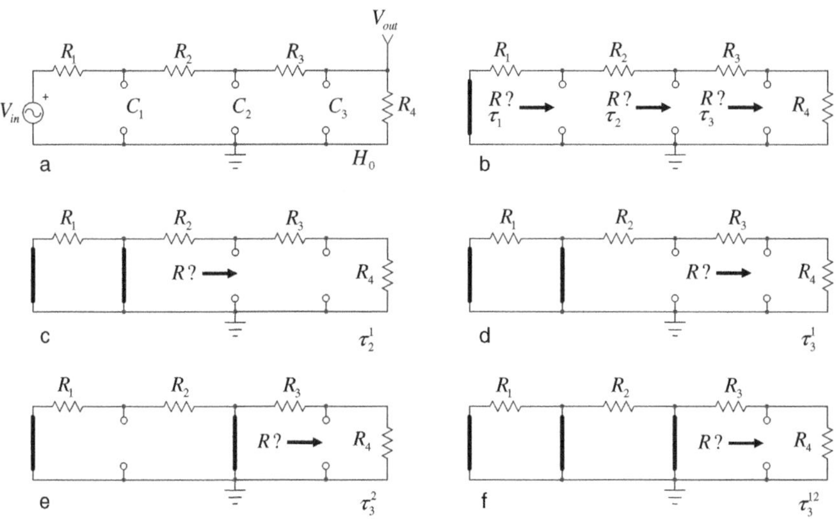

Figure 5.85 The network is split into individual pieces, helping to compute the various time constants.

The first time constants are obtained from Figure 5.85b, looking into the selected capacitor's terminals to determine the driving resistance. You should find:

$$\tau_1 = [R_1 \| (R_2 + R_3 + R_4)] C_1 \tag{5.373}$$

$$\tau_2 = [(R_1 + R_2) \| (R_3 + R_4)] C_2 \tag{5.374}$$

$$\tau_3 = [R_4 \| (R_1 + R_2 + R_3)] C_3 \tag{5.375}$$

The first coefficient b_1 can be formed:

$$b_1 = \tau_1 + \tau_2 + \tau_3 = [R_1 \| (R_2 + R_3 + R_4)] C_1 + [(R_1 + R_2) \| (R_3 + R_4)] C_2 + [R_4 \| (R_1 + R_2 + R_3)] C_3 \tag{5.376}$$

The higher-order time constants are obtained with the help of Figure 5.85c to e:

$$\tau_2^1 = [R_2 \| (R_3 + R_4)] C_2 \tag{5.377}$$

$$\tau_3^1 = [R_4 \| (R_2 + R_3)] C_3 \tag{5.378}$$

$$\tau_3^2 = (R_4 \| R_3) C_3 \tag{5.379}$$

They lead to:

$$\begin{aligned}b_2 &= \tau_1\tau_2^1+\tau_1\tau_3^1+\tau_2\tau_3^2\\ &= \left[R_1\|(R_2+R_3+R_4)\right]C_1\left[R_2\|(R_3+R_4)\right]C_2\\ &\quad+\left[R_1\|(R_2+R_3+R_4)\right]C_1\left[R_4\|(R_2+R_3)\right]C_3\\ &\quad+\left[(R_1+R_2)\|(R_3+R_4)\right]C_2\left(R_4\|R_3\right)C_3\end{aligned} \tag{5.380}$$

Finally, looking at Figure 5.85f, we have the last time constant:

$$\tau_3^{12} = \left(R_3\|R_4\right)C_3 \tag{5.381}$$

giving us the last denominator term b_3:

$$b_3 = \tau_1\tau_2^1\tau_3^{12} = \left[R_1\|(R_2+R_3+R_4)\right]C_1\left[R_2\|(R_3+R_4)\right]C_2\left(R_3\|R_4\right)C_3 \tag{5.382}$$

The denominator $D(s)$ is given by:

$$\begin{aligned}D(s) &= 1+b_1s+b_2s^2+b_3s^3\\ &= 1+s\left[\left[R_1\|(R_2+R_3+R_4)\right]C_1+\left[(R_1+R_2)\|(R_3+R_4)\right]C_2+\left[R_4\|(R_1+R_2+R_3)\right]C_3\right]\\ &\quad+s^2\begin{bmatrix}\left[R_1\|(R_2+R_3+R_4)\right]C_1\left[R_2\|(R_3+R_4)\right]C_2\\ +\left[R_1\|(R_2+R_3+R_4)\right]C_1\left[R_4\|(R_2+R_3)\right]C_3\\ +\left[(R_1+R_2)\|(R_3+R_4)\right]C_2\left(R_4\|R_3\right)C_3\end{bmatrix}\\ &\quad+s^3\left(\left[R_1\|(R_2+R_3+R_4)\right]C_1\left[R_2\|(R_3+R_4)\right]C_2\left(R_3\|R_4\right)C_3\right)\end{aligned} \tag{5.383}$$

There are no zeros in this network as placing any capacitor in its high-frequency state cancels the response. The transfer function is thus:

$$H(s) = \frac{1}{1+b_1s+b_2s^2+b_3s^3} \tag{5.384}$$

If the poles are well separated, this expression can be put under the following form:

$$H(s) \approx \frac{1}{(1+b_1s)\left(1+s\frac{b_2}{b_1}\right)\left(1+s\frac{b_3}{b_2}\right)} \tag{5.385}$$

To test these expressions, we need a reference 'raw' transfer function. Looking at Figure 5.75, we can analyze the network with two Thévenin's circuits. The first one is affected by a first output impedance:

$$R_{th1}(s) = \left(\frac{1}{sC_1}\right)\|R_1 \tag{5.386}$$

while the second generator features an impedance equal to:

$$R_{th2}(s) = (R_{th1}(s)+R_2)\|\left(\frac{1}{sC_2}\right) \tag{5.387}$$

The complete transfer function is given by:

$$H_{ref}(s) = \frac{\frac{1}{sC_1}}{\frac{1}{sC_1}+R_1}\cdot\frac{\frac{1}{sC_2}}{R_{th1}(s)+R_2+\frac{1}{sC_2}}\cdot\frac{\left(\frac{1}{sC_3}\right)\|R_4}{\left(\frac{1}{sC_3}\right)\|R_4+R_{th2}(s)+R_3} \tag{5.388}$$

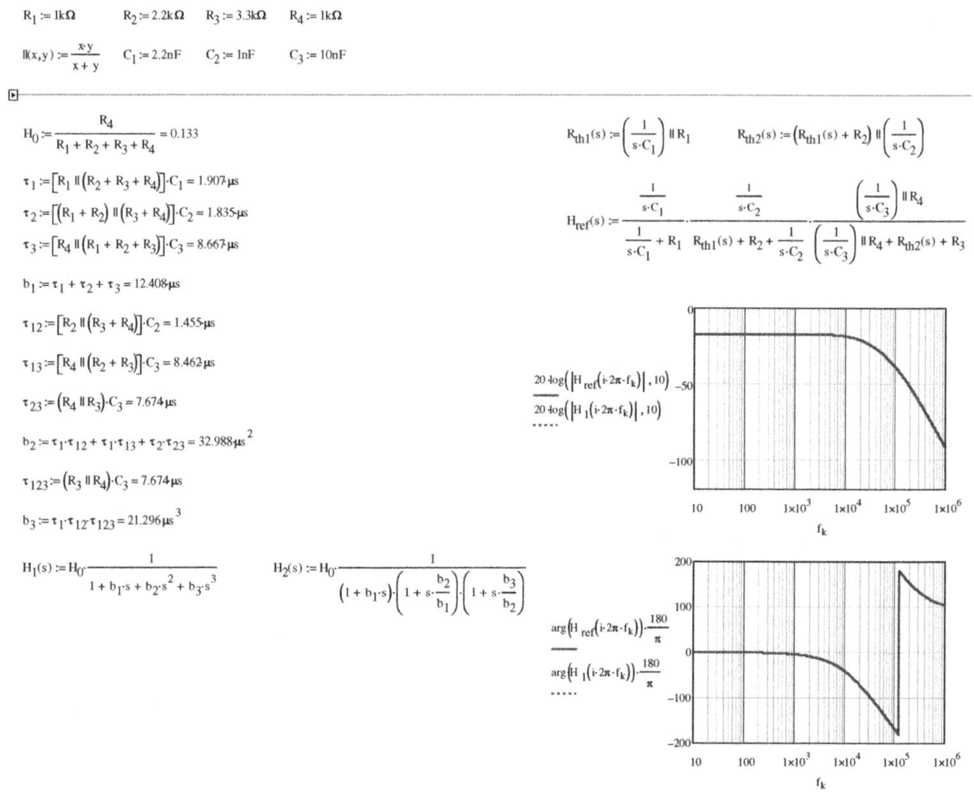

$R_1 := 1k\Omega \quad R_2 := 2.2k\Omega \quad R_3 := 3.3k\Omega \quad R_4 := 1k\Omega$

$\|(x,y) := \frac{x \cdot y}{x+y} \quad C_1 := 2.2nF \quad C_2 := 1nF \quad C_3 := 10nF$

$H_0 := \frac{R_4}{R_1 + R_2 + R_3 + R_4} = 0.133$

$\tau_1 := \left[R_1 \| (R_2 + R_3 + R_4)\right] \cdot C_1 = 1.907 \cdot \mu s$

$\tau_2 := \left[(R_1 + R_2) \| (R_3 + R_4)\right] \cdot C_2 = 1.835 \cdot \mu s$

$\tau_3 := \left[R_4 \| (R_1 + R_2 + R_3)\right] \cdot C_3 = 8.667 \cdot \mu s$

$b_1 := \tau_1 + \tau_2 + \tau_3 = 12.408 \cdot \mu s$

$\tau_{12} := \left[R_2 \| (R_3 + R_4)\right] \cdot C_2 = 1.455 \cdot \mu s$

$\tau_{13} := \left[R_4 \| (R_2 + R_3)\right] \cdot C_3 = 8.462 \cdot \mu s$

$\tau_{23} := (R_4 \| R_3) \cdot C_3 = 7.674 \cdot \mu s$

$b_2 := \tau_1 \cdot \tau_{12} + \tau_1 \cdot \tau_{13} + \tau_2 \cdot \tau_{23} = 32.988 \cdot \mu s^2$

$\tau_{123} := (R_3 \| R_4) \cdot C_3 = 7.674 \cdot \mu s$

$b_3 := \tau_1 \cdot \tau_{12} \cdot \tau_{123} = 21.296 \cdot \mu s^3$

$H_1(s) := H_0 \cdot \frac{1}{1 + b_1 \cdot s + b_2 \cdot s^2 + b_3 \cdot s^3} \qquad H_2(s) := H_0 \cdot \frac{1}{(1 + b_1 \cdot s) \cdot \left(1 + s \cdot \frac{b_2}{b_1}\right) \cdot \left(1 + s \cdot \frac{b_3}{b_2}\right)}$

$R_{th1}(s) := \left(\frac{1}{s \cdot C_1}\right) \| R_1 \qquad R_{th2}(s) := (R_{th1}(s) + R_2) \| \left(\frac{1}{s \cdot C_2}\right)$

$H_{ref}(s) := \frac{\frac{1}{s \cdot C_1}}{\frac{1}{s \cdot C_1} + R_1} \cdot \frac{\frac{1}{s \cdot C_2}}{R_{th1}(s) + R_2 + \frac{1}{s \cdot C_2}} \cdot \frac{\left(\frac{1}{s \cdot C_3}\right) \| R_4}{\left(\frac{1}{s \cdot C_3}\right) \| R_4 + R_{th2}(s) + R_3}$

Figure 5.86 Mathcad® confirms the excellent agreement between the raw transfer function and the expression we derived in (5.384).

The Mathcad® sheet appears in Figure 5.86 and confirms the good agreement between (5.384) and (5.388).

To compare response more precisely than having magnitude and phase curves pasted on a graph, you can plot the magnitude and phase differences between the functions you want to compare. If they are perfectly equal, the error is extremely small as confirmed in Figure 5.87.

Problem 2:

Figure 5.88 shows the stages to determine the network denominator of this 3rd-order filter. The quasi-static gain is simple to obtain since all inductor are shorted in dc. From sketch (a):

$$H_0 = 1 \tag{5.389}$$

The three first time constants are determined using sketches b to d and are defined asL:

$$\tau_1 = \frac{L_1}{R_1 \| R_2 \| R_3} \tag{5.390}$$

$$\tau_2 = \frac{L_2}{R_2 \| R_3} \tag{5.391}$$

$$\tau_3 = \frac{L_3}{R_3} \tag{5.392}$$

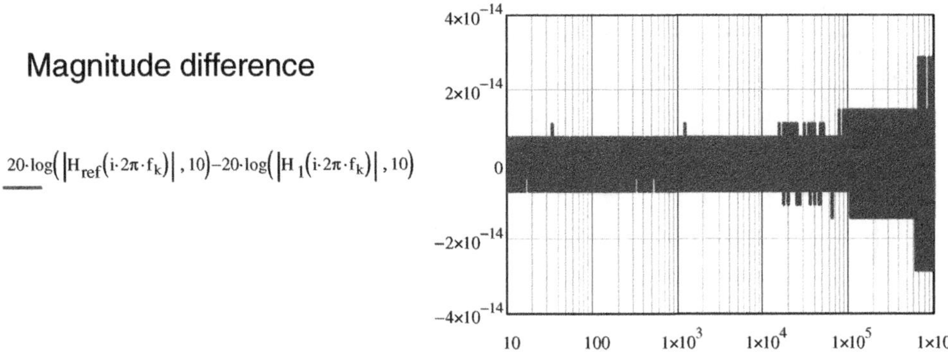

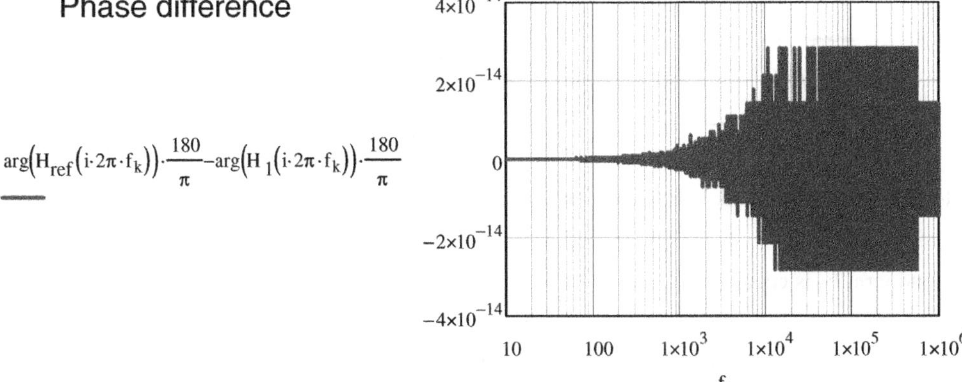

Figure 5.87 To precisely compare expressions, plotting the difference of their magnitude and phase responses is a good way to spot mismatches. Here, you can see that the error between the functions is in the solver noise.

The first term b_1 is formed by adding (5.390), (5.391) and (5.392):

$$b_1 = \tau_1 + \tau_2 + \tau_3 = \frac{L_1}{R_1 \| R_2 \| R_3} + \frac{L_2}{R_2 \| R_3} + \frac{L_3}{R_3} \tag{5.393}$$

The second term b_2 needs $\tau_2^1 \tau_3^1$ and τ_3^2. Sketches e to g should let you obtain them quite easily:

$$\tau_2^1 = \frac{L_2}{R_1 + R_2 \| R_3} \tag{5.394}$$

$$\tau_3^1 = \frac{L_3}{R_3 + R_2 \| R_1} \tag{5.395}$$

$$\tau_3^2 = \frac{L_3}{R_3 + R_2} \tag{5.396}$$

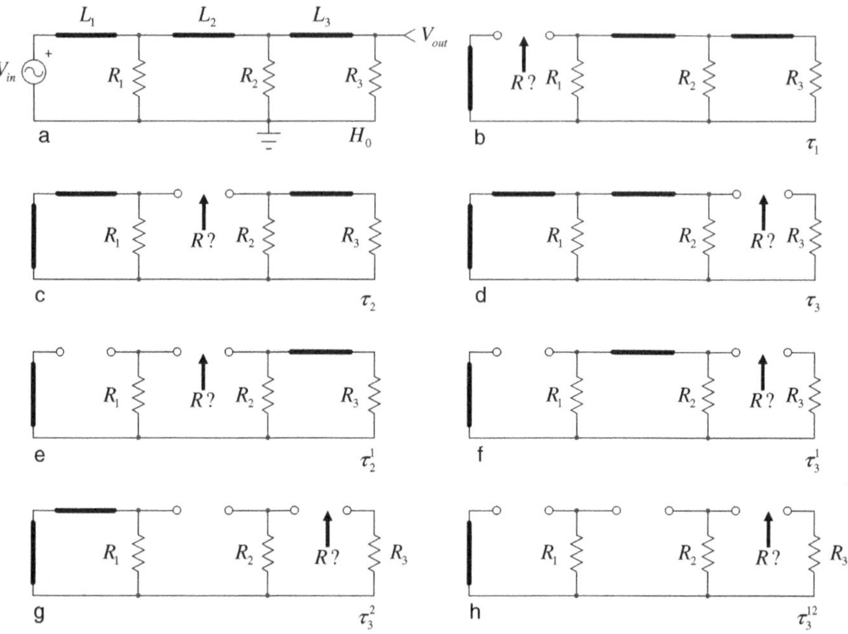

Figure 5.88 The network is split into individual pieces, helping to compute the various time constants.

Second term b_2 is equal to:

$$\begin{aligned} b_2 &= \tau_1\tau_2^1 + \tau_1\tau_3^1 + \tau_2\tau_3^2 \\ &= \frac{L_1}{R_1 \| R_2 \| R_3}\frac{L_2}{R_1 + R_2 \| R_3} + \frac{L_1}{R_1 \| R_2 \| R_3}\frac{L_3}{R_3 + R_2 \| R_1} + \frac{L_2}{R_2 \| R_3}\frac{L_3}{R_3 + R_2} \end{aligned} \tag{5.397}$$

Finally, the last time constant is obtained by looking at L_3's terminals while inductors 1 and 2 are set to their high-frequency state (open) as drawn in Figure 5.88h:

$$\tau_3^{12} = \frac{L_3}{R_3 + R_2} \tag{5.398}$$

We can now form b_3:

$$b_3 = \tau_1\tau_2^1\tau_3^{12} = \frac{L_1}{R_1 \| R_2 \| R_3}\frac{L_2}{R_1 + R_2 \| R_3}\frac{L_3}{R_3 + R_2} \tag{5.399}$$

which leads us to the final denominator expression:

$$\begin{aligned} D(s) &= 1 + b_1 s + b_2 s^2 + b_3 s^3 \\ &= 1 + s\left[\frac{L_1}{R_1 \| R_2 \| R_3} + \frac{L_2}{R_2 \| R_3} + \frac{L_3}{R_3}\right] \\ &\quad + s^2\left[\frac{L_1}{R_1 \| R_2 \| R_3}\frac{L_2}{R_1 + R_2 \| R_3} + \frac{L_1}{R_1 \| R_2 \| R_3}\frac{L_3}{R_3 + R_2 \| R_1} + \frac{L_2}{R_2 \| R_3}\frac{L_3}{R_3 + R_2}\right] \\ &\quad + s^3\left[\frac{L_1}{R_1 \| R_2 \| R_3}\frac{L_2}{R_1 + R_2 \| R_3}\frac{L_3}{R_3 + R_2}\right] \end{aligned} \tag{5.400}$$

Given the absence of zeros, the transfer function is immediate:

$$H(s) = \frac{1}{D(s)} \tag{5.401}$$

An expression like (5.400) looks complex but the way resistors are paralleled makes assumptions easier to analyze. For instance, assume R_3 goes infinite, meaning the network becomes unloaded, then you see that the L_3 contribution disappears and all terms having R_3 paralleled simplify. Try to develop (5.400) and you lose this kind of insight.

Now that we have our transfer function, we need a raw expression to confront dynamic responses. The method is the same as we applied in problem 1: define Thévenin's generators and apply an impedance divider formula. You should find:

$$R_{th1}(s) = sL_1 \| R_1 \tag{5.402}$$

$$R_{th2}(s) = (R_{th1}(s) + sL_2) \| R_2 \tag{5.403}$$

The complete 'raw' transfer function is given by:

$$H_{ref}(s) = \frac{R_1}{sL_1 + R_1} \cdot \frac{R_2}{R_{th1}(s) + R_2 + sL_2} \cdot \frac{R_3}{R_3 + R_{th2}(s) + sL_3} \tag{5.404}$$

Figure 5.89 shows what the Mathcad® sheet delivered and confirms that dynamic responses are similar.

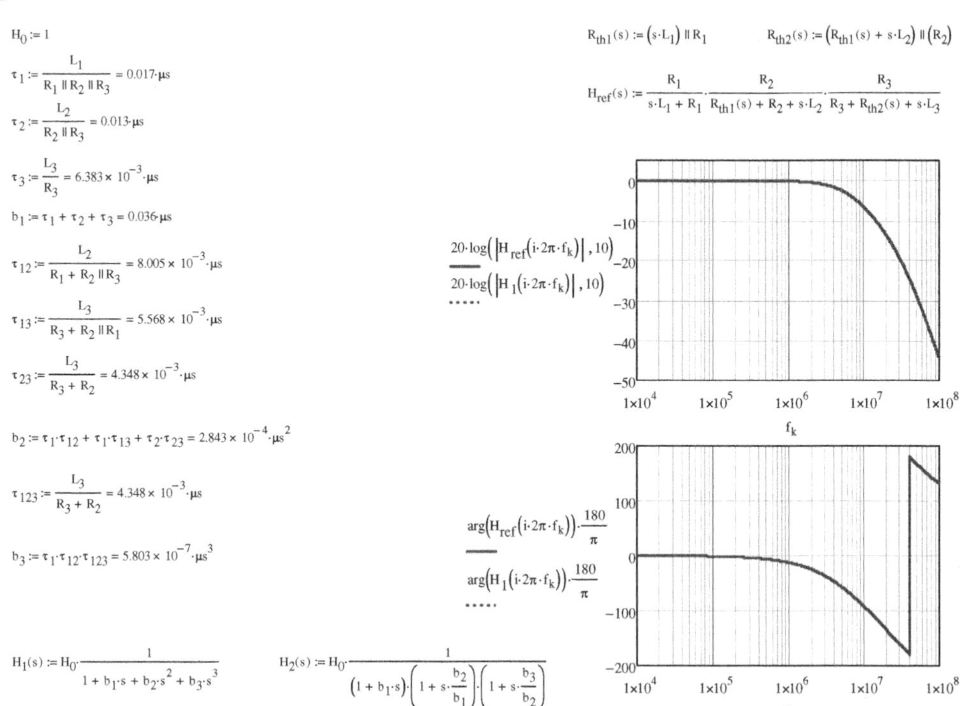

Figure 5.89 Mathcad® confirms our approach is correct, curves perfectly match.

Problem 3:

This is a tweeter filter that is supposed to cut off low frequencies. There are four energy-storing elements with independent state variables, this is a 4th-order system. As we have already done, we split the time constant sketches into two figures, Figures 5.90 and 5.91. From Figure 5.90, we see that two capacitors are in series with the input waveform so:

$$H_0 = 0 \tag{5.405}$$

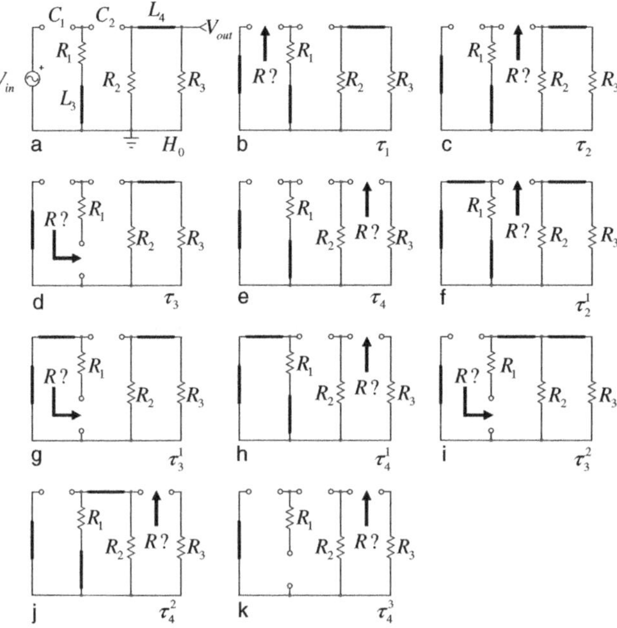

Figure 5.90 A 4th-order network requires numerous sketches but drawing them with care is the key to success.

The four time constants are obtained by observing sketches from b to e. If everything goes well, you should find:

$$\tau_1 = R_1 C_1 \tag{5.406}$$

$$\tau_2 = \left(R_1 + R_2 \| R_3\right) C_2 \tag{5.407}$$

$$\tau_3 = \frac{L_3}{\infty} = 0 \tag{5.408}$$

$$\tau_4 = \frac{L_4}{R_2 + R_3} \tag{5.409}$$

We can now form the first denominator coefficient b_1 as:

$$\begin{aligned} b_1 &= \tau_1 + \tau_2 + \tau_3 + \tau_4 \\ &= R_1 C_1 + \left(R_1 + R_2 \| R_3\right) C_2 + \frac{L_4}{R_2 + R_3} \end{aligned} \tag{5.410}$$

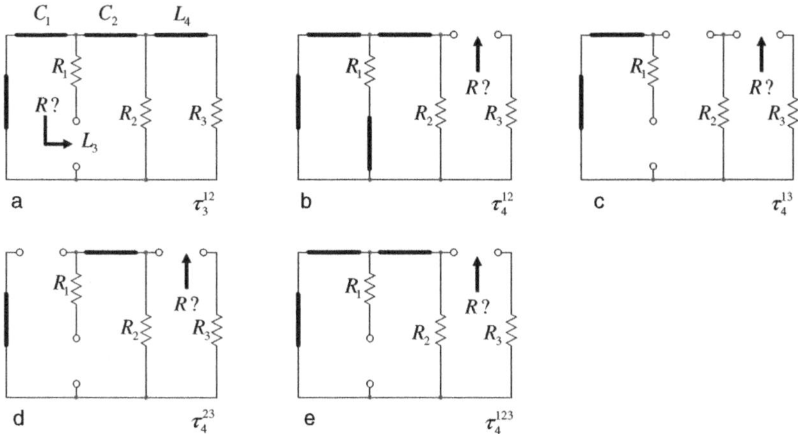

Figure 5.91 This the second part of the separate sketches for higher order coefficients.

We can now alternatively set energy-storing elements in different states as shown in the remaining sketches. Looking at the component's terminals in the various configurations gives us:

$$\tau_2^1 = (R_2 \| R_3) C_2 \tag{5.411}$$

$$\tau_3^1 = \frac{L_3}{R_1} \tag{5.412}$$

$$\tau_4^1 = \frac{L_4}{R_2 + R_3} \tag{5.413}$$

$$\tau_3^2 = \frac{L_3}{R_1 + R_2 \| R_3} \tag{5.414}$$

$$\tau_4^2 = \frac{L_4}{R_3 + R_1 \| R_2} \tag{5.415}$$

$$\tau_4^3 = \frac{L_4}{R_2 + R_3} \tag{5.416}$$

Associating these time constants leads to coefficient b_2 defined as:

$$\begin{aligned} b_2 &= \tau_1\tau_2^1 + \tau_1\tau_3^1 + \tau_1\tau_4^1 + \tau_2\tau_3^2 + \tau_2\tau_4^2 + \tau_3\tau_4^3 \\ &= R_1C_1(R_2 \| R_3)C_2 + R_1C_1\frac{L_3}{R_1} + R_1C_1\frac{L_4}{R_2 + R_3} + (R_1 + R_2 \| R_3)C_2\frac{L_3}{R_1 + R_2 \| R_3} + (R_1 + R_2 \| R_3)C_2\frac{L_4}{R_3 + R_1 \| R_2} \\ &= R_1C_1(R_2 \| R_3)C_2 + R_1C_1\frac{L_3}{R_1} + R_1C_1\frac{L_4}{R_2 + R_3} + C_2L_3 + (R_1 + R_2 \| R_3)C_2\frac{L_4}{R_3 + R_1 \| R_2} \end{aligned} \tag{5.417}$$

Now jumping to Figure 5.91, we have:

$$\tau_3^{12} = \frac{L_3}{R_1} \tag{5.418}$$

$$\tau_4^{12} = \frac{L_4}{R_3} \tag{5.419}$$

$$\tau_4^{13} = \frac{L_4}{R_2 + R_3} \tag{5.420}$$

$$\tau_4^{23} = \frac{L_4}{R_2 + R_3} \tag{5.421}$$

Coefficient b_3 is then defined by:

$$\begin{aligned} b_3 &= \tau_1\tau_2^1\tau_3^{12} + \tau_1\tau_2^1\tau_4^{12} + \tau_1\tau_3^1\tau_4^{13} + \tau_2\tau_3^2\tau_4^{23} \\ &= R_1C_1(R_2\|R_3)C_2\frac{L_3}{R_1} + R_1C_1(R_2\|R_3)C_2\frac{L_4}{R_3} + R_1C_1\frac{L_3}{R_1}\frac{L_4}{R_2+R_3} + (R_1 + R_2\|R_3)C_2\frac{L_3}{R_1 + R_2\|R_3}\frac{L_4}{R_2+R_3} \end{aligned} \tag{5.422}$$

Finally, the last coefficient is given by Figure 5.91e:

$$\tau_4^{123} = \frac{L_4}{R_3} \tag{5.423}$$

leading to b_4's definition

$$b_4 = \tau_1\tau_2^1\tau_3^{12}\tau_4^{123} = R_1C_1(R_2\|R_3)C_2\frac{L_3}{R_1}\frac{L_4}{R_3} = C_1\left(\frac{R_2}{R_2+R_3}\right)C_2L_3L_4 \tag{5.424}$$

The denominator $D(s)$ is defined by combining (5.410), (5.417), (5.422) and (5.424):

$$\begin{aligned} D(s) &= 1 + b_1s + b_2s^2 + b_3s^3 + b_4s^4 \\ &= 1 + \left(R_1C_1 + (R_1 + R_2\|R_3)C_2 + \frac{L_4}{R_2+R_3}\right)s \\ &+ \left(R_1C_1(R_2\|R_3)C_2 + R_1C_1\frac{L_3}{R_1} + R_1C_1\frac{L_4}{R_2+R_3} + C_2L_3 + (R_1 + R_2\|R_3)C_2\frac{L_4}{R_3 + R_1\|R_2}\right)s^2 \\ &+ \left(R_1C_1(R_2\|R_3)C_2\frac{L_3}{R_1} + R_1C_1(R_2\|R_3)C_2\frac{L_4}{R_3} + R_1C_1\frac{L_3}{R_1}\frac{L_4}{R_2+R_3}\right. \\ &\left. + (R_1 + R_2\|R_3)C_2\frac{L_3}{R_1 + R_2\|R_3}\frac{L_4}{R_2+R_3}\right)s^3 \\ &+ C_1\left(\frac{R_2}{R_2+R_3}\right)C_2L_3L_4s^4 \end{aligned} \tag{5.425}$$

With four energy-storing elements, the numerator is defined by the following generalized form:

$$\begin{aligned} N(s) = H_0 &+ s\left(\tau_1H^1 + \tau_2H^2 + \tau_3H^3 + \tau_4H^4\right) \\ &+ s^2\left(\tau_1\tau_2^1H^{12} + \tau_1\tau_3^1H^{13} + \tau_1\tau_4^1H^{14} + \tau_2\tau_3^2H^{23} + \tau_2\tau_4^2H^{24} + \tau_3\tau_4^3H^{34}\right) \\ &+ s^3\left(\tau_1\tau_2^1\tau_3^{12}H^{123} + \tau_1\tau_2^1\tau_4^{12}H^{124} + \tau_1\tau_3^1\tau_4^{13}H^{134} + \tau_2\tau_3^2\tau_4^{23}H^{234}\right) \\ &+ s^4\left(\tau_1\tau_2^1\tau_3^{12}\tau_4^{123}H^{1234}\right) \end{aligned} \tag{5.426}$$

However, given the two series capacitors, a lot of the dc transfer functions H will equal zero, nicely simplifying (5.426). Rather than covering all cases, we can check the non-zero transfer function when energy-storing elements are alternatively set to their high-frequency state. To ensure that dc goes through the network, C_1 and C_2 must be set to their high-frequency state (short circuit) while L_3 and L_4 remain in their dc state (short circuit). Therefore:

$$H^{12} = 1 \tag{5.427}$$

The second possibility is that C_1, C_2 are in their high-frequency state and L_3 also. As L_4 in this configuration remains in its dc state, the transfer function is:

$$H^{123} = 1 \tag{5.428}$$

All other transfer functions return 0. (5.426) greatly simplifies to:

$$N(s) = \tau_1\tau_2^1 H^{12} s^2 + \tau_1\tau_2^1\tau_3^{12} H^{123} s^3 = a_2 s^2 + a_3 s^3 \tag{5.429}$$

We can factor $a_2 s^2$ and obtain:

$$N(s) = a_2 s^2\left(1 + \frac{a_3}{a_2}s\right) = a_2 s^2\left(1 + \frac{C_1 L_3 (R_2 \| R_3) C_2}{R_1 C_1 (R_2 \| R_3) C_2}\right) = a_2 s^2\left(1 + \frac{L_3}{R_1}s\right) \tag{5.430}$$

Factoring $a_2 s^2$ in the denominator then rearranging with a_2/b_2 as the leading term, you should obtain:

$$H(s) = \frac{a_2}{b_2}\frac{1 + \frac{L_3}{R_1}s}{1 + \frac{b_1}{b_2 s} + \frac{b_3 s}{b_2} + \frac{1}{b_2 s^2} + \frac{b_4 s^2}{b_2}} \tag{5.431}$$

Before we carry on, we need a raw transfer function to check our equations. From Figure 5.77 we can apply Thévenin twice and express output resistances by:

$$R_{th1}(s) = (sL_3 + R_1) \| \left(\frac{1}{sC_1}\right) \tag{5.432}$$

$$R_{th2}(s) = \left(R_{th1}(s) + \frac{1}{sC_2}\right) \| R_2 \tag{5.433}$$

leading to the reference transfer function:

$$H_{ref}(s) = \frac{R_1 + sL_3}{sL_3 + R_1 + \frac{1}{sC_1}} \cdot \frac{R_2}{R_{th1}(s) + R_2 + \frac{1}{sC_2}} \cdot \frac{R_3}{R_3 + R_{th2}(s) + sL_4} \tag{5.434}$$

We have everything needed to graph dynamic responses and compare them in a Mathcad® sheet. Results appear in Figure 5.92 and confirm our equations are correct.

Problem 4:

Further to the input-to-output transfer function of Figure 5.77 circuit, what is the output impedance of this filter? One thing remarkable to network analysis already highlighted in Chapter 1 is that natural time constants composing the denominator $D(s)$ are solely dependent on the circuit architecture: whether you observe the output across R_3 or across R_2, excite the network with a current source across R_3 or with a voltage source in series with L_3 or C_1, time constants and therefore D remain the same. Since we already have evaluated that part in Figure 5.77, there is no need to run the exercise again for determining the circuit output impedance: $D(s)$ is common to both transfer functions $H(s)$ and $Z_{out}(s)$. On the other hand, the zeros change since they depend on where you inject and probe the response. Given the order of the circuit, we will use the generalized approach in which the numerator for an impedance Z is defined as follows:

$$\begin{aligned} N(s) = {} & R_0 + s\left(\tau_1 Z^1 + \tau_2 Z^2 + \tau_3 Z^3 + \tau_4 Z^4\right) \\ & + s^2\left(\tau_1\tau_2^1 Z^{12} + \tau_1\tau_3^1 Z^{13} + \tau_1\tau_4^1 Z^{14} + \tau_2\tau_3^2 Z^{23} + \tau_2\tau_4^2 Z^{24} + \tau_3\tau_4^3 Z^{34}\right) \\ & + s^3\left(\tau_1\tau_2^1\tau_3^{12} Z^{123} + \tau_1\tau_2^1\tau_4^{12} Z^{124} + \tau_1\tau_3^1\tau_4^{13} Z^{134} + \tau_2\tau_3^2\tau_4^{23} Z^{234}\right) \\ & + s^4\left(\tau_1\tau_2^1\tau_3^{12}\tau_4^{123} Z^{1234}\right) \end{aligned} \tag{5.435}$$

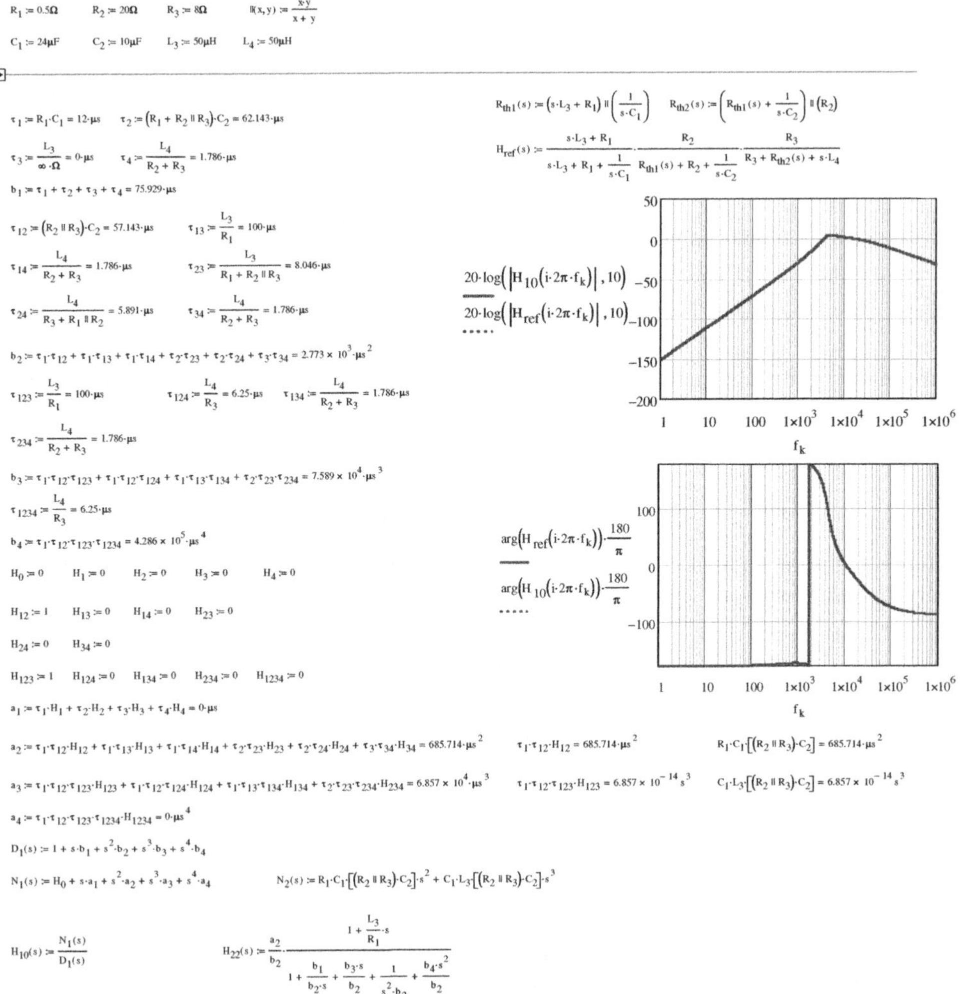

Figure 5.92 Mathcad® confirms that all our arrangements lead to similar results.

Figure 5.93 shows all individual sketches to determine the output resistance when capacitors and inductors are put in their dc or high-frequency state.

Starting from Figure 5.93a, we have:

$$R_0 = R_3 \| R_2 \tag{5.436}$$

$$Z^1 = R_3 \| R_2 \tag{5.437}$$

$$Z^2 = R_1 \| R_2 \| R_3 \tag{5.438}$$

$$Z^3 = R_3 \| R_2 \tag{5.439}$$

$$Z^4 = R_3 \tag{5.440}$$

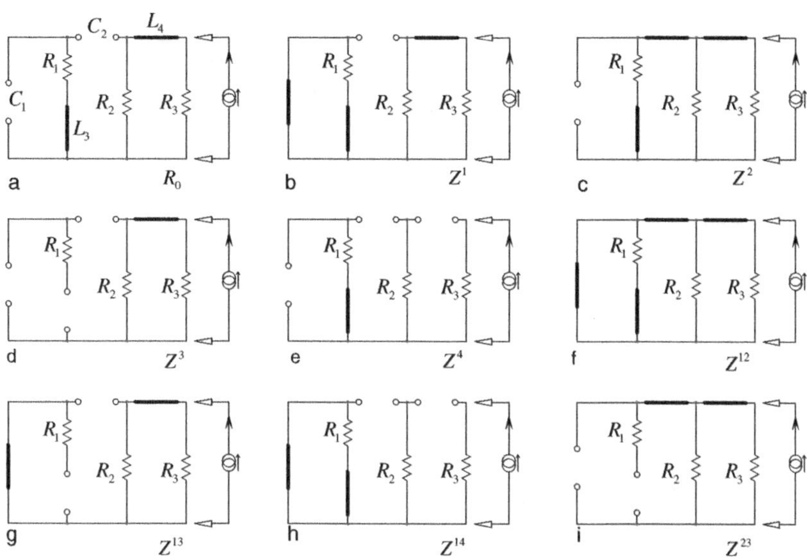

Figure 5.93 Here we evaluate the output resistance when capacitors and/or inductors are open or shorted.

We can form the first coefficient a_1 by assembling natural time constants and results from (5.437) to (5.440):

$$\begin{aligned} a_1 &= R_1C_1(R_3\|R_2) + (R_1 + R_2\|R_3)C_2(R_1\|R_2\|R_3) + 0\cdot(R_3\|R_2) + \frac{L_4}{R_2+R_3}R_3 \\ &= R_1C_1(R_3\|R_2) + (R_1 + R_2\|R_3)C_2(R_1\|R_2\|R_3) + \frac{L_4}{R_2+R_3}R_3 \end{aligned} \tag{5.441}$$

From Figure 5.93f to i, we carry and find:

$$Z^{12} = 0 \tag{5.442}$$

$$Z^{13} = R_3\|R_2 \tag{5.443}$$

$$Z^{14} = R_3 \tag{5.444}$$

$$Z^{23} = R_2\|R_3 \tag{5.445}$$

then jumping to Figure 5.94a and b:

$$Z^{24} = R_3 \tag{5.446}$$

$$Z^{34} = R_3 \tag{5.447}$$

The second coefficient a_2 can now be formed as:

$$\begin{aligned} a_2 &= \tau_1\tau_2^1Z^{12} + \tau_1\tau_3^1Z^{13} + \tau_1\tau_4^1Z^{14} + \tau_2\tau_3^2Z^{23} + \tau_2\tau_4^2Z^{24} + \tau_3\tau_4^3Z^{34} \\ &= R_1C_1(R_2\|R_3)C_2\cdot 0 + R_1C_1\frac{L_3}{R_1}(R_3\|R_2) + R_1C_1\frac{L_4}{R_2+R_3}R_3 \\ &\quad + (R_1+R_2\|R_3)C_2\frac{L_3}{R_1+R_2\|R_3}(R_3\|R_2) + (R_1+R_2\|R_3)C_2\frac{L_4}{R_3+R_1\|R_2}R_3 + \frac{L_3}{\infty}\cdot\frac{L_4}{R_2+R_3}R_3 \\ &= R_1C_1\frac{L_3}{R_1}(R_3\|R_2) + R_1C_1\frac{L_4}{R_2+R_3}R_3 + C_2L_3(R_3\|R_2) + (R_1+R_2\|R_3)C_2\frac{L_4}{R_3+R_1\|R_2}R_3 \end{aligned} \tag{5.448}$$

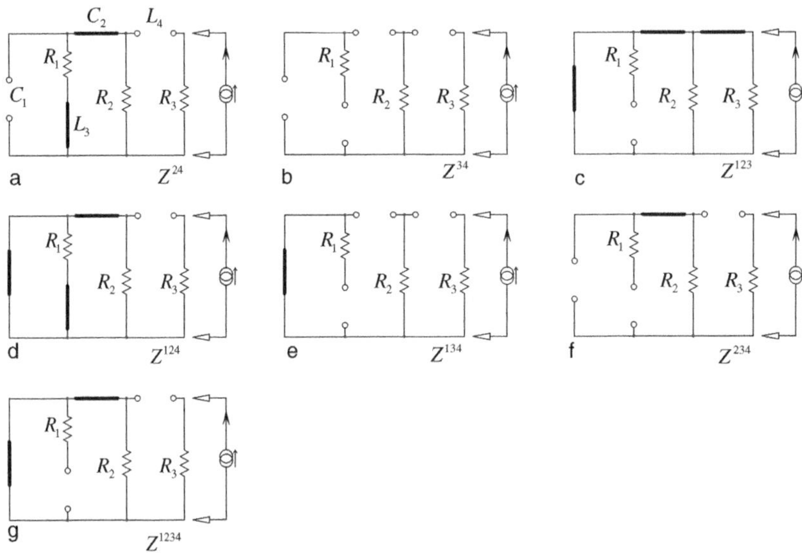

Figure 5.94 The second part of the dc gains exercise.

For a_3, we can look at Figure 5.94 to determine the terms. From sketches c to f, we have:

$$Z^{123} = 0 \tag{5.449}$$

$$Z^{124} = R_3 \tag{5.450}$$

$$Z^{134} = R_3 \tag{5.451}$$

$$Z^{234} = R_3 \tag{5.452}$$

leading to:

$$\begin{aligned} a_3 &= \tau_1\tau_2^1\tau_3^{12}Z^{123} + \tau_1\tau_2^1\tau_4^{12}Z^{124} + \tau_1\tau_3^1\tau_4^{13}Z^{134} + \tau_2\tau_3^2\tau_4^{23}Z^{234} \\ &= R_1C_1(R_2 \| R_3)C_2\frac{L_3}{R_1}\cdot 0 + R_1C_1(R_2 \| R_3)C_2\frac{L_4}{R_3}R_3 + R_1C_1\frac{L_3}{R_1}\frac{L_4}{R_2+R_3}R_3 \\ &\quad + (R_1 + R_2 \| R_3)C_2\frac{L_3}{R_1 + R_2 \| R_3}\frac{L_4}{R_2+R_3}R_3 \\ &= R_1C_1(R_2 \| R_3)C_2L_4 + R_1C_1\frac{L_3}{R_1}\frac{L_4}{R_2+R_3}R_3 + C_2L_3\frac{L_4}{R_2+R_3}R_3 \end{aligned} \tag{5.453}$$

Finally, the last term H^{1234} is obtain when all energy-storing elements are set in their high-frequency state. From Figure 5.94g, as L_4 in high-frequency state (open) isolates R_3 from the rest of the circuit, we find:

$$Z^{1234} = R_3 \tag{5.454}$$

giving

$$a_4 = C_1(R_2 \| R_3)C_2L_3\frac{L_4}{R_3} \tag{5.455}$$

Factoring R_0 in, the numerator is then defined by:

$$N(s) = 1 + \left(\frac{R_1C_1(R_3\|R_2) + (R_1 + R_2\|R_3)C_2(R_1\|R_2\|R_3) + \frac{L_4}{R_2+R_3}R_3}{R_3\|R_2}\right)s$$
$$+\left(R_1C_1\frac{L_3}{R_1}(R_3\|R_2) + R_1C_1\frac{L_4}{R_2+R_3}R_3 + C_2L_3(R_3\|R_2) + (R_1 + R_2\|R_3)C_2\frac{L_4}{R_3+R_1\|R_2}R_3\right)\frac{1}{R_3\|R_2}s^2$$
$$+\left(R_1C_1(R_2\|R_3)C_2L_4 + R_1C_1\frac{L_3}{R_1}\frac{L_4}{R_2+R_3}R_3 + C_2L_3\frac{L_4}{R_2+R_3}R_3\right)\frac{1}{R_3\|R_2}s^3$$
$$+\left(C_1(R_2\|R_3)C_2L_3\frac{L_4}{R_3}\right)\frac{1}{R_3\|R_2}s^4 \tag{5.456}$$

The transfer is then obtained by associating (5.436), (5.425) and (5.456). With a resistive leading term, its unit is ohm:

$$Z_{out}(s) = (R_3\|R_2)\frac{N(s)}{D(s)} \tag{5.457}$$

To check our calculations, we need a raw transfer function expression. Looking at Figure 5.78, we simply need to compute the following impedance expressions:

$$Z_a(s) = \left(\frac{1}{sC_1}\right)\|(R_1 + sL_3) + \frac{1}{sC_2} \tag{5.458}$$

$$Z_b(s) = sL_4 + R_2\|Z_a(s) \tag{5.459}$$

to obtain

$$Z_{ref}(s) = R_3\|Z_b(s) \tag{5.460}$$

We have captured all time constants and gains in a Mathcad® sheet which appears in Figure 5.95. As you can see, the response is quite distorted in phase and magnitude and both expressions match each other very well.

Problem 5:

This circuit reflects parasitic elements found on a printed circuit board while developing a power converter. The input is the high-voltage signal we want to observe, scaled down by the various resistors. The added capacitors are parasitic terms present on the board that affect the signal propagation to R_1. This is a 3rd-order network. We will go through the classical series of sketches to determine the denominator coefficients. Drawings appear in Figure 5.96. If we start from the upper left, Figure 5.96a, the 0-Hz gain is immediate:

$$H_0 = \frac{R_1}{R_1 + R_2 + R_3 + R_4} \tag{5.461}$$

Figure 5.96b lets us determine the three time constants involving C_2, C_3 and C_4. Looking into these capacitors terminals gives:

$$\tau_4 = \left[R_4\|(R_1 + R_2 + R_3)\right]C_4 \tag{5.462}$$

$$\tau_2 = \left[(R_1 + R_4)\|(R_3 + R_2)\right]C_2 \tag{5.463}$$

$$\tau_3 = \left[(R_1 + R_2)\|(R_3 + R_4)\right]C_3 \tag{5.464}$$

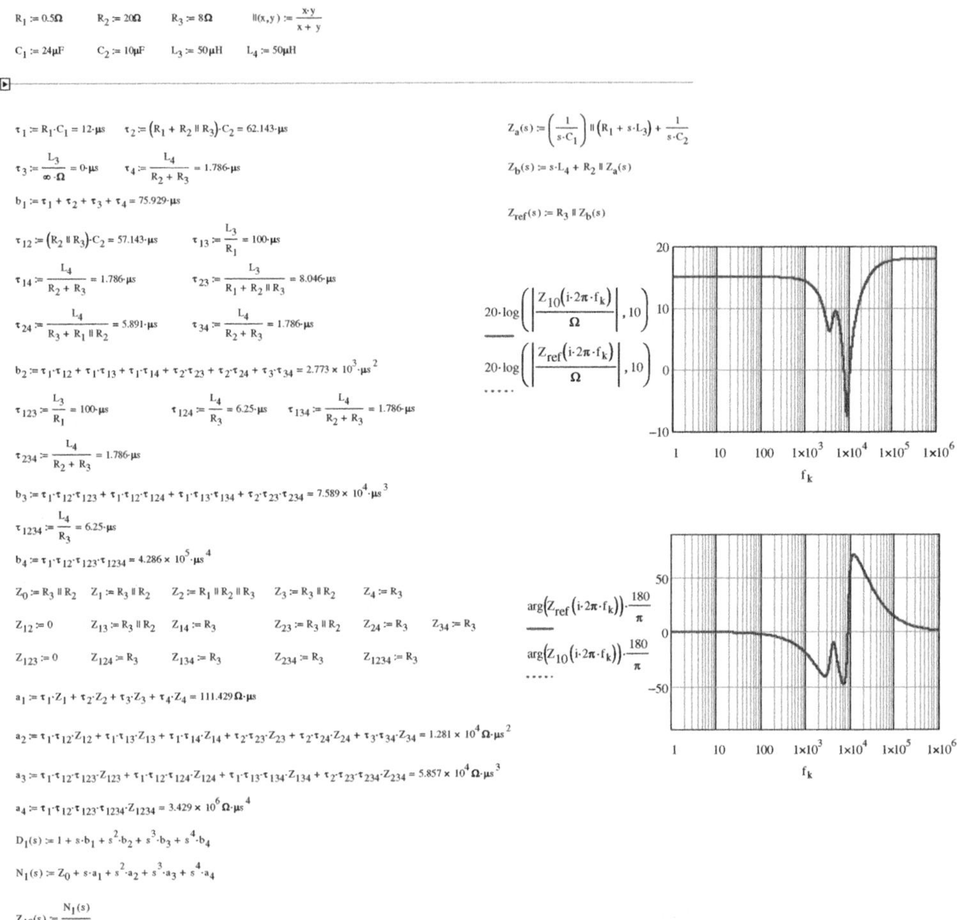

Figure 5.95 The Mathcad® sheet confirms our approach as all dynamic responses agree well with each other.

We can assemble these coefficient to form the 1st denominator term b_1:

$$b_1 = \tau_4 + \tau_2 + \tau_3 = [R_4 \| (R_1 + R_2 + R_3)]C_4 + [(R_1 + R_4) \| (R_3 + R_2)]C_2 + [(R_1 + R_2) \| (R_3 + R_4)]C_3 \tag{5.465}$$

Figure 5.96c to e show us the energy-storing elements combination to build the other time constants. We have:

$$\tau_2^4 = [R_1 \| (R_2 + R_3)]C_2 \tag{5.466}$$

$$\tau_3^4 = [R_3 \| (R_1 + R_2)]C_3 \tag{5.467}$$

$$\tau_3^2 = [R_1 \| R_4 + R_3 \| R_2]C_3 \tag{5.468}$$

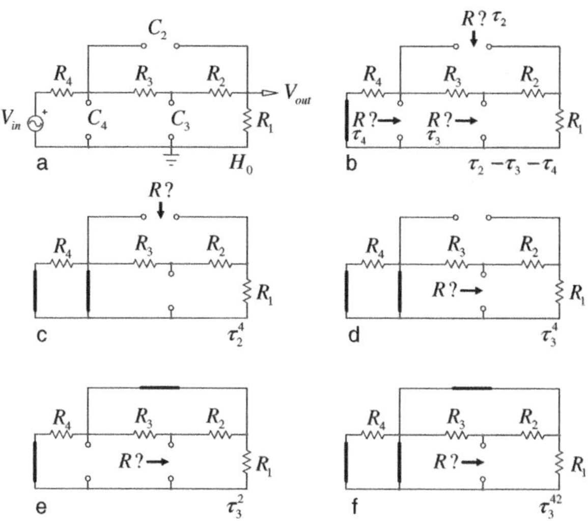

Figure 5.96 The circuit is broken in several simple sketches to help determine the denominator coefficients.

These expressions lead to the second denominator coefficient, b_2, equal to:

$$\begin{aligned} b_2 = \tau_4\tau_2^4 + \tau_4\tau_3^4 + \tau_2\tau_3^2 = &\left[R_4 \| (R_1 + R_2 + R_3)\right]C_4\left[R_1 \| (R_2 + R_3)\right]C_2 \\ &+\left[R_4 \| (R_1 + R_2 + R_3)\right]C_4\left[R_3 \| (R_1 + R_2)\right]C_3 \\ &+\left[(R_1 + R_4) \| (R_3 + R_2)\right]C_2\left[R_1 \| R_4 + R_3 \| R_2\right]C_3 \end{aligned} \tag{5.469}$$

Finally, Figure 5.96f gives the coefficient we miss to build the b_3 term:

$$\tau_3^{42} = (R_2 \| R_3)C_3 \tag{5.470}$$

leading to

$$b_3 = \left[R_4 \| (R_1 + R_2 + R_3)\right]C_4\left[R_1 \| (R_2 + R_3)\right]C_2(R_2 \| R_3)C_3 \tag{5.471}$$

The denominator $D(s)$ is then equal to:

$$\begin{aligned} D(s) &= 1 + b_1 s + b_2 s^2 + b_3 s^3 \\ &= 1 + \left(\left[R_4 \| (R_1 + R_2 + R_3)\right]C_4 + \left[(R_1 + R_4) \| (R_3 + R_2)\right]C_2 + \left[(R_1 + R_2) \| (R_3 + R_4)\right]C_3\right)s \\ &+ \left(\begin{array}{l}\left[R_4 \| (R_1 + R_2 + R_3)\right]C_4\left[R_1 \| (R_2 + R_3)\right]C_2 + \left[R_4 \| (R_1 + R_2 + R_3)\right]C_4\left[R_3 \| (R_1 + R_2)\right]C_3 \\ +\left[(R_1 + R_4) \| (R_3 + R_2)\right]C_2\left[R_1 \| R_4 + R_3 \| R_2\right]C_3\end{array}\right)s^2 \\ &+\left(\left[R_4 \| (R_1 + R_2 + R_3)\right]C_4\left[R_1 \| (R_2 + R_3)\right]C_2(R_2 \| R_3)C_3\right)s^3 \end{aligned} \tag{5.472}$$

To build the numerator, we go through the various gains calculations drawn in Figure 5.97. Sketches a to c gives us the gains values when 1 capacitor is shorted. We have:

$$H^4 = 0 \tag{5.473}$$

$$H^2 = \frac{R_1}{R_1 + R_4} \tag{5.474}$$

$$H^3 = 0 \tag{5.475}$$

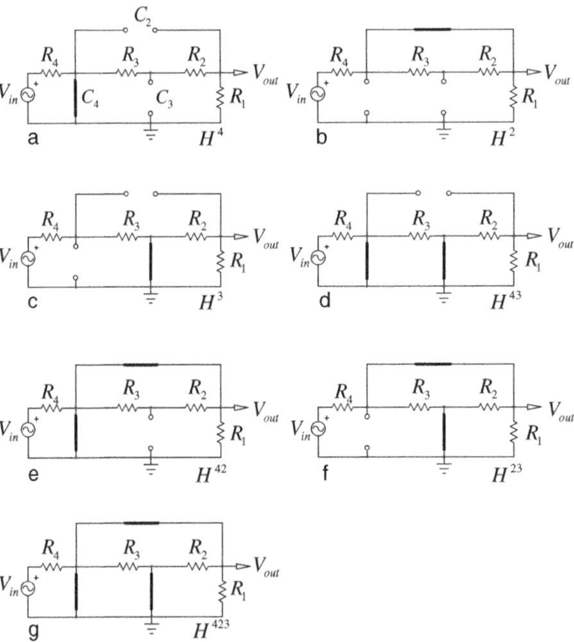

Figure 5.97 The same exercise but the input source is back in the analysis to determine the various gain values.

We have the first term a_1:

$$a_1 = \tau_4 H^4 + \tau_2 H^2 + \tau_3 H^3 = \left[(R_1 + R_4) \| (R_3 + R_2)\right] C_2 \frac{R_1}{R_1 + R_4} \tag{5.476}$$

The second term a_2 involves Figure 5.97d to f. Going through the various configurations, we find:

$$H^{42} = 0 \tag{5.477}$$

$$H^{43} = 0 \tag{5.478}$$

$$H^{23} = \frac{R_1 \| R_2 \| R_3}{R_1 \| R_2 \| R_3 + R_4} \tag{5.479}$$

The a_2 coefficient is quickly obtained as the first two gains are equal to 0:

$$a_2 = \tau_4 \tau_3^4 H^{43} + \tau_4 \tau_2^4 H^{42} + \tau_2 \tau_3^2 H^{23} = \left[(R_1 + R_4) \| (R_3 + R_2)\right] C_2 \left[R_1 \| R_4 + R_3 \| R_2\right] C_3 \frac{R_1 \| R_2 \| R_3}{R_1 \| R_2 \| R_3 + R_4} \tag{5.480}$$

The last gain is obtained when all capacitors are replaced by short circuits as in Figure 5.97g. You obtain:

$$H^{423} = 0 \tag{5.481}$$

The last term a_3 is thus equal to 0. We can now determine the numerator $N(s)$ as:

$$\begin{aligned} N(s) &= H_0 + s\tau_2 H^2 + s^2\tau_2\tau_3^2 H^{23} \\ &= \frac{R_1}{R_4+R_2+R_3+R_1} + s\left([(R_1+R_4)\|(R_3+R_2)]C_2\frac{R_1}{R_1+R_4}\right) \\ &+s^2\left([(R_1+R_4)\|(R_3+R_2)]C_2[R_1\|R_4+R_3\|R_2]C_3\frac{R_1\|R_2\|R_3}{R_1\|R_2\|R_3+R_4}\right) \end{aligned} \tag{5.482}$$

If we factor in H_0, we have:

$$\begin{aligned} N(s) &= H_0\left(1+s\tau_2\frac{H^2}{H_0}+s^2\tau_2\tau_3^2\frac{H^{23}}{H_0}\right) \\ &= H_0\left(\begin{array}{l} 1+s\left([(R_1+R_4)\|(R_3+R_2)]C_2\dfrac{R_1}{R_1+R_4}\dfrac{R_4+R_2+R_3+R_1}{R_1}\right) \\ +s^2\left([(R_1+R_4)\|(R_3+R_2)]C_2[R_1\|R_4+R_3\|R_2]C_3\dfrac{R_1\|R_2\|R_3}{R_1\|R_2\|R_3+R_4}\dfrac{R_4+R_2+R_3+R_1}{R_1}\right)\end{array}\right) \end{aligned} \tag{5.483}$$

The transfer function is then obtained by combining (5.483) and (5.472):

$$H(s) = H_0\frac{D(s)}{N(s)} \tag{5.484}$$

Then applying factorization techniques described in Chapter 2 to the numerator and the denominator, you should obtain:

$$H(s) \approx H_0\frac{\left(1+\dfrac{s}{\omega_{z_1}}\right)\left(1+\dfrac{s}{\omega_{z_2}}\right)}{\left(1+\dfrac{s}{\omega_{p_1}}\right)\left(1+\dfrac{s}{\omega_{p_2}}\right)\left(1+\dfrac{s}{\omega_{p_3}}\right)} \tag{5.485}$$

The Mathcad® sheet appears in Figure 5.98 which shows the resulting dynamic responses.

To check these results, we cannot unfortunately easily apply the superposition theorem to define a raw transfer function as we did in previous examples. This is because R_4 and C_4 affect the impedance driving C_2 and R_3 and we could not split the voltage across C_4 in two distinct nodes, alternatively set to 0. Rather, we have simulated the circuit with SPICE and ac responses perfectly superimpose with Figure 5.98 graphs, confirming the derivation we did. As an alternative approach, my colleague Dr. Capilla used a different approach, the signal-flow graph technique, to determine the transfer function of this network. I then compared my results with his by plotting the difference between magnitudes and phase responses. The result, together with the coefficients he derived, appear in Figure 5.99. The error between formulas lies in the solver resolution noise. Now assume an error is detected in Figure 5.98 flow, fixing it by tweaking one or several time constants would not be a complicated exercise. On the other hand, identifying a mistake in Figure 5.99 coefficients would lead to a tedious correction for which Marco from Tropoja would say: 'good luck' ☺

Problem 6:

This is a typical filter associated with a phase-locked loop (PLL) phase detector whose output is a current. The transfer function linking V_{out} to I_{in} is a transimpedance Z expressed in volts per amperes or ohms. You have three distinct capacitors, this is a 3rd-order network. Here, when quickly looking at the circuit, turning the excitation off will leave C_1 alone, without any dc path when all other capacitors

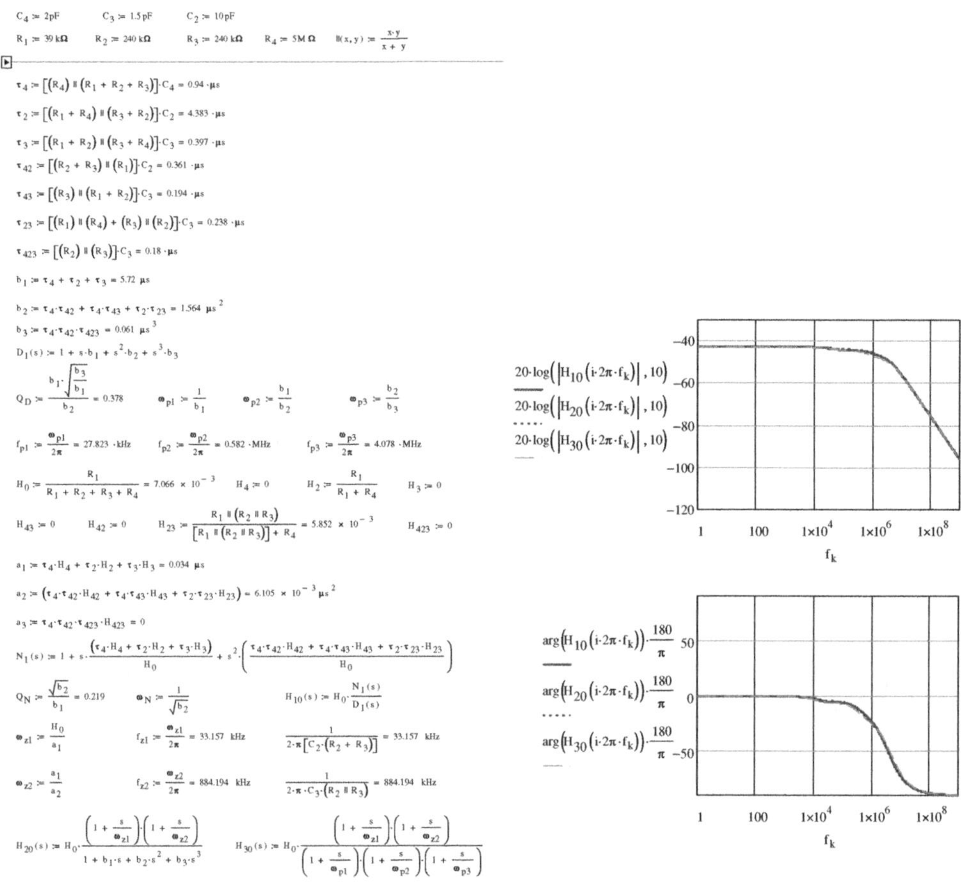

Figure 5.98 This sheet shows how the various transfer function nicely match.

are open. And you can see the same configuration happening for the other capacitors. To avoid a possibly troublesome situation, I have added an extra resistor, R_1, which mimics the current source output resistance. This resistance physically exists and bounds the maximum dc gain. At the end, we can decide that this resistance is fairly large and see how the transfer function reduces. Let's start with the classical series of sketches shown in Figure 5.100. The dc transimpedance R_0 is defined as:

$$R_0 = R_1 \tag{5.486}$$

The three time constants are obtained from Figure 5.100b. You have:

$$\tau_1 = C_1R_1 \tag{5.487}$$

$$\tau_2 = C_2(R_1 + R_2) \tag{5.488}$$

$$\tau_3 = C_3(R_1 + R_3) \tag{5.489}$$

which let us for the first term, b_1:

$$b_1 = \tau_1 + \tau_2 + \tau_3 = C_1R_1 + C_2(R_1 + R_2) + C_3(R_1 + R_3) \tag{5.490}$$

$R1 := R_1 \quad R2 := R_2 \quad R3 := R_3 \quad R4 := R_4 \quad C4 := C_4 \quad C3 := C_3 \quad C2 := C_2$

$$H_{00} := \frac{R1}{R1 + R2 + R3 + R4}$$

$$\omega_{11} := \frac{\frac{C2 \cdot R2}{2} + \frac{C2 \cdot R3}{2} + \frac{\sqrt{C2 \cdot \left(C2 \cdot R2^2 + C2 \cdot R3^2 + 2 \cdot C2 \cdot R2 \cdot R3 - 4 \cdot C3 \cdot R2 \cdot R3\right)}}{2}}{C2 \cdot C3 \cdot R2 \cdot R3} \qquad \omega_{22} := \frac{\left[\frac{C2 \cdot R2}{2} + \frac{C2 \cdot R3}{2} - \frac{\sqrt{C2 \cdot \left(C2 \cdot R2^2 + C2 \cdot R3^2 + 2 \cdot C2 \cdot R2 \cdot R3 - 4 \cdot C3 \cdot R2 \cdot R3\right)}}{2}\right]}{(C2 \cdot C3 \cdot R2 \cdot R3)}$$

$$b_{11} := \frac{(C2 \cdot R1 \cdot R2 + C2 \cdot R1 \cdot R3 + C3 \cdot R1 \cdot R3 + C2 \cdot R2 \cdot R4 + C3 \cdot R1 \cdot R4 + C3 \cdot R2 \cdot R3 + C2 \cdot R3 \cdot R4 + C3 \cdot R2 \cdot R4 + C4 \cdot R1 \cdot R4 + C4 \cdot R2 \cdot R4 + C4 \cdot R3 \cdot R4)}{(R1 + R2 + R3 + R4)} = 5.72\,\mu s$$

$$b_{22} := \frac{(C2 \cdot C3 \cdot R1 \cdot R2 \cdot R3 + C2 \cdot C3 \cdot R1 \cdot R2 \cdot R4 + C2 \cdot C3 \cdot R1 \cdot R3 \cdot R4 + C2 \cdot C4 \cdot R1 \cdot R2 \cdot R4 + C2 \cdot C3 \cdot R2 \cdot R3 \cdot R4 + C2 \cdot C4 \cdot R1 \cdot R3 \cdot R4 + C3 \cdot C4 \cdot R1 \cdot R3 \cdot R4 + C3 \cdot C4 \cdot R2 \cdot R3 \cdot R4)}{(R1 + R2 + R3 + R4)} = 1.564\,\mu s^2$$

$$b_{33} := \left(\frac{C2 \cdot C3 \cdot C4 \cdot R1 \cdot R2 \cdot R3 \cdot R4}{R1 + R2 + R3 + R4}\right) = 0.061\,\mu s^3$$

$$Tjose(s) := \frac{R1}{R1 + R2 + R3 + R4} \cdot \frac{\left(1 + \frac{s}{\omega_{11}}\right) \cdot \left(1 + \frac{s}{\omega_{22}}\right)}{b_{33} \cdot s^3 + b_{22} \cdot s^2 + b_{11} \cdot s + 1}$$

Magnitude difference

$$20 \cdot \log\left(\left|Tjose\left(i \cdot 2\pi \cdot f_k\right)\right|, 10\right) - 20 \cdot \log\left(\left|H_{10}\left(i \cdot 2\pi \cdot f_k\right)\right|, 10\right)$$

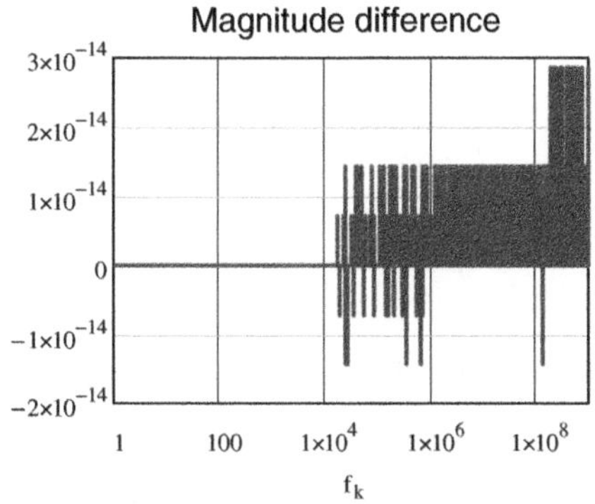

Phase difference

$$\arg\left(Tjose\left(i \cdot 2\pi \cdot f_k\right)\right) \cdot \frac{180}{\pi} - \arg\left(H_{10}\left(i \cdot 2\pi \cdot f_k\right)\right) \cdot \frac{180}{\pi}$$

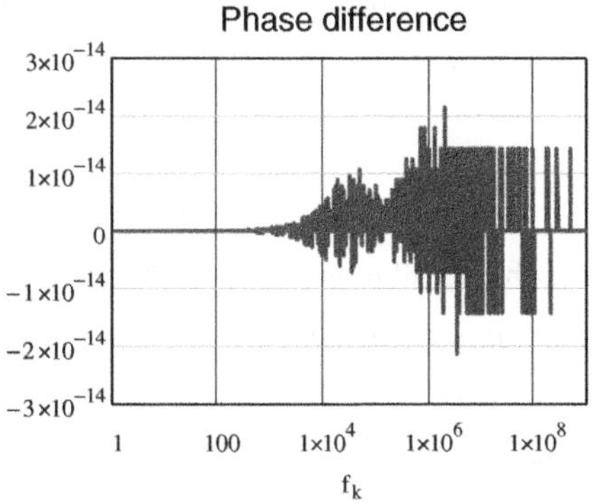

Figure 5.99 The transfer function obtained with the FACTs is identical to that obtained with the signal flow graph technique.

From Figure 5.100c to e, we have all we need to determine b_2:

$$\tau_2^1 = C_2 R_2 \tag{5.491}$$

$$\tau_3^1 = C_3 R_3 \tag{5.492}$$

$$\tau_3^2 = C_3 \left(R_2 \| R_1 + R_3\right) \tag{5.493}$$

which is finally equal to:

$$b_2 = \tau_1 \tau_2^1 + \tau_1 \tau_3^1 + \tau_2 \tau_3^2 = C_1 C_2 R_1 R_2 + C_1 R_1 C_3 R_3 + C_2 (R_1 + R_2) C_3 \left(R_2 \| R_1 + R_3\right) \tag{5.494}$$

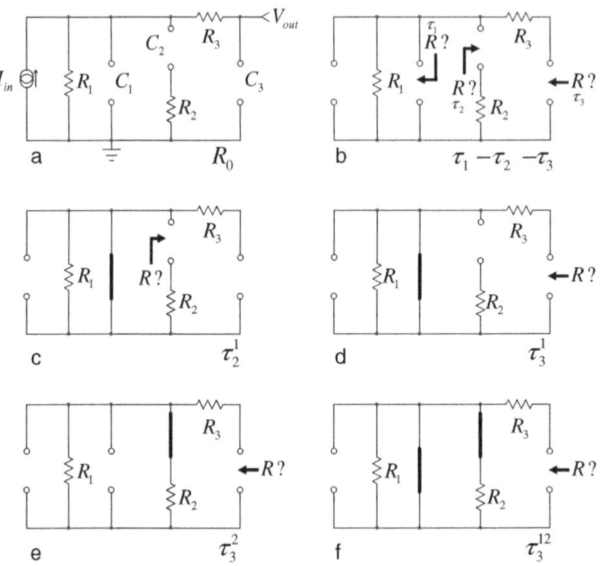

Figure 5.100 Six small sketches lead you to the natural time constant definitions.

The last time constant is obtained with the help of Figure 5.100f:

$$\tau_3^{12} = C_3 R_3 \tag{5.495}$$

leading to b_3:

$$b_3 = \tau_1 \tau_2^1 \tau_3^{12} = C_1 C_2 C_3 R_1 R_2 R_3 \tag{5.496}$$

If you now assemble up all terms from (5.490), (5.494) and (5.496), you should find:

$$\begin{aligned} D(s) = {} & 1 + s[C_1 R_1 + C_2(R_1 + R_2) + C_3(R_1 + R_3)] \\ & + s^2\left[C_1 C_2 R_1 R_2 + C_1 R_1 C_3 R_3 + C_2(R_1 + R_2) C_3\left(R_2 \| R_1 + R_3\right)\right] + s^3 C_1 C_2 C_3 R_1 R_2 R_3 \end{aligned} \tag{5.497}$$

If you now factor R_1, you obtain:

$$D(s) = R_1 \left(\begin{aligned} & \frac{1}{R_1} + s\left[C_1 + C_2 \frac{R_1 + R_2}{R_1} + C_3 \frac{R_1 + R_3}{R_1}\right] \\ & + s^2 \left[C_1 C_2 R_2 + C_1 C_3 R_3 + C_2 \left(\frac{R_1 + R_2}{R_1}\right) C_3 \left(R_2 \| R_1 + R_3\right)\right] + s^3 C_1 C_2 C_3 R_2 R_3 \end{aligned} \right) \tag{5.498}$$

and if R_1 is a fairly large value, $D(s)$ simplifies to:

$$D(s) \approx R_1\left(s[C_1 + C_2 + C_3] + s^2[C_1(C_2 R_2 + C_3 R_3) + C_2 C_3(R_2 + R_3)] + s^3 C_1 C_2 C_3 R_2 R_3\right) \tag{5.499}$$

Now, if we jump to Figure 5.101, we see that every time C_1 or C_3 are set to their high-frequency state (a short circuit), the response is 0 and Z also. Actually, one gain is different from 0 and it is that of Figure 5.101b:

$$Z^2 = R_1 \| R_2 \tag{5.500}$$

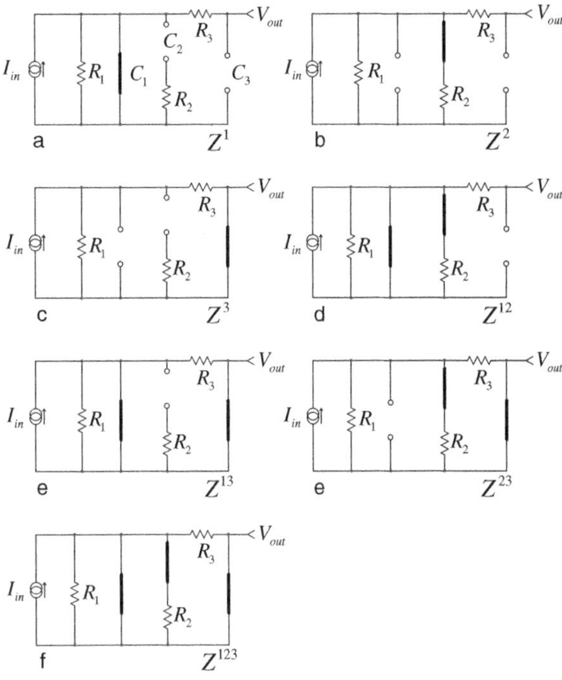

Figure 5.101 Another batch of sketches for the gain definitions and you are done.

The denominator $N(s)$ is thus extremely simple and is defined by:

$$N(s) = R_0 + s\tau_2 Z^2 = R_0\left(1 + s\tau_2 \frac{Z^2}{R_0}\right) = R_1(1 + sR_2C_2) \tag{5.501}$$

The transfer function is obtained when dividing (5.501) by (5.499):

$$Z(s) = \frac{R_1(1 + sR_2C_2)}{R_1(s[C_1 + C_2 + C_3] + s^2[C_1(C_2R_2 + C_3R_3) + C_2C_3(R_2 + R_3)] + s^3C_1C_2C_3R_2R_3)} \tag{5.502}$$

We can of course simplify by R_1 and factor sR_2C_2 to form an inverted zero. The transfer function then becomes:

$$Z(s) = \frac{R_2C_2}{C_1 + C_2 + C_3} \frac{1 + \dfrac{1}{sR_2C_2}}{1 + s\dfrac{C_1(C_2R_2 + R_3C_3) + C_2C_3(R_2 + R_3)}{C_1 + C_2 + C_3} + s^2\dfrac{C_1C_2C_3R_2R_3}{C_1 + C_2 + C_3}} \tag{5.503}$$

We can also transform the 2nd-order form into cascaded poles provided the quality factor is low. In this case:

$$Z(s) \approx \frac{R_2C_2}{C_1 + C_2 + C_3} \frac{1 + \dfrac{1}{sR_2C_2}}{\left(1 + s\dfrac{C_1(C_2R_2 + R_3C_3) + C_2C_3(R_2 + R_3)}{C_1 + C_2 + C_3}\right)\left(1 + s\dfrac{C_1C_2C_3R_2R_3}{C_1C_2R_2 + C_1C_3R_3 + C_2C_3R_2 + C_2C_3R_3}\right)} \tag{5.504}$$

The reference transfer function is obtained by paralleling impedances made of R_2-C_2 and R_3-C_3 then applying an impedance divider made of C_3 and R_3:

$$Z_{ref}(s) = \left(R_1 \| Z_1(s) \| \left(\frac{1}{sC_1}\right) \| Z_2(s)\right) \frac{1}{1 + sR_3C_3} \tag{5.505}$$

with

$$Z_1(s) = R_2 + \frac{1}{sC_2} \tag{5.506}$$

$$Z_2(s) = R_3 + \frac{1}{sC_3} \tag{5.507}$$

Figure 5.102 confirms our calculations are correct and the rearranged transfer function in (5.504) matches the reference function quite well.

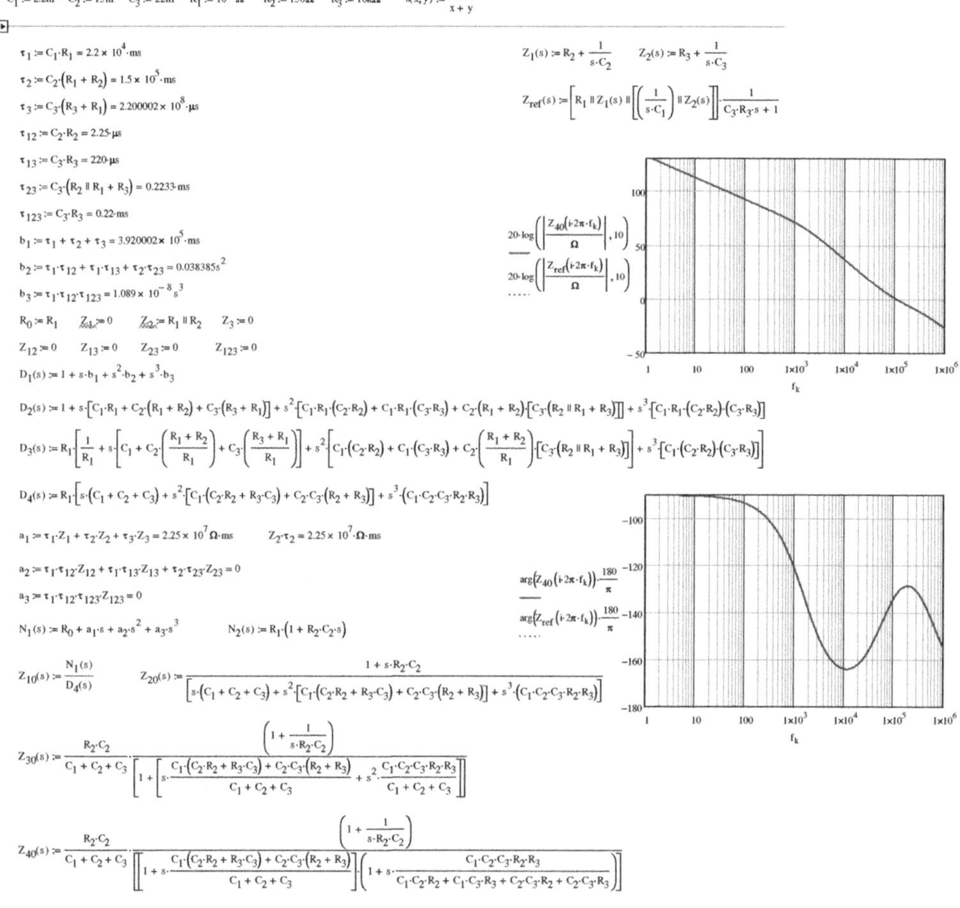

Figure 5.102 The Mathcad® sheet confirms that our equations are correct when dynamic responses including the reference transimpedance are in agreement.

Problem 7:

This circuit represents one possible model of quartz crystal. In this configuration, it is likely that the absence of ohmic losses prevents us from easily forming time constants. The approach we adopted before can be duplicated here by adding an extra resistor we will call R_{inf} across the crystal. The first transfer function is the quasi-static gain R_0 obtained from Figure 5.103a:

$$R_0 = R_{\text{inf}} \tag{5.508}$$

Then, the time constants are obtained from Figure 5.103b and c:

$$\tau_1 = R_{\text{inf}} C_1 \tag{5.509}$$

$$\tau_2 = R_{\text{inf}} C_2 \tag{5.510}$$

$$\tau_3 = \frac{L_3}{\infty} = 0 \tag{5.511}$$

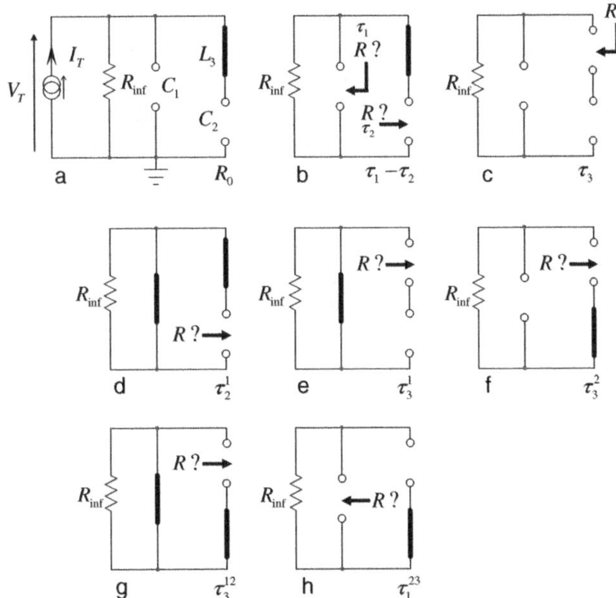

Figure 5.103 A series of sketches helps define time constants but indeterminacies are likely to happen.

The first term b_1 is obtained by adding all time constants:

$$b_1 = \tau_1 + \tau_2 + \tau_3 = R_{\text{inf}}(C_1 + C_2) \tag{5.512}$$

From Figure 5.103d to f, we have:

$$\tau_2^1 = C_2 \cdot 0 \tag{5.513}$$

$$\tau_3^1 = \frac{L_3}{\infty} \tag{5.514}$$

$$\tau_3^2 = \frac{L_3}{R_{\text{inf}}} \tag{5.515}$$

If we now assemble these time constants with those from (5.509) to (5.511), we have:

$$b_2 = \tau_1\tau_2^1 + \tau_1\tau_3^1 + \tau_2\tau_3^2 = R_{inf}C_1 \cdot 0 \cdot C_2 + R_{inf}C_1\frac{L_3}{\infty} + R_{inf}C_2\frac{L_3}{R_{inf}} = C_2L_3 \tag{5.516}$$

A common mistake here would be to consider terms factored with (5.511), (5.513) and (5.514) equal to 0 then decide to immediately omit them in the writing of b_2 because they are null. You must multiply the time constants together making sure no indeterminacies are formed in the factors. Here you see that without R_{inf}, you would have many indeterminacies when replacing it by infinity. As there are no issues in (5.516) then simplifications occur.

The last term is obtained from Figure 5.103g in which capacitors C_1 and C_2 are set to their high-frequency state. In this case, inductor L_3 sees a short circuit leading to:

$$\tau_3^{12} = \frac{L_3}{0} \tag{5.517}$$

This is a nice indeterminacy. How do we get rid of it? By reshuffling the time constants into a different combination. We have the choice between two, τ_1^{23} or τ_1^{32}. You may correctly object that these two expressions are similar as both C_2 and L_3 are set to their high-frequency state while we look into C_1's terminals. However, in the first case, the preceding factor will be $\tau_2\tau_3^2$ or $\tau_3\tau_2^3$ in the second option.τ_2^3 gives another indeterminacy as looking into C_2's terminals while the inductor is removed gives infinity. On the other hand, determining τ_3^2 is already available from (5.515) and Figure 5.103g gives us:

$$\tau_1^{23} = C_1R_{inf} \tag{5.518}$$

Associating (5.518), (5.515) and (5.510) leads to:

$$b_3 = \tau_2\tau_3^2\tau_1^{23} = R_{inf}C_2\frac{L_3}{R_{inf}}C_1R_{inf} = R_{inf}C_1C_2L_3 \tag{5.519}$$

What if the indeterminacy had not been resolved by reshuffling? In this case, adding a resistor across L_3 would have offered a dc path for (5.517). The pain is that you have modified the structured by adding this extra component (even if you make it approach infinity later on) and you must recompute the previous time constants accounting for this extra component.

The denominator can be written associating the time constants we have derived:

$$D(s) = 1 + R_{inf}(C_1 + C_2)s + C_2L_3s^2 + R_{inf}C_2C_1L_3s^3 \tag{5.520}$$

Factoring R_{inf}, we write:

$$D(s) = R_{inf}\left(\frac{1}{R_{inf}} + s(C_1 + C_2) + s^2\frac{C_2L_3}{R_{inf}} + s^3C_1C_2L_3\right) \tag{5.521}$$

Considering R_{inf} a high-value resistor, the above expression simplifies to:

$$D(s) \approx R_{inf}\left[s(C_1 + C_2) + s^3C_2C_1L_3\right] \tag{5.522}$$

The denominator is obtained, we can look at the various gains to determine the zeros. The sketches are in Figure 5.104. When C_1 is replaced by a short circuit, the response is 0. The situation repeats for C_2 in high frequency state and in series with L_3 set in its dc state. The rest of the configurations return R_{inf}. As a summary:

$$Z^1 = 0 \tag{5.523}$$

$$Z^2 = 0 \tag{5.524}$$

$$Z^3 = R_{inf} \tag{5.525}$$

With these expressions on hand, we can form a_1:

$$a_1 = \tau_1 Z^1 + \tau_2 Z^2 + \tau_3 Z^3 = R_{\mathrm{inf}} C_1 \cdot 0 + R_{\mathrm{inf}} C_2 \cdot 0 + \frac{L_3}{\infty} \cdot R_{\mathrm{inf}} = 0 \quad (5.526)$$

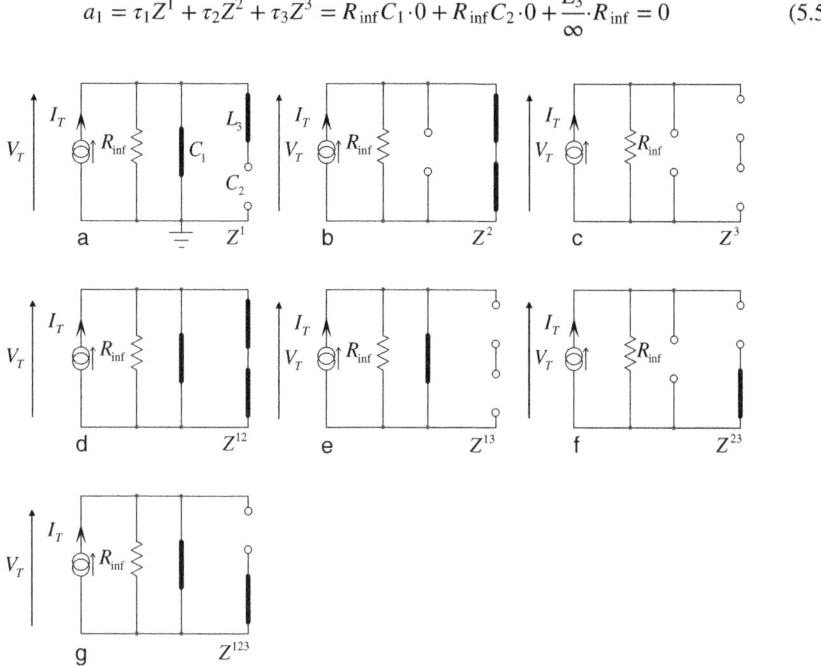

Figure 5.104 The zeros are obtained by looking at the gains obtained in various conditions.

Proceeding with Figure 5.104d to f, we have:

$$Z^{12} = 0 \quad (5.527)$$

$$Z^{13} = 0 \quad (5.528)$$

$$Z^{23} = R_{\mathrm{inf}} \quad (5.529)$$

Leading to the following a_2 definition:

$$a_2 = \tau_1 \tau_2^1 Z^{12} + \tau_1 \tau_3^1 Z^{13} + \tau_2 \tau_3^2 Z^{23} = R_{\mathrm{inf}} C_1 \cdot 0 \cdot C_2 \cdot 0 + R_{\mathrm{inf}} C_1 \frac{L_3}{\infty} \cdot 0 + R_{\mathrm{inf}} C_2 \frac{L_3}{R_{\mathrm{inf}}} R_{\mathrm{inf}} = R_{\mathrm{inf}} C_2 L_3 \quad (5.530)$$

In sketch (g), the response is also 0:

$$Z^{123} = 0 \quad (5.531)$$

which implies that

$$a_3 = \tau_2 \tau_3^2 \tau_1^{32} Z^{123} = R_{\mathrm{inf}} C_2 \frac{L_3}{R_{\mathrm{inf}}} C_1 R_{\mathrm{inf}} \cdot 0 = 0 \quad (5.532)$$

Combining (5.526), (5.530) and (5.532), the denominator is equal to:

$$N(s) = R_0 + a_1 s + a_2 s^2 + a_3 s^3 = R_{\mathrm{inf}} \left(1 + L_3 C_2 s^2\right) \quad (5.533)$$

and the final transfer function is given by:

$$Z(s) = \frac{R_{\text{inf}}\left(1 + s^2 L_3 C_2\right)}{R_{\text{inf}}[s(C_1 + C_2) + s^3 C_2 C_1 L_3]} = \frac{1 + s^2 L_3 C_2}{s(C_1 + C_2) + s^3 C_2 C_1 L_3} \tag{5.534}$$

The raw transfer function is easy and equal to:

$$Z_{ref}(s) = \left(\frac{1}{sC_1}\right) \| \left(sL_3 + \frac{1}{sC_2}\right) \tag{5.535}$$

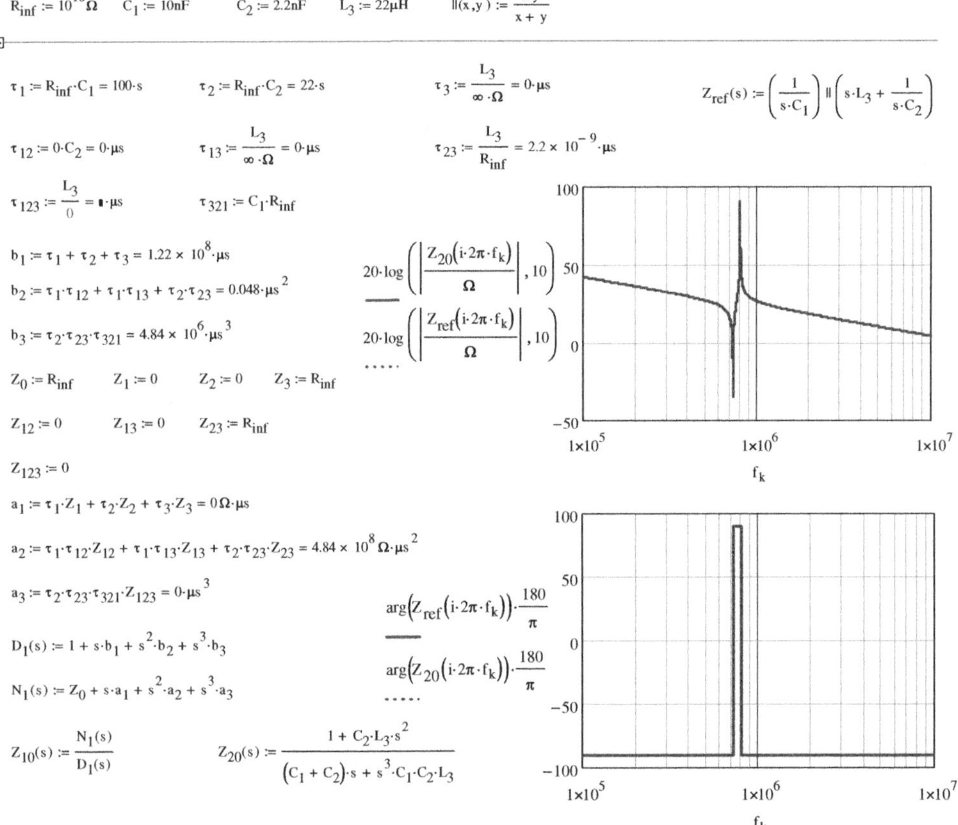

Figure 5.105 Mathcad® confirms that all our equations are correct.

All curves in Figure 5.105 match and confirm our approach. Now, it is important to realize that doing the maths in your head or asking Mathcad® to simplify the below equation:

$$Z = \frac{\frac{1}{s \cdot C_{.1}} \cdot \left(s \cdot L_{.3} + \frac{1}{s \cdot C_{.2}}\right)}{\frac{1}{s \cdot C_{.1}} + \left(s \cdot L_{.3} + \frac{1}{s \cdot C_{.2}}\right)}$$

immediately gives

$$Z = \frac{C_{.2} \cdot L_{.3} \cdot s^2 + 1}{C_{.1} \cdot s + C_{.2} \cdot s + C_{.1} \cdot C_{.2} \cdot L_{.3} \cdot s^3}$$

which is way faster than our fast analytical approach! Engineering judgment is thus extremely important to select the right tool in relationship to the network complexity: do not choose a hammer to smash the fly. Nevertheless, solving this quartz impedance taught us how to reshuffle time constants and efficiently remove indeterminacies.

Problem 8:

This structure describes the simplified representation of a LLC resonant switching converter detailed in [1] and [2]. R_1 is often labeled R_{ac} and represents the load supplied through a diode bridge reflected to the primary-side of an isolation transformer affected by a turns ratio $N = N_p/N_s$. L_3 is the transformer's magnetizing inductance while L_2 models the leakage inductance. Both are usually linked via a fixed ratio (e.g. $L_3 = 2L_2$) to bring specific performance. We can first ignore the transformer turns ratio N and bring it back at the end. This is a 3rd-order network and we can already draw sketches corresponding to various energy-storing elements states combinations. They appear in Figure 5.106. The dc gain H_0 is immediate and equals:

$$H_0 = 0 \tag{5.536}$$

The three 1st-order time constants are obtained from Figure 5.106b to c:

$$\tau_1 = 0 \cdot C_1 \tag{5.537}$$

$$\tau_2 = \frac{L_2}{\infty} \tag{5.538}$$

$$\tau_3 = \frac{L_3}{R_1} \tag{5.539}$$

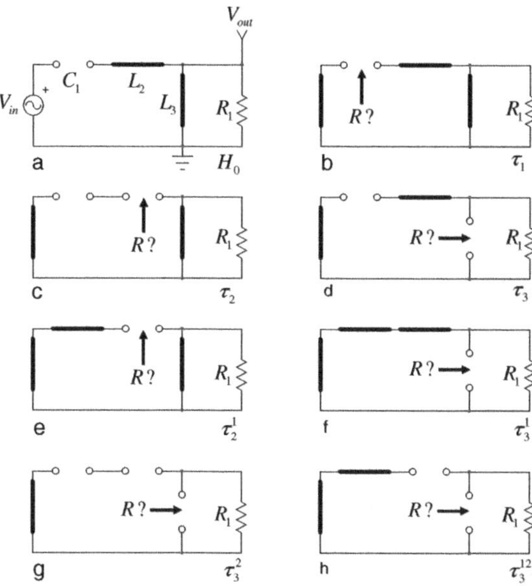

Figure 5.106 The LLC circuit host three energy-storing elements and its natural time constants can be determined with these small sketches:

We can form b_1 by adding (5.537), (5.538) and (5.539):

$$b_1 = \tau_1 + \tau_2 + \tau_3 = \frac{L_3}{R_1} \tag{5.540}$$

If we carry on using Figure 5.106d to f, we have:

$$\tau_2^1 = \frac{L_2}{0} \tag{5.541}$$

$$\tau_3^1 = \frac{L_3}{0} \tag{5.542}$$

and:

$$\tau_3^2 = \frac{L_3}{R_1} \tag{5.543}$$

Looking at (5.541) and (5.542), we have two nice infinite time constants which combined with the previous values are likely to form indeterminacies. What matters is the lack of ohmic path with C_1 and L_2. To temporarily solve this issue, we can install a small resistance R_d in series between C_1 and L_2 and update all precedent equations:

$$\tau_1 = R_d C_1 \tag{5.544}$$

$$\tau_2 = \frac{L_2}{\infty} \tag{5.545}$$

$$\tau_3 = \frac{L_3}{R_1} \tag{5.546}$$

b_1 is updated to:

$$b_1 = \tau_1 + \tau_2 + \tau_3 = R_d C_1 + \frac{L_3}{R_1} \tag{5.547}$$

Considering $R_d \to 0$, (5.547) simplifies to:

$$b_1 = \frac{L_3}{R_1} \tag{5.548}$$

and

$$\tau_2^1 = \frac{L_2}{R_d} \tag{5.549}$$

$$\tau_3^1 = \frac{L_3}{R_d \| R_1} \tag{5.550}$$

$$\tau_3^2 = \frac{L_3}{R_1} \tag{5.551}$$

b_2 is now determined by:

$$b_2 = \tau_1\tau_2^1 + \tau_1\tau_3^1 + \tau_2\tau_3^2 = R_d C_1 \frac{L_2}{R_d} + R_d C_1 \frac{L_3}{R_d \| R_1} + \frac{L_2}{\infty}\frac{L_3}{R_1} = C_1 L_2 + C_1 L_3 \left(\frac{R_1 + R_d}{R_1}\right) \tag{5.552}$$

Considering $R_d \to 0$, (5.552) simplifies to:

$$b_2 = C_1(L_2 + L_3) \tag{5.553}$$

The last term is obtained from Figure 5.106h:

$$\tau_3^{12} = \frac{L_3}{R_1} \tag{5.554}$$

and leads to the definition of b_3:

$$b_3 = \tau_1\tau_2^1\tau_3^{12} = R_d C_1 \frac{L_2}{R_d}\frac{L_3}{R_1} = \frac{C_1 L_2 L_3}{R_1} \tag{5.555}$$

The complete denominator definition is thus:

$$D(s) = 1 + b_1 s + b_2 s^2 + b_3 s^3 = 1 + s\frac{L_3}{R_1} + s^2[C_1(L_2 + L_3)] + s^3\frac{C_1 L_2 L_3}{R_1} \tag{5.556}$$

The numerator is obtained by looking at all gain sketches gathered in Figure 5.107. Here, it is easy, all transfer functions return 0 except H^{13} which is 1. Therefore, we have:

$$a_1 = \tau_1 H^1 + \tau_2 H^2 + \tau_3 H^3 = 0 \tag{5.557}$$

$$a_2 = \tau_1\tau_2^1 H^{12} + \tau_1\tau_3^1 H^{13} + \tau_2\tau_3^2 H^{23} = R_d C_1 \frac{L_3}{R_d \| R_1} \tag{5.558}$$

Considering $R_d \to 0$, (5.558) simplifies to:

$$a_2 = C_1 L_3 \tag{5.559}$$

The last term is also 0:

$$a_3 = \tau_1\tau_2^1\tau_3^{12} H^{123} = 0 \tag{5.560}$$

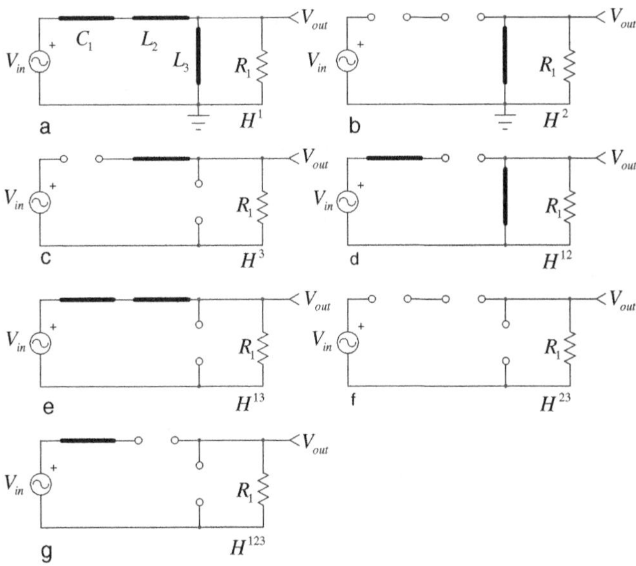

Figure 5.107 All transfer functions return 0 except one, H^{13}, naturally associating C_1 and L_3 in a zero definition.

The numerator $N(s)$ is equal to 1 single term:

$$N(s) = H_0 + a_1 s + a_2 s^2 + a_3 s^3 = s^2 C_1 L_3 \tag{5.561}$$

We now have our final transfer function by dividing (5.561) by (5.556) and including N to reach V_{out} in Figure 5.82:

$$H(s) = \frac{1}{N} \frac{s^2 C_1 L_3}{1 + s\frac{L_3}{R_1} + s^2[C_1(L_2 + L_3)] + s^3 \frac{C_1 L_2 L_3}{R_1}} \tag{5.562}$$

If you look back to Figure 5.82, R_1 is in parallel with L_3. In high-output current conditions, R_1 is very small and short circuits L_3. The resonant frequency involving L_2 and C_1 can be defined as:

$$\omega_s = \frac{1}{\sqrt{L_2 C_1}} \tag{5.563}$$

When the output current decreases, or when the converter becomes lightly loaded, L_3 comes back in series with L_2 to form a second resonant frequency expressed as:

$$\omega_m = \frac{1}{\sqrt{C_1(L_2 + L_3)}} \tag{5.564}$$

If we define a quality factor Q by:

$$Q = R_1 \sqrt{\frac{C_1}{L_2}} \tag{5.565}$$

Then after several manipulations, (5.562) can be rewritten as:

$$H(s) = \frac{1}{N} \frac{\frac{L_3}{L_2}\left(\frac{s}{\omega_s}\right)^2}{1 + s\frac{L_3}{R_1} + \left(\frac{s}{\omega_m}\right)^2 + s^3 \frac{L_3}{L_2} \frac{1}{Q{\omega_s}^3}} \tag{5.566}$$

As a side remark, independent from this exercise, you can use this expression to plot the dc transfer function of the LLC converter. It is not to be mixed up with its small-signal response which cannot easily be analytically determined. In this case, R_1 (also labeled as R_{ac} in the literature) must be replaced by:

$$R_1 = \frac{8}{\pi^2} N^2 R_L \tag{5.567}$$

Please see [1] to understand how (5.567) is derived.

Q is updated to:

$$Q = N^2 R_L \sqrt{\frac{C_1}{L_2}} \tag{5.568}$$

in which R_L is the resistance loading the LLC converter. The updated expression becomes:

$$H(s) = \frac{1}{N} \frac{\frac{L_3}{L_2}\left(\frac{s}{\omega_s}\right)^2}{1 + s\frac{\pi^2 L_3}{8 \cdot N^2 R_L} + \left(\frac{s}{\omega_m}\right)^2 + s^3 \frac{L_3}{L_2} \frac{\pi^2}{8 \cdot Q{\omega_s}^3}} \tag{5.569}$$

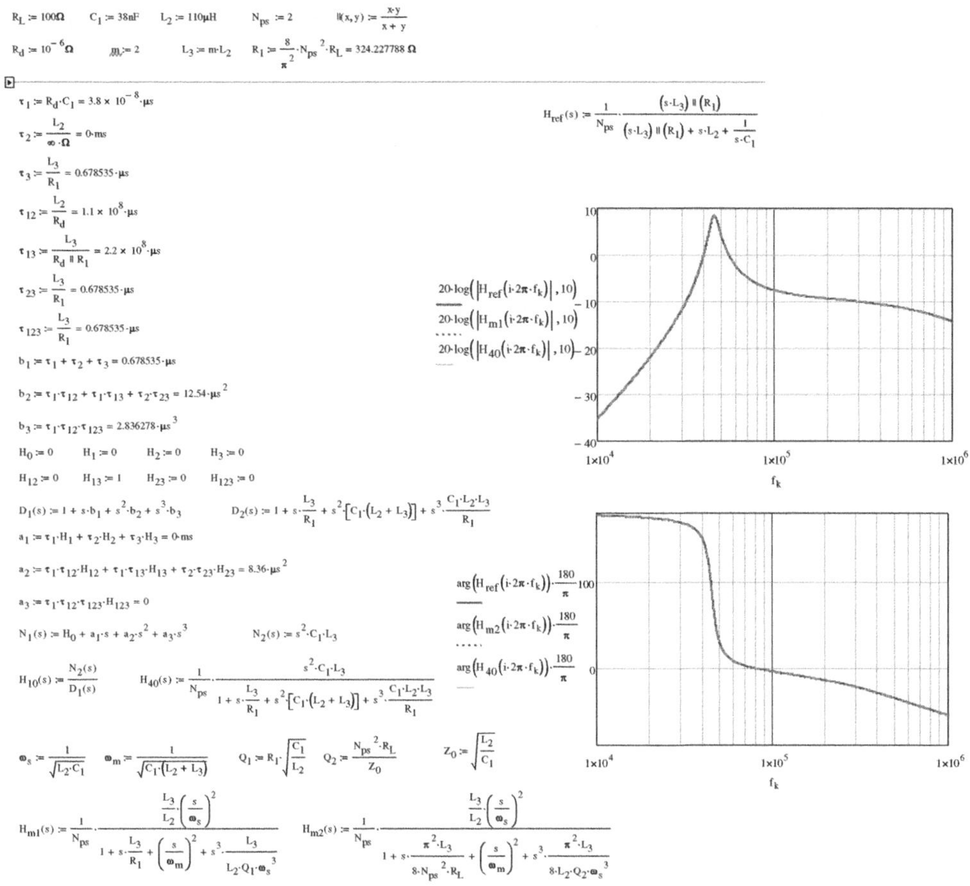

Figure 5.108 The Mathcad® sheet shows that all transfer functions give the same ac dynamic response.

The reference transfer function is simple to find and equals:

$$H_{ref}(s) = \frac{1}{N} \frac{sL_3 \| R_1}{sL_3 \| R_1 + sL_2 + \dfrac{1}{sC_1}} \tag{5.570}$$

All these curves gathered in Figure 5.108 confirm that our approach is correct. (5.566) is plotted under the formula labeled H_{m1} while H_{m2} describes the same formula but depending on the LLC load, R_L involving (5.567) and (5.568).

Problem 9:

This circuit describes a loudspeaker model detailed in [3]. To determine an impedance, the driving variable is a current I_T while the response is V_T across the network. When you set the excitation to 0 A, the circuit shows that L_1's upper terminal is floating, leading to a 0-valued time constant obtained with the infinite symbol in the denominator. This situation is likely to generate indeterminacies and we may need to add another resistance across the network or reshuffle time constants at some points. We will see. We start with Figure 5.109a where the dc resistance R_0 offered by the network is determined:

$$R_0 = R_1 \tag{5.571}$$

The 3 other time constants are obtained with Figure 5.109b to e:

$$\tau_1 = \frac{L_1}{\infty} \tag{5.572}$$

$$\tau_2 = \frac{L_2}{R_2} \tag{5.573}$$

$$\tau_3 = C_3 \cdot 0 \tag{5.574}$$

$$\tau_4 = \frac{L_4}{R_3} \tag{5.575}$$

We can form the first term, b_1:

$$b_1 = \tau_1 + \tau_2 + \tau_3 + \tau_4 = \frac{L_2}{R_2} + \frac{L_4}{R_3} \tag{5.576}$$

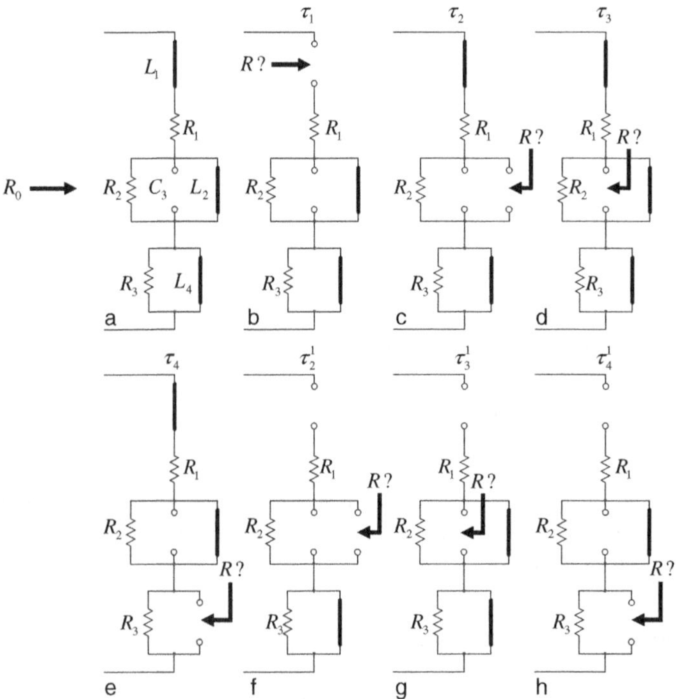

Figure 5.109 Sketches are helpful to derive the various time constants. Here the first part of the natural time constants determination.

The next batch of time constants use Figure 5.109f to h and Figure 5.110a to c:

$$\tau_2^1 = \frac{L_2}{R_2} \tag{5.577}$$

$$\tau_3^1 = C_3 \cdot 0 \tag{5.578}$$

$$\tau_4^1 = \frac{L_4}{R_3} \tag{5.579}$$

$$\tau_3^2 = C_3 R_2 \tag{5.580}$$

$$\tau_4^2 = \frac{L_4}{R_3} \tag{5.581}$$

$$\tau_4^3 = \frac{L_4}{R_3} \tag{5.582}$$

The second term b_2 is equal to:

$$\begin{aligned} b_2 &= \tau_1\tau_2^1 + \tau_1\tau_3^1 + \tau_1\tau_4^1 + \tau_2\tau_3^2 + \tau_2\tau_4^2 + \tau_3\tau_4^3 \\ &= \frac{L_1}{\infty}\frac{L_2}{R_2} + \frac{L_1}{\infty} \cdot 0 \cdot C_3 + \frac{L_1}{\infty}\frac{L_4}{R_3} + \frac{L_2}{R_2} C_3 R_2 + \frac{L_2}{R_2}\frac{L_4}{R_3} + C_3 \cdot 0 \cdot \frac{L_4}{R_3} \\ &= \frac{L_2}{R_2} C_3 R_2 + \frac{L_2}{R_2}\frac{L_4}{R_3} = L_2 C_3 + \frac{L_2}{R_2}\frac{L_4}{R_3} \end{aligned} \tag{5.583}$$

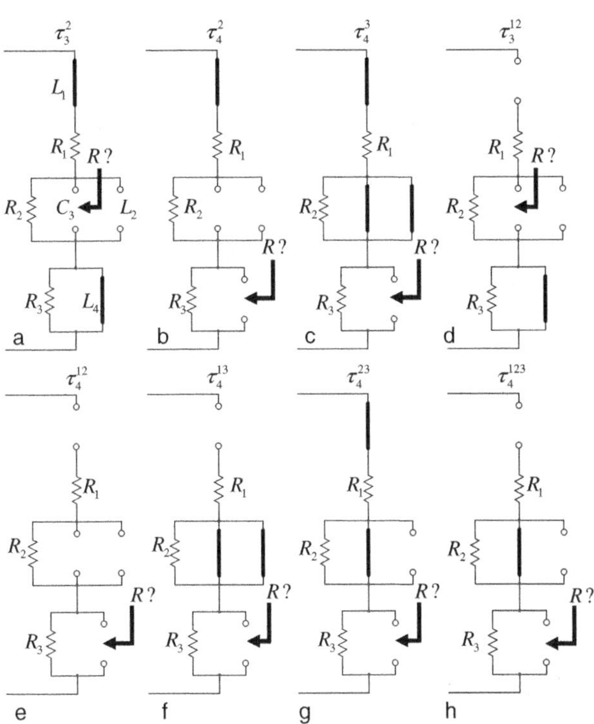

Figure 5.110 This picture is the second part of the natural time constants determination.

Now from Figure 5.110d to g, we can findL:

$$\tau_3^{12} = C_3 R_2 \tag{5.584}$$

$$\tau_4^{12} = \frac{L_4}{R_3} \tag{5.585}$$

$$\tau_4^{13} = \frac{L_4}{R_3} \tag{5.586}$$

$$\tau_4^{23} = \frac{L_4}{R_3} \tag{5.587}$$

Assembling these elements with the previous time constants leads to:

$$\begin{aligned} b_3 &= \tau_1 \tau_2^1 \tau_3^{12} + \tau_1 \tau_2^1 \tau_4^{12} + \tau_1 \tau_3^1 \tau_4^{13} + \tau_2 \tau_3^2 \tau_4^{23} \\ &= \frac{L_1}{\infty}\frac{L_2}{R_2} C_3 R_2 + \frac{L_1}{\infty}\frac{L_2}{R_2}\frac{L_4}{R_3} + \frac{L_1}{\infty} \cdot 0 \cdot C_3 \frac{L_4}{R_3} + \frac{L_2}{R_2} C_3 R_2 \frac{L_4}{R_3} \\ &= \frac{L_2 C_3 L_4}{R_3} \end{aligned} \tag{5.588}$$

Finally, with the help of Figure 5.110h, we have:

$$\tau_4^{123} = \frac{L_4}{R_3} \tag{5.589}$$

giving us

$$b_4 = \tau_1 \tau_2^1 \tau_3^{12} \tau_4^{123} = \frac{L_1}{\infty}\frac{L_2}{R_2} C_2 R_2 \frac{L_4}{R_3} = 0 \tag{5.590}$$

With all these definitions on hand, we can express the denominator D as:

$$D(s) = 1 + b_1 s + b_2 s^2 + b_3 s^3 + b_4 s^4 = 1 + s\left(\frac{L_2}{R_2} + \frac{L_4}{R_3}\right) + s^2\left(L_2 C_3 + \frac{L_2}{R_2}\frac{L_4}{R_3}\right) + s^3 \frac{L_2 C_3 L_4}{R_3} \tag{5.591}$$

Now that we have our denominator, we need to determine the numerator N. NDI is usually a more complex approach than the generalized formula except when dealing with impedance. This is because nulling the response in Figure 5.83 is a degenerate case in which the current generator is replaced by a short circuit. Applying a NDI to this network is thus quite simple as shown in Figure 5.111 and Figure 5.112. We start with Figure 5.111a to d:

$$\tau_{1N} = \frac{L_1}{R_1} \tag{5.592}$$

$$\tau_{2N} = \frac{L_2}{R_1 \| R_2} \tag{5.593}$$

$$\tau_{3N} = C_3 \cdot 0 \tag{5.594}$$

$$\tau_{4N} = \frac{L_4}{R_1 \| R_3} \tag{5.595}$$

We have our first numerator coefficient, a_1:

$$a_1 = \tau_{1N} + \tau_{2N} + \tau_{3N} + \tau_{4N} = \frac{L_1}{R_1} + \frac{L_2}{R_1 \| R_2} + C_3 \cdot 0 + \frac{L_4}{R_1 \| R_3} = \frac{L_1}{R_1} + \frac{L_2}{R_1 \| R_2} + \frac{L_4}{R_1 \| R_3} \tag{5.596}$$

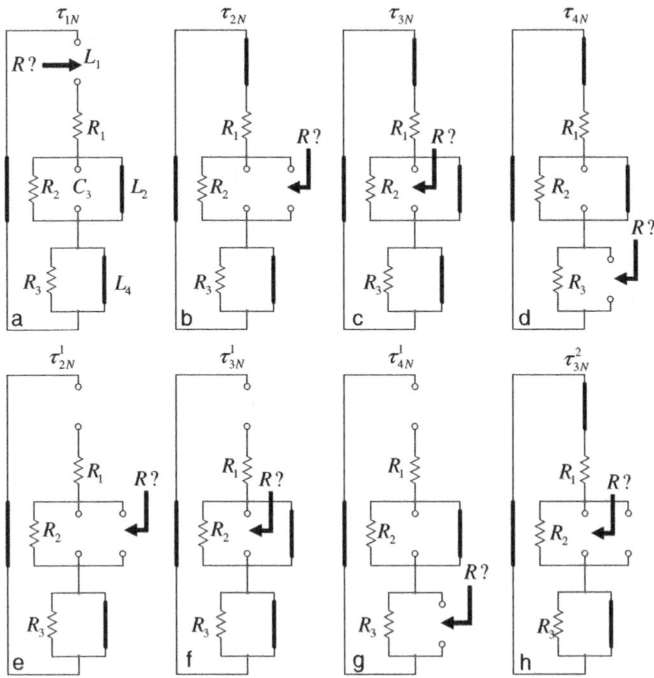

Figure 5.111 Zeros can be obtained using NDI quite easily given the current generator replaced by a short circuit.

From Figure 5.111e to Figure 5.112b, we have:

$$\tau^1_{2N} = \frac{L_2}{R_2} \tag{5.597}$$

$$\tau^1_{3N} = C_3 \cdot 0 \tag{5.598}$$

$$\tau^1_{4N} = \frac{L_4}{R_3} \tag{5.599}$$

$$\tau^2_{3N} = C_3 (R_2 \| R_1) \tag{5.600}$$

$$\tau^2_{4N} = \frac{L_4}{R_3 \| (R_1 + R_2)} \tag{5.601}$$

$$\tau^3_{4N} = \frac{L_4}{R_3 \| R_1} \tag{5.602}$$

We can express a_2 as:

$$\begin{aligned} a_2 &= \tau_{1N}\tau^1_{2N} + \tau_{1N}\tau^1_{3N} + \tau_{1N}\tau^1_{4N} + \tau_{2N}\tau^2_{3N} + \tau_{2N}\tau^2_{4N} + \tau_{3N}\tau^3_{4N} \\ &= \frac{L_1}{R_1}\frac{L_2}{R_2} + \frac{L_1}{R_1} C_3 \cdot 0 + \frac{L_1}{R_1}\frac{L_4}{R_3} + \frac{L_2}{R_1 \| R_2} C_3 (R_2 \| R_1) + \frac{L_2}{R_1 \| R_2}\frac{L_4}{R_3 \| (R_1 + R_2)} + C_3 \cdot 0 \frac{L_4}{R_3 \| R_1} \\ &= \frac{L_1}{R_1}\frac{L_2}{R_2} + \frac{L_1}{R_1}\frac{L_4}{R_3} + L_2 C_3 + \frac{L_2}{R_1 \| R_2}\frac{L_4}{R_3 \| (R_1 + R_2)} \end{aligned} \tag{5.603}$$

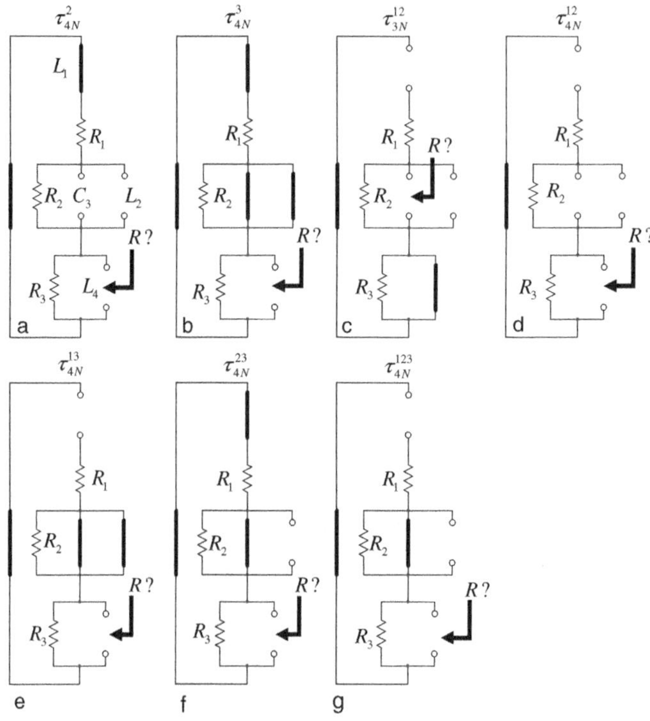

Figure 5.112 This picture is the second part of the NDI exercise for the numerator determination.

From Figure 5.112c to f, we can determine a_3 terms as:

$$\tau_{3N}^{12} = C_3 R_2 \tag{5.604}$$

$$\tau_{4N}^{12} = \frac{L_4}{R_3} \tag{5.605}$$

$$\tau_{4N}^{13} = \frac{L_4}{R_3} \tag{5.606}$$

$$\tau_{4N}^{23} = \frac{L_4}{R_3 \| R_1} \tag{5.607}$$

which lead us to forming a_3 as:

$$\begin{aligned} a_3 &= \tau_{1N}\tau_{2N}^{1}\tau_{3N}^{12} + \tau_{1N}\tau_{2N}^{1}\tau_{4N}^{12} + \tau_{1N}\tau_{3N}^{1}\tau_{4N}^{13} + \tau_{2N}\tau_{3N}^{2}\tau_{4N}^{23} \\ &= \frac{L_1}{R_1}\frac{L_2}{R_2}C_3 R_2 + \frac{L_1}{R_1}\frac{L_2}{R_2}\frac{L_4}{R_3} + \frac{L_1}{R_1}C_3 \cdot 0 \frac{L_4}{R_3} + \frac{L_2}{R_1 \| R_2} C_3 \left(R_2 \| R_1\right) \frac{L_4}{R_3 \| R_1} \\ &= \frac{L_1 L_2}{R_1} C_3 + \frac{L_1}{R_1}\frac{L_2}{R_2}\frac{L_4}{R_3} + L_2 C_3 \frac{L_4}{R_3 \| R_1} \end{aligned} \tag{5.608}$$

The last term is obtained from Figure 5.112g:

$$\tau_{4N}^{123} = \frac{L_4}{R_3} \tag{5.609}$$

which leads us to

$$a_4 = \tau_{1N}\tau_{2N}^{1}\tau_{3N}^{12}\tau_{4N}^{123} = \frac{L_1}{R_1}\frac{L_2}{R_2}C_3R_2\frac{L_4}{R_3} = \frac{L_1L_2L_4C_3}{R_1R_3} \tag{5.610}$$

The complete numerator N is expressed as:

$$\begin{aligned} N(s) &= 1 + sa_1 + s^2a_2 + s^3a_3 + s^4a_4 \\ &= 1 + s\left(\frac{L_1}{R_1} + \frac{L_2}{R_1 \| R_2} + \frac{L_4}{R_1 \| R_3}\right) + s^2\left(\frac{L_1}{R_1}\frac{L_2}{R_2} + \frac{L_1}{R_1}\frac{L_4}{R_3} + L_2C_3 + \frac{L_2}{R_1 \| R_2}\frac{L_4}{R_3 \| (R_1+R_2)}\right) \\ &+ s^3\left(\frac{L_1L_2}{R_1}C_3 + \frac{L_1}{R_1}\frac{L_2}{R_2}\frac{L_4}{R_3} + L_2C_3\frac{L_4}{R_3 \| R_1}\right) + s^4\frac{L_1L_2L_4C_3}{R_1R_3} \end{aligned} \tag{5.611}$$

and the transfer function we want is:

$$Z(s) = R_0\frac{N(s)}{D(s)} = R_1 \cdot \frac{N(s)}{1 + s\left(\frac{L_2}{R_2} + \frac{L_4}{R_3}\right) + s^2\left(L_2C_3 + \frac{L_2}{R_2}\frac{L_4}{R_3}\right) + s^3\frac{L_2C_3L_4}{R_3}} \tag{5.612}$$

The raw transfer function is the series connection of several impedances:

$$Z_{ref}(s) = sL_1 + R_1 + Z_1(s) + Z_2(s) \tag{5.613}$$

in which

$$Z_1(s) = R_2 \| \left(\frac{1}{sC_3}\right) \| sL_2 \tag{5.614}$$

$$Z_2(s) = R_3 \| sL_4 \tag{5.615}$$

We have captured all these formulas in Figure 5.113 and all dynamic responses perfectly superimpose.

Problem 10:
This is an inverting biquad filter as presented in [4]. We will determine this transfer function using the classical series of sketches to determine the natural time constants and, later on, the transfer function gains. From Figure 5.114a, we have:

$$H_0 = 0 \tag{5.616}$$

Then, from Figure 5.114b, we can determine the three time constants:

$$\tau_1 = C_1R_1 \tag{5.617}$$

$$\tau_2 = C_2R_1 \tag{5.618}$$

$$\tau_3 = C_3R_1 \tag{5.619}$$

which lead to a simple first denominator coefficient b_1:

$$b_1 = \tau_1 + \tau_2 + \tau_3 = R_1(C_1 + C_2 + C_3) \tag{5.620}$$

We now carry on with sketches from Figure 5.114c to e. In sketch (c), the output is zero volt, there is no driving signal and the negative pin is also 0, grounding R_1. Therefore:

$$\tau_2^1 = C_2R_2 \tag{5.621}$$

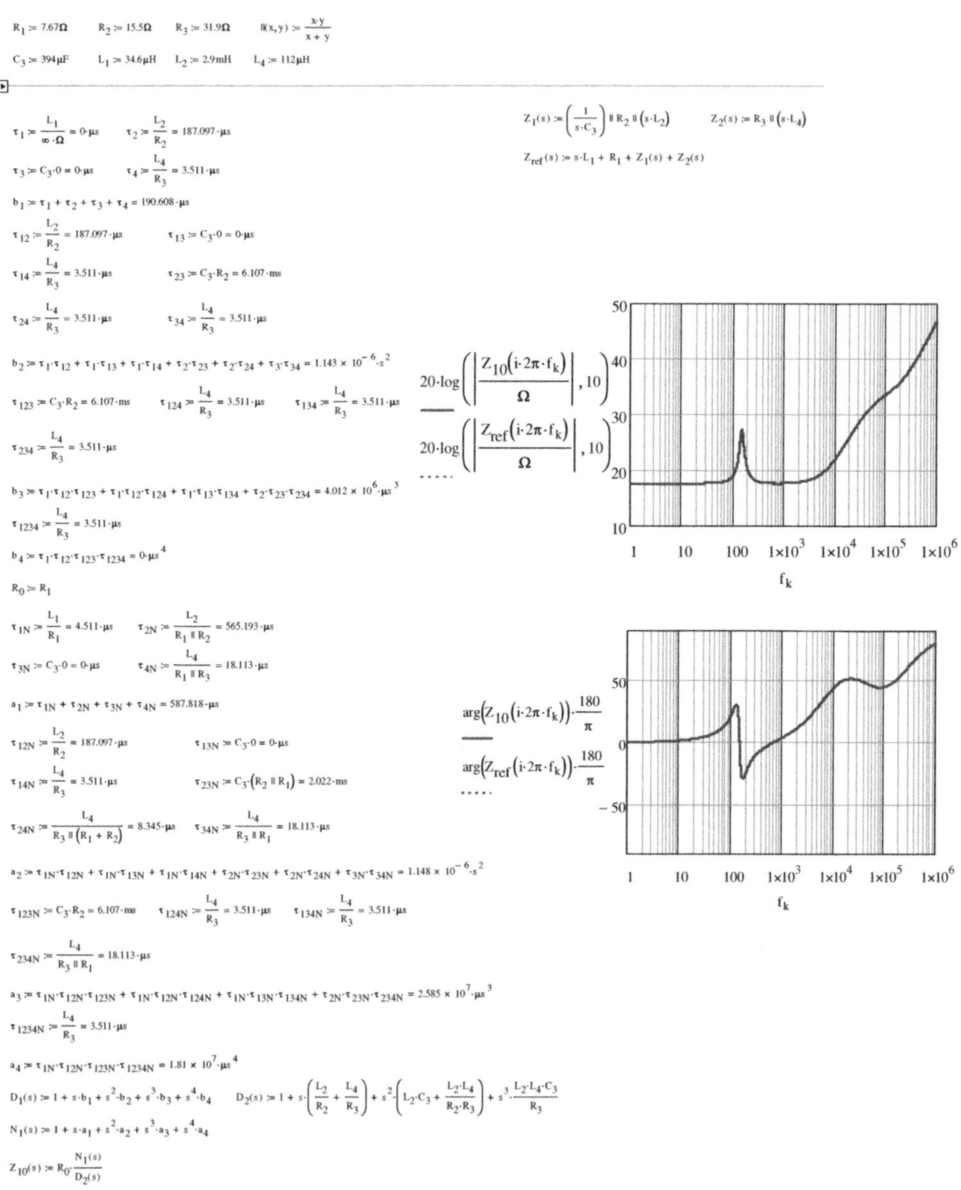

Figure 5.113 The Mathcad® sheet confirms our approach is correct.

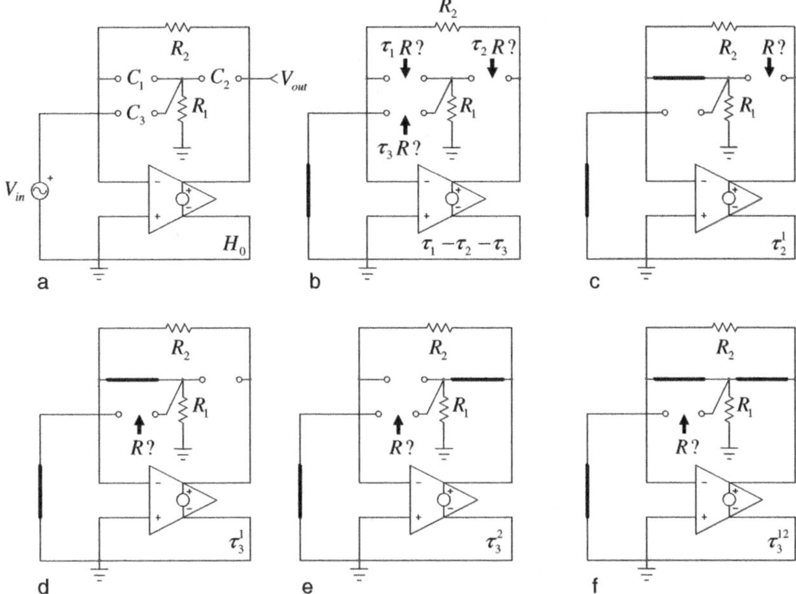

Figure 5.114 A series of sketches will lead us to the determination of the denominator D.

In sketch (d), an intermediate stage is necessary as represented in Figure 5.115. The test current I_T is the sum of I_1 and I_2:

$$I_T = I_1 + I_2 \tag{5.622}$$

The first current I_1 is the negative voltage pin over R_1. The negative voltage pin is V_{out} divided by the op amp open-loop gain A_{OL}:

$$I_1 = -\frac{V_{out}}{A_{OL}R_1} \tag{5.623}$$

The second current, I_2, is equal to the inverting pin voltage minus the output voltage and divided by R_2:

$$I_2 = \frac{V_{(-)} - V_{out}}{R_2} = \frac{-\frac{V_{out}}{A_{OL}} - V_{out}}{R_2} = -\frac{\left(\frac{1}{A_{OL}} + 1\right)}{R_2} V_{out} \tag{5.624}$$

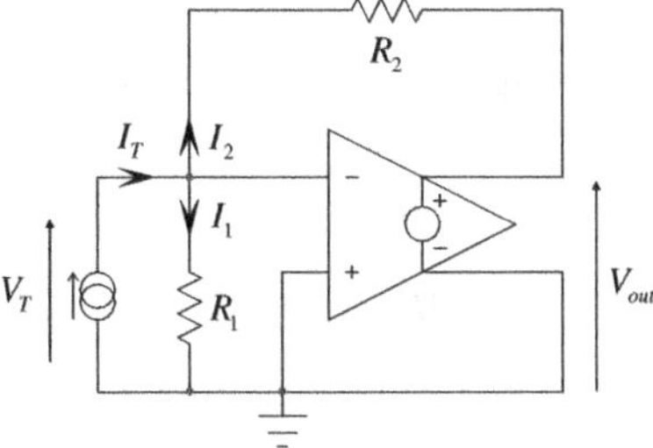

Figure 5.115 This intermediate sketch will help determine τ_3^1.

The voltage V_T is that of the negative pin:

$$V_T = -\frac{V_{out}}{A_{OL}} \tag{5.625}$$

The resistance seen from C_3's terminals is thus:

$$\frac{V_T}{I_T} = \frac{-\dfrac{V_{out}}{A_{OL}}}{-\dfrac{\left(\dfrac{1}{A_{OL}}+1\right)}{R_2}V_{out} - \dfrac{V_{out}}{A_{OL}R_1}} = \frac{R_1R_2}{R_1+R_2+A_{OL}R_1} \tag{5.626}$$

The time constant associated with C_3 in this configuration is:

$$\tau_3^1 = C_3\frac{R_1R_2}{R_1+R_2+A_{OL}R_1} \tag{5.627}$$

We proceed with sketch (e) where we find:

$$\tau_3^2 = C_3 \cdot 0 \tag{5.628}$$

The second term b_2 is then equal to:

$$\begin{aligned} b_2 &= \tau_1\tau_2^1 + \tau_1\tau_3^1 + \tau_2\tau_3^2 = C_1R_1C_2R_2 + C_1R_1C_3\frac{R_1R_2}{R_1+R_2+A_{OL}R_1} + C_2R_1\cdot 0\cdot C_3 \\ &= C_1R_1C_2R_2 + C_1R_1C_3\frac{R_1R_2}{R_1+R_2+A_{OL}R_1} \end{aligned} \tag{5.629}$$

When the open loop gain A_{OL} approaches infinity, (5.629) simplifies to:

$$b_2 = C_1C_2R_1R_2 \tag{5.630}$$

τ_3^{12} is obtained with Figure 5.114f:v

$$\tau_3^{12} = C_3 \cdot 0 \tag{5.631}$$

leading to the 3rd coefficient definition:

$$b_3 = \tau_1\tau_2^1\tau_3^{12} = C_1R_1C_2R_2\cdot 0\cdot C_3 = 0 \tag{5.632}$$

The denominator D can be expressed as:

$$D(s) = 1 + b_1s + b_2s^2 + b_3s^3 = 1 + sR_1(C_1+C_2+C_3) + s^2C_1C_2R_1R_2 \tag{5.633}$$

To obtain the numerator definition, we will go through a series of gain configurations as shown in Figure 5.116.

If you carefully study all these configurations, they all return 0, except H^{13} in sketch (f) which returns $-A_{OL}$:

$$H_0 = H^1 = H^2 = H^3 = H^{12} = H^{23} = H^{123} = 0 \tag{5.634}$$

$$H^{13} = -A_{OL} \tag{5.635}$$

With these gains definitions, only $\tau_1\tau_3^1H^{13}$ is different from zero. Therefore:

$$a_1 = \tau_1H^1 + \tau_2H^2 + \tau_3H^3 = C_1R_1\cdot 0 + C_2R_1\cdot 0 + C_3R_1\cdot 0 = 0 \tag{5.636}$$

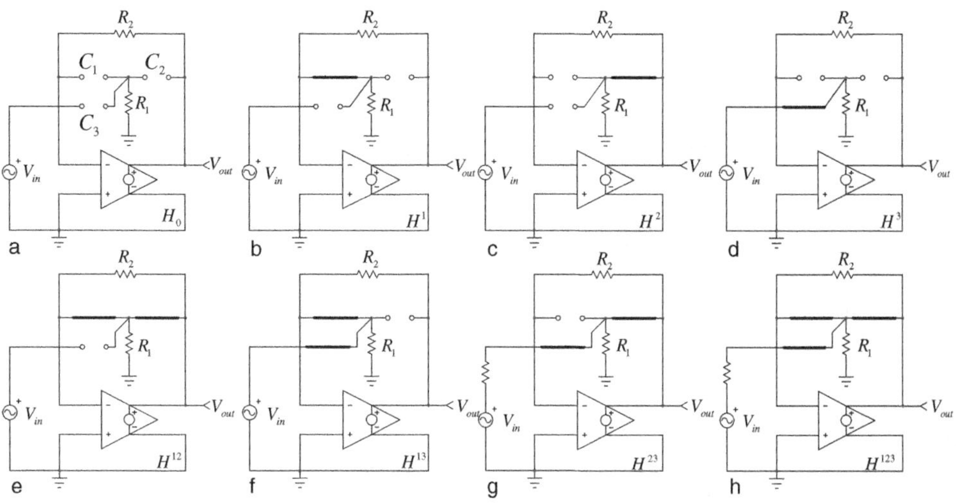

Figure 5.116 The gains are easy to determine in these configurations.

$$a_2 = \tau_1\tau_2^1 H^{12} + \tau_1\tau_3^1 H^{13} + \tau_2\tau_3^2 H^{23} = C_1R_1C_2R_2 \cdot 0 - C_1R_1C_3\left(\frac{R_1R_2}{R_1+R_2+A_{OL}R_1}\right)A_{OL} + C_2R_1C_3 \cdot 0 \cdot 0$$
$$= -C_1R_1C_3\left(\frac{R_1R_2}{R_1+R_2+A_{OL}R_1}\right)A_{OL} \tag{5.637}$$

$$a_3 = \tau_1\tau_2^1\tau_3^{12} H^{123} = R_1C_1R_2C_2 \cdot 0 \cdot C_3 \cdot 0 = 0 \tag{5.638}$$

If we now look at (5.637) and study the expression when A_{OL} approaches infinity, the term simplifies to:

$$a_2 = \lim_{A_{OL}\to\infty} -C_1R_1C_3\left(\frac{R_1R_2}{\frac{R_1}{A_{OL}}+\frac{R_2}{A_{OL}}+R_1}\right) = -C_1C_3R_1R_2 \tag{5.639}$$

The numerator $N(s)$ is thus defined by a single term:

$$N(s) = -s^2C_1C_3R_1R_2 = -\left(\frac{s}{\omega_0}\right)^2 \tag{5.640}$$

with

$$\omega_0 = \frac{1}{\sqrt{C_1C_3R_1R_2}} \tag{5.641}$$

If all capacitors are equal to C, then (5.641) simplifies to:

$$\omega_0 = \frac{1}{C\sqrt{R_1R_2}} \tag{5.642}$$

We now have our final transfer function obtained by dividing (5.640) by (5.633):

$$H(s) = -\frac{s^2 C_1 C_3 R_1 R_2}{1 + sR_1(C_1 + C_2 + C_3) + s^2 C_1 C_2 R_1 R_2} \tag{5.643}$$

Based on the Chapter 2 definition, we can rewrite the fraction by expressing a quality factor and a resonant frequency

$$Q = \frac{\sqrt{b_2}}{b_1} = \frac{\sqrt{C_1 C_2 R_1 R_2}}{(C_1 + C_2 + C_3)R_1} = \frac{1}{C_1 + C_2 + C_3}\sqrt{\frac{C_1 C_2 R_2}{R_1}} \tag{5.644}$$

If all capacitors are equal to C, then (5.644) simplifies to:

$$Q = \frac{1}{3}\sqrt{\frac{R_2}{R_1}} \tag{5.645}$$

The resonant angular frequency is obtained by:

$$\omega_0 = \frac{1}{\sqrt{b_2}} = \frac{1}{\sqrt{C_1 C_3 R_1 R_2}} \tag{5.646}$$

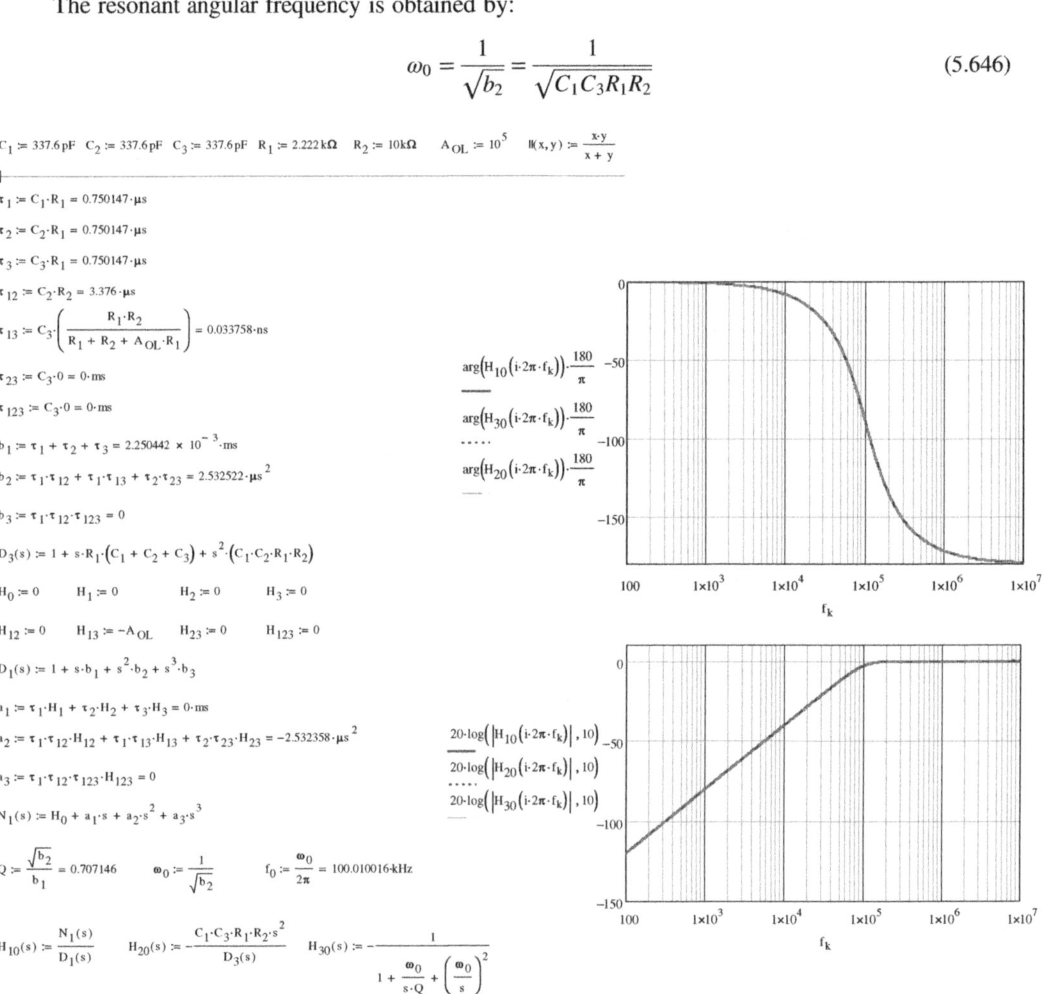

Figure 5.117 All transfer function lead to identical dynamic responses.

If all capacitors equal C, then (5.646) also simplifies to:

$$\omega_0 = \frac{1}{C\sqrt{R_1 R_2}} \tag{5.647}$$

Using these expressions (5.643) can be rewritten as:

$$H(s) = -\frac{\left(\frac{s}{\omega_0}\right)^2}{1 + \frac{s}{\omega_0 Q} + \left(\frac{s}{\omega_0}\right)^2} \tag{5.648}$$

This transfer function can be rearranged following Chapter 2 guidelines in a more compact form:

$$H(s) = -H_\infty \frac{1}{1 + \frac{\omega_0}{sQ} + \left(\frac{\omega_0}{s}\right)^2} \tag{5.649}$$

in which H_∞ is the gain when s approaches infinity, 1 or 0 dB.

References

1. C. Basso. Understanding the LLC Structure in Resonant Application. ON Semiconductor application note AND8311, http://www.onsemi.com/pub_link/Collateral/AND8311-D.PDF (last accessed 12/12/2015).
2. C. Basso. A Simple Dc SPICE Model for the LLC Converter. ON Semiconductor application note AND8255, http://www.onsemi.com/pub_link/Collateral/AND8255-D.PDF (last accessed 12/12/2015).
3. https://gasstationwithoutpumps.wordpress.com/2013/02/15/seventeenth-day-of-circuits-class-inductors-and-gnuplot-tutorial/ (last accessed 12/12/2015).
4. http://www.filter-solutions.com/active.html (last accessed 12/12/2015).

Conclusion

Chapter 5 ends this book dedicated to fast analytical circuits techniques. I thought I would not document ten problems on 3rd and 4th order networks but I finally did it! I hope you have enjoyed the ride, I truly did when writing this book. I can tell you that there is nothing more gratifying than capturing the last coefficient and seeing a graph that perfectly matches simulation results or the raw transfer function dynamic response! If you follow the flow I adopted in this book and solve simple problems first – *petit à petit* in French – then there is no reason to fail when confronting the method with more complex circuits. Actually with the help of Mathcad® and a SPICE simulator, you have all the tools to verify your calculations with a zeroed excitation or with a nulled response. This is my contribution to the technique, giving the reader the ability to assist calculations with a simulator in the simplest way. And it works: I have been trapped several times and was able to identify my mistakes with the small individual sketches and immediately correct the guilty tau in the spreadsheet. With an arm-long expression mixing all sorts of terms, I would have needed to restart from scratch or throw away the chapter pages against the wall! The references I gave at the end of each chapter are worth considering if you wish to extend your skills to the field of micro-electronics and control loop analysis. Dr. Vorpérian's book and Dr. Middlebrook's papers/training foils are certainly documents to read as they include a lot of demonstrations and proofs I could not include in this document. After all, this book can be seen as a stepping stone before considering higher level readings.

I sincerely hope my book modestly contributes to a wider adoption of these analytical techniques. The tools I described and how to exercise them should help the student and the engineer fulfill their daily tasks whether we talk about homework or designing a new product. As usual, I will be happy to hear comments, corrections or suggestions from you all. Please note that all Mathcad files are available for download from my web page http://cbasso.pagesperso-orange.fr/Spice.htm

Christophe Basso
May 2015

Linear Circuit Transfer Functions: An Introduction to Fast Analytical Techniques, First Edition. Christophe P. Basso.

Glossary of Terms

Ac	Alternating current. This abbreviation defines either a *bipolar* or *varying* current. An ac gain or an ac voltage now looks strange, as ac originally defines a current. The term *ac response* will be used by extension in this book to designate a Bode plot. The term can be interchanged with *dynamic* or *frequency* response. Please note that ac is never capitalized except when it starts a sentence.
Bias point	A bias point also called a dc or static operating point describes all currents and voltages in a circuit without excitation. In SPICE, a static operating point (keyword .OP) is systematically determined before any simulation is run. Around this operating point, all non-linear terms are linearized before simulation begins.
Bode plot	Also referred to as a frequency response plot, a Bode plot (named after Hendrik Bode) is the graphical representation of a complex transfer function in which the phase or the argument (in degrees) and the magnitude (in dB) are vertically plotted versus a horizontal log-compressed frequency axis.
Butterworth	A Butterworth filter is often referred to as a maximally-flat magnitude response. The denominator of Butterworth filters can be expressed under a normalized form ensuring the maximally-flat response for different orders.
Compensator	A compensator is an active (operational amplifier-based for instance) or passive circuit used in a control system. It helps realizing a precise and stable closed-loop system by forcing a selected crossover frequency and ensuring adequate phase and gain margins.
Crossover	In a control system operated in open-loop conditions (the feedback or return path is open), the crossover frequency f_c is the frequency point at which the transfer function magnitude linking the output variable to the control variable is equal to 1 or 0 dB.
Dc	Direct current. This abbreviation originally defining either a *constant* or *unipolar* current is often used to qualify other variables such as gains or voltages. However, a dc voltage or a dc gain now sounds contradictory when you recall the original definition. A quasi-static gain is a better word to designate the amplification or attenuation at a 0-Hz frequency. Please note that dc is never capitalized except when it starts a sentence.

Linear Circuit Transfer Functions: An Introduction to Fast Analytical Techniques, First Edition. Christophe P. Basso.

DPI — Driving Point Impedance. This term implies that excitation and response are observed at the same port. For instance, biasing a port with a 1-A ac current source and determining the voltage across that port gives you the impedance offered by the port. An impedance is one kind of transfer function among the six available definitions.

EET — The Extra Element Theorem defined by R.D. Middlebrook. In a linear circuit featuring an element designated as *extra* (usually an element that complicates the analysis when present), the EET states that the transfer function of the said circuit is the transfer function determined without the extra element (replaced by a short circuit or physically removed) affected by a correction term. The EET's underlying theory paves the way for the FACTs.

Excitation — This is a driving waveform applied to the circuit under study. It can be applied to the circuit input but it can also be an independent source driving a reactance port. A current excitation is usually applied across a circuit's element while a voltage excitation is usually inserted in series with a circuit's element.

Entropy — Entropy in thermodynamics characterizes the degree of disorder in a system. By analogy, a *high-entropy* transfer function implies that the result is disorganized and does not give any insight into what the dynamic response is. On the contrary, a *low-entropy* transfer function lets you figure out its asymptotic response by expressing poles, zeros and gains in a clear and ordered form.

FACTs — Fast Analytical Circuit Techniques. They are techniques developed to let you determine transfer functions in the most efficient way by determining the circuit time constants in two conditions: when the excitation is turned off and when the response is nulled.

Gain margin — In the Bode plot describing the frequency response of a compensated control system operated in open-loop conditions, the gain margin (GM) defines the distance of the magnitude curve to the 0-dB line at a frequency where the argument or the phase of the loop gain is 0°.

GIC — General Impedance Converter. This particular op amp-based structure is employed in filtering applications where robustness to component values variations is wanted.

Inspection — Rather than deriving the transfer function using KVL or KCL, you can, in certain cases, observe or *inspect* the electrical diagram and infer the transfer function without writing a single line of algebra. It is often the case for the zeros as shown in many of the available examples. Inspection usually leads to the simplest possible form of transfer functions.

KCL — Kirchhoff's Current Law. A relationship linking currents at any node in a circuit: the sum of currents flowing into a node is equal to the sum of currents leaving that node.

KVL — Kirchhoff's Voltage Law. The algebraic sum of voltages in a loop or mesh is always zero.

Mathcad® — This is a mathematical solver that lets you capture equations, solve systems of equations numerically or symbolically, plot transfer functions and many other useful things. I have documented most of the examples in version 15 and the files are available for download in my personal webpage, http://cbasso.pagesperso-orange.fr/Spice.htm

NDI	Null Double Injection. As the word implies, the double injection involves the excitation associated with a second source – usually a current generator I_T – adjusted to null the output response. NDI is employed to determine the zeros in a transfer function. Chapters abound with examples in which a SPICE template lets you simulate the null response of a given circuit with a voltage-controlled current source.
Notch filter	A notch filter is a type of band-stop filter – a rejector – affected by a high quality factor.
Null	A null in the response simply means that you create specific conditions in the circuit under study so that the excitation does not propagate to produce an output response (see NDI above). The output variable is nulled at 0 V or 0 A. It is different than an output short circuit as explained in Chapter 2.
Order	The order of a circuit is given by the polynomial degree of the denominator. That degree depends on the number of energy-storing elements in the said circuit. More precisely, the degree depends on the number of individual state variables you have identified. Degenerate cases exist when you have a capacitive loop or a pure inductive node. See Chapter 2 for more details.
Phase margin	In the Bode plot describing the frequency response of a compensated control system operated in open-loop conditions, the phase margin (PM) corresponds to the difference between the argument (the phase) measured at the crossover frequency f_c and the $-360°$ or $0°$ limit. Phase margin affects transient response and is usually greater than 45°.
Port	A port describes a pair of connecting terminals sharing a common current. They are the connecting terminals which let you drive a circuit under study or observe a response signal at any place. The input/output ports are commonly employed terms. A port can be dynamically created when a reactance – a capacitor or an inductor – is temporarily removed from the circuit creating a new observation window through its open terminals.
Quasi-static	A quasi-static gain designates a gain obtained as f approaches 0 Hz. However, it is not V_{out} over V_{in} in which input and output would be static bias points, but is ΔV_{out} over ΔV_{in}. ΔV_{out} is actually V_{out2}–V_{out1} respectively obtained for V_{in2} and V_{in1} whose difference forms ΔV_{in}. The term quasi-static gain is advantageously employed over *dc gain* which sounds inappropriate as dc designates a direct – read continuous – current.
Reactance	The reactance is the imaginary component of an impedance. It is positive for an inductor, negative for a capacitor and equal to 0 for a resistance. By extension in the book, a reactance designates an energy-storing element, a capacitor (capacitive reactance) or an inductor (inductive reactance).
Response	This is the waveform observed in a circuit as an effect of excitation. It can be monitored at the circuit output but also at any other node in the circuit. Changing the point in the circuit at which the response is observed affects the position of the zeros (if any) and modifies the numerator N but does not alter the natural time constants: the denominator D is unchanged.
Small-signal	A small-signal or *incremental* model describes the behavior of a non-linear element linearized across an operating point. By linearized, we approximate the behavior of that element by linear expressions. Physically, a small-signal excitation implies a driving signal of sufficiently low amplitude to maintain

the circuit in a linear mode while studying its response to a stimulus. A non-distorted output variable, the response, confirms a linear operation.

SPICE — Simulation Program with Integrated Circuit Emphasis. It is a software program which allows you to simulate an electrical circuit in static conditions, in harmonic analysis or in transient conditions. You can find several free packages such as editor's demonstration versions (OrCAD's PSpice, Intusoft's IsSpice, Design Soft's Tina) or totally free versions like Linear Technology's LTSpice.

State variables — A state variable defines the state of the circuit at $t = 0$. If you want to know the node voltage of a circuit involving a capacitor or an inductor after the power-on sequence ($t > 0$), you need to know if these elements were already charged or energized at the beginning of the analysis, $t = 0$. These *initial conditions* (abbreviated IC in SPICE) are the circuit *state variables*. One usually designates the capacitor voltage by x_2 while the inductor current is x_1.

Storage — Capacitors and inductors are energy-storing elements. A capacitor C stores $\frac{1}{2}CV^2$ with V being the voltage across the capacitor. An inductor L stores $\frac{1}{2}LI^2$, with I being the current flowing in the inductor.

Switching converter — A power converter involving time-discontinuous currents and voltages as opposed to a linear converter in which these variables remain continuous in time. Switching converters or dc-dc converters are more compact and efficient than their linear counterparts. However, they are usually noisier in terms of electromagnetic signature. A switching converter must be linearized before FACTs can be applied to determine its transfer functions.

Time constant — The time constant τ ("tau") involving a capacitor C with a resistance R is defined as $\tau = RC$. When an inductor L is involved, the time constant definition becomes $\tau = L/R$. The unit is time in seconds [s]. You determine the circuit natural time constants when the excitation is turned off. Assembling these time constants leads to expressing the transfer function denominator.

Transfer function — It is a mathematical relationship linking a response to an excitation. There are six types of transfer functions as detailed in Chapter 1: voltage or current gains, impedance or transimpedance, admittance or transadmittance.

Index

Printed and bound by CPI Group (UK) Ltd, Croydon, CR0 4YY

17/03/2025

14640913-0002